JN438858

2012

INTERNATIONAL DESIGN EXCHANGE PROJECT

The 3rd Hidden Space Project

HIGH STREET

Aldgate in London

01
LONDON METROPOLITAN University
Sir John Cass Department of Art, Media and Design
Interior Design and Technology BA (Hons)
UK

Tehelan-ro in Seoul

02
GACHON University
Department of Interior Architecture
Rep. of KOREA

Flyovers in Seoul

03
DONGYANG MIRAE University
Department of Interior Design
Rep. of KOREA

The 3rd Hidden Space Project

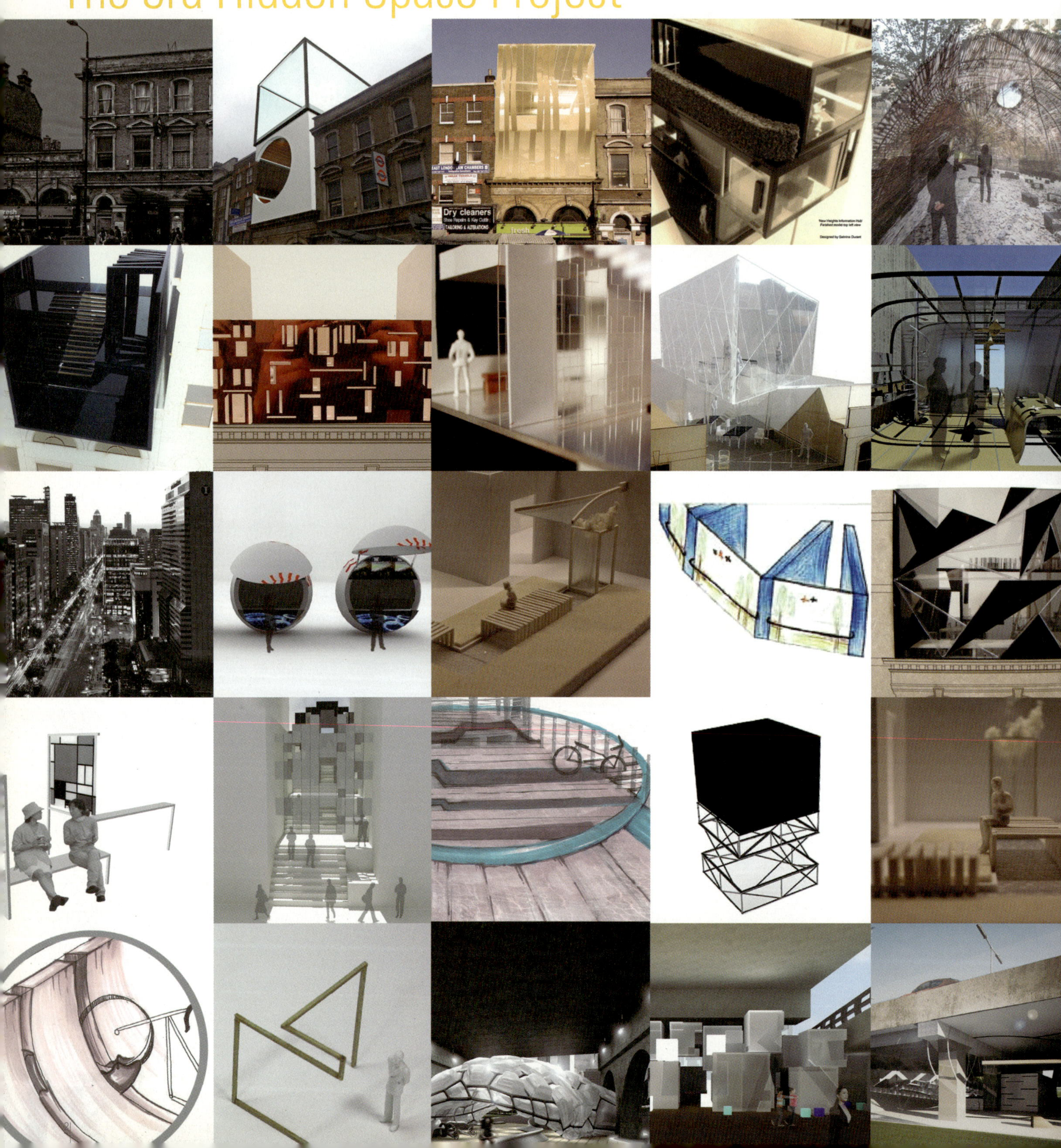

HIGH STREET_37 Projects

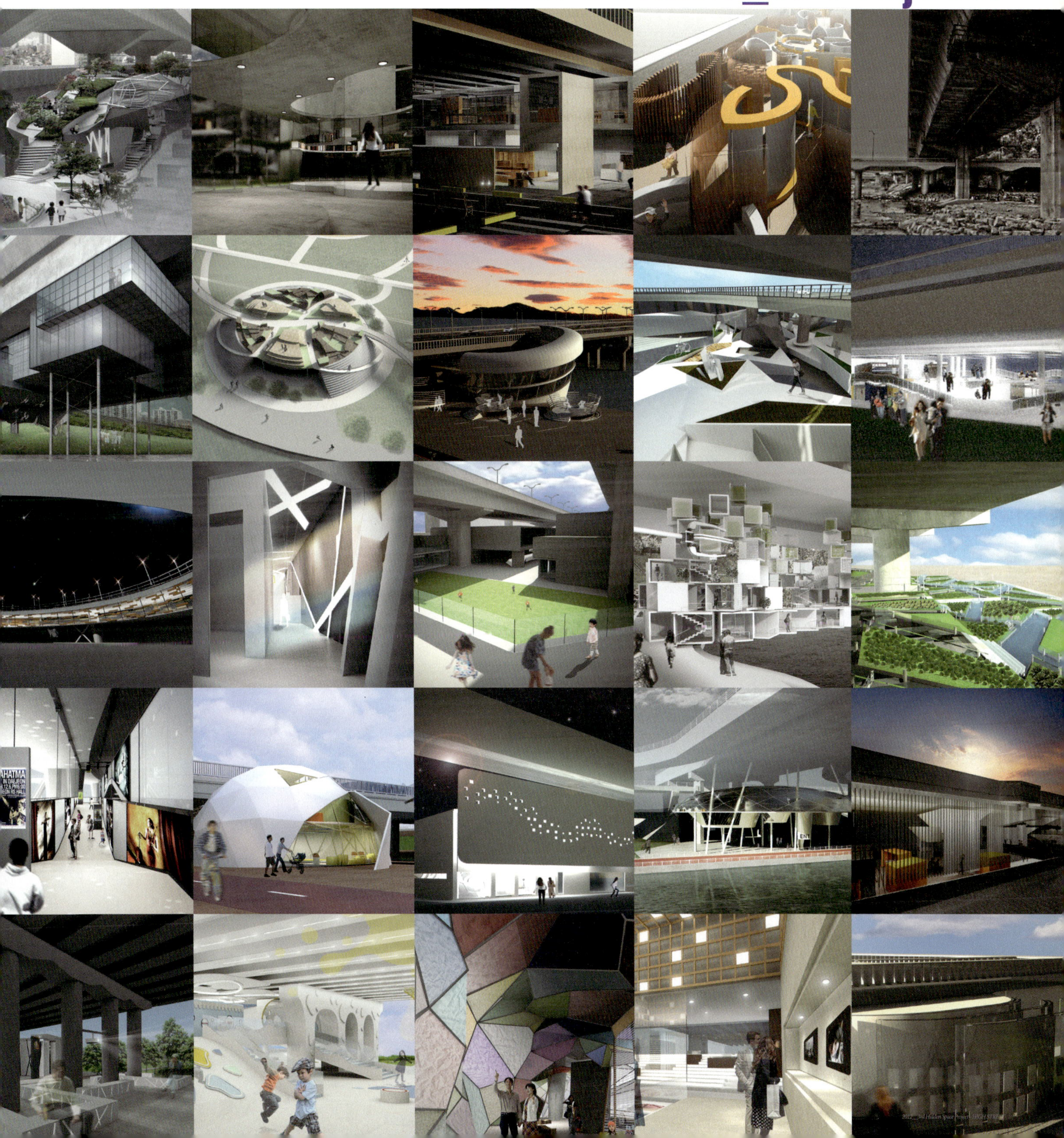

contents

01 LONDON METROPOLITAN University _ UK

INTRODUCTION

PROJECTS

02 GACHON University _ Rep. of KOREA

INTRODUCTION

PROJECTS

03 DONGYANG MIRAE University _Rep. of KOREA

INTRODUCTION

PROJECTS

The 3rd
Hidden Space Project
HIGHSTREET

LONDON Metropolitan University
Sir John Cass Department of Art, Media and Design
Interior Design and Technology BA (Hons)

Sir John Cass Department of Art, Media and Design
41-47 Commercial Road
Lonodon E1 1LA
Switchboard 020 7423 0000
www.londonmet.ac.uk

LONDON Metropolitan Univ
Sir John Cass Department o
Interior Design and Technol
Modelling Interior Structures
DE2014C
2 0 1 2 HIGH STREET

rsity
Art, Media and Design
gy BA (Hons)

The Workshop focuses on the area around Whitechapel – a fringe of the city swayed gently by autonomous development, property speculation and the steady decline in industry.

These fluctuations have left the area with a strange accumulation of leftover space, service networks and incomplete urban narratives. By drawing the actuality of this Bataillean 'dust', the students will first compile a matrix of spatial fragments, and focus one particular site of there choice to speculate on their socio-spatial potential whilst reinterpreting the city as a catalogue of beautifully incomplete objects.

Hidden space Workshop/London Metropolitan University/Ecole Boulle/12-16/12/2011

Michael Griffiths, AA project

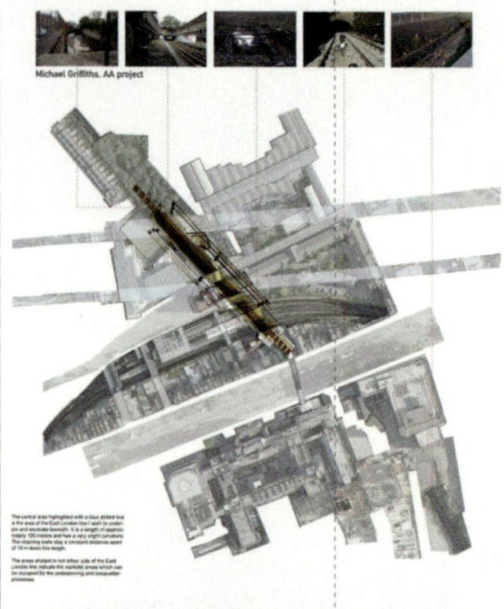

Whitechapel Road/London

Hidden space Workshop/London Metropolitan University/Ecole Boulle/12-16/12/2011

LondonMet

Tutors
Students

Kaye Newman, Director of the program

Jannette Harris, Lead tutor

Jérôme Picard, Lecturer

Hidden space Workshop/London Metropolitan University/Ecole Boulle/12-16/12/2011

Agathe Oury, Christina Rech and Maya Fabri

Camilla Chung

Lois Woodgate and Christopher Crawford Kelly

Samaneh Khazaei

Sara Mitrova and Sabrina Durant

Shahzad Akhtar and Gabriella Ramacciotti

Lost in Whitechapel road Market, but here for the workshop :
Buket Kutlu
Lina Navickaite
Veronika Lovasova
Juna Margariti
Farida Hasan
Monika Krzyzanowska

Hidden space Workshop/London Metropolitan University/Ecole Boulle/12-16/12/2011

Ecole Boulle

Tutors
Students

Valerie de Calignon

Erwan Lebourdonnec

Bruno Rosenzweig

Roland Lemoine

Christelle Bernard

Hidden space Workshop/London Metropolitan University/Ecole Boulle/12-16/12/2011

DSAA1 AI

DSAA2 AI

Hidden space Workshop/London Metropolitan University/Ecole Boulle/12-16/12/2011

Tom Jones

As an Associate Principal of Populous, Tom has worked on a number of projects, including the design proposals for the Viv Richards Stadium in Antigua as part of the 2007 ICC Cricket World Cup. He has also played an important role in the design of the Emirates Stadium, the new home ground for Arsenal FC. After leading the design and tender phases of the new Headquarters building for the club, Tom was responsible for delivering the final construction stages of the main stadium – on budget and ahead of time.

Tom joined the London Olympic Stadium team, taking responsibility for design delivery and securing planning permission for the project. This involved coordinating the work of a large design team, as well as undertaking extensive public consultation on the design proposals.

Tom is now responsible for managing business development in the London office of Populous and has spoken at a range of conferences and events around the world. He is a guest tutor at London Metropolitan University and is driven by an ambition to design and realise responsible, dynamic architecture - based on specialist research that he has undertaken in environmental design and sustainable architecture.

Peter Sheard

Peter Sheard is the practice area leader for the Gensler landscaping studio. He has over 25 years professional experience in landscape architecture and urban design and forms part of the multidisciplinary master planning team. He works closely with Gensler's architects and workplace designers to develop schemes for public realm spaces including public squares, streetscapes, parks, atria, roof gardens and interior spaces. Peter's in-depth knowledge of the landscape planning process and sustainability ensures that his designs for architectural developments are fully integrated into their natural surroundings.

Enrico Caruso

As a Regional Design Leader for Europe, Enrico makes ongoing contributions to Gensler's Firmwide Design Leadership group to develop best design practices across the firm. He has led teams on major interior design projects for multinational corporations in the United Kingdom, Canada, Europe and the Middle East.

He is a guest tutor at the London Metropolitan University and sits on the Industry Advisory Board for Ravensbourne College of Design and Communication. He is a valued design mentor for junior staff and actively involved in the recruiting and development of some of the best talent available in the industry

Jerome Picard

Jerome Picard is currently working with S333 where he is developing his experience of architect/urban planner, after completing a Master in Housing and Urbanism at the Architectural Association, focusing on spatial strategies enabling Innovation environment in complex urban context.

Between 2007 and 2009, Jerome worked as an architectural designer with Dominique Perrault Architecture where he was involved in multiple scale projects. His collaboration includes major educational and public commissions, such as the EWHA University campus in Seoul, the European Court of Justice in Luxembourg and the Tennis Centre in Madrid. Conjointly, He led several design creations such as Lemma chair and Quartetto vases for Sawaya and Moroni, edited for the Biennale of Milan 2008 and 2009.

Hidden space Workshop / London Metropolitan University / Ecole Boulle / 12-16/12/2011

Site visit and Tour
Workshop
Review
Free/Gathering

	Monday	Tuesday	Wednesday	Thursday	Friday
09		Archiwalk 1 Richter at the Tate Modern National Theater Denny Lasdun	Workshop Exploring concept	Archiwalk 2 OMA Progress Barbican Centre	Final critic With Guest critic
10					
12	Arriving/Drop and Go				
02	Site visit Whitechapel Road	Meet Lecturers	Workshop Pushing forward	Workshop Final development	Final critic
	Meet Tom jones	Workshop Thematic approach Research group Rapid making exploration			Departure
06	Welcome drink				
08			Mid-week review		

Hidden space Workshop / London Metropolitan University / Ecole Boulle / 12-16/12/2011

Archiwalk Panorama

The Tate and the South Bank

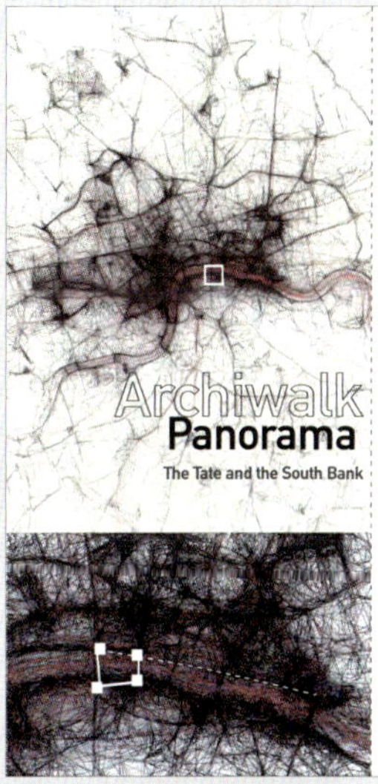

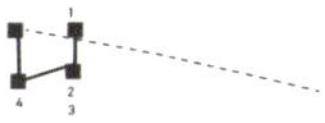

1-Meeting at Saint Paul 9.30am

1-Tate Modern

3-Gerhard Richter: Panorama

4-Lasdun National Theatre

Notes :

Hidden space Workshop / London Metropolitan University / Ecole Boulle / 12-16/12/2011

Archiwalk OMA Progress Barbican

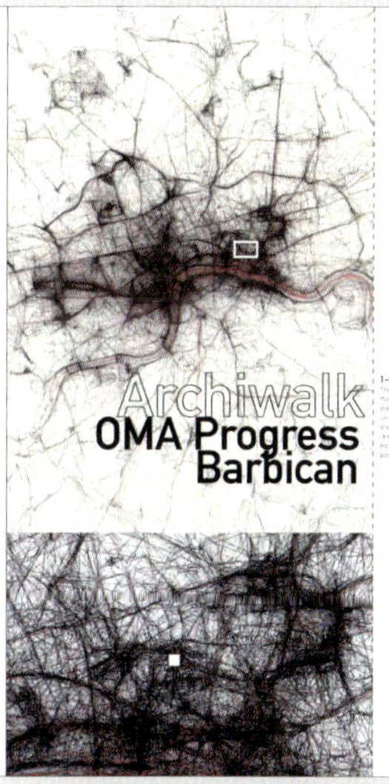

1-OMA Progress

2-Barbican centre

Notes :

Hidden space Workshop / London Metropolitan University / Ecole Boulle / 12-16/12/2011

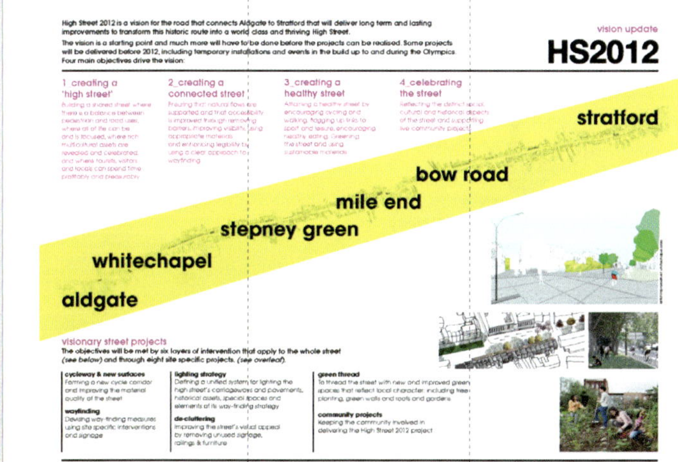

Hidden space Workshop / London Metropolitan University / Ecole Boulle / 12-16/12/2011

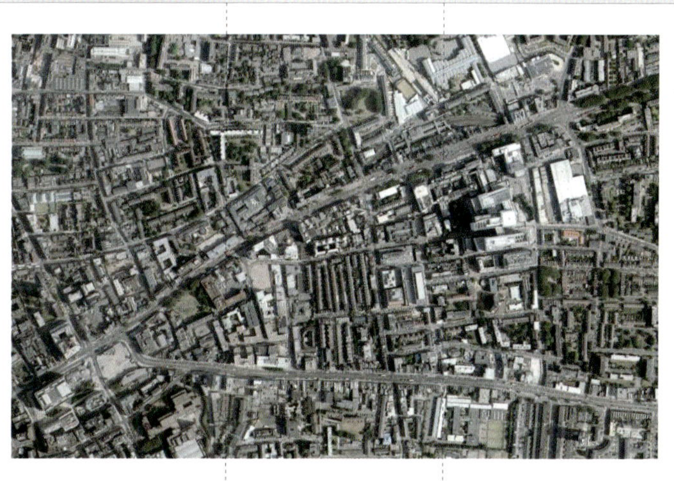

Hidden space Workshop / London Metropolitan University / Ecole Boulle / 12-16/12/2011

GROUP PROJECTS: GROUP 1

"OPEN HOUSE"

Three main platforms within the structure represent spaces to gather, relax, eat, sleep, work on a laptop. Hidden from the street they bring intimacy to this public space. Infused with light, energy and positive spirit, it will be a magnet for the local community!

In the garden ...

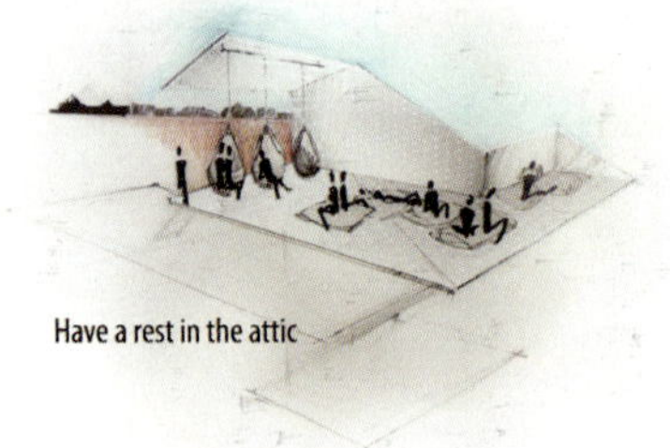

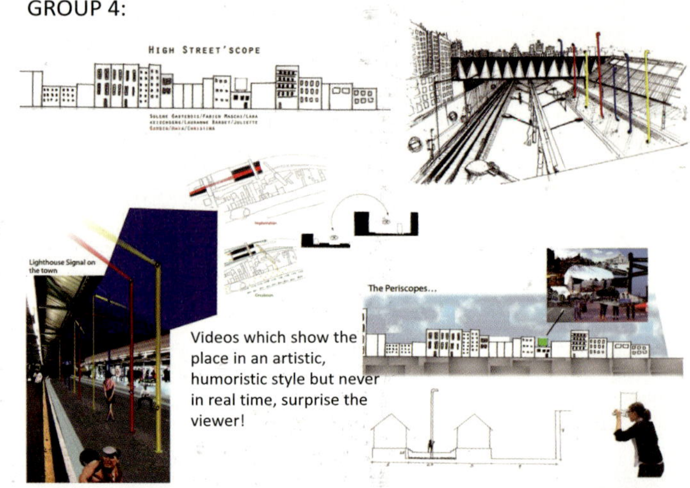

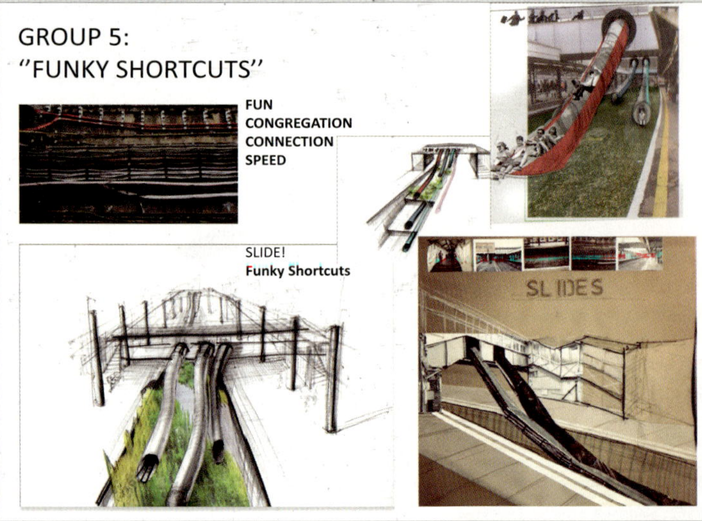

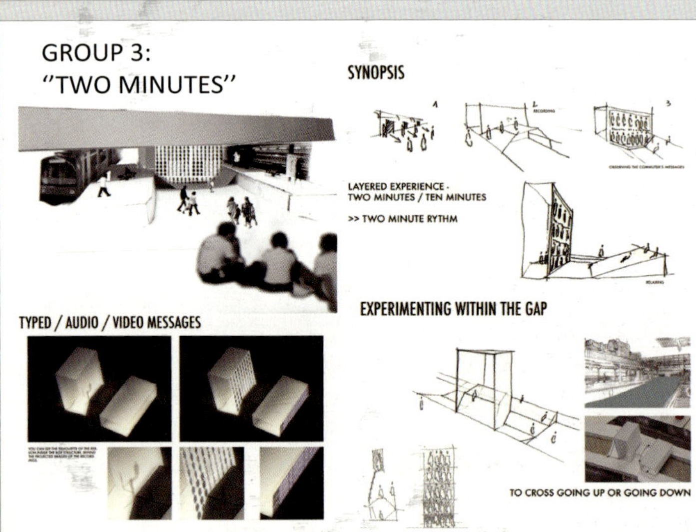

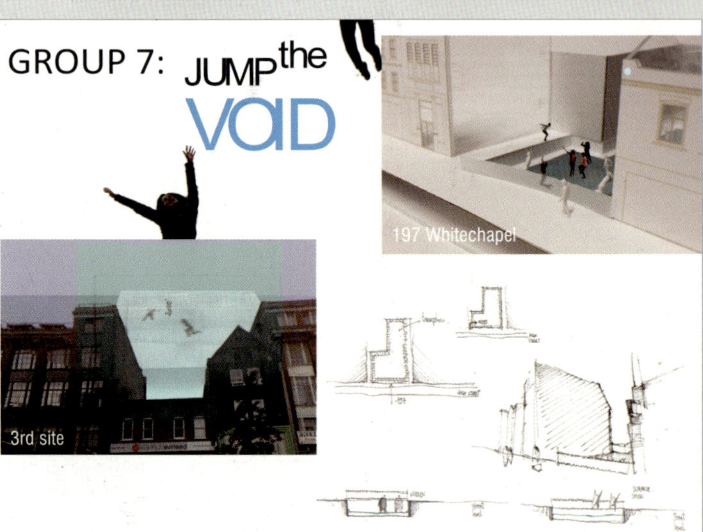

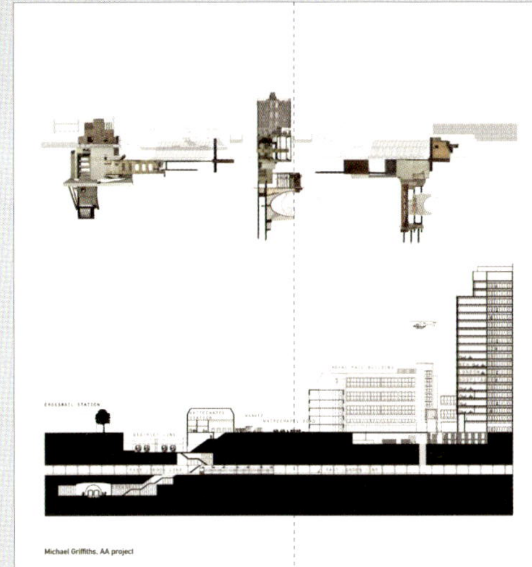

Michael Griffiths, AA project

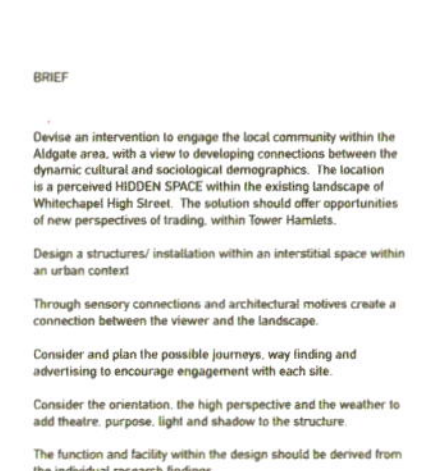

BRIEF

Devise an intervention to engage the local community within the Aldgate area, with a view to developing connections between the dynamic cultural and sociological demographics. The location is a perceived HIDDEN SPACE within the existing landscape of Whitechapel High Street. The solution should offer opportunities of new perspectives of trading, within Tower Hamlets.

Design a structures/ installation within an interstitial space within an urban context

Through sensory connections and architectural motives create a connection between the viewer and the landscape.

Consider and plan the possible journeys, way finding and advertising to encourage engagement with each site.

Consider the orientation, the high perspective and the weather to add theatre, purpose, light and shadow to the structure.

The function and facility within the design should be derived from the individual research findings

Hidden space Workshop / London Metropolitan University/Ecole Boulle / 12-16/12/2011

Hidden Space Day 1 12/12/2011

Site visit/Lecture by Tom Jones and Jerome Picard/Team introduction at the MET

Follow us on http://hiddenspacewhitechapel.blogspot.com

Hidden Space Whitechapel / London Metropolitan University / Ecole Boulle / 12-16/12/2011
The workshop newsletter #1

Hidden space Workshop / London Metropolitan University/Ecole Boulle / 12-16/12/2011

Hidden Space Day 5 16/12/2011

Final Presentation

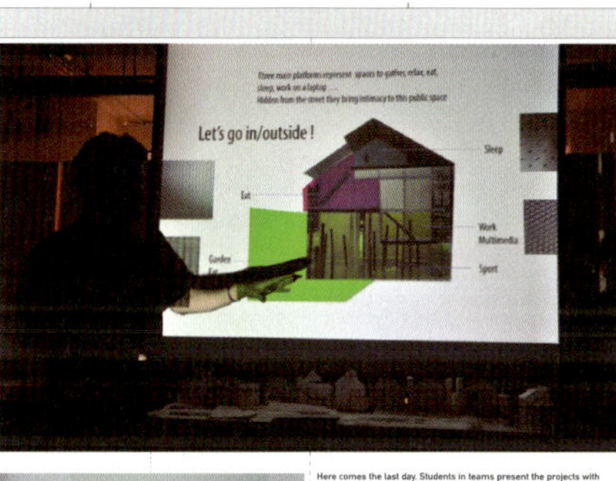

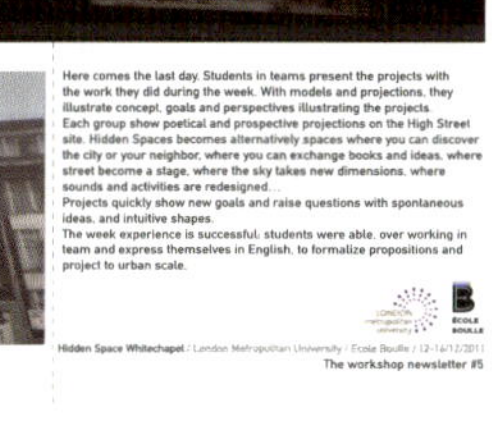

Here comes the last day. Students in teams present the projects with the work they did during the week. With models and projections, they illustrate concept, goals and perspectives illustrating the projects. Each group show poetical and prospective projections on the High Street site. Hidden Spaces becomes alternatively spaces where you can discover the city or your neighbor, where you can exchange books and ideas, where street become a stage, where the sky takes new dimensions, where sounds and activities are redesigned...
Projects quickly show new goals and raise questions with spontaneous ideas, and intuitive shapes.
The week experience is successful: students were able, over working in team and express themselves in English, to formalize propositions and project to urban scale.

Hidden Space Whitechapel / London Metropolitan University / Ecole Boulle / 12-16/12/2011
The workshop newsletter #5

Hidden space Workshop / London Metropolitan University/Ecole Boulle / 12-16/12/2011

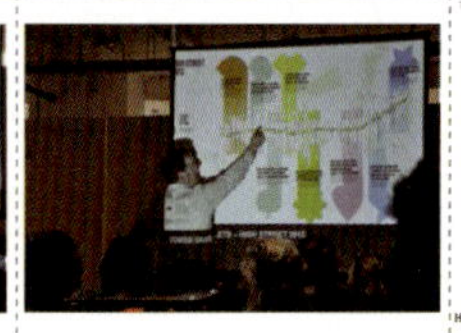

Hidden Space Whitechapel / London Metropolitan University / Ecole Boulle / 12-16/12/2011
The workshop newsletter #1

Robert Mull
Dean of Sir John Cass Faculty of Art, Architecture and Design

This year is a year of new beginnings. I was appointed Dean in 2011 and have since formed a new Faculty which now encompasses Art, Architecture and Design and a brand new identity known as the Sir John Cass Faculty of Art, Architecture and Design which will be launched this year at our Summer Shows.
The merger has offered us an opportunity to talk through our ethos, the courses and indeed the changing needs of students. Whilst change can be a little uncomfortable the new structure will be more sympathetic to the needs of the creative disciplines and the response from both lecturers and students has been fantastic and I very much look forward to seeing the progress and culmination of innovative, inspired and the inventive projects next year 2013.
In a similar way the focus of this book looks at the Hidden Space, where space has been overlooked, forgotten and therefore is not able to participate or integrate into fabric of everyday life. When offered the opportunity to look with fresh eyes the possibilities of new allegiances and synergies abound, wonderfully exciting and truly exhilarating springing us forward to work with a sense vocation and relevance to both our local neighbourhoods and our global community.

올해는 의미있는 새로운 시작의 해라고 할 수 있다. 나는 2011년 학과장이 되어 그때부터 미술, 건축, 그리고 디자인을 포함한 새로운 교수진과 함께, 이번 여름 전시에서 선보이게 될 Sir John Cass Faculty of Art, Architecture and Design (런던 매트로폴리탄대학 미술디자인학부)의 새로운 아이덴티티를 만들어냈기 때문이다.
이러한 학과간의 합병은 새로운 수업과정과 학생들에게 필요한 것이 무엇인지 그리고 교수진은 어떤 생각들을 가져야하는지 등에 관해 논의할 기회를 주었다. 변화를 갖는다는 것이 쉽지만은 않았지만 이 새로운 구조는 앞으로 학생들과 교수들의 창의적 요구와 발전에 더 부응할 수 있을 것이다. 교수진과 학생들의 반응과 결과물들은 매우 괄목할만한 것이었기에 2013년에 보여 주게 될 진보와 획기적이면서도 탁월하며 독창적인 프로젝트들과의 만남이 매우 기대된다.
이 책은 Hidden Space, 즉 간과되거나 잊혀져 버려서 더 이상 일상생활의 일부가 되지 못하는 그런 공간들에 집중하여 바라본 것이라고 할 수 있다.
한국의 훌륭한 교류학교들과 함께 시너지 효과를 가지고 우리 주변의 커뮤니티와 나아가 세계를 새로운 시선으로 바라보며 작업을 할 수 있게 된 것에 크게 기대가 되는 바이다.

Kaye Newman

Course Organizer
Interior Design and Technology BA (Hons)
London Metropolitan University
Sir John Cass Department of Art, Media and Design
Interior Design and Technology BA (Hons)

Hidden Space 2012

Our cities now have to rethink and redevelop the use of space more creatively and ingeniously than ever before. Space is precious in any context but the extra pressure cities'face with the ever growing need to accommodate increasing populations mean that everyopen area is now a new possibility for spatial development.

One of the most recent and fascinating is discovering interstitial space, gaps in-between.Their appeal lies in their surprise and undiscovered disposition. Their purpose might involve a large community or an individual but because of its hidden or exposed position, providing an adventure, or unexpected experience. It provides opportunities for older established buildings to convey a new brief and therefore a new spirit, encouraging creativity and positive thinking, using a different energy.

The urban master planners and the local communities have to consider a more responsible composition of ideas to our environmental needs. Sustainable energy use and material choice are to be rewarded but how does this connect with the individual and the surroundings? What is our sensory involvement? Actions that involve breathing fresh air, feeling the breeze, hearing the cars rumble by, bathing in direct light, seem fundamentally prearranged but is the experience total? How do we make this encounter poignant and memorable?

The Aldgate area, where our Hidden Space is located, sits on the City fringe adjoining multi-national corporate businesses to some of the poorest areas in London which exist in its East End. The 2012 Olympic Games was secured on the basis that the East End was one of the poorest areas in Europe and in need of global support to redevelop and renew confidence amongst its diverse and multicultural inhabitants. The Olympics has in turn created opportunities to think about its legacy and the future of the East End and projects such as High Street 2012 have been written to investigate and explore the needs of the communities and their environments. From this point, the Faculty of Art, Architecture and Design took on the Aldgate project and have been developing ideas surrounding the issues of the area.

The Interior students in particular looked at trade and employment as an over-arching theme. Why does this community from Whitechapel fear the corporate establishments? How do we bridge the gap between the steel and glass architecture to simple community housing estates and try to establish a positive journey guiding people to employment and career opportunities? Three sites have been identified and used to develop ideas of structural language, developing familiar and recognizable spaces to dispel the fear and allow the visitor to acclimatize in stages to the city environments.

Each of the students has looked at forms and materials that incorporate ideas of confidence building, openness and inclusivity. The interventions allows for views and vistas from the street and from within spanning the depth of London across the city giving the viewer a sense of inclusion and belonging.

Kaye Newman
Course Organizer - 코스 기획자

우리의 도시는 공간을 어떻게 훨씬 더 창의적으로 이용하는 주제에 대해서 재고하고 재발전시켜야 하는 필요성이 그 어느 때 보다도 대두되고 있다. 공간이란 언제나 중요하게 여겨져 온 주제이지만 계속 불어나는 인구의 압박들 때문에라도 도시의 오픈 스페이스는 그 어디라도 새로운 공간디자인의 가능성을 타진해 보는 대상이 될 수 있다.
최근 가장 고무적인 것은 도시의 틈바구니에 끼어있거나 남겨진 사이 공간들 혹은 틈새 공간들의 발견이라고 할 수 있다. 이러한 공간들은 개인을 위한 사적인 공간일 수도 있고 커뮤니티를 위한 공간일 수도 있으나 숨겨져 있는 이유로 인해 우리에게 모험과 의외의 경험을 하게 해준다. 또한 오래된 건물들이 무언가 다른 에너지를 수혈 받고 새로운 용도를 가지게 되어 창의적이고 긍정적인 모습으로 새롭게 변할수 있는 기회를 제공한다.
도시설계자들과 지역 커뮤니티는 우리의 환경에서 필요로 하는 것에 대한 구상들을 보다 책임감 있게 다루어야 한다. 지속 가능한 에너지 사용과 재료의 선택도 좋지만, 어떻게 개인과 주변을 연결시킬 것인가? 그리고 우리 감각과는 어떻게 관련이 있는가? 신선한 공기를 마셔들이는 행위, 바람을 느끼는 것, 지나가는 자동차 소리를 듣는 것, 직접 햇살을 쬐는 깃 등의 감각적 경험은 정해지고 예정된 경험처럼 보이지만 과연 그 경험은 완전체일까? 어떻게 이 접촉을 더 기억에 남게 할 수는 없는 것일까 ?

우리의 HIDDEN SPACE인 Aldgate지역은 런던도심의 가장자리에 있으며, 다국적 기업의 사무실들과 런던에서 가장 가난한 East End 지역들이 맞닿아 있다. 2012 올림픽 경기는 East End가 유럽에서 가장 가난한 지역 중 하나이며 지역의 재개발과 여러 국적의 이주민들의 자신감 회복을 위해 국제적인 지지가 필요하다는 이유로 런던에서 하게 되었다.
올림픽은 East End의 미래를 생각할 수 있는 기회를 가져왔다. 커뮤니티와 환경을 위해 필요한 것들을 조사하기 위해 HIGH STREET와 같은 프로젝트들이 기획되었고, 더불어 런던의 미술, 건축, 그리고 디자인 학부들은 해당 지역의 둘러싸고 있는 문제들을 풀기 위한 아이디어를 생각해내기 시작했다.
실내건축 학생들은 상거래와 고용을 테마로 바라보았다. 왜 Whitechapel의 주민들은 기업화된 시장이 침투하는 것을 두려워할까? 어떻게 하면 철과 유리로 된 신축건축물들과 기존의 소박한 지역의 주택들 사이의 간극을 줄이면서 동시에 사람들을 취업의 길을 만들어내며 긍정적인 여정을 만들어낼 수 있을까?
친근하며 알기 쉬운 공간을 통해 도시에 대한 두려움 없이 방문객이 도시 환경에 단계적으로 익숙해질 수 있는 구조적인 언어를 개발하기 위해 세 곳을 선정하였다.
학생들 개개인은 열림과 포함이라는 주제와 연결되기에 적합한 형태와 재료를 찾아보았다. 이러한 인터벤션은 런던이라는 도시의 어느 곳에서든지 자연스러운 풍경을 볼 수 있게 하여 도시에 새로운 깊이감을 심어주면서, 도시를 보는 사람으로 하여금 소속감을 안겨주게 된다.

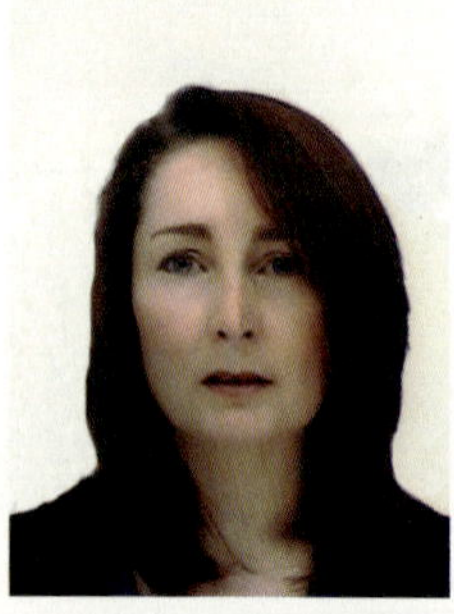

Janette Harris MALTHE, SEDA

Senior Lecturer
Module Leader
Interior Design and Technology BA (Hons)
London Metropolitan University
Sir John Cass Department of Art, Media and Design
Interior Design and Technology BA (Hons)

Modelling Interior Structures DE2069C Year 2
Semester 1
DE2014C

Module Summary

The module is concerned with the exploration and realisation of structure and materials within a space. Workshop and class based activities encourage a wide range of skills, highlighting sensitivity to design matters, reflective practice, informal debate, peer assessment and presentation. Software is used to produce orthographic details, visuals and drawings for the production of components. Scale is explored in the context of human interaction. Spatial concepts are considered through modelling in three dimensions, using a wide range of machinery and hand tools, allowing the student to fully explore all modelling possibilities. Process is developed using proven and experimental techniques. The module promotes the growth of existing skills combined with the exploration of the potentials of technology.

The Pedagogy

The students undertake this project in the first semester of their second year. The project lasts 14 weeks, with the last 5 weeks producing a professional model and visualisation. The emphasis is to encourage the students to design through making, rather than producing a 3D output from a drawing. This experiential process utilises haptic interaction with materials and process, allowing for a more informed approach and an appreciation of spatial concepts. In turn this approach promotes reflection, challenging students to take risks with design concepts, within the confines of the brief. Activities within taught sessions such as peer and self-assessment supports this methodology.

The Hidden Space project posed a number of challenges

- Connecting the site to the context and those who work within the building
- Understanding the aspect of the site through environmental analysis
- Understanding the demographic of the area through analysis and diagnostic
- Understanding the materiality of the existing building and it's surroundings
- Developing a concept that dealt with cultural and sociological issues and responding through appropri ate design solutions.

Constructivist View of Learning

The students are taught using constructivist learning theories that develop a learning community through communication and cooperation through activity, based on work by Vygostsky, influenced by Lave & Wenger 1991, ideas of Learning Communities.

Activity

Develops knowledge construction

Learning is developed by

- building on existing knowledge
- Students being allowed time to incubate, allowing ideas to evolve
- Tasks and exercises are provided to keep students on track and within the time frame of the module
- Interaction with professional practice through critique and mentored sessions with the workshop
- Developing skills through experimentation, interaction with peers and reflecting on process and practice

Activities of construction understanding

- Interaction with materials, understanding their limitations, possibilities and appropriate solutions, in regard to modelling and the reality.
- Interaction with professional practice
- Interaction in order to understand the individuals level of understanding with their peers

Design Principles for constructivist teaching and learning activities

- Ownership of the task
- Coaching and modelling of thinking skills
- Scaffolding
- Guided discovery
- Opportunity for reflection
- Ill-structured problems

To enable students with the face-to-face sessions an online learning environment BlackBoard is used. Week by week the students are given data, task and are asked to form a community of learning to promote independent and meaningful learning.

- Discussion and debate through a blog
- Students encouraged to develop a resource library for the module for the whole class to use
- Students set up study groups to share practice and process developing a community of practice
- Students to develop reflective practice by using an online portfolio
- Feedback given to students through mail within WebLearn and writing their personal portfolio's
- Feedback from the students to the tutor through an anonymous online proforma
- Peer and Self assessment.

Students are asked to present their ideas throughout the semester to the class, professional practice and to other courses within the Faculty. The aim is to build student confidence to stand up for their ideas, and to answer questions that may prove challenging. The students then exhibit their work within Clerkenwell Design Week, The Association of Shop Fitters Annual Design Awards and London Festival of Architecture, as part of the Korean exhibit. The interaction with professional practice develops key skills of communication, connections and community.
The collaboration with Gachon University and DongYang Mirae University in Seoul South Korea, has given the students opportunity to demonstrate their perspective on the Hidden Space brief, and this book offers us an insight to three Universities approaches. We are very proud to be part of this project, and thank all the participants that have made this possible.

모듈(교육과정)개요

본 모듈은 어떤 공간 속에서 구조물과 재료들을 탐색하고 구체화시키는 데 역점을 두고 있다. 워크샵과 수업시간 동안의 활동들을 통해 다양한 기술들을 익힐 수 있는데, 디자인 소재에 대한 감수성, 격의 없는 토론, 동료 학생에 대한 평가, 그리고 발표 등이 대표적이다. 시각이미지와 드로잉 등을 처리할 수 있는 소프트웨어가 활용된다. 사람들 사이의 상호작용을 최대한 고려하며, 공간 컨셉트는 3차원 모델링을 통해 이루어진다. 다양한 기계 및 수공구를 활용하며, 이를 통해 학생들은 모델링과 관련된 다양한 가능성을 탐색할 수 있다. 기존의 기술들 뿐만 아니라 실험적인 기술들도 차용된다. 본 모듈을 통해 기존 기술들의 향상을 도모하는 동시에 기술의 새로운 가능성을 탐색하고자 한다.

Janette Harris
Senior Lecturer
Module Leader – 교과목 지도교수

페다고지(수업방법)

12주로 이뤄진 본 모듈은 2학년 1학기에 진행된다. 이 즈음이 되면 학생들은 3D 스튜디오 맥스, 포토샵, 그리고 오토캐드 등에 대한 충분한 지식을 갖게 되지만, 수업 초기에는 주로 드로잉, 스케치 모델링, 그리고 사진촬영 등에 주안점을 둠으로써, 학생들이 시행착오를 통해 다양한 컨셉트를 탐구할 수 있도록 한다. 이러한 접근법을 통해 학생들의 기존 생각을 되짚어보는 한편 토론과 논의의 장을 마련함으로써 반성적 사고와 활동을 발전시킬 수 있다. 모듈 내의 각 수업별로 활동을 분산배치함으로써 학생들이 지식의 폭을 넓혀나갈 수 있도록 한다.

학생들은 학기 동안 자신의 반과 다른 반, 그리고 전문가들에게 아이디어를 발표하게 하였다. 이는 학생의 자신감 구축과 함께 질문 답변 능력을 길러주기 위함이었다. 그리고 학생들은 Clerkenwell Design Week, 국가 점포 설치 협회 연간 디자인 상, 그리고 런던 건축 축제 때 한국 전시의 일부로 작품을 전시하였다. 이러한 전문가와의 접촉은 학생의 소통 능력, 관련성, 그리고 사회성에 큰 도움이 된다.
한국 서울의 가천대학교와 동양미래대학교의 공동 작업은 학생들에게 HIDDEN SPACE에 대한 자신의 관점을 보여줄 수 있었고, 이 책은 세 대학의 접근에 대한 이해를 도와준다. 우리는 이 프로젝트의 일원이 된 것이 자랑스러우며, 이 프로젝트를 가능케 한 모든 이들에게 감사를 표한다.

christopher crawford kelly
rombedingo77@hotmail.com
www.christophercrawfordkelly.tumblr.com

The Arc Building

THE BRIEF

To devise a series of architectural and interior solutions to engage the local community with local business within the Aldgate area, with a view to future employment.

요약

Aldgate 지역의 주민들을 지역 비즈니스와 연관시키고 미래 취업과 연관되는 실외와 실내 건축 방안을 찾아보는 것.

AIM

The location of the structures should encourage the customers and visitors to travel to each of the three locations (Braham Park, Spital Yard and Whitechapel Station) whilst offering opportunities of new perspectives within Tower Hamlets.

목적

고객과 방문객이 세 공간 (Braham Park, Spital Yard, 그리고 Whitechapel Station)에서 방문객이나 소비자들의 유입을 장려하는 방향 이어야하며 Tower Hamlets 지역의 새로운 가능성들을 제시해 준다.

INTENTION

Design a structure/ installation within an interstitial space within an urban context and, through sensory connections and architectural motives, create a connection between the viewer and the landscape.

의도

틈새 공간 안에 도시적인 건축물/설치물을 만들어 감각적인 연결점과 건축적 모티브를 통해 사람과 풍경 사이에 연관성을 형성한다.

SPITAL YARD

This site will be where the more specific and detailed events happen. There will be lectures and seminars happening here. There will be opportunities for the people from the corporate businesses which surround the area to give advice to the local people of the Whitechapel area who are looking for new ways towards future employment. There will be workshops, events, lectures and exhibitions here to involve the diverse group of people to encourage interaction and integration with one another, putting them in an environment where they would otherwise not come together.

BRAHAM PARK

This site will be for people passing by. As it stands, the site is more like a pathway rather than a place to sit and relax. The installation will be something that will capture the attention of the passers-by and plant the seed of curiosity into them so that they stop to investigate what is inside. As there are tall office buildings surrounding the site, the installation will have an impact when viewed from above as well as when experienced at street level.

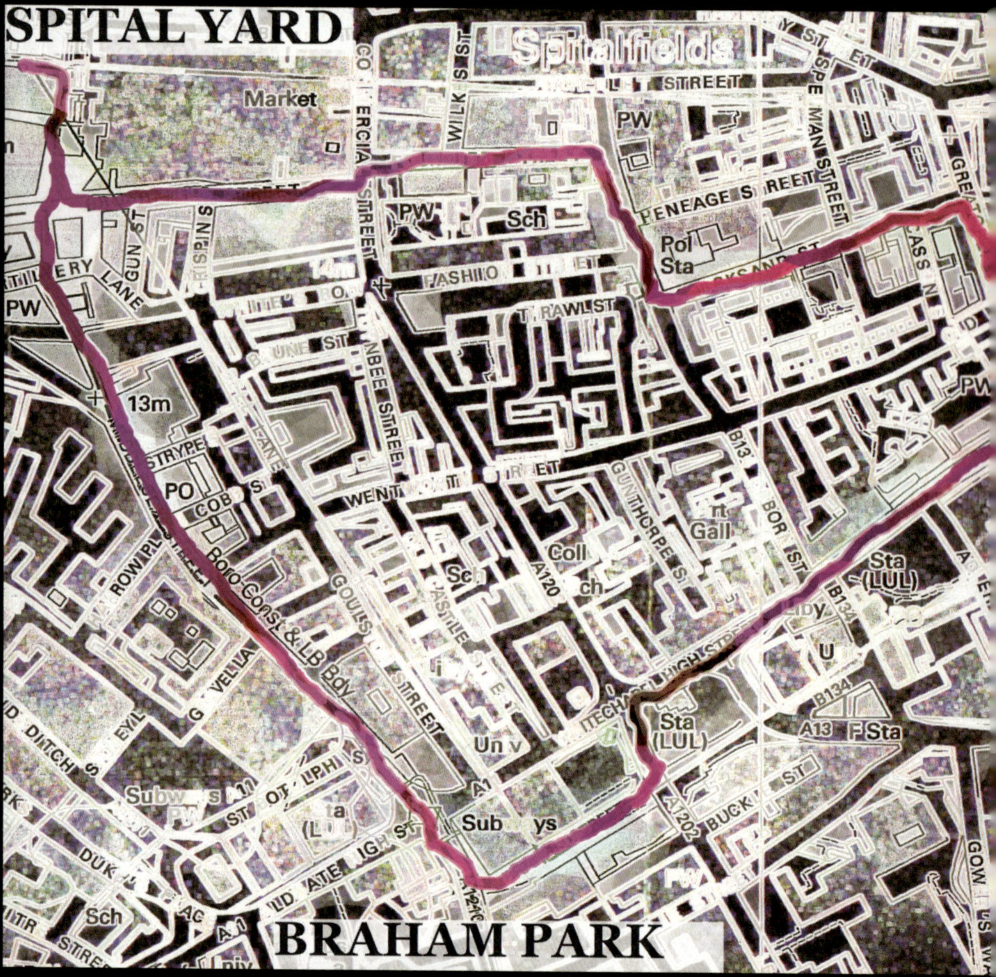

SPITAL YARD

이 사이트는 더 특정한 목적의 섬세한 이벤트가 열리는 곳이다. 강의와 세미나를 할 수 있는 곳이다. 주변 회사의 기업인들이 취업을 찾는 Whitechapel 지역 주민들에게 일에 대한 기대가 생겨날 수 있도록 조언할 수 있는 기회를 제공하는 공간이기도 하다. 워크샵, 이벤트, 강의와 전시가 열려 평상시라면 모이지 않을 다양한 사람들을 모아 서로간의 교류와 연대를 장려할 것이다.

BRAHAM PARK

이 사이트는 지나가는 행인을 위한 곳이다. 이 사이트는 앉아서 쉬는 공간보다는 길에 가깝다. 이곳에는 행인의 눈길을 사로잡을 무언가를 설치하고 그들에게 호기심을 불러일으켜 안에 무엇이 있는지 알아내기 위해 안으로 들어오게 유도 할 것이다. 주변에 높은 회사 건물들이 있기 때문에 길거리에서 체험하는 것과 동시에 다른 건물의 위에서 봤을 때에도 효과가 있을 것이다.

WHITECHAPEL STATION

Observational drawings of Whitechapel Station

THE CONCEPT

The concept is centrally focussed upon the notion of breaking the circle. It is apparent that there is a distinct problem with employment in the area. This is a consequence of people who are stuck in a pattern of behaviour and situations which are curtailing their chances of employment. It may seem fairly obvious and a generic statement to make, but the circle of behaviour which has exacerbated the situation, rather than improving it, must be broken.

The concept is multifaceted in response to a convoluted and complex issue within the community. Consequently, there will be other aspects such as ascension or elevation with a kinetic energy and a sense of movement to convey the progression gained from the experience within the sites and the scheme. From this, there will be the notion of new vistas, new platforms and new perspectives.

The final aspect of the concept involves a feeling of comfort and consideration to show how the scheme is there to provide opportunities in a positive way rather than judgementally. From this, there will be embracing circles throughout to also show that a connection is being made – a sense of contact.

THE CONCEPT

이 컨셉은 일상적 태도를 바꾸는 데에 집중하고 있다. 이 지역에 취업 문제가 있음은 분명하다. 이것은 사람들의 생활 습관과 상황이 취업의 기회를 줄어들게 하기 때문이다. 당연한 일반론처럼 들릴 수도 있지만, 상황이 나아지는 대신 오히려 악화되도록 만드는 일상의 행동패턴과 사고들은 깨져 마땅하다고 본다.

컨셉은 커뮤니티안의 복잡한 문제거리들에 응대하기위해서 다면적인 성격을 가진 것이다. 운동 에너지와 움직임을 이용해 올라가는 부분 역시 발전의 느낌을 주고 싶었다. 여기서 새로운 풍경, 새로운 단계, 새로운 관점을 느낄 수 있을 것이다.

컨셉의 마지막 시도로 편안함과 배려의 느낌을 느끼게 해주고자했는데 비판적인 지적만이 아니고 긍정적인 방법으로도 기회를 줄 수 있다는 걸 알리고자 했다. 서로가 서로를 품은 원들의 의미는 무언가가 이어지고 있다는 것 – 서로 접촉이 있음을 보여주는 것이다.

BRAHAM PARK

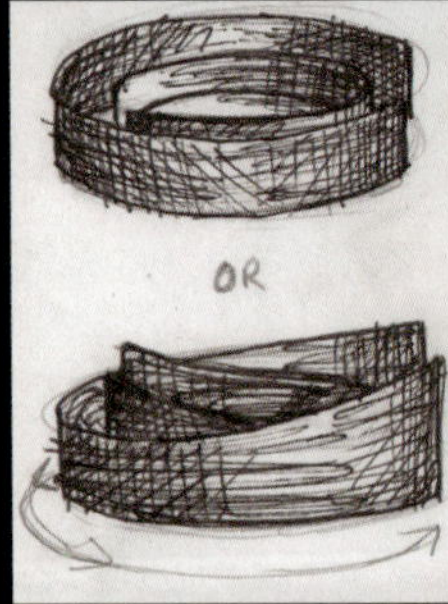

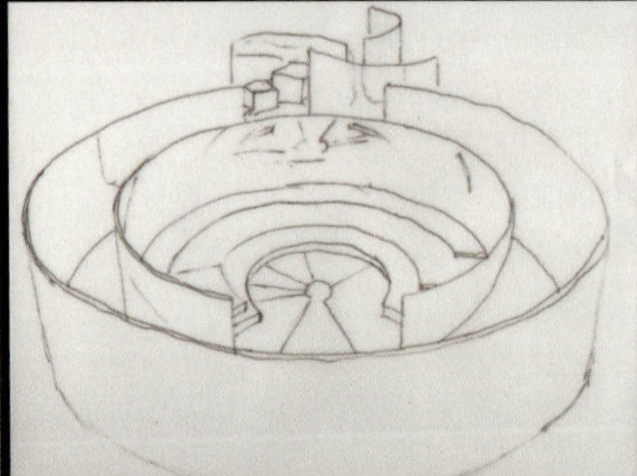

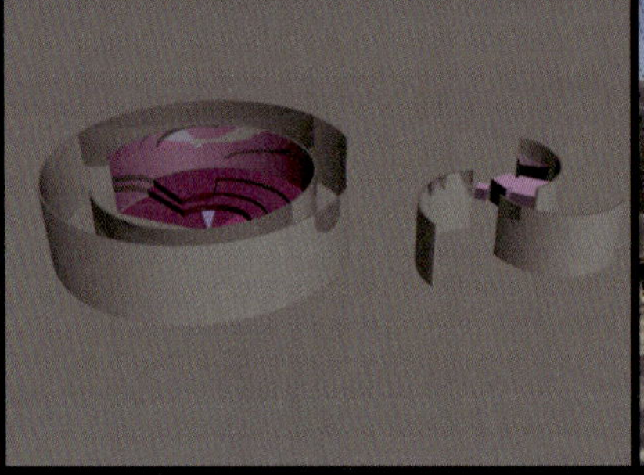

Development of idea

This site will be an auditorium of sorts, but on a smaller scale. It incorporates the notion of embracing circles. The steps within are formed out of broken circles, which aid a sense of progression and ascension.
It will be a place to sit and contemplate. A place with a sense of privacy. Within this area there will be a screen which provides basic information about the scheme and what the intention of the three sites is.
Between the three sites, the logo will be placed on lamp posts, metallic plaques could be placed in the paving stones to act as breadcrumbs leading people who are intrigued to the other two sites.

이 사이트는 작은 강당들이 들어설 것이다. 이것은 원의 형태를 적용하였다. 안의 계단들은 깨진 원들로 이뤄져 있는데 이것은 발전과 진보의 느낌을 준다.
사적인 느낌이 드는 이곳은 앉아서 생각하는 곳이 될 것이다. 이 안에는 세 공간에 대한 안내정보와 공간의 의도가 무엇인지 보여주는 스크린이 있을 것이다.
세 사이트 사이, 가로등에 금속의 로고를 둘 것이다. 이 금속 로고는 사람들을 다른 두 사이트로 인도하는 헨델과 그레텔의 빵 조각 같은 역할을 할 수 있을 것이다.

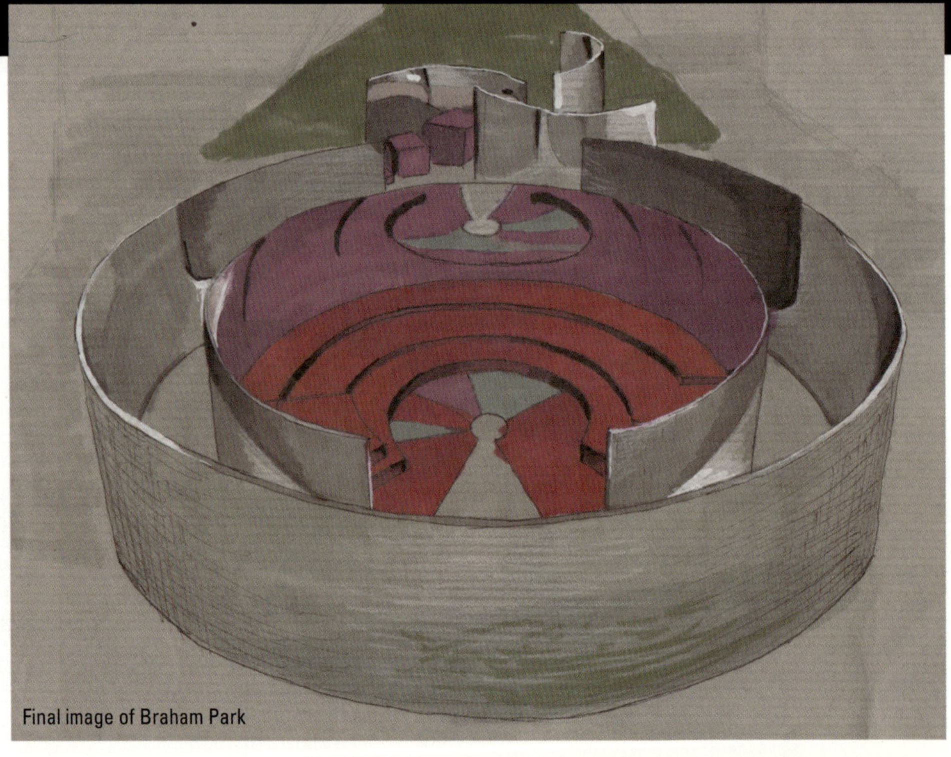

Final image of Braham Park

SPITAL YARD

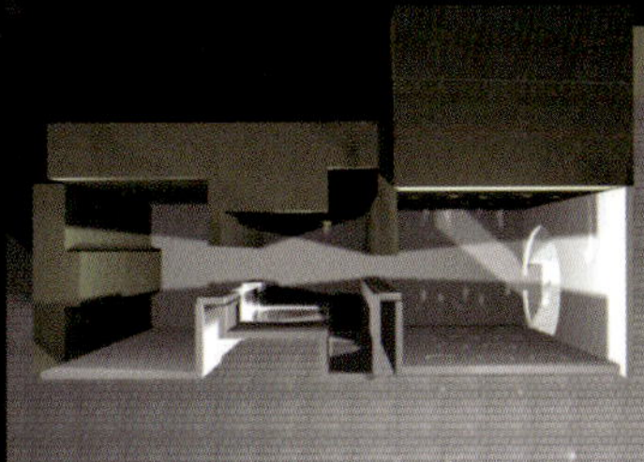

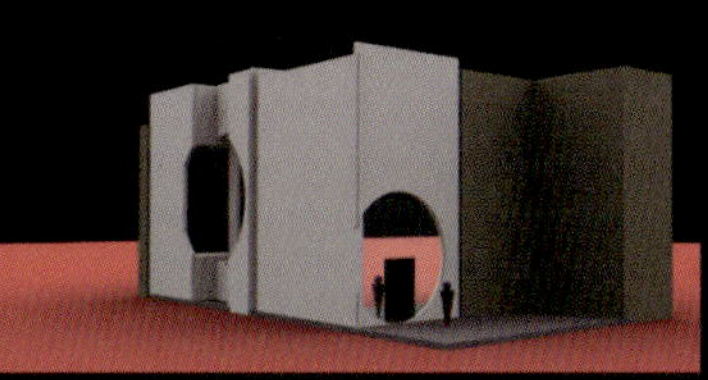

Development of idea

The shape of the building is directly linked to the geometry of the site. The jutting nooks and crannies are mirrored to create a dynamic interior with defined spaces for specific activities. From the exterior, the various facades form a circle from a specific perspective. The strong arc shapes give a clear link to the brand logo and the main hub in Whitechapel.
The Corian facade also forms a strong link to the Whitechapel site and also creates a refreshing sense of identity and boldness within its surroundings.

건물의 모양은 사이트의 지형적 모양에서 직설적으로 가져온 것이다. 튀어나온 모서리들은 역동적인 실내 공간을 표현한 것이다. 실외에서는 여러모습의 표면들이 특정 시점에서는 원을 형성한다. 강한 호 모양의 선은 브랜드 로고이면서 이곳이 Whitechapel의 중심 허브라는 개념을 연결시켜 준다.
Corian(DuPont 회사가 만든 표면 물질)의 표면 역시 Whitechapel의 사이트와 연결되고 주변환경을 안에서 대담하게 새로운 정체성을 표현한다.

WHITECHAPEL STATION

Development drawings of idea for Whitechapel

Structure set back in context

Structure set flush in context

Structure set forward in context

WHITECHAPEL STATION

This site will be there for general advice about employment opportunities and prospects. There will be computer stations for research and trained advisors with whom appointments can be booked for career advice etc. Events could be held there to celebrate cultural occasions to include people of minority groups who are suffering from unemployment.

이 사이트는 취업에 대한 조언을 구할 수 있는 곳이다. 컴퓨터가 있어 조사를 할 수 있고, 또 예약을 통해 취업과 관련해서 여러 충고와 조언을 구할 수 있거나 자문 위원들을 만나볼 수 있는 공간이 될 것이다. 문화적 이벤트를 열어 취업으로 고생을 하고 있는 소수 집단들을 위로할 수도 있는 공간이 될 것이다.

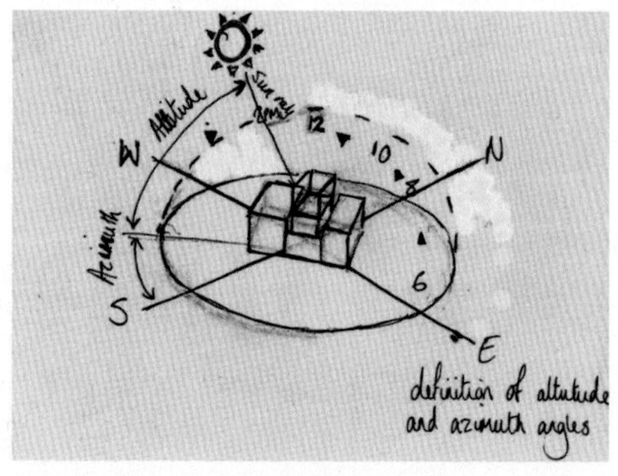

Diagram of sun's position to the structure

The building will have a glass box surrounding it. When the site is closed, the glass box lowers completely over the building, thus blocking the doorways. When it is open, the glass box rises up to symbolise ascension, but it also creates a vista to view the local area from a different perspective. The glass box will be very expansive and seemingly borderless to symbolise limitless potential. The movement also incorporates the idea of vibrancy and dynamism through kinetic energy.

빌딩은 유리 상자가 감싸고 있을 것이다. 사이트가 닫혔을 때에는 유리 상자가 내려와 출입구를 막을 것이다. 열렸을 때에는 상향을 상징하기 위해, 그리고 사이트에 대한 새로운 시점의 풍경을 자아내기 위해 상자가 위로 올라갈 것이다. 유리 상자는 크고, 또 보기에는 경계가 없어 운동성을 통한 역동성과 활기참을 표현할 것이다.

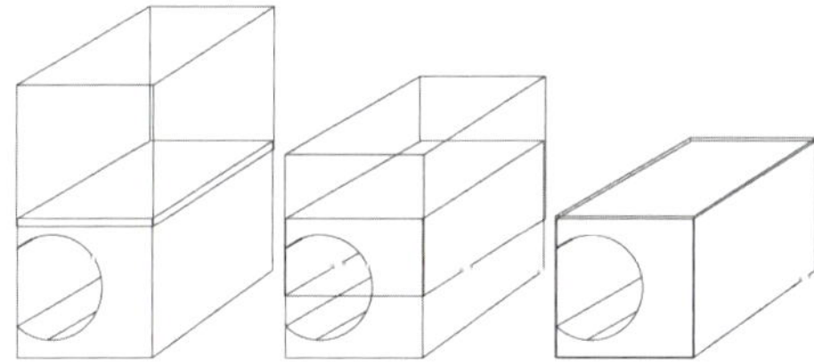

Glass box rising to open

Inspirational image Glass structure rising Studio Tjep; Amsterdam Airport.

Sketch model testing out light qualities and where the structure should be placed within the site.

MATERIALS

Initially the materials for this intervention were going to be dictated by the local materials in order to blend in. This did not suit the idea of this building breaking the circle and the pattern. It should stand out.

The height and proportions of the building are sympathetic to the surroundings, but the materials should be refreshing and distinct. The main building will be clad in white Corian and the glass structure will be clean and bright. Precedence for this can be taken from the 'Ideas Store' which is a glass building located a few buildings down on the same street.

재료

처음에는 주변 환경과 어울리는 재료를 선택하려고 했으나, 이는 생활습관을 바꾸는 컨셉과 어울리지 않았다. 환경과 어울리기보다는 눈에 띄는 것을 원했다.

빌딩의 높이와 비율은 주변 환경과 어울리되, 재료는 새롭고 뚜렷해야 한다. 주 빌딩은 하얀 corian을 이용하여 유리로 된 조형물이 깨끗하고 밝을 것이다. 이것은 같은 거리에 몇 블록 떨어진 곳에 있는 유리 빌딩인 'Ideas Store' 에서 얻어온 아이디어이다.

Final image of the intervention on site

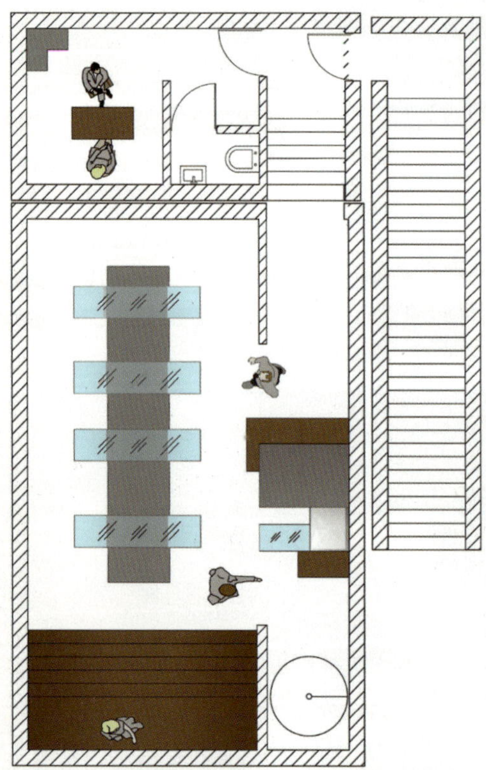

Plan

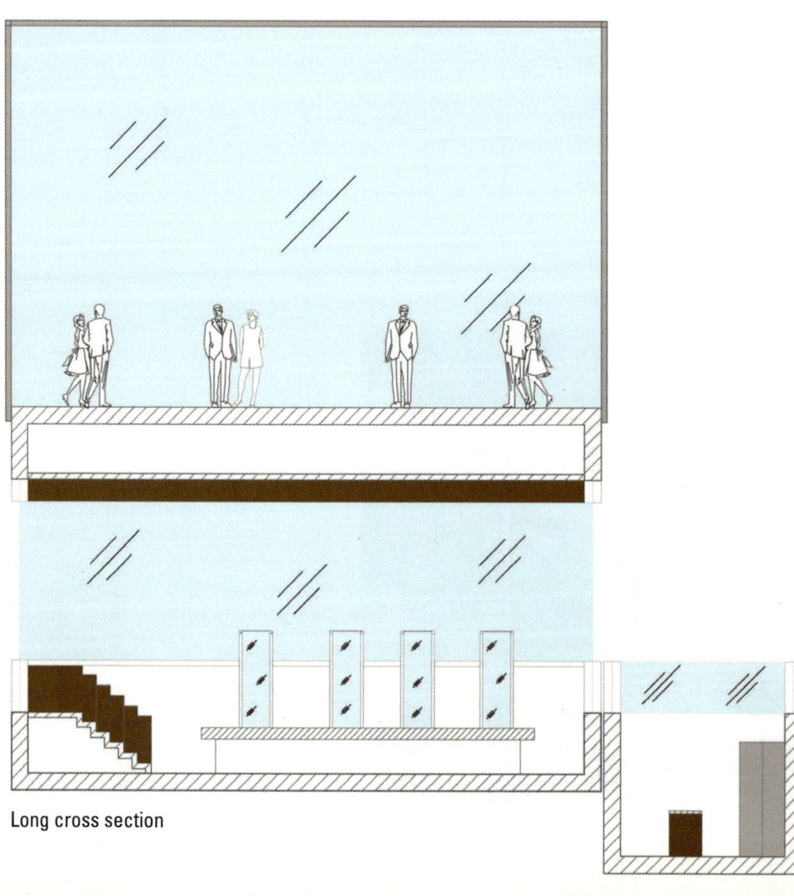

Long cross section

External elevation of Whitechapel Site

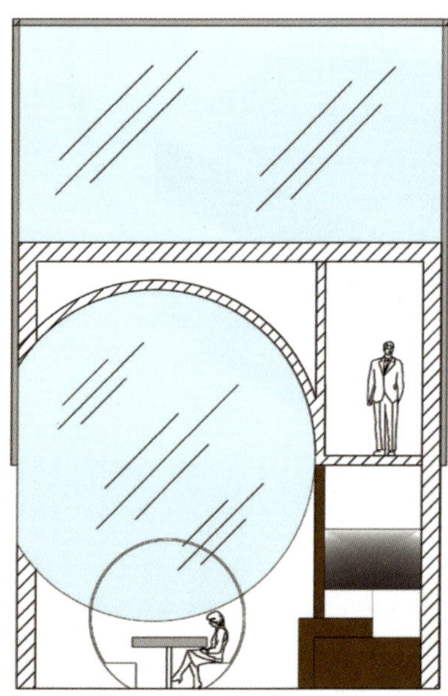

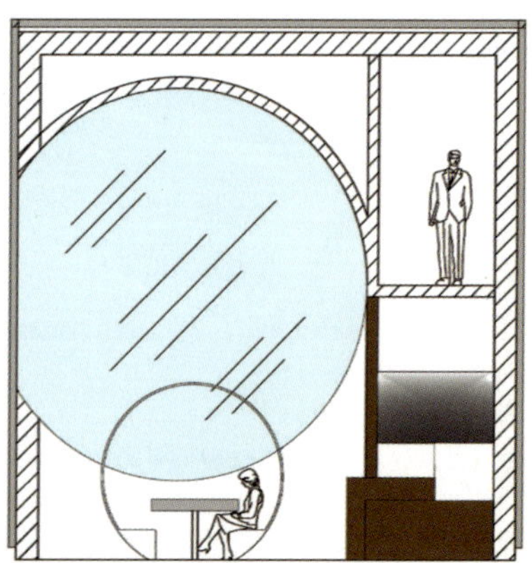

Short cross sections showing glass box at various stages of ascention

THE INTERIOR

The interior of the site includes a lower level and an upper level on the first floor. The lower level is not part of the structure seen from the front. It includes and office for one on one meetings as well as a WC. The upper level contains a reception desk, a modular seating system which slides across the length of a concrete table. At the front of this level is a curved platform which sits at the window for an interesting view outside. There is a spiral staircase, which leads to the glass vista on the second floor for a 360 degree view of the surrounding area.

실내

사이트 실내의 아래층에는 두 개의 단이 있을 것이다. 아랫단은 앞에서 봤을 때 건물의 일부가 아닌 것처럼 보인다. 일대일 미팅을 위한 사무실과 화장실이 있을 것이다. 윗단은 접수 데스크와 앉을 수 있는 공간을 마련해 둘 것이다. 이 공간의 앞쪽에는 굽은 단이 있어 밖에서 봤을 때 흥미로운 광경이 될 것이다. 나선형 계단으로 유리 상자가 있는 2층으로 올라갈 수 있다.

Gabriella Ramacciotti
Email: gabs.ram@hotmail.com

URBAN NEST

" a nest for dreaming.." Gaston Bachelard

THE BRIEF

The project brief was to create a series of structures in three specifically selected sites, to engage the community with local business. The principal design, the 'hub' rooftop space in Whitechapel, would connect with the subsidiary sites Braham park in Aldgate, Spital yard in Spitalfields through a journey motivated by employment. Promoting the advancement of education and the acquisition of skills was central to the concept, Job Wanted.

요약

프로젝트 목적은 세 곳의 특정한 구역에 시리즈로 지역 사회를 주변 기업들과 연결시키기 위한 건축물을 세우는 것이었다. Whitechapel의 중심지의 디자인은 Braham Park와 Spital Yard의 보조 건축물들은 고용이라는 주제를 가지고 동선을 구성한다. '구직' 이라는 컨셉에는 교육의 발전과 기술의 습득을 홍보하는 것이 중요했다.

BRAHAM PARK

Despite its rich and colourful heritage, social, economic and cultural barriers have blighted the East End of London for centuries. Today the area is characterised by pockets of culture and a diverse, thriving population, but poverty and transitional immigrant communities have also created a sense of alienation and detachment. There is an invisible 'glass divide' where the affluent highly-skilled and well-educated professionals of the city sit in marked contrast to the poverty, lack of skills and unemployment of the local community.

The challenge is to encourage access for the local community to the opportunities and prospects in Spitalfields and the nearby City. The solution will need to create a fertile ground to foster dialogue, cultivate a sense of belonging and help to create strong communities that work effectively together.

런던의 East End는 풍부하고 다양한 배경에도 불구하고 사회적, 경제적, 그리고 문화적 경계로 인해 수세기 동안 낙후된 채 있었다. 오늘의 East End 지역은 다양한 문화가 모여 있는 곳이자 번성하는 곳으로 알려져 있지만, 가난과 이주민의 유입으로 이민으로 인해 소외감과 형성되었다. 보이지 않는 벽은 도시의 고학력의 전문가들과 가난한 무직의 지역 주민들 사이에 알게 모르게 생겨났다.

도전은 지역 사회에 Spitalfields와 주변 도시에 있는 기회와 가능성에 대해 알리고 매개해 주는 것이다. 해결책은 소통을 하고 소속감을 키워주며 함께 일할 수 있는 경쟁력 있는 지역사회를 만들어주는 것이다.

SPITAL YARD

THE CONCEPT

" .....the discovery of a nest is always a source of fresh emotion"
- Gaston Bachelard

The concept of the Urban Nest evolved through looking at parallels in the animal kingdom; successful ecosystems where communities work together in unity and harmony.
Ants, the archetypal workers, provided the ideal metaphor; highly organised, they live in structured nest communities, negotiate complex relationships and have perfected the art of social networking.

The Urban Nest is an inviting space, a welcome refuge, but also a centre of activity and somewhere you return to; this cyclical networking process will ensure business and the local community engage and build fruitful bonds to secure the future of the younger generation.

At the same time, the ethereal space can perhaps recapture something of the surprise, magic and naive wonder we used to feel in our childhood at finding a nest unexpectedly..?

컨셉

"둥지의 발견은 언제나 새로운 감정의 원천이 된다."

– 가스통 바슐라르

도시적 '둥지'의 개념은 여러 사회가 힘을 합쳐 조화를 이루는 성공적인 생태계 즉, 자연에서 가져온 개념이다.
전형적인 일꾼의 유형인 개미에서 이상적인 메타포로 가져왔다. 그들은 '둥지' 사회에 살며, 계획적이고 또 복잡한 관계 사이에서 서로 타협하며 완벽한 사회 연결망을 구성했다.
URBAN NEST는 사람을 환영하는 공간임과 동시에 활동의 중심지이고 돌아올 수 있는 곳이다. 이 순환적인 연결 과정은 지역 사회와 기업들이 미래의 세대를 위해 유대감을 형성하는 데에 적극적으로 참여하도록 독려한다.

동시에 위쪽의 공간으로 인해 무언가 놀라운, 우연히 찾아낸 새 둥지 같은 마술 같고 순수한 어릴 적의 감성을 되찾아 질지도 모른다.

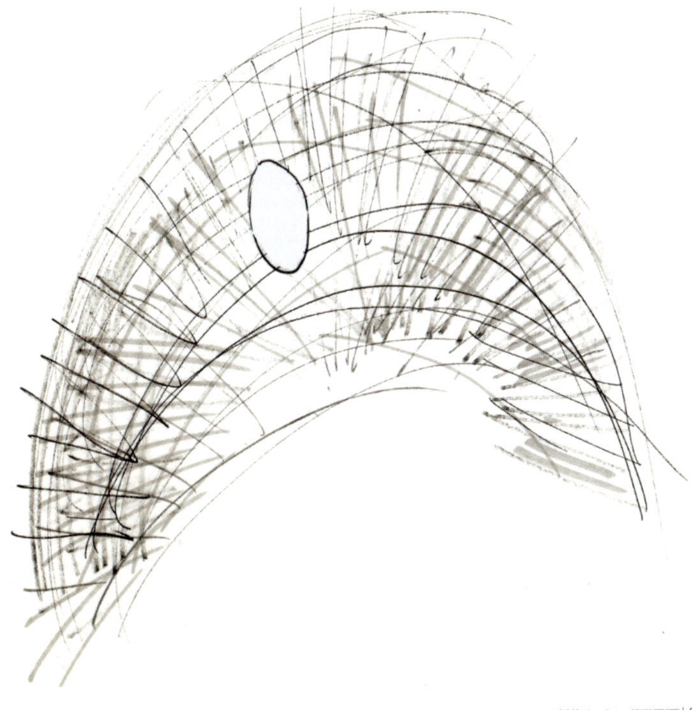

Taking its rhythm from the neighbouring features, the 'woven' facade suggests the forms of traditional nest-building, yet with an urban flavour.

주변의 모습을 본떠 엮인 모양의 파사드가 특징인 이 건축물은 전형적인 새둥지를 도시적인 감성으로 만들 것이다.

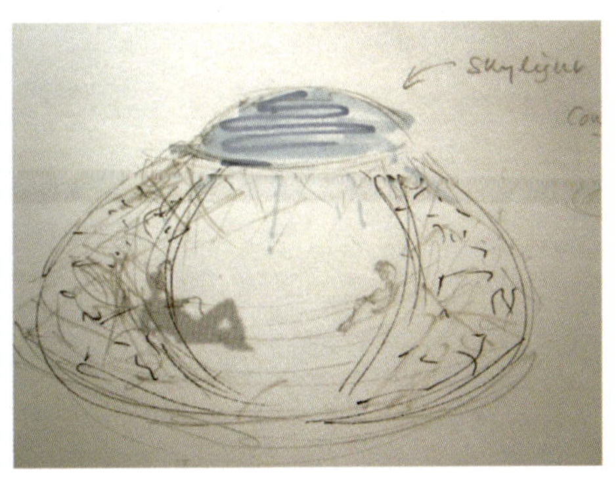

From soft to hard, rigid to fluid, the forms and materials of the exterior design to the interior, mirror the emotional journey of the visitor, taking a leap of faith into the unknown. From a position of uncertainty with perhaps a feeling of fear and apprehension, upon entering these give way to a sense of pleasure and comfort. The light, ethereal space with open porous, translucent soft materials and delicate spun, woven textures are inviting, comforting, nurturing.

부드러운 것에서 딱딱한 것, 규칙적인 것에서 유동적인 것으로 이어져 방문자의 모르는 곳으로 뛰어드는 기분을 보여주는 재료와 형태를 선택했다. 불확실하고 어쩌면 무서울 수도 있는 곳에서 편안하고 안도감을 주는 곳으로의 이동을 표현하고 싶었다. 가벼운 천상의 공간은 구멍이 많고 반투명한 재료와 함께 섬세하게 짜인 직물의 사용은 초대적이고, 편안하고, 진정되는 느낌을 준다.

A reference to the silk-weaving tradition prevalent in the East End, experimentation with woven mesh material wall structures, seating and domes.

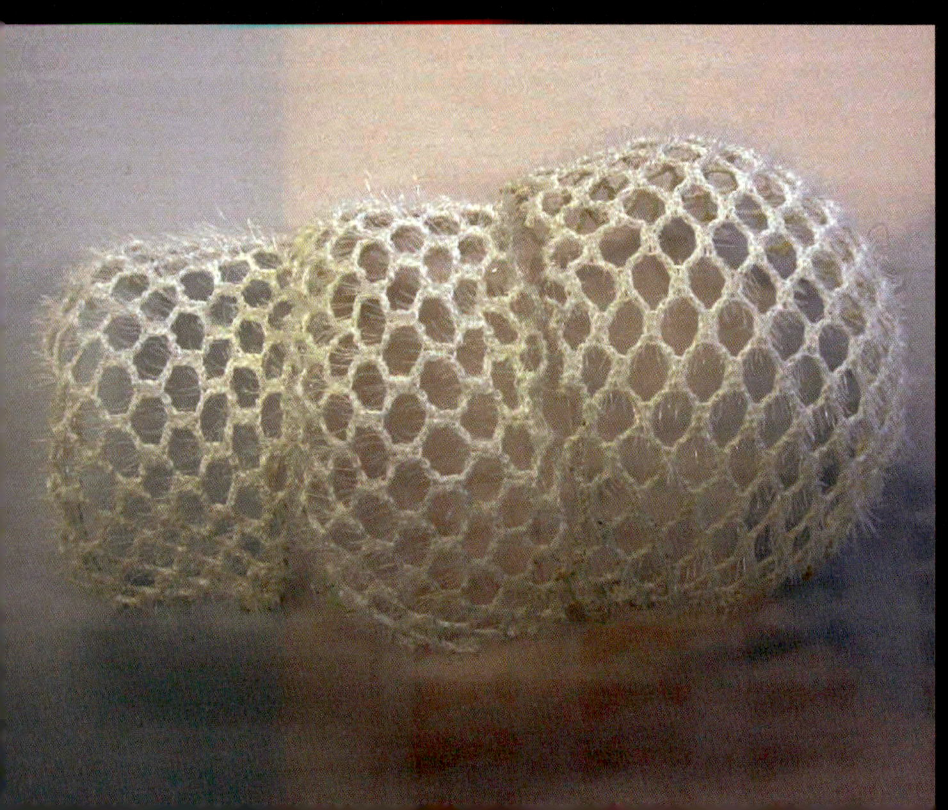

Once inside, access to the first floor is via the glass spiral staircase, where visitors enter the inverted nest-like mesh chambers to meet and greet. This is a platform for intimate one to one discussion, and the exchange of information.
They then proceed to the upper gallery, a large, communal space for sharing, talking, dreaming..
The woven rooftop creates captivating shadows as visitors relax and reflect, with a small opening to the sky..a dream world removed from the bustle of the street market outside. Hopefully they will leave inspired and motivated to take the next step on their journey!
Sustainable materials have been used throughout the design: fsc wood (facade, rooftop detail) recycled glass (main structure of rooftop) and pva synthetic material (moulded chambers, soft undulating walls & seating.

안에 들어가면 첫 번째 층은 유리로 된 나선형 계단으로 들어가며, 이곳에는 둥지 모양의 방들이 있어 사람을 만날 수 있다. 이곳은 일대일 대화를 할 수 있는 친밀한 공간이고, 또 정보를 교환할 수 있는 공간이다.
한 층 위의 갤러리는 커다랗게 트인 공간이며, 사람들이 서로 나누고, 말하고, 꿈꿀 수 있는 공간이다.
방문자들이 휴식을 취할 때 엮인 모양의 옥상을 흥미로운 모양의 그림자를 만들어낸다. 하늘의 작은 조각이 보이는 이곳은 외부의 정신없는 시장 거리와 단절되어있다. 아마 사람들은 이곳에서 영감과 자극을 받아 떠날 수 있지 않을까.재료들은 환경 친화적인 FSC 나무 (모양, 천장), 재활용 유리 (천장), 그리고 PVA 인공 재료 (방, 파도 모양의 벽과 의자) 등을 골랐다.

The images of the model capture the magic of the Urban Nest.. The striking patterns of the facade and rooftop emerged with a distinctly Islamic feel in sympathy and harmony with the local community.

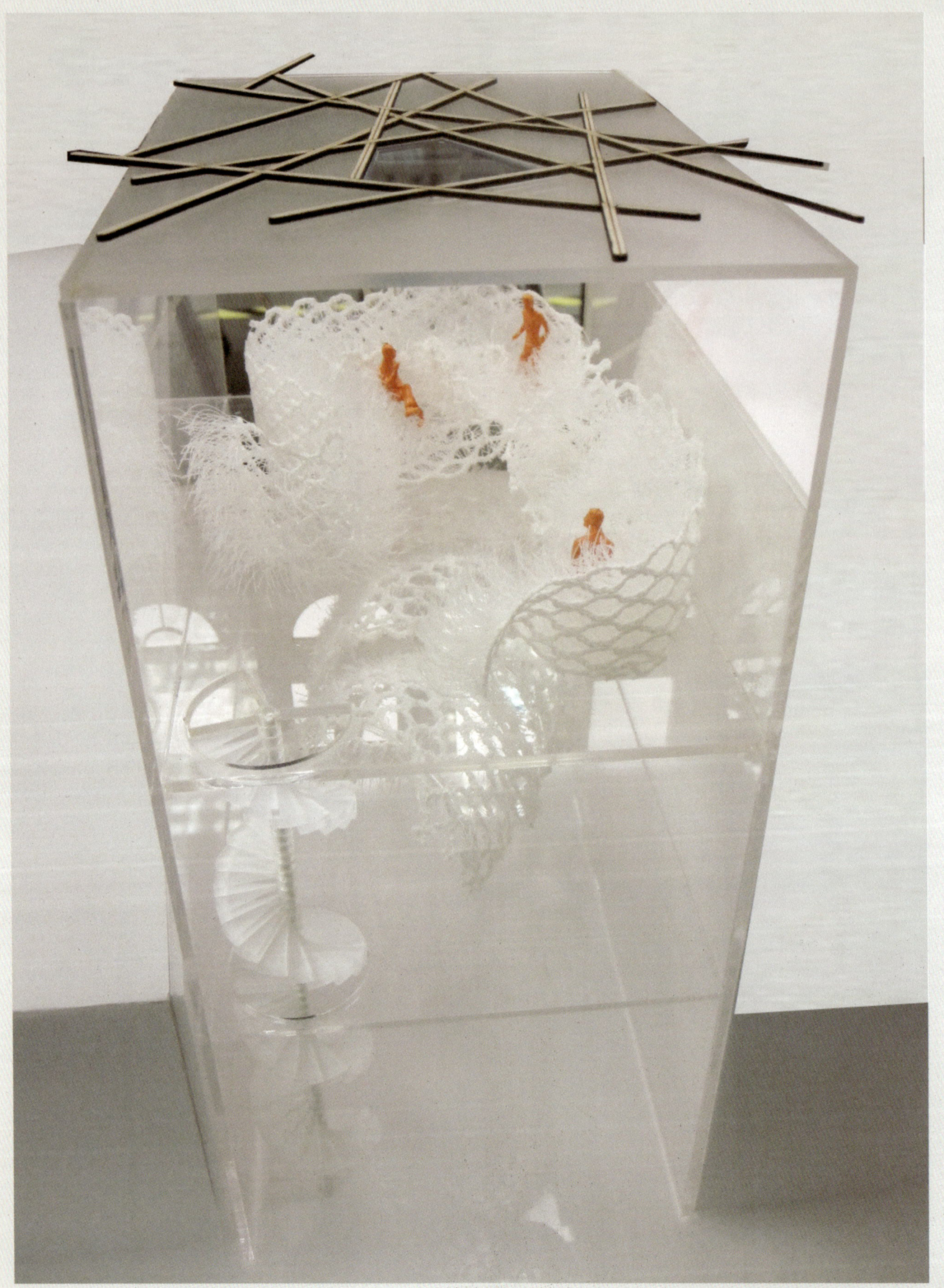

AW CHAMBERS
n Consultants
Fax: 020 7247 4722
fresh
SURIYA & Co
SOLICITORS

BRAHAM PARK

SPITAL YARD

Lina Navickaite
lnavickait@googlemail.com

Visible Vibes

SITE PLAN

SITE IMAGES

PROPOSAL

There are so many places or areas that could be called non-existent in London. Some of them are just forgotten others are being ignored. That could be because the businesses they are lead by are not powerful enough to make them outstand, so even communities based there there usually are left behind. One of such an areas could be a triangle from Whitechapel passing Braham Park and ending up in Spitafields.

AIM

Interior designers and architects were given a symbolical triangle on the map: Braham Park, Whitechapel, Spitafields. The idea is to create spaces where people could gather and speak about the particular subject, help each other and would feel free to break the wall and enter the different areas around, so that would be a big transition and step forward to this development of communities and businesses.

The aim of a project is to design three spaces based on their areas needs (depending on businesses around, people and their distinguishing features) focussing on employability through connecting different communities, encouraging them to share, support each other, help and improve together.

제안

런던에는 '존재하지 않는다' 고 할 수 있는 공간이 상당히 많다. 그저 잊혀 지나갈 수 있는 곳일 수도 있고, 어떤 곳은 외면당할 수도 있다. 어쩌면 그 이유는 그곳을 이끌어갈 만한 사업적 경쟁력이 부족해서 지역 사회까지도 그 경쟁력의 뒤편에 소외되는 것일지도 모른다. 이렇게 남겨지거나 외면당하는 곳들 중 하나가 Whitechapel에서 Braham Park를 지나 Spitafields까지 이어지는 삼각지대이다.

목적

실내 디자인과 건축학과 학생들에게 Braham Park, Whitechapel, 그리고 Spitafields이 삼각형으로 표기된 지도를 정했다. 목직은 사람들이 모이고 특정 주제에 대해 이야기를 나누며 서로를 도와 그 곳을 둘러싼 보이지 않는 벽을 깨고 주변의 다른 지역들과 연계 될 수 있는 기회를 마련, 즉 지역 사회와 기업들에게 발전으로 이어질 수 있도록 환기시켜줄 수 있는 공간을 마련하는 것이다.

이 프로젝트의 목적은 주변 기업과 지역 주민의 필요성인 취업에 집중해서 다른 지역과의 교류를 만들어주고 서로 협력과 나눔을 통해 발전을 장려하는 것이다.

SPITAFIELDS

WHITECHAPEL

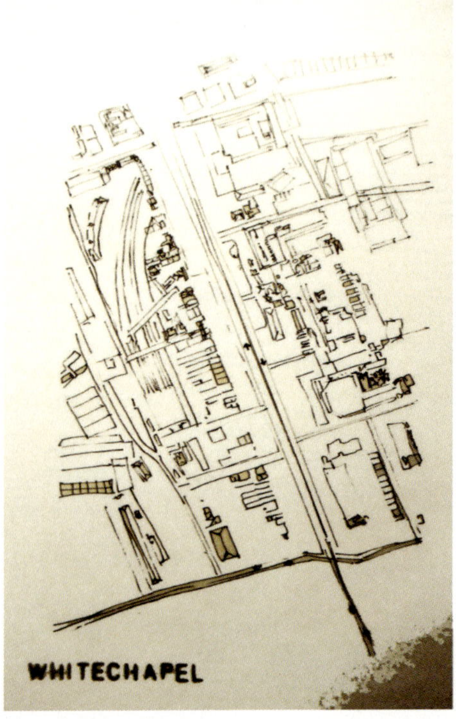

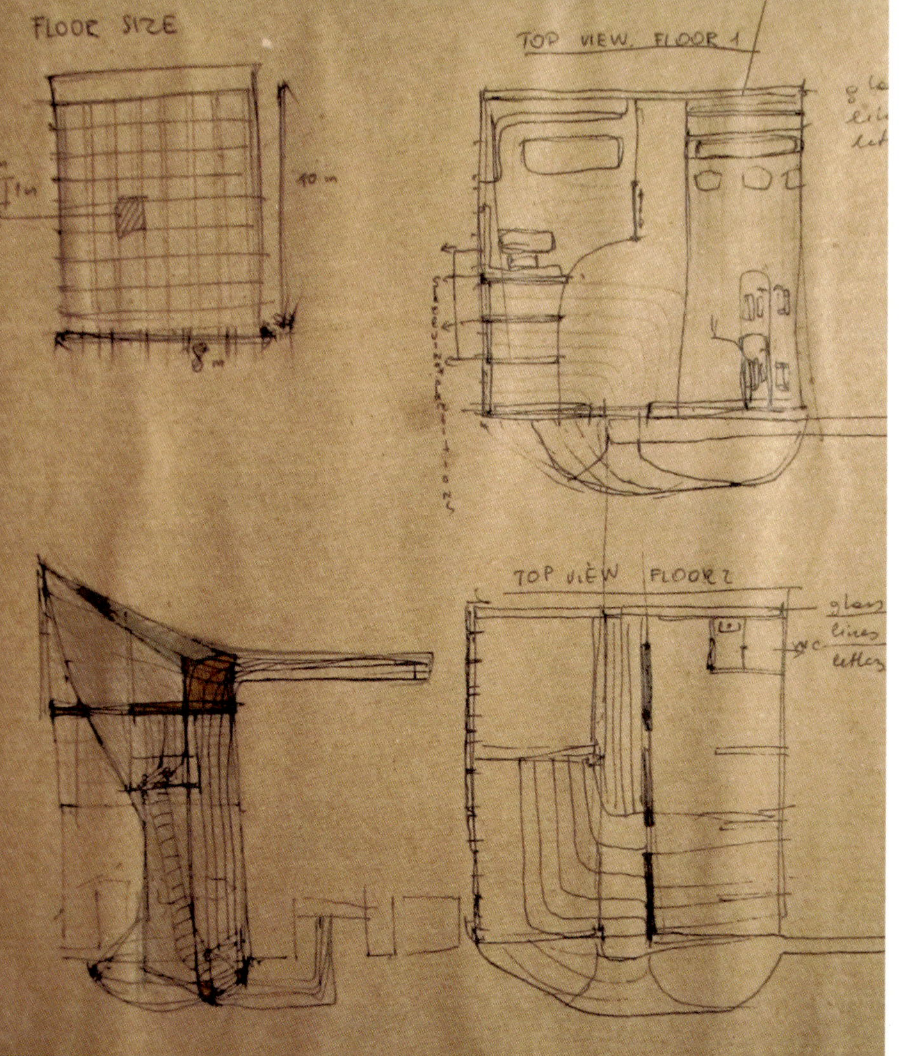

VISIBLE VIBES

VIBE is a metaphorical word for links connections and spreading. As interior designers we need to give a visual effect and experience and held the idea from imaginary life to the reality. We need an interior or exterior to reflect the main idea, that is the main affection from our side. So, we make it VISIBLE. HISTORY, EDUCATE, CHANGE, ENCOURAGE, BELIEF, LIFE, FUTURE - are the words used in interior design and LINE which flows, spreads and runs around interior too. These words are being kept in the walls of history, but the lines transfers it to the next level - preset or future. All these symbols leads the passenger througout the space providing a clue of a metaphorical meaning of his journey.

VISIBLE VIBES

VIBE라는 말은 서로의 연결과 확장, 확산에 대한 은유적 표현이다. 실내 디자이너로서 우리는 시각적 효과를 통한 경험을 주고 상상과 현실을 이어준다. 우리는 주된 생각을 투영할 실내 혹은 실외가 필요하고, 그것은 우리 쪽에서는 가장 중요한 요소들이기에 애착이 가는 것들이다. 그래서 우리는 그것들을 시각적으로 보이게 만든다. 역사, 교육, 변화, 격려, 신뢰, 삶, 미래 – 이것들이 실내 디자인에서 쓸 단어들이다. 흘러가고, 뻗어 나가고, 실내를 따라가는 선 역시 있다. 이 단어들은 역사의 벽에 갇혀있지만, 선들이 이것들을 다음 단계로 끌어낸다. – 현재 혹은 미래로. 이 모든 상징들은 방문객의 짧은 여행의 의미에 대한 은유적 표현이 된다.

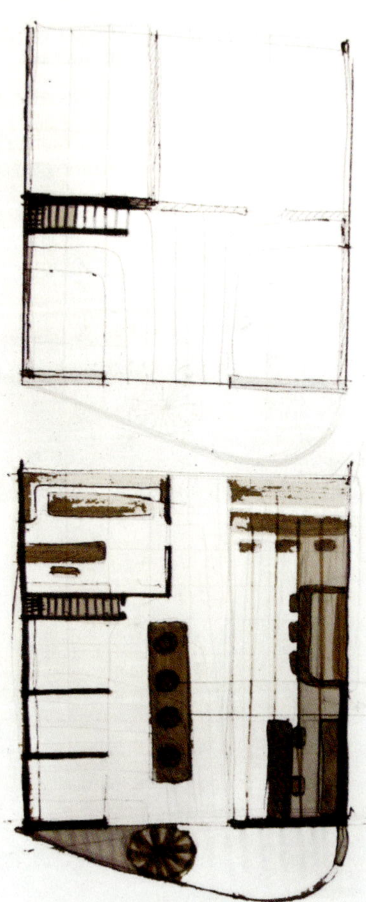

SPACE PLANNING SKETCH

SKETCHING INTERIOR

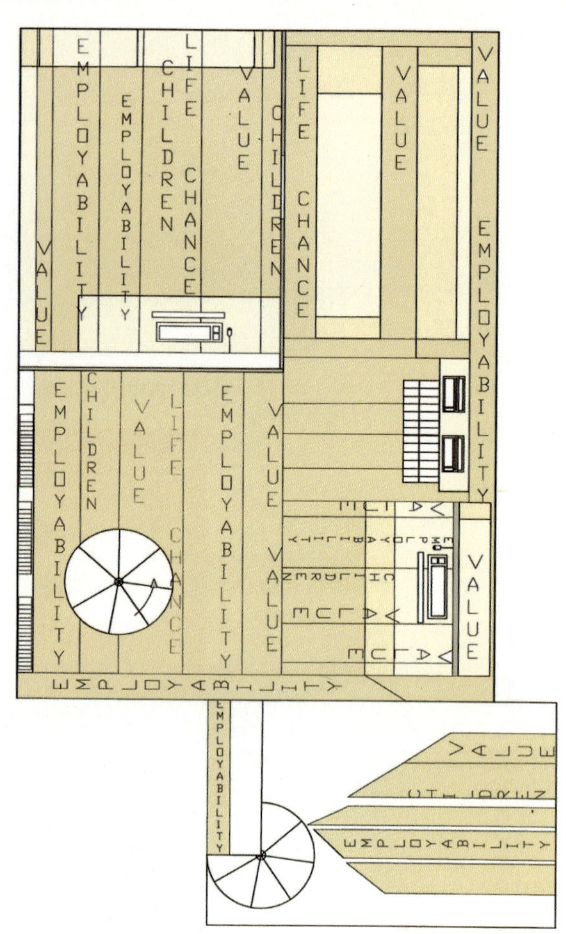

WHITECHAPEL FLOOR 1

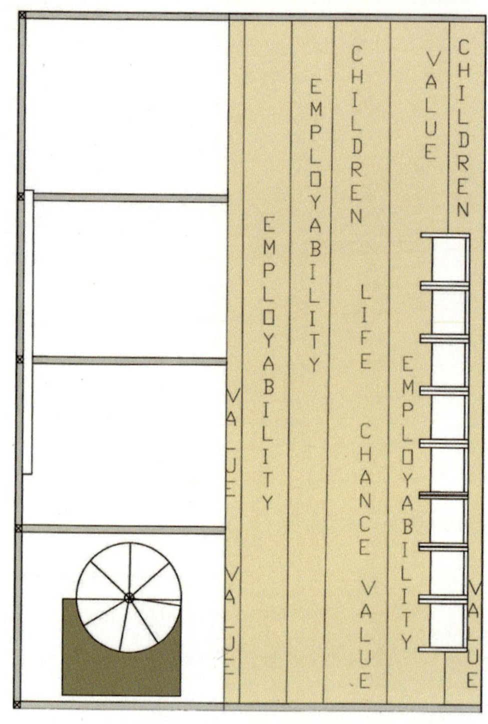

WHITECHAPEL FLOOR 2

WHITECHAPEL ELEVATION

WHITECHAPEL INTERIOR & EXTERIOR SKETCHES

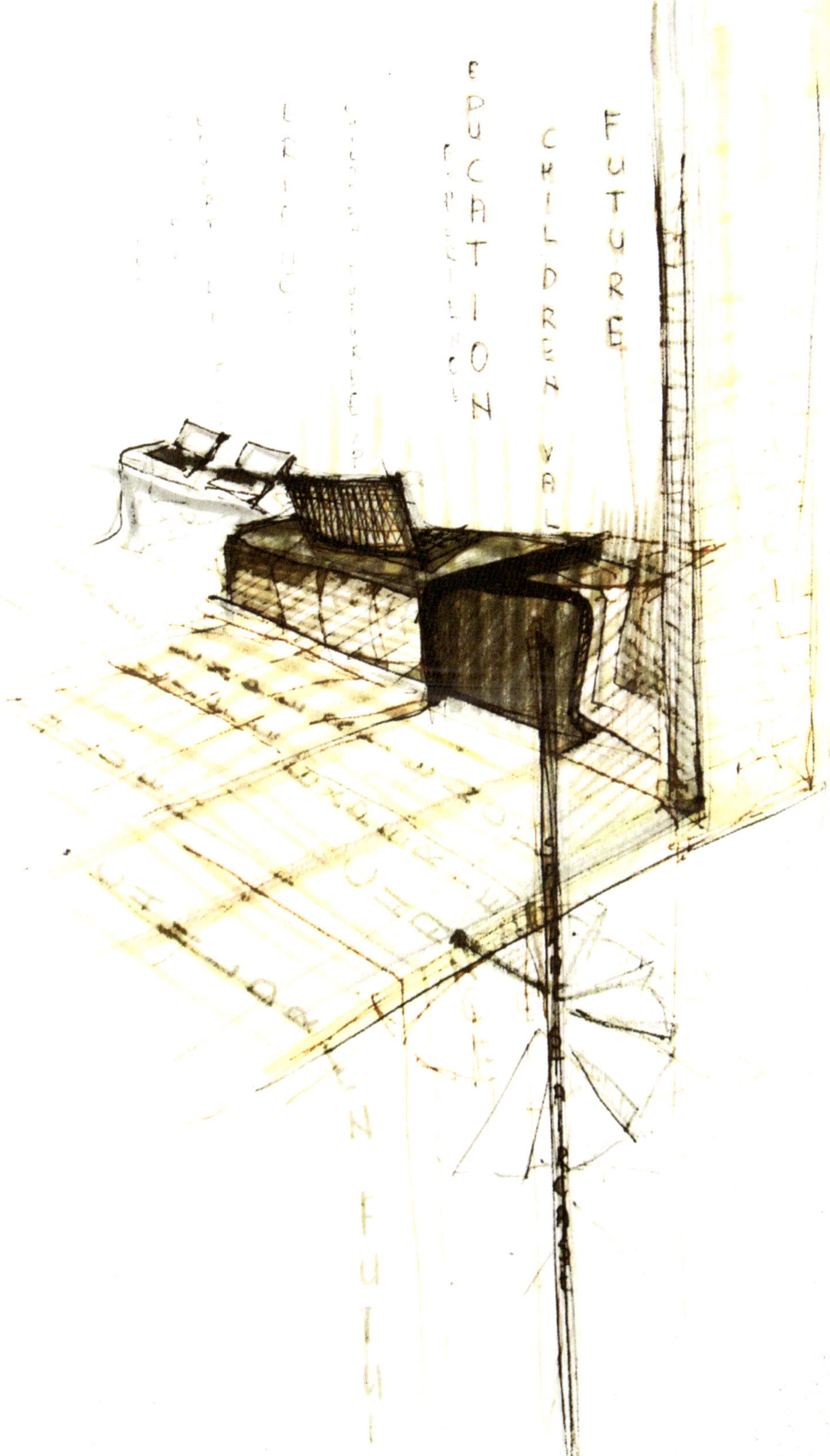

FUTURE LIFE CHANCE
EMPLOYABILITY
CHANCE
EMPLOYABILITY
VALUE
ABILITY
LIFE

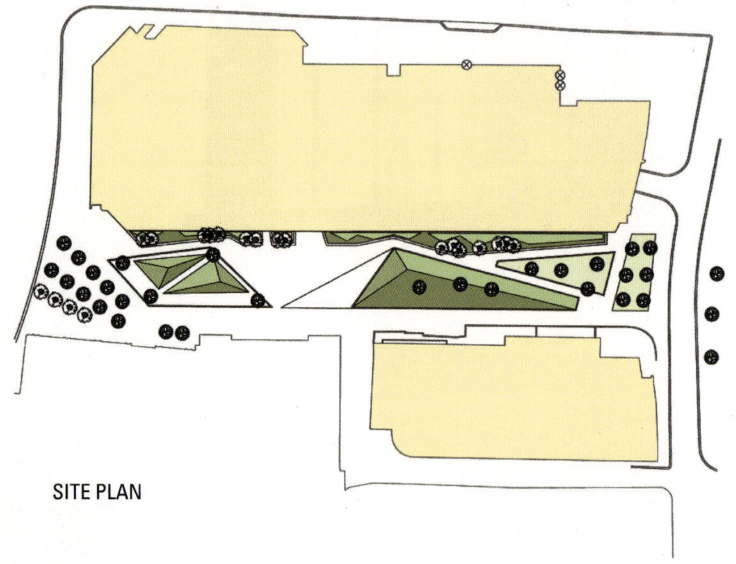

SITE PLAN

Braham Park is a place in between. It is a reminder or a place to develop your thoughts. It encourages taking the next step to the future – Spitafields, where The new life begins.

Braham Park는 사이에 있는 공간이다. 이곳은 과거를 상기시켜 주는 곳일 수도 있고, 생각을 하게 해주는 공간일 수도 있다. 이곳 미래, 즉 Spitafields로의 다음 한 걸음을 격려해준다.

Transfer point from Whitechapel to Spitalfields: Information Board

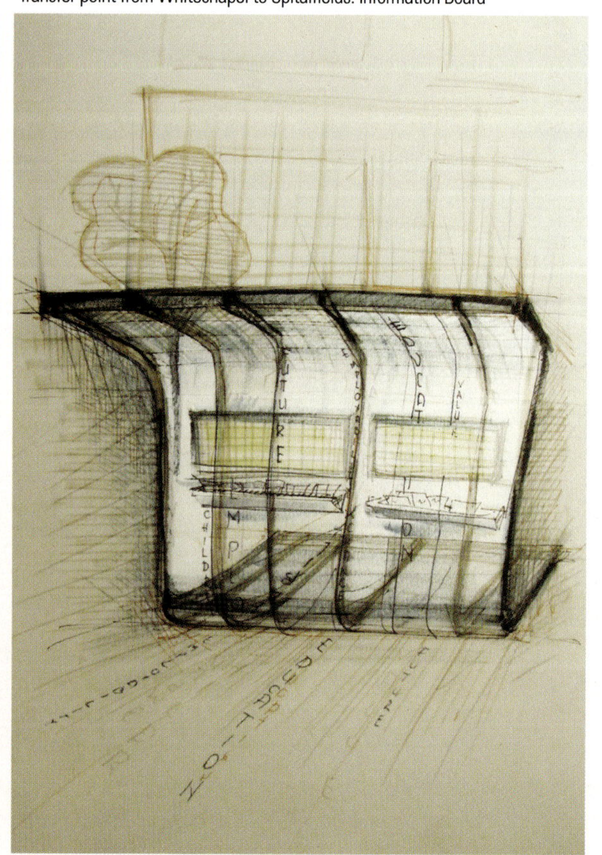

Idea Development

The Last point of a triangle:
Space to have interviews and meetings about the community employment opportunities

Lois Woodgate
loiswoodgate@hotmail.co.uk

Hidden Spaces

SPITALFIELDS

WHITECHAPEL

THE CONCEPT

The concept acts as a guide for the community from Whitechapel through a series of journeys to overcome a perceived barrier, which is stopping them from entering Spitalfields. This barrier, even though is a hypothetical one, seems to be opaque and stopping the community from seeing the opportunities on the other side.

The journey they will travel through will guide them from the opaque to the transparent, helping them to gain confidence in themselves and push boundaries, whilst having the comforts around them for support and inspiration. These journeys will be placed into the Whitechapel, Braham Park and Spitalfield as stepping-stones towards moving through this barrier. They will be positive areas, places of opportunity and hope allowing a transition through the space with security, whilst changing the visitor's perception, bridging the gap between the two communities.

이 컨셉은 Whitechapel의 주민들이 Spitalfields로 가는 벽을 넘어서기 위한 지도가 된다. 이 벽은 비록 가상적인 것이지만 불투명하여 지역 사회가 반대편에 있는 기회를 보는 것을 막는 듯하다.

이들이 떠날 여정은 이들을 불투명한 것에서 투명한 것으로 인도하며, 자신을 찾고 경계를 깨는 것을 도와줌과 동시에, 주변에는 익숙한 요소들이 있어 영감과 안정감을 줄 것이다. 이 여정들은 Whitechapel에서 Braham Park, 그리고 Spitalfield까지 이어지는 징검다리가 될 것이다. 긍정적인 공간이 되고, 기회의 공간이 되어서 사람들에게는 변화를 주기도 하지만 안심되는 공간이기도 해서 두 공간을 이어주는 공간이 될 것이다.

Braham Park

Window examples in Whitechapel

Community Hall in Whitechapel

Whilst on an educational walk around Aldgate I learnt that the community does not stray away from their comfort zone, they stay within certain areas and tend to have repetitive motions such as the journeys they take, shops they visit and people they meet. From this I looked for repetition within the Whitechapel landscape.

The concept of 'windows' has been chosen not only because they are used to provide light into the interior and create interesting reflections into the landscape. They are also connected to the phrase "window of opportunity" and therefore will aim to provide this in the design proposal.

Aldgate 주변을 걸어 다니면서 그곳의 주민들은 행동 범위가 그다지 넓지 않다는 것을 깨달았다. 특정 공간 안에 머물고, 만나는 사람, 들리는 가게, 떠나는 여행마저도 늘 같은 패턴이다. 이것에서 출발하여 Whitechapel 안에서 반복되는 것을 찾았다.

창문이라는 소재는 실내에 빛을 들이고 반사광을 만들 뿐만 아니라, 'Window of opportunity' (절호의 기회) 라는 말에도 사용된다.

East London Mosque

Sculpture in Aldgate

Windows within the landscape

Experimentation with window positions

Memorial in Altab Ali Park

The Royal London Hospital

Opposite Whitechapel Station

View from Whitechapel Highstreet

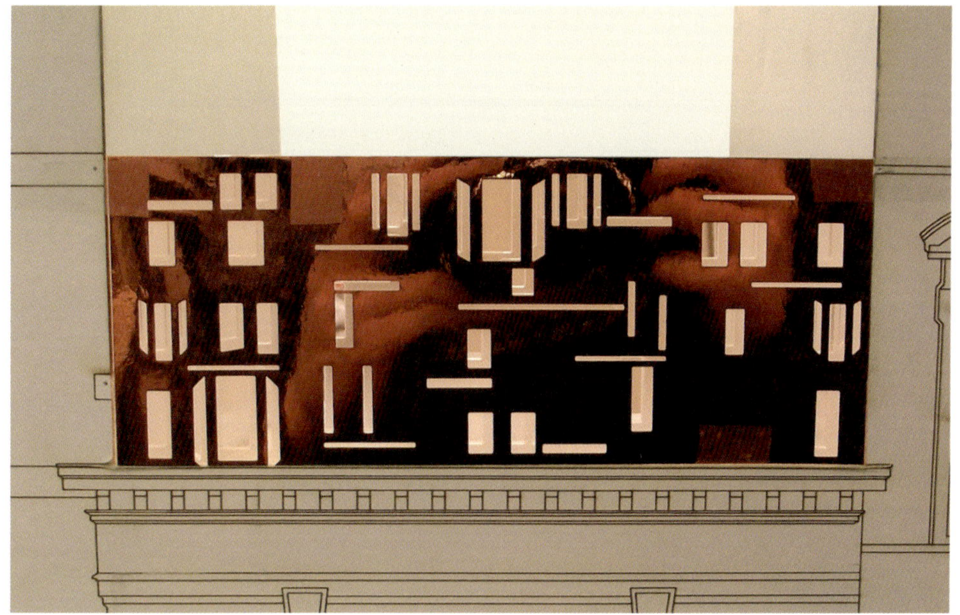

Copper Facade

Instead of having a brick facade copper plate panelling has been proposed to blend in with the warm tones of the brick surrounding it but also will still have enough of an impact to draw people's attention upwards, beyond the sight-line of the shop facades so they are aware of its presence and invite them up into the building.

벽돌 모양 대신에 구리판을 이용하여 주변 벽돌의 따듯한 색감과 어우러지지만 그래도 사람들의 눈길을 끌 수 있을 정도로 가게들 보다는 높게 설치할 것이다.

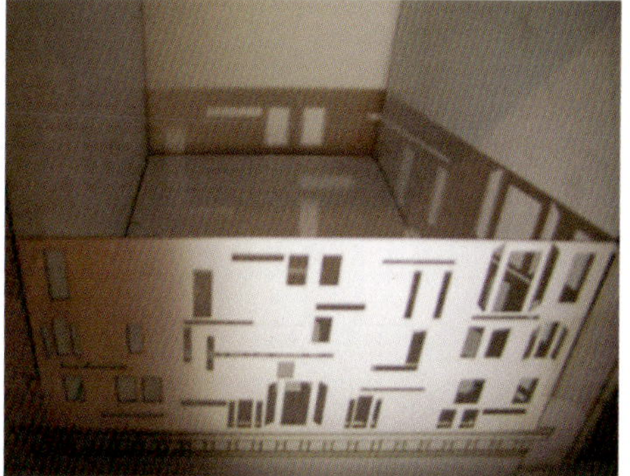

Experimentation With Window effects

The concept will alter the nature of the repetitive behaviour and architecture.
The direction and sizes of the 'windows' have been adjusted and have been layered and slightly moved off centre to adjust the slight-line and 'framing' of what people can see.
The frosted layer will make the viewer think about what it is they are really seeing. Do they have a clear vision of what they want or is it translucent?
It also softens the shadows, which fall in the interior helping the area look less threatening.

Light and shadow within the building

Whitechapel Elevation

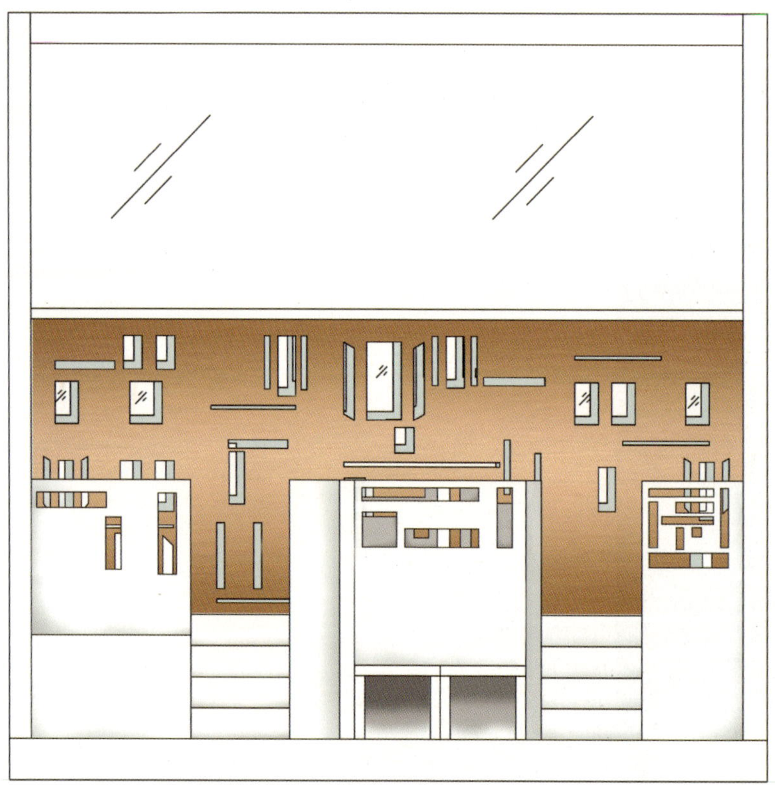

Whitechapel Section

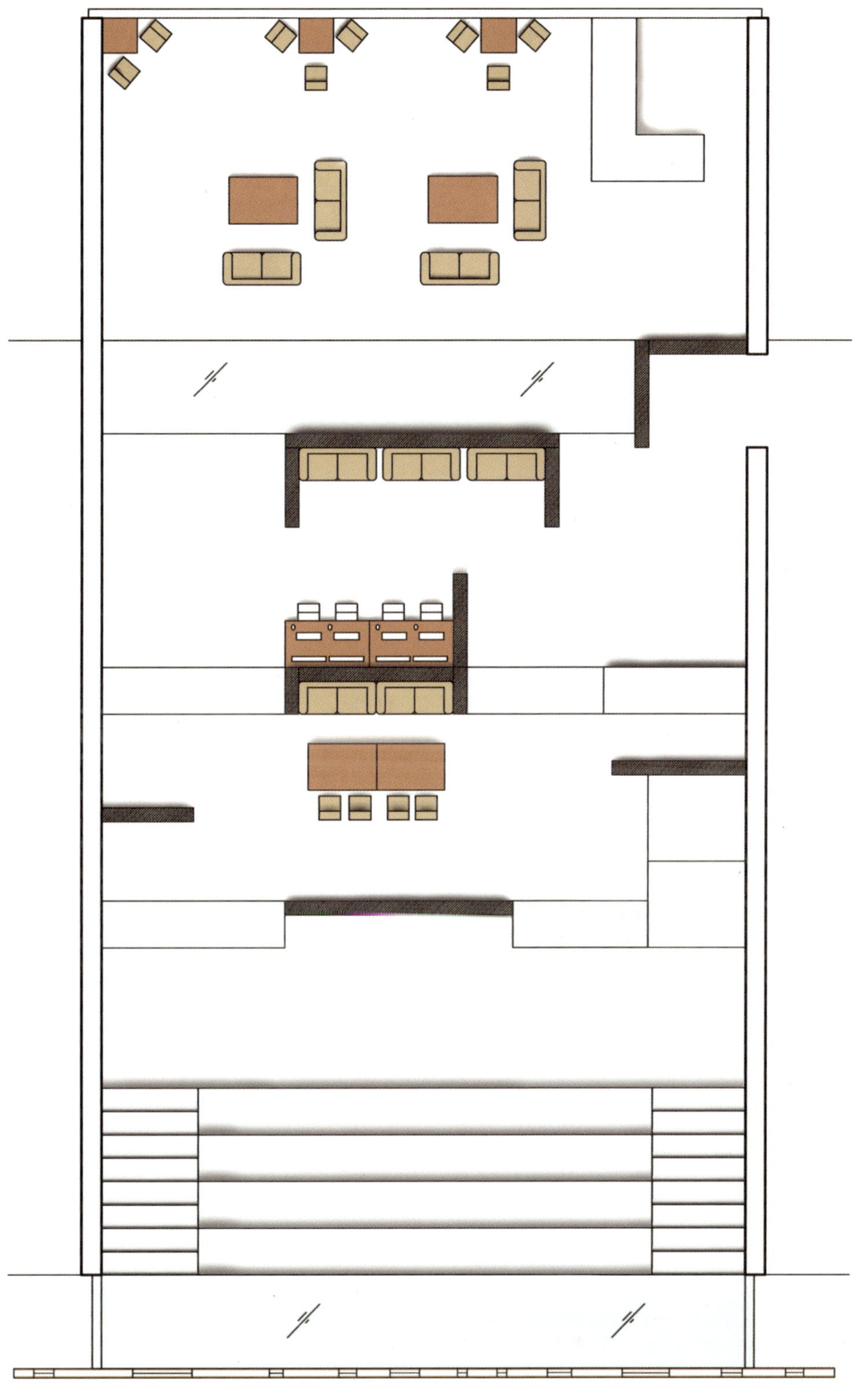

Whitechapel Plan

View from the platform

The consultation Area

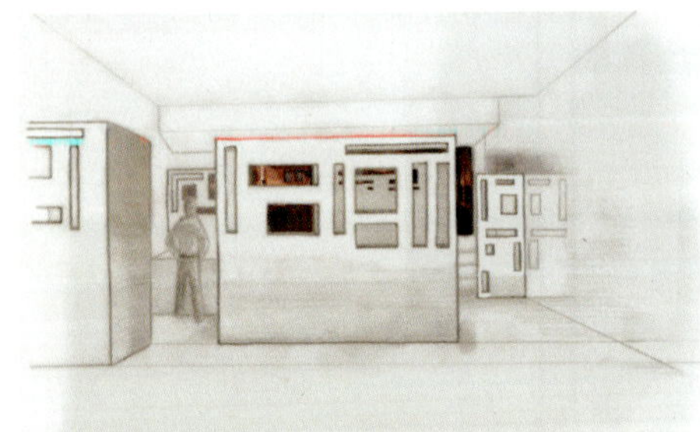

View of the front, from the rear

The front of the building and steps/auditorium

Final Images

For the interior, the journey intends to be an educational one, with metaphorical boundaries to cross while they are inside, such as a change in materials, lighting and height.
The series of walls reflects the facade as they are positioned in layers with 'windows', which will frame parts of the space as well as allowing light to travel further into the building. The windows will also create shapes of shadows and light, which will move around the building as the sun moves, reflecting the idea of time and change. I've chosen to have a clear top to open up the space and add light into the room. It also opposes the restriction of a window and it's frame as they are completely open to the sky and beyond.
The rear of the space has been extended out further to create one final boundry for the visitors which is to walk over a clear glass strip where they can see down to the ground. Beyond this final boundry, there will be an area surrounded by glass instead of brick walls, to represent the open possibilities. In this space they will be able to talk with advisors and gain the knowledge to make the next step to Braham Park and onto Spitalfields.

내부 인테리어는 메타포적 경계를 표현하는데 있어, 예를 들어 재료나 빛이나 높이의 변화가 있을 것이다.
벽들은 창들을 겹겹이 설치해 주변 모양을 반영하고, 공간의 부분부분을 틀에 넣기도 하고 실내를 더 비춰줄 것이다. 또한 창들은 시간이 감에 따라 달라지는 그림자와 빛의 모양을 만들 것이고, 이는 시간의 변화라는 테마에 어울린다. 천장은 투명해서 공간을 열어주고 빛을 더 많이 들어오게 한다. 그리고 창틀의 제한과는 반대되며 사람들이 하늘과 그 이상을 볼 수 있다.
공간의 뒤쪽은 확대되어 방문자들에게 마지막 경계를 제공한다. 바로 바닥에 땅이 보이도록 설치된 가늘고 긴 유리 조각을 건너는 것이다. 이 마지막 경계를 넘으면 벽돌로 된 벽 대신 유리로 된 공간이 기다리고 있을 것이다. 이는 열린 기회들을 상징한다. 이곳에서 그들은 조언자들과 이야기를 하며 Braham Park와 Spitalfields로의 다음 단계를 거치게 된다

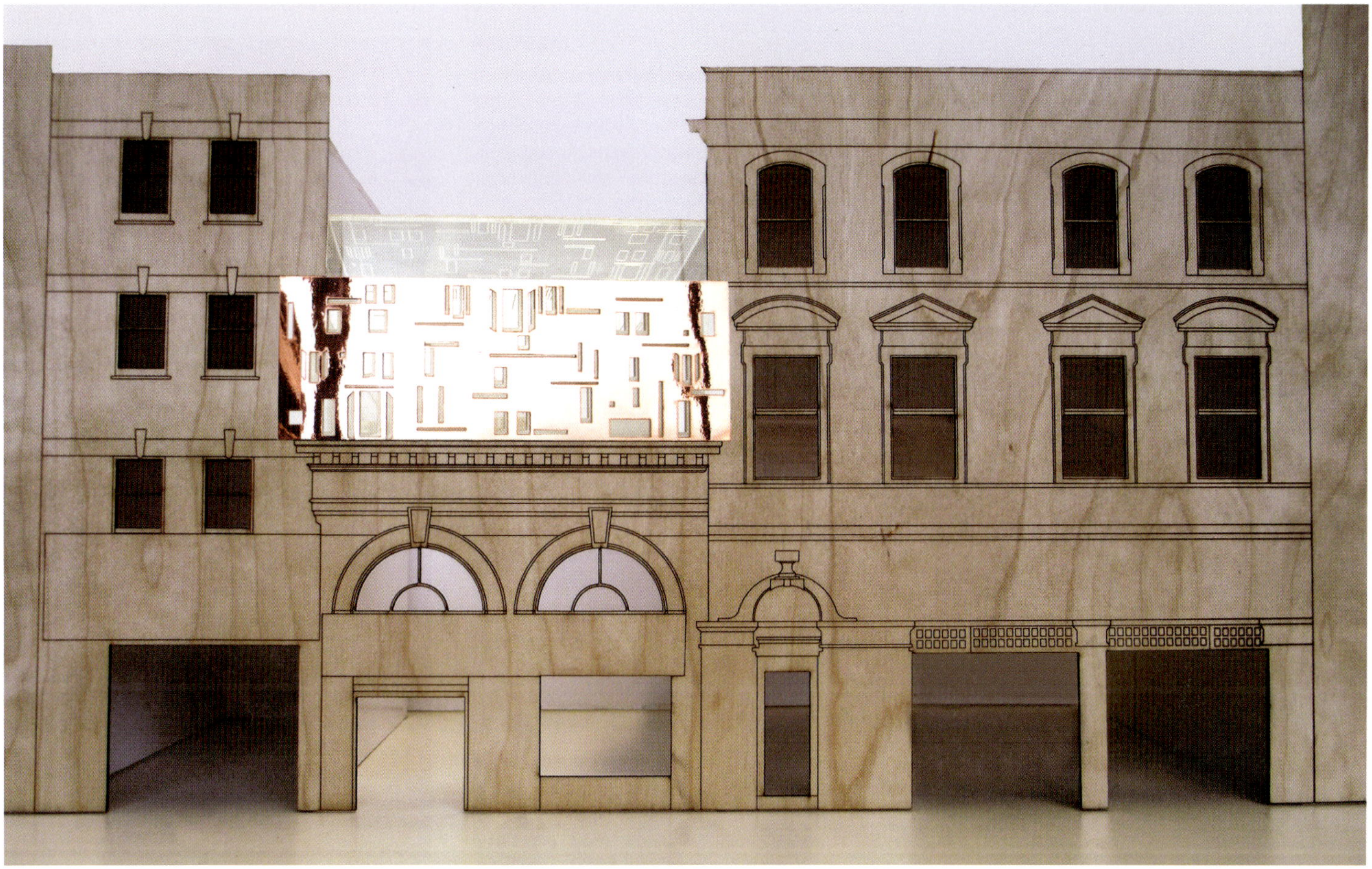

The Final Model

Elevation of the Model

View inside the model of the desk from the rear of the space

The glass boundry from the rear of the space

View through the model incuding the Copper Facade

BRAHAM PARK

Braham Park has been kept uncomplicated, as it would be an pause point between Whitechapel and Spitalfields. The walls from within the interior of the Whitechapel space have been placed in Braham Park as something familiar reminder of their first step.
The visitors can walk through the windowed planes as they travel from one side of the park to the other.

Braham Park는 다른 두 공간 사이에서 잠시 멈추는 공간이기에 간단하게 두었다. 첫 번째 단계에 대한 기억을 되새기기 위해 Whitechapel의 공간에 있는 벽 Braham Park에도 설치했다.
방문자들은 공원을 거닐며 창에 설치된 면들 사이를 걸어갈 수 있다.

The windows in the boards increase the sense of the context as light which is reflected off of the glass on the srrounding buildings either side would create interesting shadows and light patterns as the sun moves through the space.
There would be light and shadow cast on on both sides of each board representing the spreading of knowledge. This idea of light and shadow as well as time and change is a direct reflection of the Whitechapel space to make the transition from one space to the other an easy one.

판에 설치된 창들은 주변 빌딩들의 유리에서 태양의 움직임에 따라 변화하는 반사된 빛을 받아들이기에 의미를 더욱 강화시킬 것이다.
이렇게 실내에 들어온 빛과 그림자는 양쪽으로 퍼져 지식의 확대를 상징할 것이다. 빛과 그림자, 그리고 시간의 흐름은 Whitechapel 공간에서 직접 본떴으므로 한 공간에서 다른 공간으로의 이동을 더 쉽게 만들어줄 것이다.

SPITAL YARD

The design for Spitalfields is very different from Whitechapel because the two areas are very different in materials and character. Similarly to Whitechapel the design reflects the landscape, which surrounds it. The ground level reflects the building clad in wood, and the window frames are based on the window of the converted church, but are exaggerated. Once the visitor enters the building, they will enter a light and welcoming space.

Spitalfields의 디자인은 Whitechapel의 디자인과 많이 다르다. 두 지역은 성격이나 환경이 다르기 때문이다. Whitechapel과 마찬가지로 모양이 주변 환경을 본떴다. 아래층은 나무로 된 빌딩을 상징하고, 창틀들은 개조된 교회의 창에서 따왔지만, 과장되었다. 방문자가 안으로 들어올 때면 언제나 환영받을 수 있는 공간일 것이다.

In this space the visitors will be able get information and help to decide on the career they want to pursue and advice on qualifications and applications for the career they have chosen. They can return to the space for advice and can obtain help with every step, until they succeed in getting the career they desire.

이 공간 안에서 사람들은 자신이 원하는 직업에 대한 정보를 얻을 수 있다. 또한 언제든지 돌아와서 조언을 구할 수 있고, 성공할 때까지 도움을 받을 것이다.

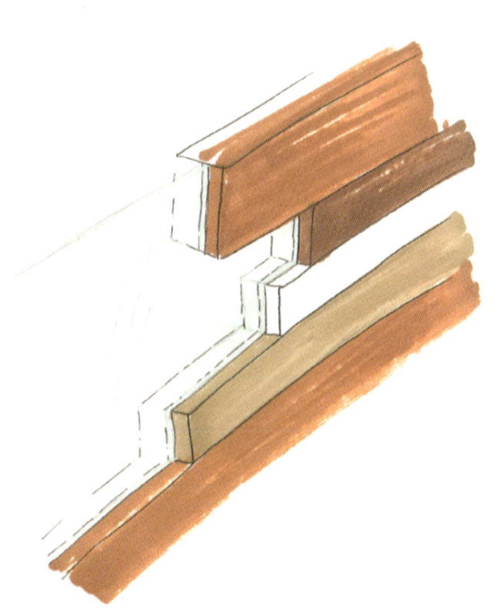

Maria Cristina Rech
cri.rech@gmail.com

Whitechapel Arts Center

SPITAL YARD

WHITECHAPEL STATION

BRAMHAM PARK

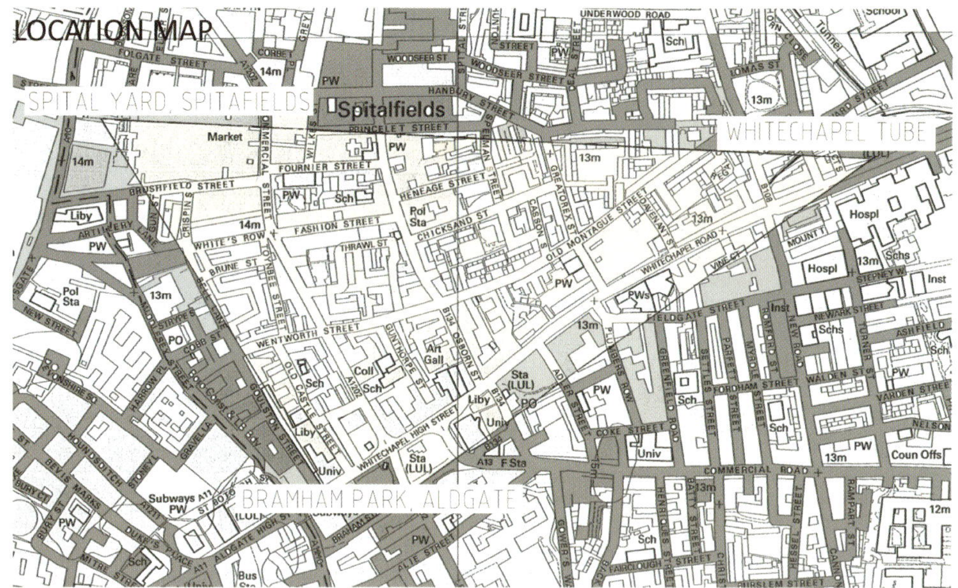

THE BRIEF

The aim of this project is to redevelop three sites: 277 Whitechapel Road, Spital Yard and Bramham Park, in order to engage all of the community with local business and employability within the Aldgate area and beyond.

The project highlights and brings alive, hidden or exposed interstitial space, gaps in-between, to help engage the community. In some areas, such as spitalfields, beautiful and vibrant locations and history are blocked out by large scale contemporary buildings. All these physical boundaries lack a visual connection, to merge cultures, classes and the community's diversities. The intervention aims to enable and offer, through the design solution, future perspectives and employment opportunities, by encouraging the local community into the environments.

요약

이 프로젝트의 목적은 Whitechapel, Spital Yard, 그리고 Braham Park의 세 공간을 지역 사회와 기업들을 연계시켜 재개발하기 위함이다.

프로젝트의 가장 큰 목적은 틈새 공간을 이용해 지역 사회와의 연계를 이끌어 내는 것이다. 예를 들어 Spitalfields에는 커다란 현대식 빌딩들이 아름다운 역사적 공간들을 막아내고 있다. 이러한 경계는 시각적인 소통과 공동체와 지역, 문화의 다양성이 통합되는 것을 방해한다. 이번 계획은 디자인을 통해 지역 공동체가 지역 환경에 참여함으로써 미래에 대한 전망과 취업 기회가 제공되는 것을 목표로 하고 있다.

SUNPATH ANALYSIS

FRONT BUILDING

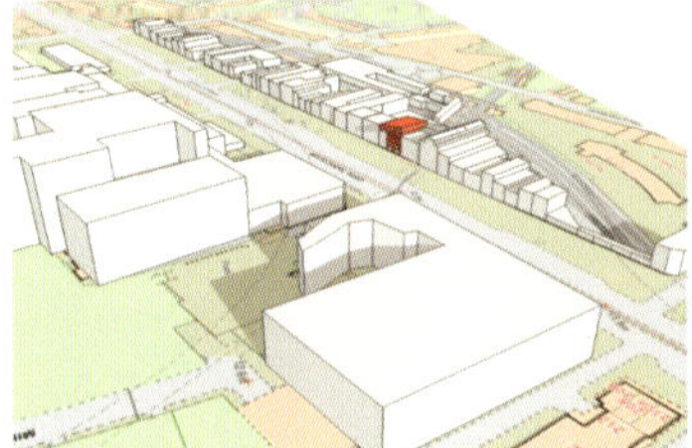

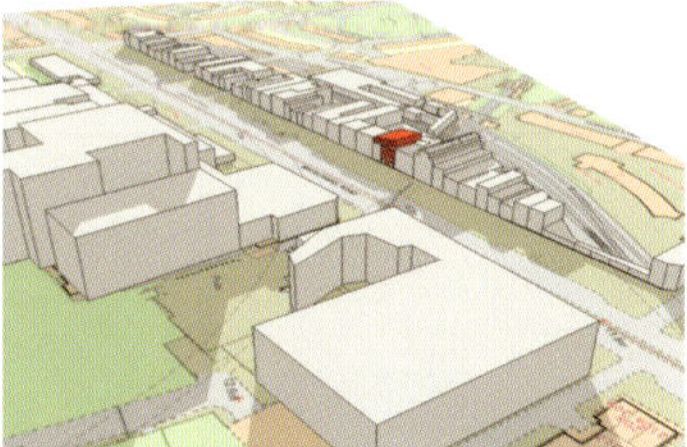

REAR OF THE BUILDING

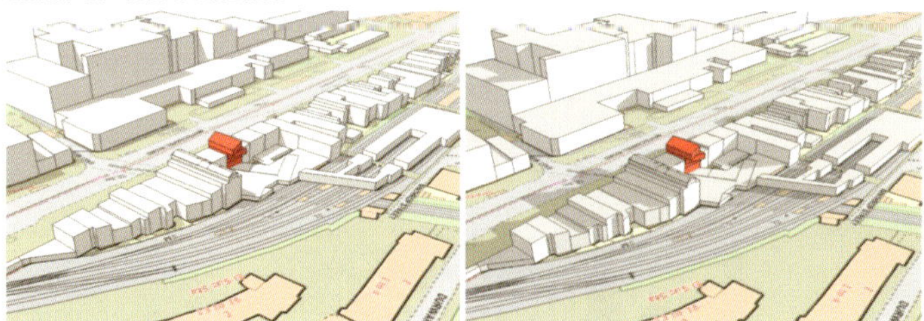

View of the Royal London Hospital

VIEW FROM THE ROYAL LONDON HOSPITAL TO WHITECHAPEL ARTS CENTER

THE CONCEPT

Through the concept of visual contact, comes the idea to design a building that engages participants to develop familiarity with the surrounding area and encourage them to explore it. The visual connection reflects the design of the building, in fact it will be a vehicle to looking anew and makes it feel as if it were a part of the future change. A unique shape that shows a spatial continuity and subtle articulation of space to represent the union and the diversity of the community; it will encourage people to explore around and develop a positive attitude with a new sense of belonging, in fact when the visitors start to explore and know more about surrounding areas, it will increase a more confident feeling and connection with the other local communities.
The shape and the different sizes represent the local community; portraying different characteristics, but linking them together they become a special unique shape.

컨셉

시각적 접근을 통한 컨셉을 통해 주변을 둘러싼 지역과 조사, 친숙함을 발전시켜 빌딩을 디자인하는데 이용하려 한다. 시각적 접촉은 빌딩의 디자인이 보다 새롭게 보이고 미래 변화의 일부로 느껴지게 된다. 공간의 연속성과 지속성을 보여주는 독창적인 형태는 통해 지역 사회의 다양성의 통합을 보여 준다. 이것은 사람들에게 소속감을 느끼게 해주어 긍정적인 마음을 가지게 하고, 방문객들이 주변을 탐사하면서 다른 지역 사회들과의 연결도 자신 있게 느낄 수 있을 것이다.
다양한 형태와 크기는 지역 공동체의 서로 다른 성격을 반영하는 것이지만 독창적인 형태가 모여 그들을 연결해 시켜준다.

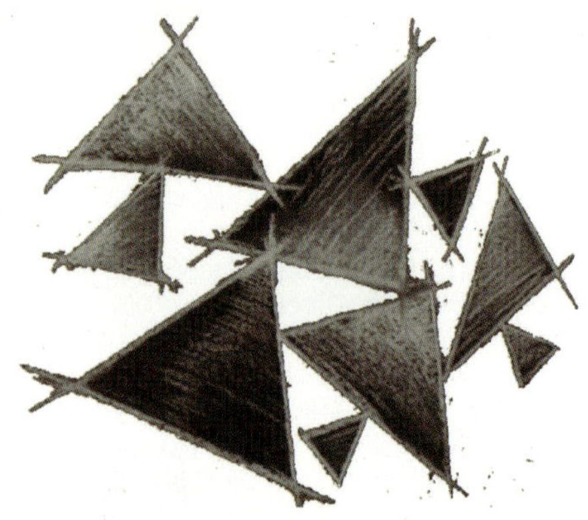

The building is design to create a place where the community around can meet and enjoy the time spend together through arts courses, but also a place where it is possible to organize business meeting or courses for a preparatory base for a future work.
It will increase the aspiration in people, and also it will be a place where the people will engages participants to develop familiarity with the surrounding area and it will encourage them to explore and makes them fell part of the future change. The building's architecture and the shape will provide different view from all the corner of the street, created small hidden space with different shadow and light, it will encourage to come inside and make a journey.

건물은 주변 사람들이 모여서 미술 수업 등을 통해 함께 시간을 보내는 곳이자, 더 나아가 비즈니스 회의 등을 준비할 수 있는 곳이기도 하다.
사람들의 포부를 길러주며, 서로 교류를 하게하고, 또 주변 환경과의 친밀도를 높여 미래를 향한 변화에 자신들도 일부분이라는 것을 깨닫게 해줄 것이다. 건물의 형태는 사방에서 다른 모양으로 보일 것이고, 다각도의 빛과 그림자에 의해 서로 다른 Hidden Space가 만들어지고, 사람들이 이 공간 안에서 느끼고 경험하기를 바란다.

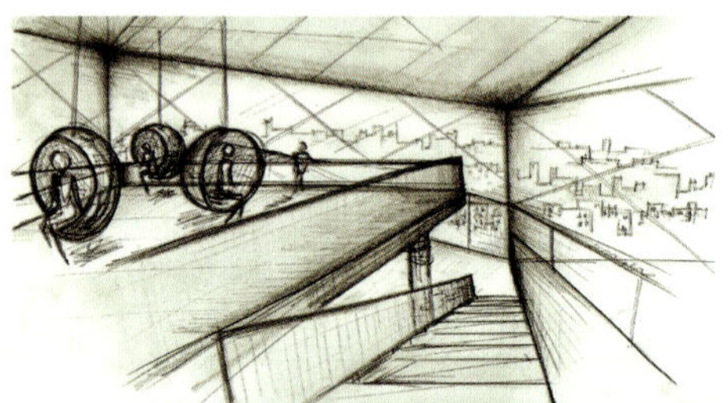

Third floor - platform from where the visitors can see the surround environment

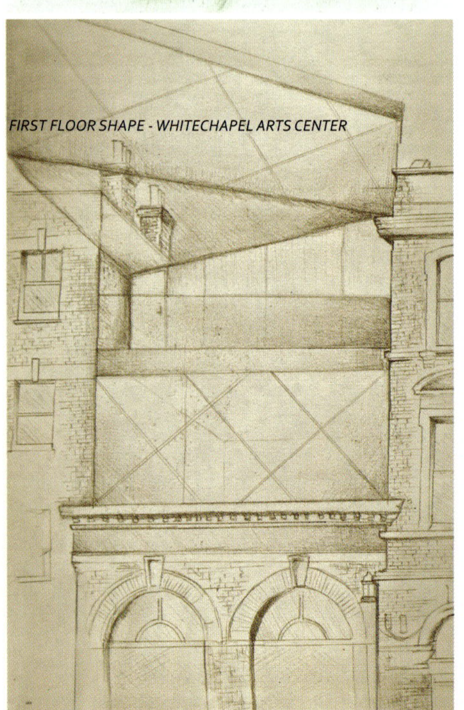

Concept development

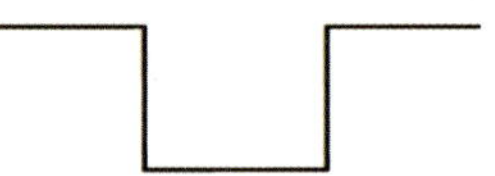

Gap between the building in whitechapel tube station

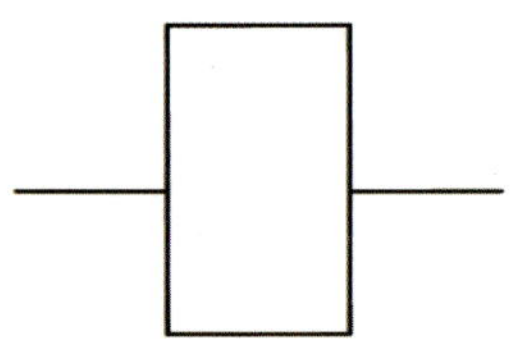

From the shape of the Whitechapel Arts centre

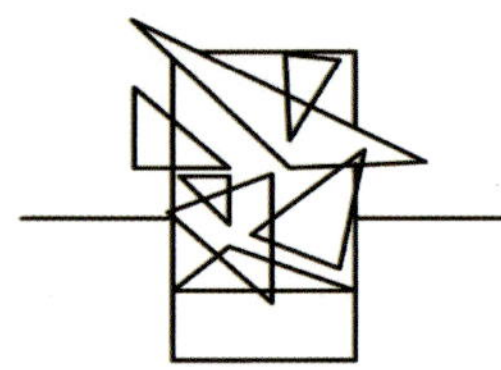

The different shapes represent the diversity character in the Community...

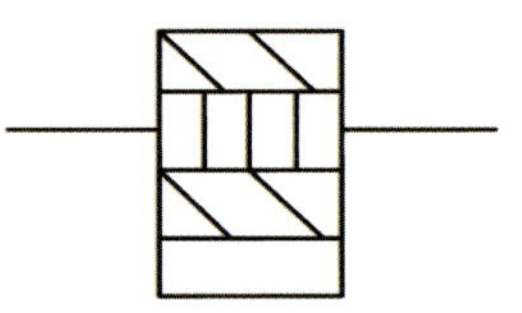

...That fit together create a unique building's structure

Whitechapel arts centre facade

Final model of the whitechapel Arts centre

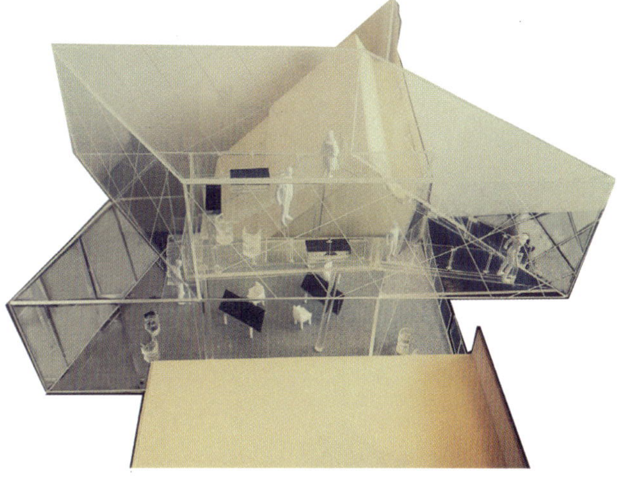

Left side view of the whitechapel Arts centre's model

Model of the second and third floor - different shape to create a unique building with a varity of segment

Front Elevation Whitechapel Arts Centre

Section AA - Whitechapel Arts Centre

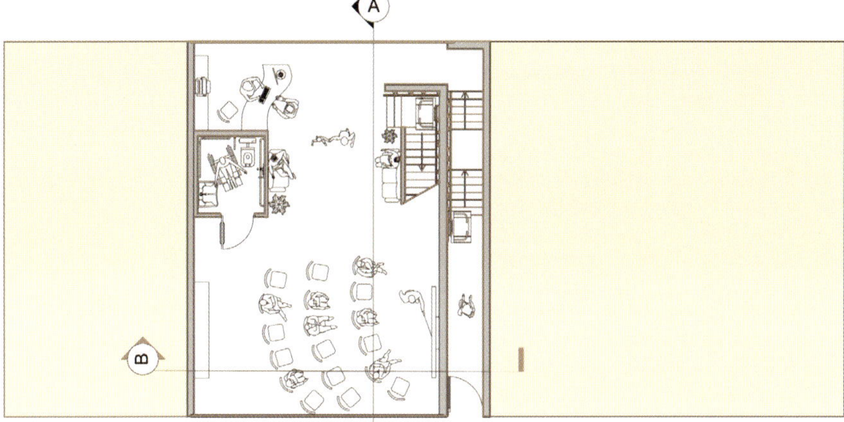

First floor plan - reception and meeting room

The space of the building is spread on three-floor plans. Starting from the first floor level, the visitors will be welcomed by a unique open space that contains the reception, waiting area, and an area dedicated to lectures or a meeting area. The first impression that the visitors will have is the uncomplicated space will make them feel welcome and free to start their journey. All the spaces are flexible and in all of them it's possible to change function and design.

건물의 공간은 세 개의 층으로 나뉘어있다. 첫 번째 층에는 탁 트인 공간에 접수처, 대기실, 그리고 강의 및 회의실이 있다. 이 공간에서 사람들은 환영 받고 언제든지 자기 행동의 자유로움을 느낌을 받을 수 있다. 모든 공간은 융통성 있게 모양과 기능을 바꿀 수 있다.

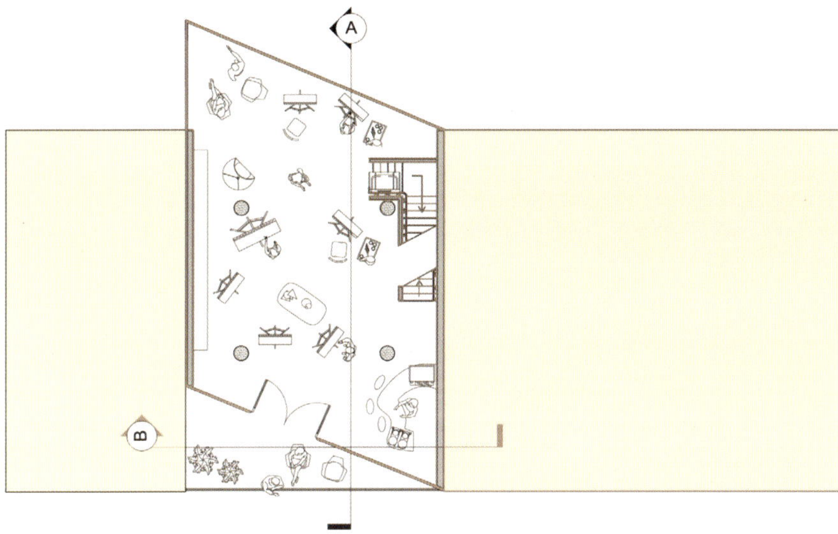

Second floor plan - arts course class , coffee shop and terrace

The second floor will be designed to the workshop area, some painting and drawing courses where the people will use this space to express themselves, feeling comforted by the surrounding locations and lighting coming from the glass walls.
This floor will provide a small cafeteria where the people can relax and enjoy the view around...In fact the strong facade's shapes that become a focus of the unclosed areas.

두 번째 층에는 미술 작업실과 워크샵 등이 있어 사람들이 자신을 표현할 수 있고, 유리를 통해 보이는 밖과 들어오는 빛에 안도감을 느낄 수 있을 것이다.
이 층에는 작은 카페테리아를 두어 사람들이 휴식을 취하고 주변의 광경을 만끽할 수 있게 할 것이다. 또한 건물의 독특한 모양은 열린 공간에서 주목할 점이 될 것이다.

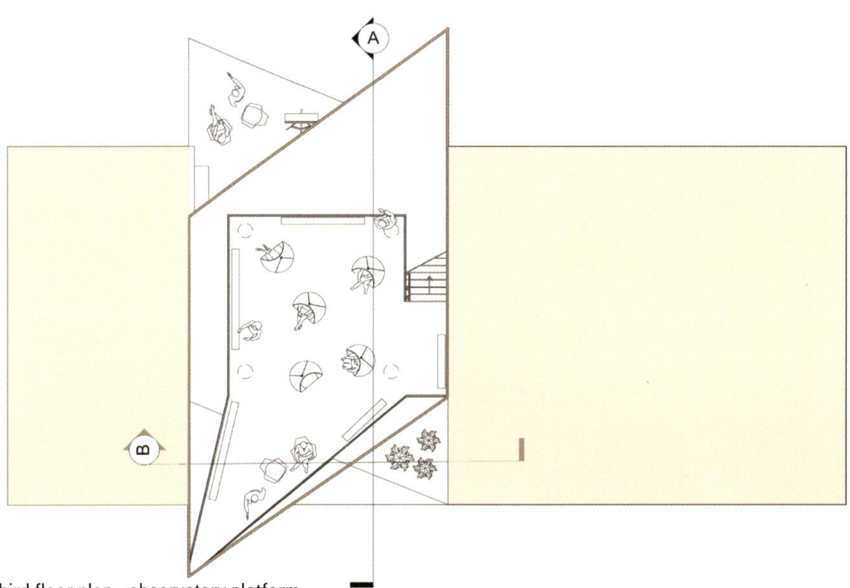

Third floor plan - observatory platform

The third floor will be a small relaxing platform where the visitors will have few moments to dedicated to their thoughts...Like an observatory, mapping and understanding the surrounding area.
This open space room creates a visual point that encourages exploration and increases the people's aspiration.The feeling that this building wants to give is curiosity and interest in learning about the communitiy identity (through courses, lectures and view and explanation of the surround area).

세 번째 층에는 방문객들이 여유로운 휴식을 취할 수 있는 공간이 있다. 이곳은 마치 관측소처럼 외부 주변 지역을 관찰할 수 있는 곳이다.
이렇게 트인 공간은 창의적인 시점에 대한 관찰을 제시하고 사람들의 열린 공간은 사람들에게 탐험을 격려하고 사람들의 목표의식을 길러준다. 이 건물은 호기심과 흥미를 느끼게 해줬으면 좋겠고, 강연이나 수업을 통해 지역 사회의 정체성에 대해 배울 수 있을 것이다.

RENEWABLE ENERGY

GLASS: The proposed building design suggests the use of Low-e coating glazed glass filled with Krypton gas for this façade instead of the Triple glazed gas infill for the others (from Velux Company). The design solution has been influenced by the façade's greater solar exposure which would affect the internal comfort increasing the need of a cooling energy.

ROOF: The design of the building will be provide an energy reduction system through renewable energy as solar PV facades, trans¬parent solar-photovoltaics (solar cells that are built into transparent triple glazing units that can easily be integrated into roofs). The lighting system is based on the concept of daylight harvesting, as anticipated above the western façade has the greater solar lighting exposure and to limit the solar heat gain Low-e glasses has been employed in the design. As artificial light will be use LED lightings (75% less energy), in fact quali¬fied LED lightings produce virtually no heat; their design are highly durable and provides optimal light colour for any environment.

WALLS: High performance envelope and 350mm insulation in walls (in the first and part of the second floor's walls) and it will be installed below the concrete slab, so the thermal capacity of the building will be increased and the concrete floor structure will absorb heat and release it into the rooms gradually over a period of time; while in some areas they will be inserted into ventilation strips behind the façade, using Katherm HK – kampmann (Heating or cooling with EC crossflow fan assistance).

CEILING: PCM (Phase Change Material) plasterboard to maintain the heat inside the rooms.

WATER: On the ground floor a Polypipe Rainstream will be installed, for a complete water re-use and filtration system (toilet/urinal flushing, grounds ir¬rigation, commercial wash down etc)

MATERIAL: the building is based in a eco-friendly design, so all the element of the structure will be reusable, from the high recycled-content steel structure, to the transparent solar-photovoltaics glass (aggregate, concrete item, drainage...) and to the furniture (in beech recycled hardwood from UK). To preserve the environmental friendly idea of this building's design and concept, the people working inside will rely in the Eviroof-fice (Environ¬mentally Friendly Office Supplies Company) for recycled office's products.

재활용 에너지

유리: 건물의 디자인은 크립톤 가스가 든 로이유리를 사용해 모양을 낼 것이다. 트리플 가스 대신 크립톤을 쓰는 것이 냉난방 비용을 절감할 수 있어 고른 소재이다.

지붕: 태양에너지를 이용하기 위해 투명한 전자판으로 천장을 만들 것이다. 햇빛의 에너지를 모아두었다가 밤에 사용하는 시스템이다. 인공 빛으로는 발열이 적고 실내를 효과적으로 비출 수 있는 LED 전구를 사용할 것이다.

벽: 350mm 단열 벽을 콘크리트 아래 설치하여 건물의 열용량을 증가시키고 콘크리트가 열을 흡수했다가 나중에 내보내도록 하고 1층 외에는 별도의 환기 장치를 설치하여 냉난방을 이용하게 할 것이다.

천장: PCM을 이용한 단열.

물: 빗물 등의 재활용을 위한 순환 시스템

재료: 이 건물은 친환경적 목적으로 만들어졌으며, 건물의 뼈대부터 가구까지 모두 재활용이 가능할 것이다. 또한 직원들 역시 재활용된 사무용품을 사용할 것이다.

Whitechapel Arts Centre, second and third floors

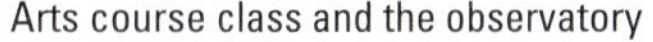
Arts course class and the observatory

BRAMHAM PARK

Cor-ten steel installation that bridges to across

This local park is located within the aldgate gyratory traffic island within the city fringe area and whitechapel. The site is surrounding by leisure building, a bank and several victorian buildings. A green open space with water features and play area in hard and soft landscaping. The proposed is to create an installation that engage the local people with the the park, so here comes the idea to apply some corten square installation. T he shape and the position (one next to the others) of the installations will allow the visitors to walk on it, and across the straight canal positionated in the middle along the park. This area is accessible to many, and this create a natural journeys to engaged the community with the local facility, from the whitechapel art gallery to whitechapel library or areas as bricklane.

지역 공원은 Aldgate의 교통안전지대를 선회하는 도로 부근에 위치해 있다. 사이트 주변에는 여가활동을 위한 빌딩, 은행, 그리고 몇 채의 빅토리아 시대 건물이 있다. 열린 공간은 딱딱함과 부드러움이 공존할 수 있다. 여기에 조형물을 설치에 조형물 위로 또 사이로 다닐 수 있게 해 사람과 공원 사이 소통을 유도한다. 이 공간은 접근성이 좋기 때문에 주민들이 도서관과 전시관 등 지역 시설을 자연스럽게 이용하게 유도할 수 있다.

Cor-ten steel installation and area Where visitors can sit and enjoy the park

CONCEPT DEVELOPMENT

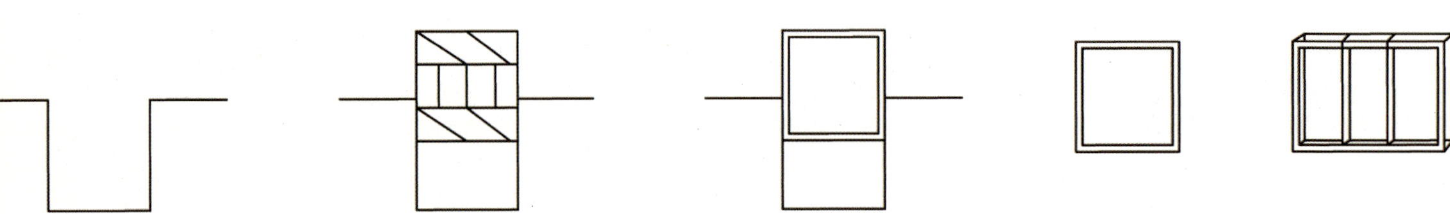

Gap between the building in Whitechapel tube station
From the shape of the Whitechapel Arts centre

Working the Installation 's shape
Thinking about the union of the diversity

The redevelopment of this area has generated unseen barriers within the existing infrastructure. The corporate buildings in spitalfields create a perceived difference between business and the surrounded community. That got already a totally different character and identity. The scale of the buildings in bishop square are extremely different from the remaining area caused an alienation of the local community. As you go through the area is possible notice how the used of materials can change and generated several feeling...from natural shadow and reflection, to different colours graduation. The de¬sign of the two buildings create a connection with the surround environment, through the use of wood material mix with glass ,and the position of the buildings. Based on the idea of creating a connection through the view between the community and the local areas , the buildings will be provide of two small roof garden where the people will enjoy the contradistinct view of the place. The space will be dedicated for an information point, where the people will be free to ask question about the area, the community events (it will be also be beneficial for some advertising for the whitechapel site and brahman park), but also a place where the visitors can ask for a job preparation (for a job's interview or to do a cv etc...).

지역의 재개발은 기반 체계에 보이지 않는 경계를 가져왔다. Bishop Square와 다른 곳의 건물들은 그 생김새가 확연히 다르고, 이는 곧 정체성의 차이로 이어진다. 재료의 차이 역시 마찬가지이지만, 새로운 디자인을 통해 관련성을 확보하길 원한다. 따라서 건축물과 주변 환경의 공통점을 찾아 지역사회 사이를 연결시킬 것이다. 또 하나의 작은 즐거움으로 건축물 옥상에는 정원을 만들어 사람들이 두 개의 다른 풍경을 즐길 수 있도록 할 것이다.

SPITAL YARD, SPITAFIELDS

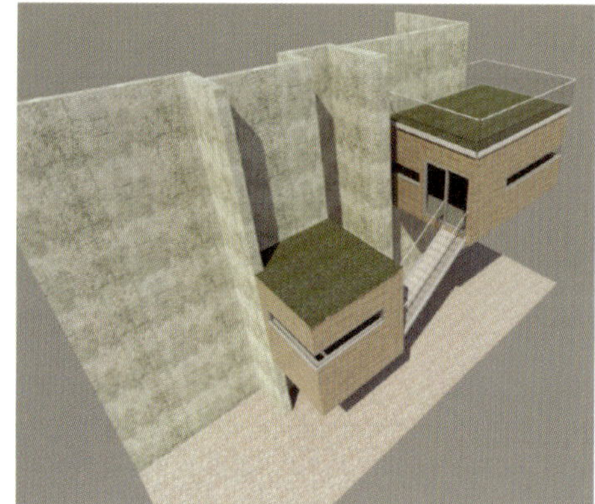

Sketch of the information centre

View from above Roof garden development

Image of the two wood blocks

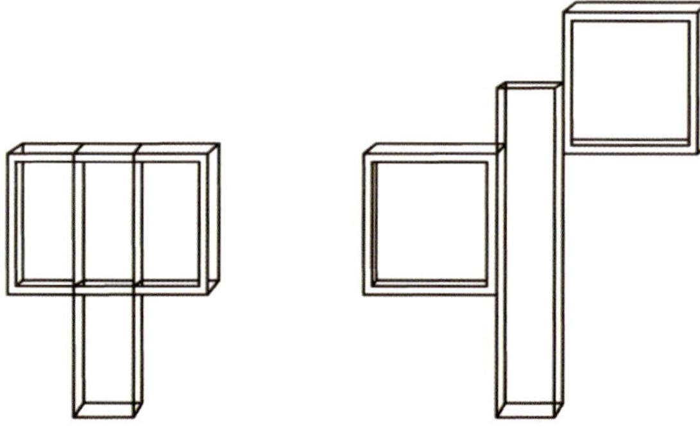

Create a balance between the different community

Maya Fabri
mayafabri@hotmail.it

Magnet

THE SITE

Whitechapel is a built-up inner city district in the London Borough of Tower Hamlets, the resident population consists of mostly families from the West Indies, India, Pakistan and Bangladesh. Its proximity to the arty enclaves of Hoxton and Shoreditch means that fashion and trend is very important.

사이트

Whitechapel은 건물이 많이 들어선 런던의 한 구역이다. 인도, 파키스탄 등에서 이민 온 사람이 대부분이다. 또한 Hoxton과 Shoreditch과 가깝기 때문에 패션과 유행이 중요한 곳이기도 하다.

3 Locations...

SPITAL YARD

SPITALFIELS (Buisness)

BISHOPS GATE

BRAHMAN PARK (Employment)

CONTEXT

Our cities now have to rethink and redevelop the use of space more creatively and ingeniously than ever growing need to accommodate increasing populations mean that every open area is now a new possibility for spatial development.
One of the most recent and fascinating discovering interstital space, gaps inbetween. Their appeal lies in their surprise and undiscovered disposition.

Their purpose might involve a large community or an individual but because of its hidden or exposed position, providing an adventure, or unexpected experience.
It provides opportunities for older established buildings to convey a new brief and therefore a new spirit, encouraging creativity and positive thinking, using a different energy.

배경

우리 도시들은 점점 불어나는 인구를 수용하기 위해 공간을 보다 더 창의적이고 재치 있게 재개발하고 사용방법을 다시 생각해보아야 한다. 따라서 열려있는 모든 공간은 개발의 대상이 될 수 있다. 최근의 가장 흥미로운 발견은 넘쳐나는 도시의 공간 속에 생겨나는 '틈새 공간' 의 발견이다. 틈새 공간의 매력은 숨겨져 있는 공간의 발견을 통해 기대하지 못한 새로운 경험을 할 수 있다는 것이다. 이를 통해 오래된 건물은 새로운 경험과 기분을 이끌어내고, 창의적이고 긍정적 사고의 변화의 에너지를 가져올 것이다.

WHITE CHAPEL (Comunity)

The urban master planners and the local communities have to consider a more responsible composition of ideas to our environmental needs. Sustainable energy use and material choice are to be rewarded but how does this connect with the individual and the surroundings? What is our sensory involvement? Actions that involve breathing fresh air, feeling the breeze, hearing the cars rumble by, bathing in direct light, seem fundamentally prearranged but is the experience total? How do we make this encounter poignant and memorable?

도시설계자들과 지역 커뮤니티는 우리의 환경에서 필요로 하는 것에 대한 구상들을 보다 책임감 있게 다루어야 한다. 지속 가능한 에너지 사용과 재료의 선택도 좋지만, 어떻게 개인과 주변을 연결시킬 것인가? 그리고 우리 감각과는 어떻게 관련이 있는가? 신선한 공기를 마셔 들이는 행위, 바람을 느끼는 것, 지나가는 자동차 소리를 듣는 것, 직접 햇살을 쬐는 것 등의 감각적 경험은 정해지고 예정된 경험처럼 보이지만 과연 그 경험은 완전체일까? 어떻게 이 접촉을 더 기억에 남게 할 수 는 없는 것일까 ?

WHY MAGNET?

A magnet is a material or object that produces a magnetic field. This magnetic field is invisible but is responsible for the most notable property of a magnet: a force that pulls on other ferromagnetic materials, such as iron, and attracts or repels other magnets.

The fact that the three sites are so close to each other but don't interact, make it feel like if un invisible wall was placed in the middle do divide them, like un unknown force...

In a magnet, when two same poles are closed to each other, they seem to repel, but the simple act of turning one of them make them join.
The mission in this project is to Brake that wall by simply "swopping the poles" in order to make a connection, and to do it I will use the fascinating behaviour of the magnet to emplane the cause, to find a solution and finally to find a nice and original way to design the sites.

Developing this project with the magnet's resistance and also the level of risk that exposure to static fields can have on people. Magnetic elements such as: attraction, division, connection, repulsion, movement and energy will also be employed in the concept.

왜 자석인가?

자석은 자기장을 형성하는 물질 혹은 물건이다. 이 자기장은 비록 눈에 보이지 않지만 자석의 가장 중요한 성질로써 바로 금속과 같이 자석의 영향을 받는 물체를 끌어당기거나 밀어내는 특성을 가지고 있다.

세 구역의 거리가 가까운데도 불구하고 교류의 부재는 마치 중간에 보이지 않는 미지의 힘이 존재하는 벽이 있는 것 같다.

자석의 같은 극을 가까이 하면 서로 밀어내려 하지만 한쪽 극을 돌려놓는 간단한 방법으로 두 자석을 이어지게 할 수 있다. 이번 프로젝트의 목적은 이와 같이 간단히 '극을 바꾸는' 것으로 보이지 않는 벽을 허무는 것이다. 그리고 이 같은 자석의 원리를 이용해 원인을 찾아내고 독창적인 디자인을 만들어낼 것이다.

이 프로젝트에 이용될 자석의 원리에는 끌림, 나눠짐, 밀어냄, 움직임, 그리고 에너지가 있다.

Development

CONCEPT

The concept is derived from social issues concerning the London borough of Tower Hamlet.
The space is for the community who are not well integrated within their society. People who feel alone and marginalized due to their lack of education or skills. People who would like to change their status and make some order in their lives in order to find satisfaction and stability and have a better future.
The division between the two areas (Spitalfield and Whitechapel) which is both socially and economically destructive for the borough stands a metephor for this project.
The MAGNET illustrates the situation and works towards solving the problem at the same time.

컨셉

이 컨셉은 Tower Hamlet의 문제점을 보고 고안해냈다.
이 공간은 전체 사회와 어우러지지 못하고 있는 지역사회를 위한 공간이다. 교육과 기술의 부족함으로 인해 소외되고 차별받는 사람들, 그리고 만족감과 안정감과 나은 미래를 위해 삶의 수준을 높이고 만족을 얻고 싶어 하는 사람들의 공간이다. 사회적, 경제적으로 분열하는 Spitalfield와 Whitechapel 사이의 문제를 이번 프로젝트를 통해 상징적으로 보여준다.
MAGNET은 이 상황을 보여주는 동시에 해결 방안을 제시할 수 있을 것이다.

Final outcome...

Journey Whitechapel

At the entrance, the floor will tip slightly forward to help people to enter and to complicate their way out.
This slope will be surrounded by a tunnel made of squares which will be placed irregularly to give a spiral effect which will "absorb" them in to the lower level. This spiral tunnel will be used to give the idea of movement like a magnet as well as leading them inside.
The visitors will then arrive at the reception desk which will appear to be suspended as it will be attached to a lateral wall to give a floating, surprise effect that will underline an important element of the magnet which is energy.

Whitechapel 여정

입구에서 바닥까지 기울어져 있어 진입은 쉽고 퇴장은 어렵게 할 것이다.
이러한 경사는 불규칙적으로 놓아서 나선 효과를 내는 사각으로 만들어진 터널로 쌓여있어 사람들을 아래층으로 끌어당길 것이다. 나선 터널은 자석과 같은 움직임을 보여주고 사람들을 안으로 유도한다.
방문객들은 접수처에 왔을 때 측면의 벽의 설치가 마치 떠있는 느낌으로 주어 놀라움을 가져다 줄 것이다. 이것은 자석의 중요한 성질을 강조한 것이다.

Finally visitors will go down to the main floor. People will find a nice, comfortable and welcoming waiting area on their right where they can sit waiting until it is time for them to be seen while reading a leaflet with a map and directions to get from Whitechapel to Spitalfields with a note saying "how can you say you live if you don't try.
The whole wall in front of them will be a nicely designed image of the City to make them realize that they are getting closer to their dream.

On their left, there will be a massive "floating" desk where people will get personal help from professionals who will make sure that they are aware of all the courses and possibilities that London has to offer. This will give them the feeling of union and connection and also make them wonder how such a big table can float, leaving them with the sensation that nothing is impossible if you try.

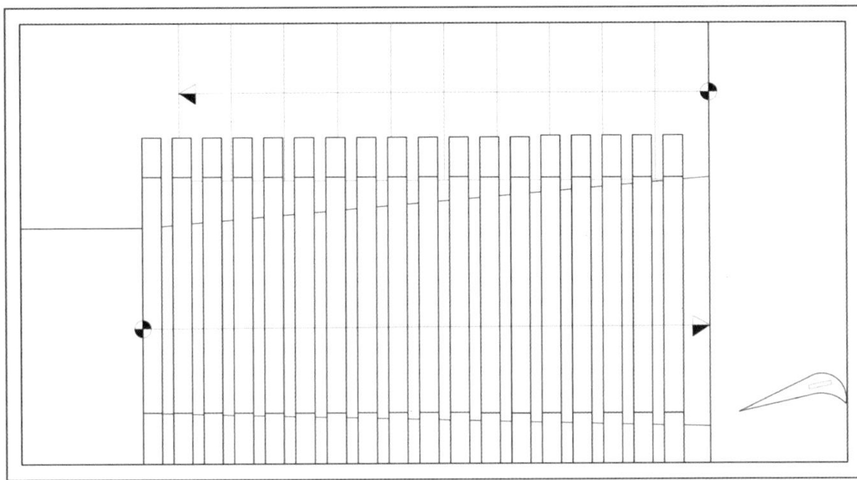

PLAN

메인 층의 오른쪽에는 쾌적하고 편안한 대기실이 있을 것이다. 사람들은 여기서 자기 순서를 기다리며 '도전해보지 않고 어떻게 살아있다고 할 수 있는가' 라는 문구가 적힌, Whitechapel에서 Spitalfield까지 가는 지도가 그려진 팜플렛을 읽고 있을 것이다.
한쪽 벽 전면에는 그들이 바람에 가까워지고 있음을 알려주기 위한 도시 디자인의 한 이미지가 있을 것이다.

왼쪽에는 전문가와 상담을 할 수 있는 '떠있는' 책상이 있을 것이다. 이곳에서 사람들은 런던에서 잡을 수 있는 기회에 대해 들으며 협동심과 소통을 느끼게 될 것이고, 동시에 어떻게 이렇게 큰 책상이 떠있는지 궁금해 할 것이다. 따라서 노력하면 불가능한 것은 없다고 느낄 수 있을 것이다.

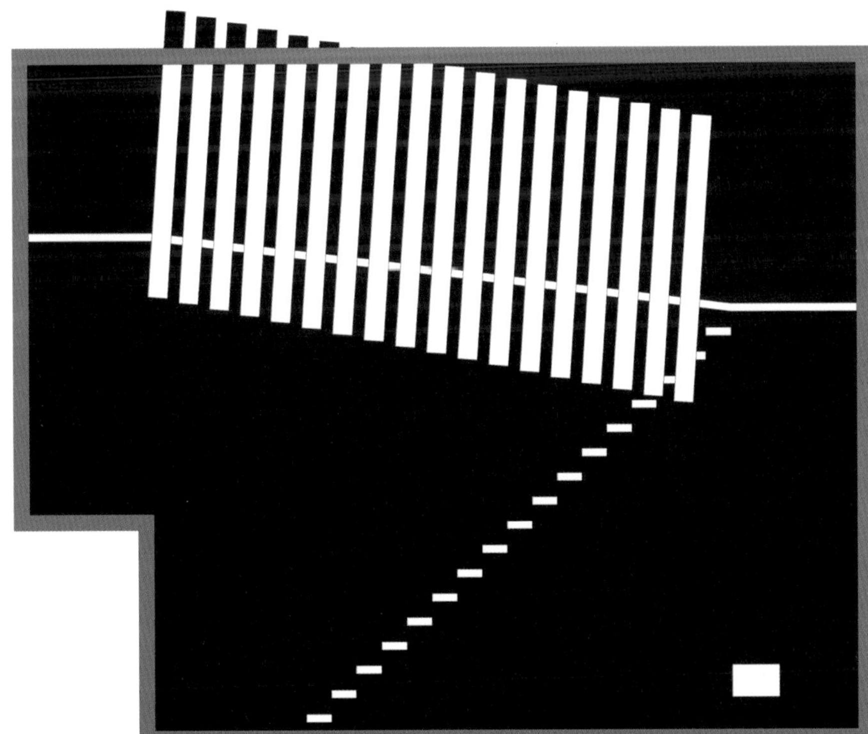

Monika Vanagaite
plenciux@yahoo.co.uk

Links

THE BRIEF

The task was to design a series of structures that encouraged and engaged the local community with employment, beyond their immediate locality, linking and developing connections. Three locations Whitechapel tube station, Braham Park and Spitafields market where provided as pivoting points, all within 15-minute walk from each other, forming a triangle on the map.

요약

우리의 과제는 지역사회를 취업하도록 격려하고 이끌 수 있는 건축물을 구상하는 것이다. 결국에는 단발성게 그치는 관심이 아니라 지속적인 교류를 형성하는 것을 말한다. 우리가 프로젝트를 할 곳은 Whitechapel, Braham Park, 그리고 Spitafields의, 서로 도보 15분 이내인 지역들의 삼각지대이다.

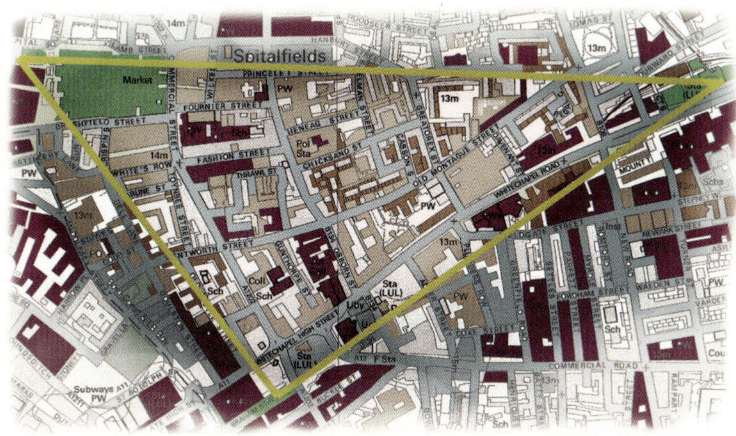

Location map

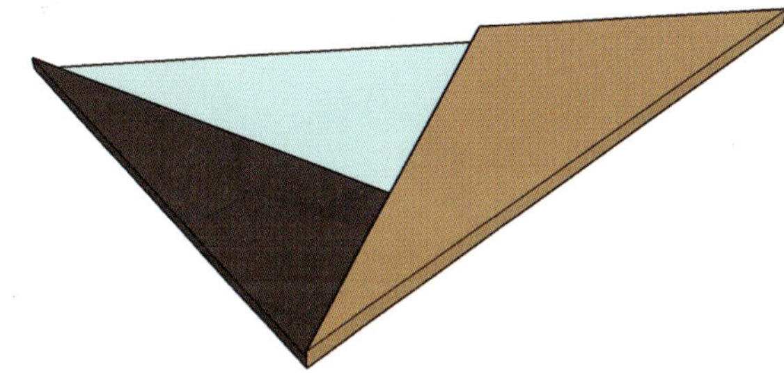

Site above whitechapel tube station

Whitechapel High Street

A place where people could visit and start their journey towards brighter future

Internal Elevation

First Floor Plan

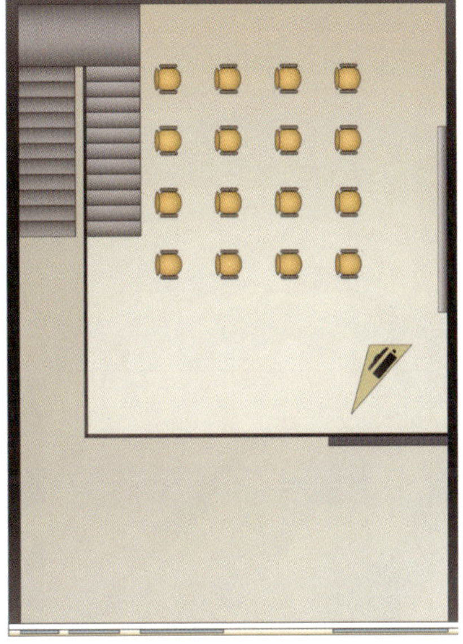

Second Floor Plan

Section

Taking the connecting lines within the landscape, to and from each of the sites as the basis for the concept, led to the metaphor of 'linking' buisiness to the community and community to employment.

프로젝트의 목적은 주변 지역들을 서로 이어주는 선에서 기업을 지역 주민들에게 이어주는 것을 고안하는 것이다.

The outcome is built within a perceived hidden space above and adjacent to Whitechapel tube station has a contemporary feel, reflected in a façade formed of glass triangles in different planes, representing opportunity and diversity. Shadow and light plays within the interior opening up and closing down space throughout the day, providing meaningful environments to engage with future possibilities.

Whitechapel 지하철 역의 hidden space는 현대적인 느낌이 든다. 서로 다른 면의 유리로 된 삼각형들이 모양을 만들어내는데, 이것은 기회와 다양성을 표현한다. 빛과 그림자가 시간에 따라 열리고 닫히는 실내에서 돌아다녀 미래의 기회에 대해 의미를 부여한다.

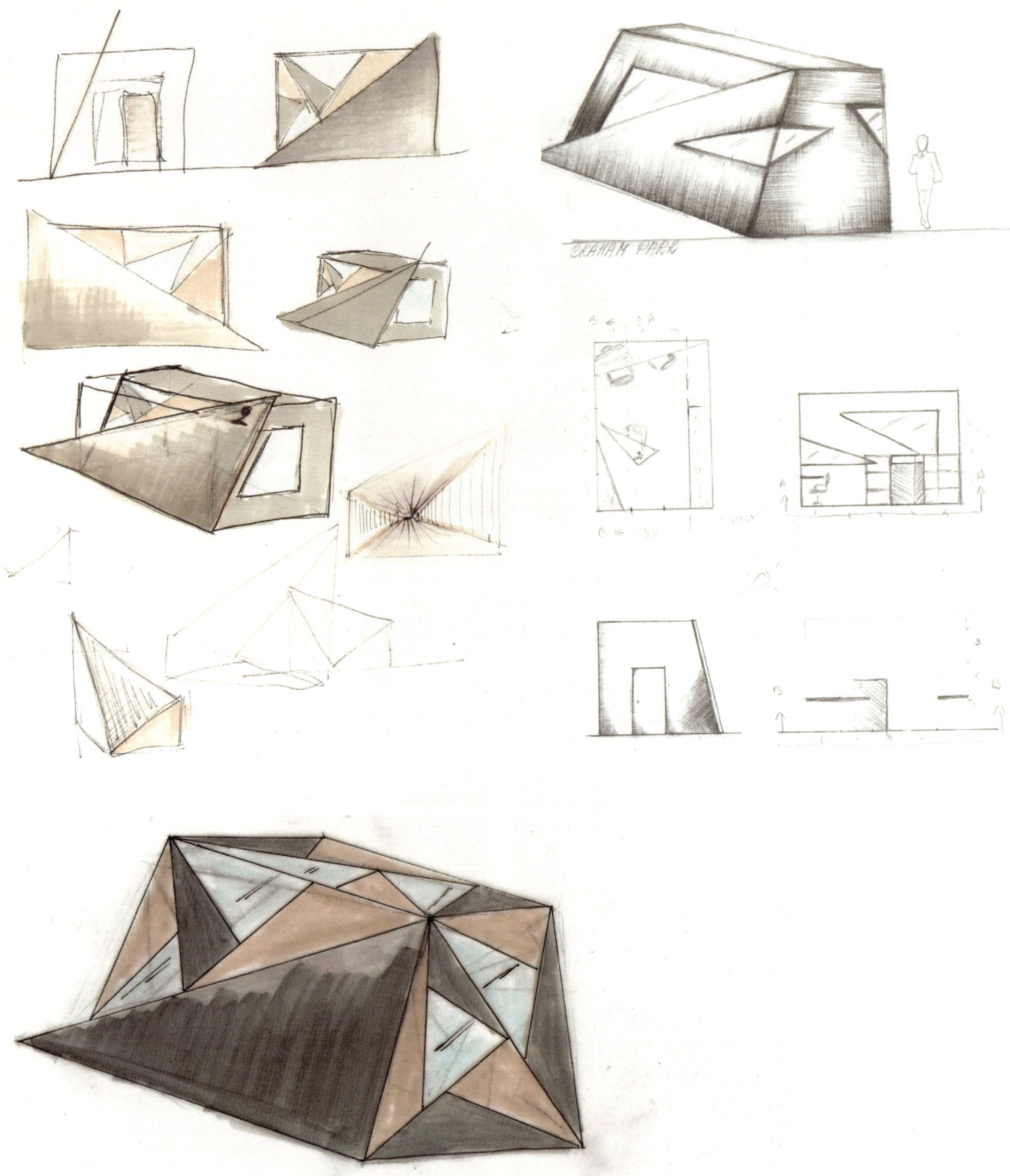
BRAHAM PARK

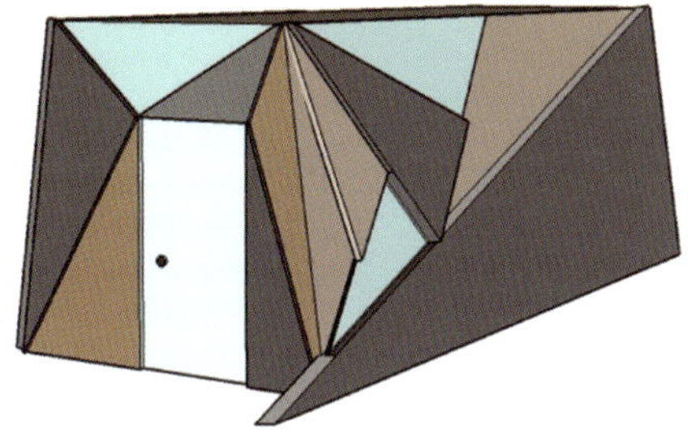

BRAHAM PARK

A small pop up bunker for people to drop of their cv , book an appointments for interviews, for a jobs or training courses and 1to 1 help for practising speech before interview. Also huge triangle peace wall pointing towards direction of Spitafields.

BRAHAM PARK

이력서를 넣거나, 직업훈련 또는 면접을 예약할 수 있는 작은 벙커. 그리고 Spitafields 쪽을 가리키는 커다란 삼각 모형

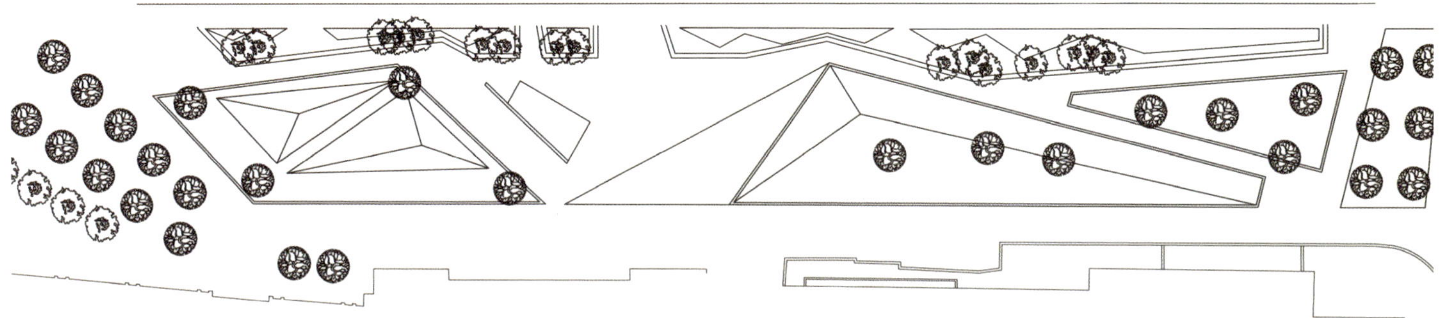

SPITALFIEDLS

The biggest building of all three located in between the old market and newly developed business complex. This place is also designed using the same style of the triangle and adds up a final piece in the triangular map. It is combines of the reception area, small help desks, and huge lecture theatre used for all purpose training, advice and teaching.

SPITALFIELDS

오래된 시장과 높은 기업 건물들 사이, 셋 중 가장 큰 건축물. 같은 스타일의 작은 삼각형들로 이뤄져 결국에는 커다란 삼각형 지도가 된다. 안에는 접수처와 작은 업무 지원 센터, 그리고 다용도의 커다란 강의실이 있다.

Sabrina Yvonne Durant
imdesignsbysd@gmail.com / adesignersnb.wordpress.com

The Domino Effect

Hidden Space: Whitechapel (main site)

Location: Tower Hamlets Borough

THE BRIEF

The Three Sites in Tower Hamlets:

Brahame Park

Whitechapel

Main Site

Spitalfields

Photographic Diagnostic of the hidden space and surroundings at Whitechapel site.

사진을 통한 Whitechapel의 지역의 Hidden space와 주변 지역

Diagnositic of Braham Park and Spitalfields site

Our idea is to bridge the gap between local community and local businesses in the borough of tower hamlets in the areas of Whitechapel, Aldgate and Spitalfields. Our plan is to devise a series of architectural and interior solutions with a view for future employment. The location of structures should encourage its customers and visitors to travel to each of these three locations, offering opportunities of new perspectives within Tower Hamlets.

지역사회와 주민 커뮤니티, 기업들을 매개해 주는 것이 취지이다. Whitechapel, Aldgate와 Spitalfields에 새로운 실내외를 건축과 공간 구성을 통해 장래에 취업에 대한 전망을 보여줄 것이다. 건축물의 위치는 방문자들에게 각기 세 곳을 돌아다니도록 유도하고 Tower Hamlets 내의 새로운 기회와 시각을 제공한다.

The Whitechapel Site

Pictures by Sabrina Durant

Initiate & Create Centre
Spitalfields Site

Pictures by Sabrina Durant

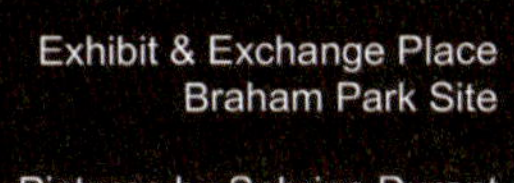

Exhibit & Exchange Place
Braham Park Site

Pictures by Sabrina Durant

Sketch ideas of hub space in whitechapel

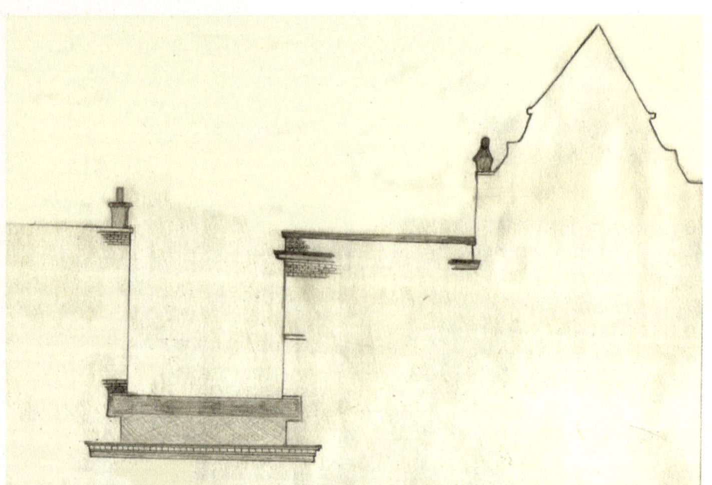

Front outline of site

The Whitechapel site is located above a shop on a busy high street, Surrounded by market stools, shops and small businesses. Home to historic buildings, Whitechapel has begun its facelift for High street 2012 preparing for the 2012 Olympic Games in London.

바쁜 도로에 위치한 가게 위의 공간이다. 이곳은 작은 회사들과 다른 가게에 둘러싸여있다. 주변에는 역사적인 건물들도 있다. 2012 런던 올림픽을 맞이하며 지역을 꾸미기 시작했다.

working with different floors levels

Sketching Ideas

Final design Overlooking the back Side of the New Heights Information Hub
최종 디자인은 New Heights Information Hub을 내려다보는 위치이다.

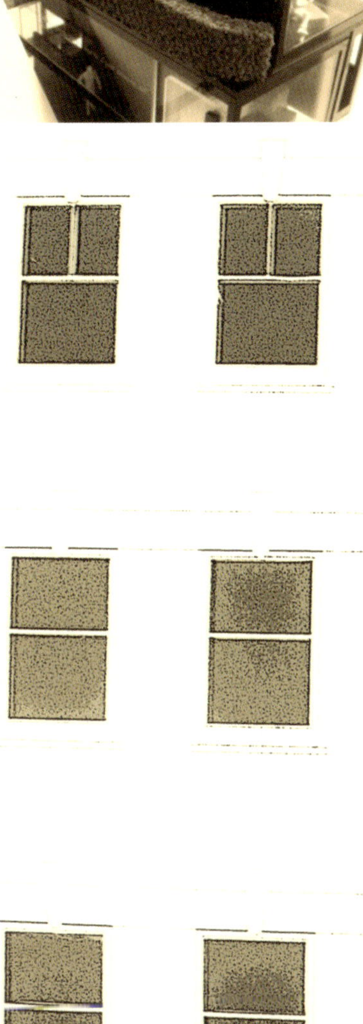

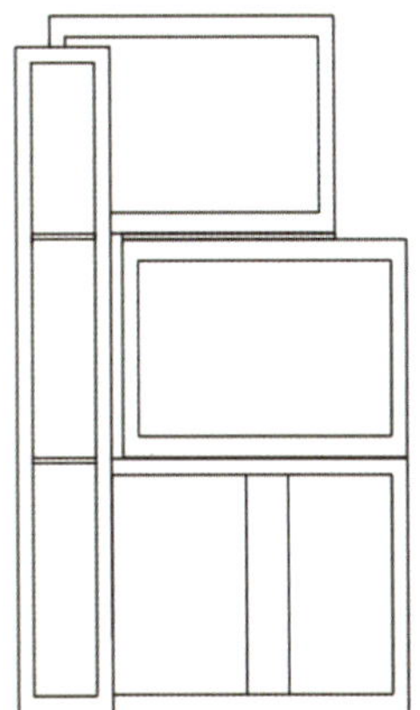

Front Elevation

Front Section

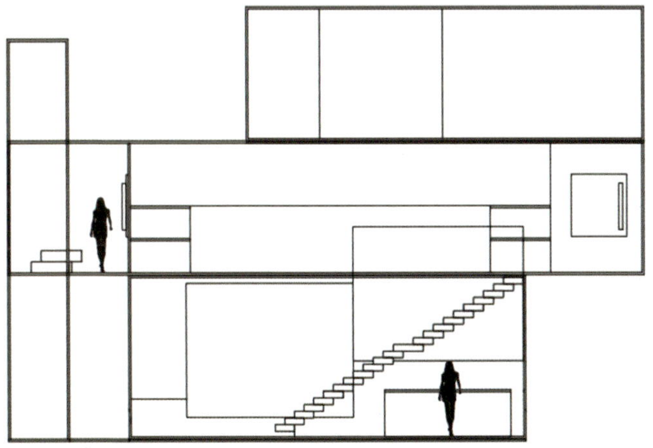

Right side Elevation

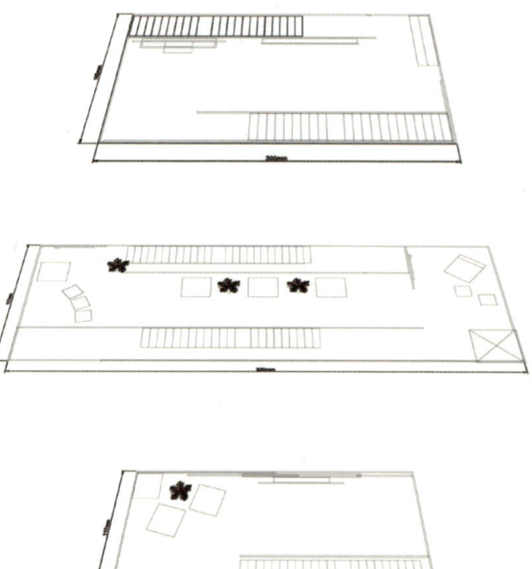

Ground, first and second floor plans

Reception area on the ground floor

Walkway through to lift

'New Heights Information Hub'
Second floor interior

Second floor interior furniture

'New Heights Information Hub'
Right side: windows with inspirational quotes

Designed by Sabrina Durant

INSPIRE...
make IDEAS happen

A view from whitechapel high street

An inspiring structure, designed to open up the mind of those who walk through the space. It has been made up of different scaled blocks over three floors. The second floor protudes over the edge, giving the sense that the world is your oyster and barriers are crossed over in a new and positive way throughout this unique design. It has been named 'The New Heights Information hub'. A space of imagination, inspiration, freedom and connection.

이 자극적인 조형물은 지나가는 사람들의 마음을 열어주기 위해 설계됐다. 3개 층이 각기 다른 크기의 육면체로 이루어져있고, 2층은 튀어나와있어 무엇이든 가능하다고 생각되게 하며, 긍정적이고 새로운 길이 열린다는 느낌을 준다. 이곳에는 "The New Heights Information Hub" 이라는 이름을 붙였으며, 상상과 영감과 자유와 연결의 장소이다.

Samaneh Khazaei
samadin2001@yahoo.co.uk

Fusion

Location: White chapel Station

White chapel High Street is on the three location that have been identified as instinctive space for an intervention that encourage employability for the local community.

장소: Whitechapel 역

Whitechapel 도로는 취업을 권장하기에 적합한 세 곳 중 한 곳으로 선정되었다.

1. Spital Yard

Christ Church

Bishop square

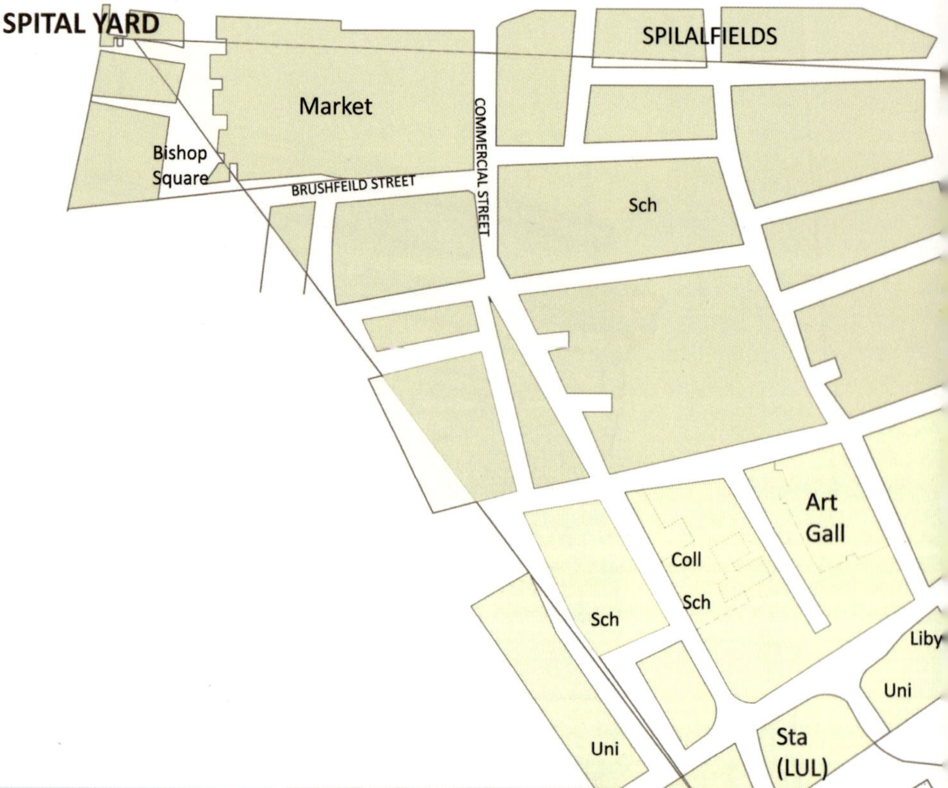

SPITAL YARD
SPILALFIELDS
Market
Bishop Square
BRUSHFEILD STREET
COMMERCIAL STREET
Sch
Art Gall
Coll
Sch
Sch
Liby
Uni
Uni
Sta (LUL)
BRAHAM PARK

THE BRIEF

Our cities are now having to rethink and redeveloped the use of space more creatively and ingeniously than ever before. Space is precious in any context, but the extra pressure cities' face with the ever growing need to accommodate increasing populations, means that every gap in the environment has the possibility for an intervention.

요약

우리의 도시는 공간을 어떻게 훨씬 더 창의적으로 이용하는 주제에 대해서 재고하고 재발전시켜야 하는 필요성이 그 어느 때 보다도 대두되고 있다. 공간이란 언제나 중요하게 여겨져온 주제이지만 계속 불어나는 인구의 압박을 때문에라도 도시의 오픈 스페이스는 그 어디라도 새로운 공간디자인의 가능성을 타진해 보는 대상이 될 수 있다.

2. Braham park

Braham park

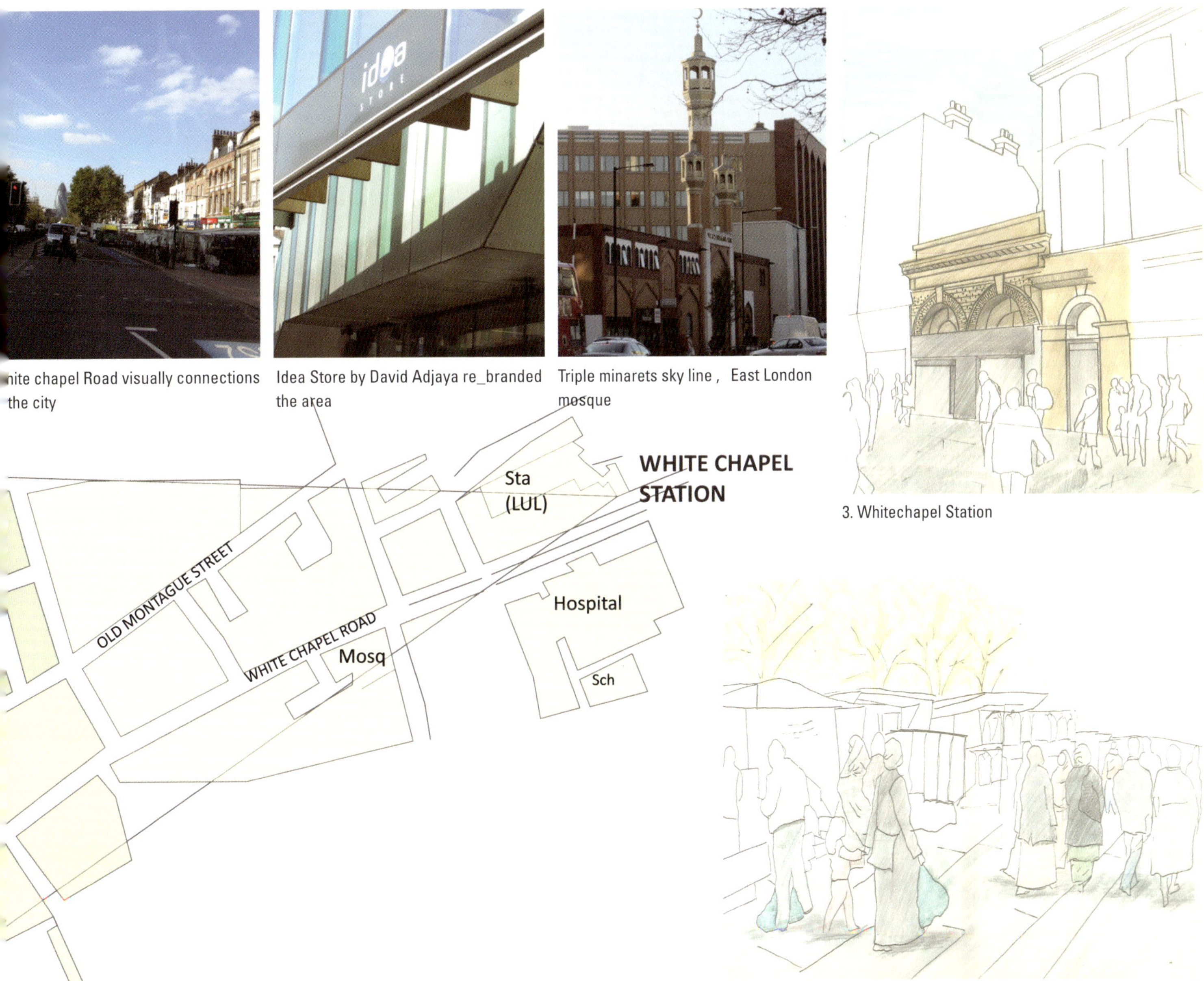

nite chapel Road visually connections the city

Idea Store by David Adjaya re_branded the area

Triple minarets sky line , East London mosque

3. Whitechapel Station

Community pattern at White chapel

THE CONCEPT

What we see at high street level in many urban neighborhoods does not reflect their aspiration for themselves and society. In particular if we want a well educated and informed society, how can we develop physical and cognitive space that is inclusive, accecible and enjoyable? How can we calling people for fusion of good principal in mind which local community see themeselves and seen by others and re-call a past to create the fusion in present and connect to the future?

The intervention is the underpining to develop a innovative shift to see a fusion of possibilities for the future.

The aim is to provide support that can help community over come the personal barriers through the design intentions.

컨셉

High Stree에서 보았을 때, 도시의 많은 동네에서는 주민 자신과 그 사회를 위한 포부가 보이지 않는다. 만약에 우리가 교육과 정보전달이 잘 된 사회를 원한다면, 포괄적이고, 접근성이 용이하고, 즐거운 공간을 만들어야 한다. 또한 주민들이 스스로가 보는 시선과 남들이 보는 시선을 긍정적으로 바꿀 필요가 있다.

이러한 개입은 미래의 변화 가능성을 통합하여 보고 변화를 가져오기 위한 것이다.

DESIGN DEVELOPMENT

Aldgate Fusion has been inspired by the local narrative, reflecting the site analysis and brief, to engage the local community with local business, with view for future employment. The expression of basic physics, utilizing the metaphor of energy needed to create Fusion with in all three sites.

디자인 개발

Aldgate Fusion은 지역의 이야기와 장소의 분석이 반영되었다. 장래의 취업 전망을 위해 지역사회와 지역 기업들은 연결되어야 하는데 이를 위해 간단한 물리적 표현과 메타포의 필요성을 보여주기 위해 세 개 장소의 조화를 통해 창의적으로 만들 것이다.

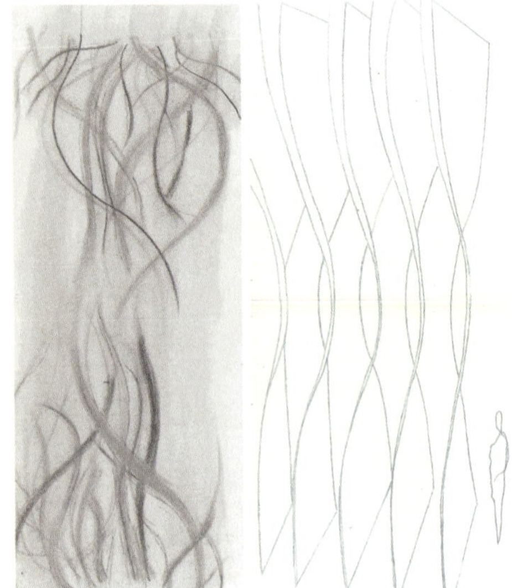

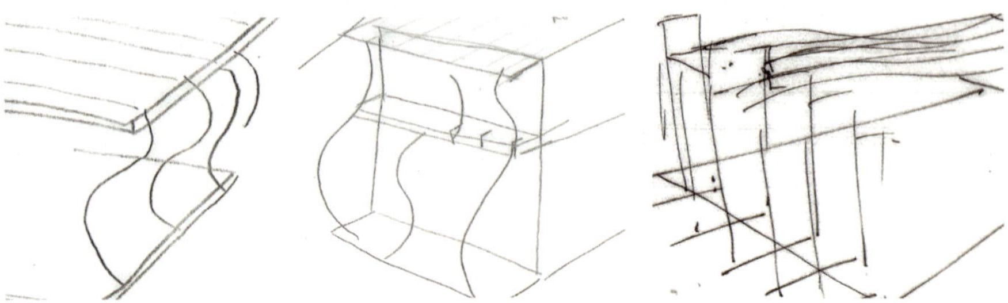

Development of the structure and forms, create the visual architectural narrative

Through the concept development a structure configuration has been formed.

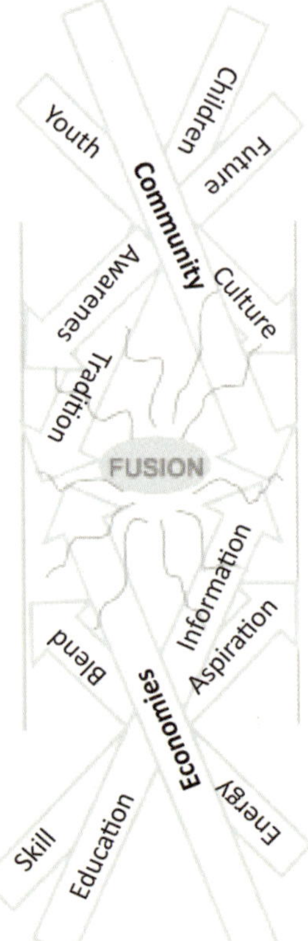

INTERIOR LAYOUT

The metaphor of fusion is reflected through, structure form and texture with the interior environment.

FURNITURE

Flexible element adapt to several positions where the number of these may form part of a group with many compositions. The result is surfaces that have different scales, adapting to different scenarios thus changing the personality of space where necessary.

내부 레이아웃

비유의 접합 점으로 구조의 형태와 내부 환경의 질감을 반영한다.

가구

여러 개가 함께 있을 경우에는 잘 어우러지도록 융통적인 면이 있어야 한다. 비율이 다른 몇 개의 표면을 이용해 필요에 따라 다른 성격을 띠게 했다.

Second floor and mezzanine Visual

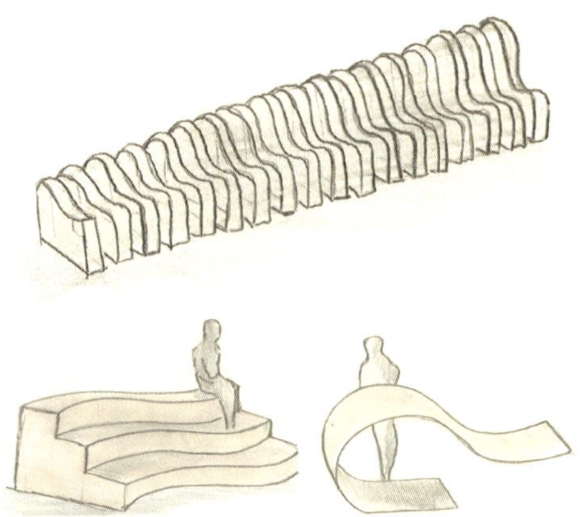

Mezzanine

Reception desk
(Glacier white corian)

The second floor can has the idea of flexible space allows for exhibition and events. Part of this floor has double high ceiling. The glass roof will bring the natural light with in space. The mezzanine floor is the inspiration point through communication and overview to exhibition.

2층은 유연한 공간으로 전시회, 이벤트 등이 가능하다. 어떤 부분은 두 배로 높은 천장이 있고, 유리로 된 지붕이 자연광이 들어오도록 한다. 중이층이 커뮤니케이션과 전시를 통해 얻은 주요 포인트이다.

GLASSING: Thermobel SunEwat

Thermobel SunEwat is a glazing that generate power using photovoltaic technology. The main environmental benefit is their contribution to the development of renewable energy sources that drastically reduce CO2 emission.

유리

태양에너지를 사용할 수 있는 전자판을 설치한다. 이는 재생 가능 에너지를 이용하고 이산화탄소가 감소된다.

Facade Elevation From White chapel Road

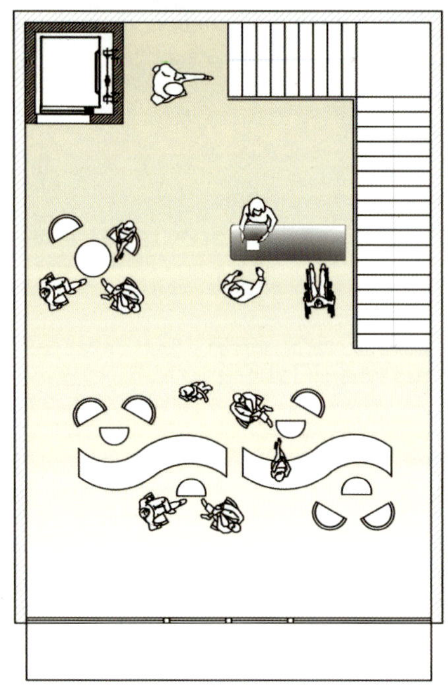
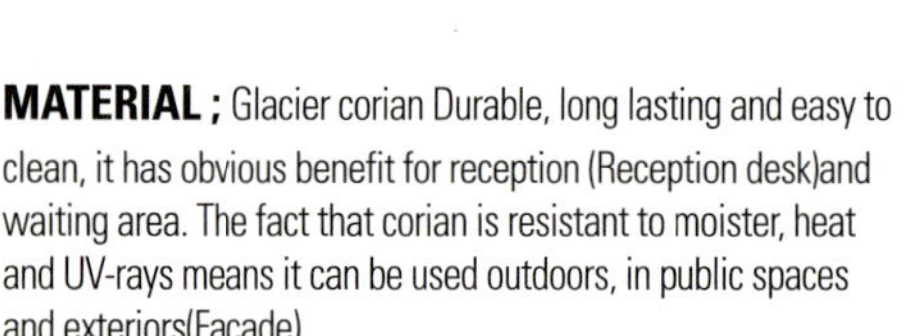
First floor plan

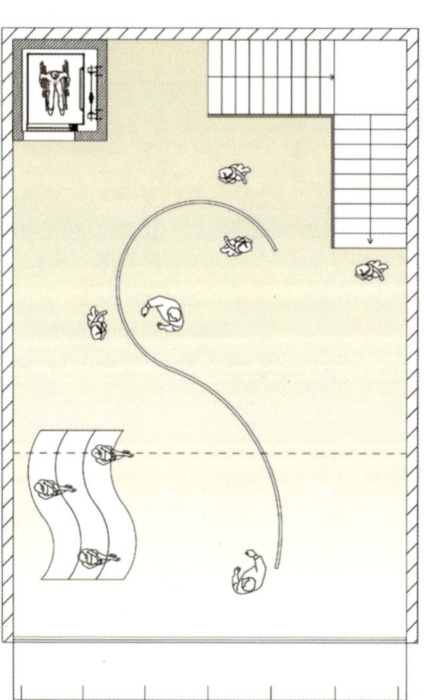
Second floor plan

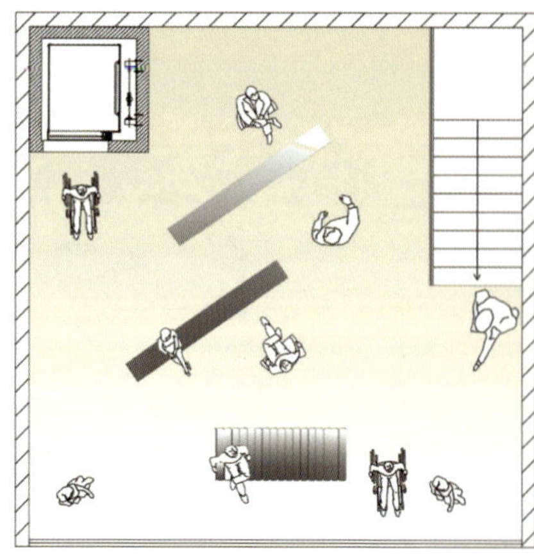
Mezzanine floor plan

MATERIAL ; Glacier corian Durable, long lasting and easy to clean, it has obvious benefit for reception (Reception desk)and waiting area. The fact that corian is resistant to moister, heat and UV-rays means it can be used outdoors, in public spaces and exteriors(Facade).

FLOORING: The Pavegens Floor Tiles Flex (Kinetic Energy Generation)

The Pavegens floor tiles flex a slight 5 millimetres when stepped on, capturing kinetic energy which is either stored in lithium polymer batteries beneath its surface or converted in to 2.1 watt-hours of electricity and distributed through out surrounding lights. The centre of the tile illuminates when stepped upon, not only informing the passerby of their contribution to the environment but also encouraging the continuation of sustainable awareness and decision.

재료 닦기 쉽고 잘 마모되지 않는 코리안을 사용. 접수처와 대기실에서 용이하게 쓸 수 있다. 습기, 열, 그리고 자외선에 강해서 실외에서도 쓸 수 있다.

바닥 운동 에너지를 저장할 수 있는 신소재를 사용하여 사람들이 걸어 다니는 행위로부터 에너지를 얻고 저장하며 나중에 쓸 수 있도록 배터리에 저장하는 시스템이다. 이 역시 환경 친화적인 방법이다.

BRAHAM PARK DESIGN PROPOSAL

This site located between two main communities. The small cube structure designed in open air and will bring the Wide variety of information and opportunities in public setting.
The design provide public space for lecture, public meeting and film screening.

Braham Park 디자인

이 사이트는 두 개 사회 사이에 위치해있다. 작은 정육면체 조형물이 정보를 제공하는 역할을 할 것이다. 이 조형물은 공공 장소에서 강연, 회의, 그리고 영화 감상을 할 수 있게 해준다.

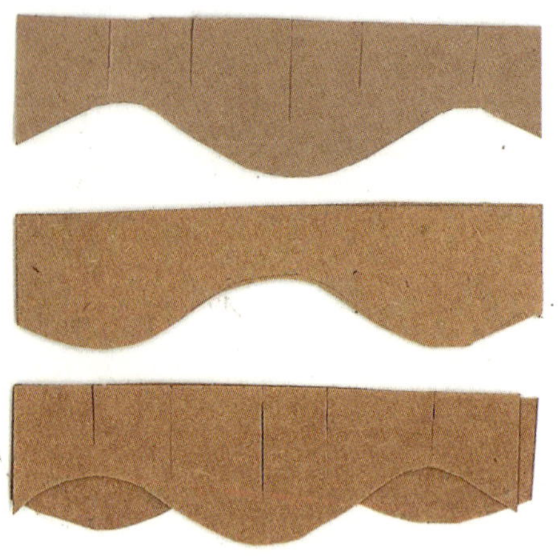

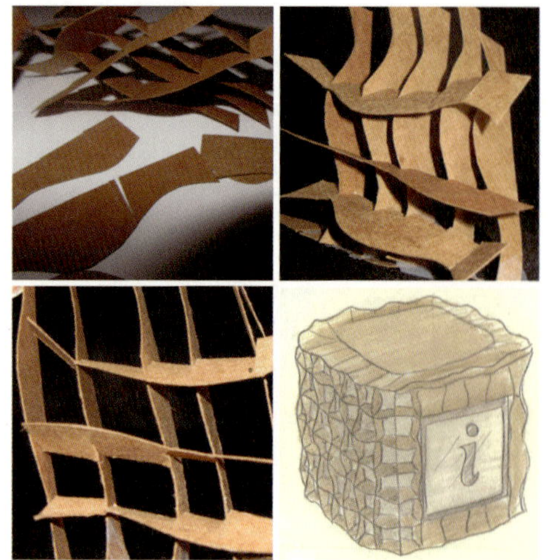

Development of the structure and forms, create the visual architectural narrative.

SPITAL YARD

The constructed of Spitalfields site ask the design response to nestle a use in to the existing architecture.

SPITAL YARD

Spitalfields에 이미 존재하는 건물들을 사이의 틈새 공간을 이용하는 것이다.

DESIGN PROPOSAL 1

The structure has the idea of flexible space allows for workshop to build the skills.
The structure can adapt to different scenarios.(Exhibition, art gallery, meeting area, small event,... .)

디자인 제안 1

기술을 배울 수 있는 워크샵을 열 수 있는 건축물. 여러 가지 상황에 쓸 수 있다 (예: 전시, 회의실 등).

SPITAL YARD DESIGN PROPOSAL 2

Climbing wall

Climbing wall designed to reflect the concept theory in practical, which will provide the fusion of personal energy to achieve goal in enjoyable and convivial way.

디자인 제안 2

벽을 등반하는 것은 개인의 노력에 의해 목적을 이룬다는 점에서 재미있고 유쾌한 놀이이다.

Sara Mitrova
mitrova.sara@gmail.com
http://saraamitrova.tumblr.com/

The Whitechapel Foundation

The Royal London Hospital, situated just across the Whitechapel Station

INTRODUCTION

Our cities now have to rethink and redevelop the use of space more creatively and ingeniously than ever before. Space is precious in any context but the extra pressure cities' face with the ever growing need to accommodate increasing populations mean that every open area is now a new possibility for spatial development.

소개

우리 도시들은 이제 공간의 사용을 더 창의적으로 생각해야 한다. 공간이란 언제나 중요하지만 도시는 계속 불어나는 인구를 수용할 수 있어야 하는 압박을 받고 있기 때문에 열린 공간이라면 어디라도 개발을 할 수 있는 곳이 된다.

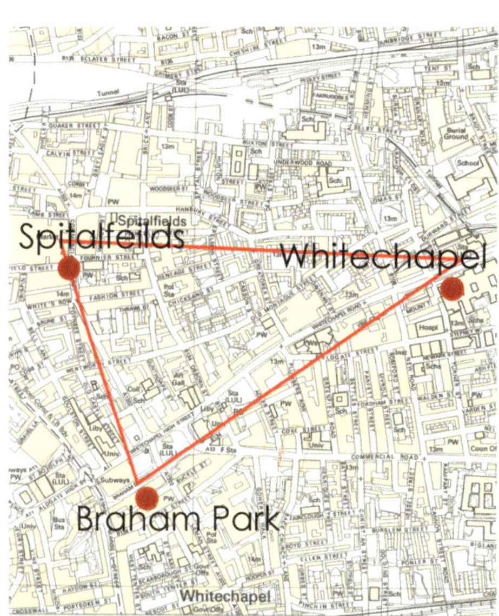

Whitechapel Station panoramic picture

images showing trade in the Whitechapel area

THE BRIEF

The brief was to devise a series of architectural and interior solution to engage local community with local business within the Aldgate area, with a view to future employment. The location of the structures would have to encourage the customers and visitors to travel to each of the three locations, offering opportunities of new perspectives, within Tower Hamlets.

요약

지역사회와 지역 기업들 사이에 다리를 놓는 것이 취지이다. Whitechapel, Aldgate와 Spitalfields에 새로운 실내외를 건축하여 미래 취업에 대한 전망을 보여줄 것이다. 건축물의 위치는 세 곳을 돌아다니도록 격려하고 Tower Hamlets 내 새로운 기회와 시선을 제공한다.

The main focus was to provide a series of structures that engage the local community with local business, within Tower Hamlets, with a view of employment. The project asked us to create a series of environments on three sites with the hub or pivoting point nestling on the roof of Whitechapel Station.

An important inspiration for my project is the Whitechapel Gallery, and the solution that was found to link the old structure with a contemporary elements.

이 프로젝트에 중요한 영감을 준 것은 Whitechapel 갤러리이고, 해결 방안은 오래된 건물을 새로운 요소들과 결합하는 데에서 나왔다.

The Whitechapel Gallery

The Whitechapel Station

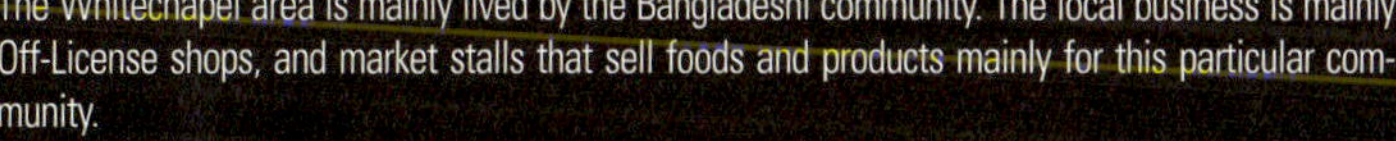
photographic diagnostics of architecture in the Whitechapel area

The Whitechapel area is mainly lived by the Bangladeshi community. The local business is mainly Off-License shops, and market stalls that sell foods and products mainly for this particular community.
The main material used for the houses around this area is London stock brick; not being an advantaged part of London, little changed since these houses were built.

Whitechapel 지역은 방글라데시 사람들이 많다. 이 지역은 주로 불법으로 운영되는 가게들과 포장마차 장사로 돌아간다.
집들을 짓는데 가장 흔히 이용된 것은 벽돌이고, 이 집들이 지어진 이후로 변한 것은 별로 없다.

architecture on the Whitechapel High street

sketch showing concept development

THE CONCEPT

The concept is called "mind the gap" inspired by the space between the two buildings at Whitechapel Station. However, I want to give it a positive outlook, and see the space as a challenge, a void that has the opportunity of message and impact through architectural experience. It is the message of awareness.

컨셉

이 컨셉은 '틈을 주의하시오' 라는 제목이다. Whitechapel 역의 두 건물 사이의 틈에서 영감을 얻었다. 나는 이 틈을 도전으로 보고 싶다. 건축을 통해 교훈과 영향을 줄 수 있는 공간이다.

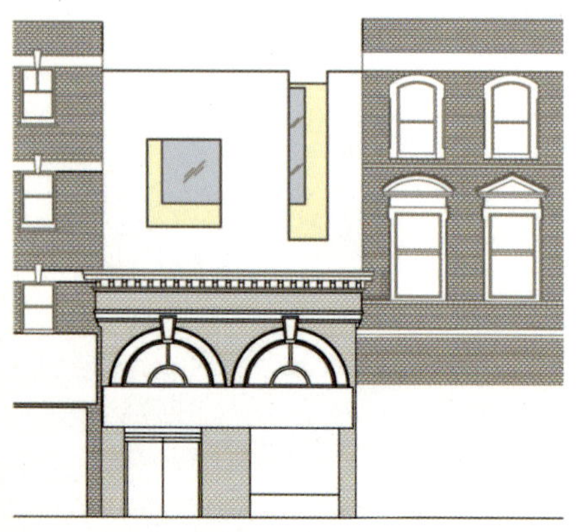

concept development

The structures used for the facade as well as the interior of the space have a note of newness and freshness, but their effect is not too sharp in order to respect local architectural principles. The layered facade also challenges people to engage and start thinking of things in a new different way – whether they are inside or outside the space.

모양을 내기 위해 이용한 조형물과 실내를 꾸미기 위한 조형물은 새로움과 신선함이 있지만, 지역 특유의 건물들을 의식해서 너무 날카롭게 만들지는 않았다. 여러 겹으로 된 모양은 사람들이 새로운 방식으로 생각하기를 권한다 – 안에 있든 밖에 있든 상관 없이.

Sketch showing interior development

model pictures showing facade development

The concept is all about encouraging people to create links within the community, links between Spitalfields and Whitechapel. The space would have a creative purpose, but an inclusive and practical one. There would be creative short courses (such as painting, photography etc) that would engage students, youth, ordinary locals and local artists. The work produced would be then exhibited in the other building in Spitalfields. Through this a new, more open way of thinking would be created, people would be encouraged to learn about arts, and explore new cultural values. The space would gain dynamic informational and educational purpose, but also new links and bridges between communities.

컨셉은 사람들에게 지역 사회 안에서 연결점을 만드는 것을 격려하는 게 목적이다. 공간은 창의적 목적을 가졌지만, 동시에 포함적이고 실용적이다. 회화, 사진 등을 배울 수 있는 짧은 수업들이 있고, 학생, 젊은이, 지역 주민, 그리고 주변의 예술가들을 불러들일 수 있다. 자신들이 만든 작품을 Spitalfields의 다른 건물에서 전시할 수 있을 것이다. 이를 통해 새롭고 더 열린 방식의 사고가 만들어질 수 있고, 사람들은 예술에 대해 호기심이 생기고 문화적 가치를 한번 더 생각해볼 수 있을 것이다. 이 공간은 활동적으로 정보를 공유할 수 있고, 지역사회 사이에 새로운 관계와 다리들이 생길 것이다.

CONCEPT: FACADE

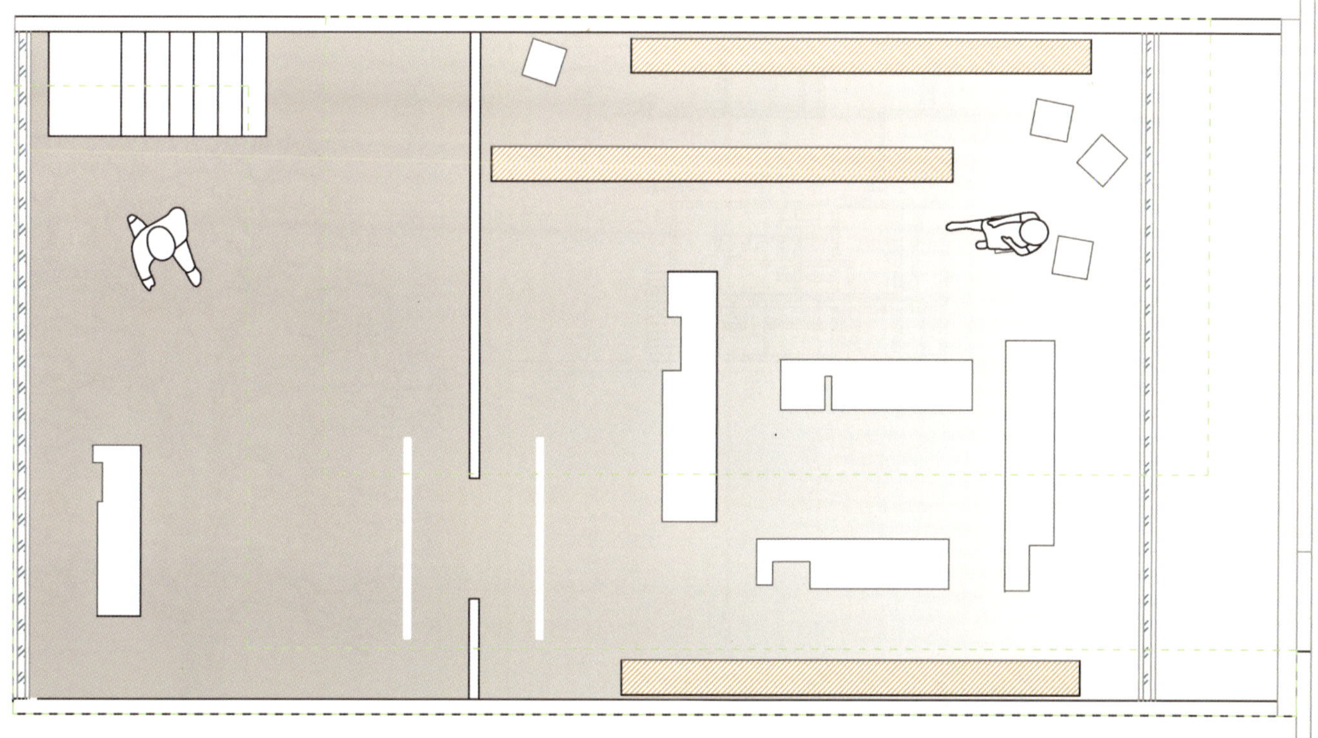

INTERIOR PLAN 1

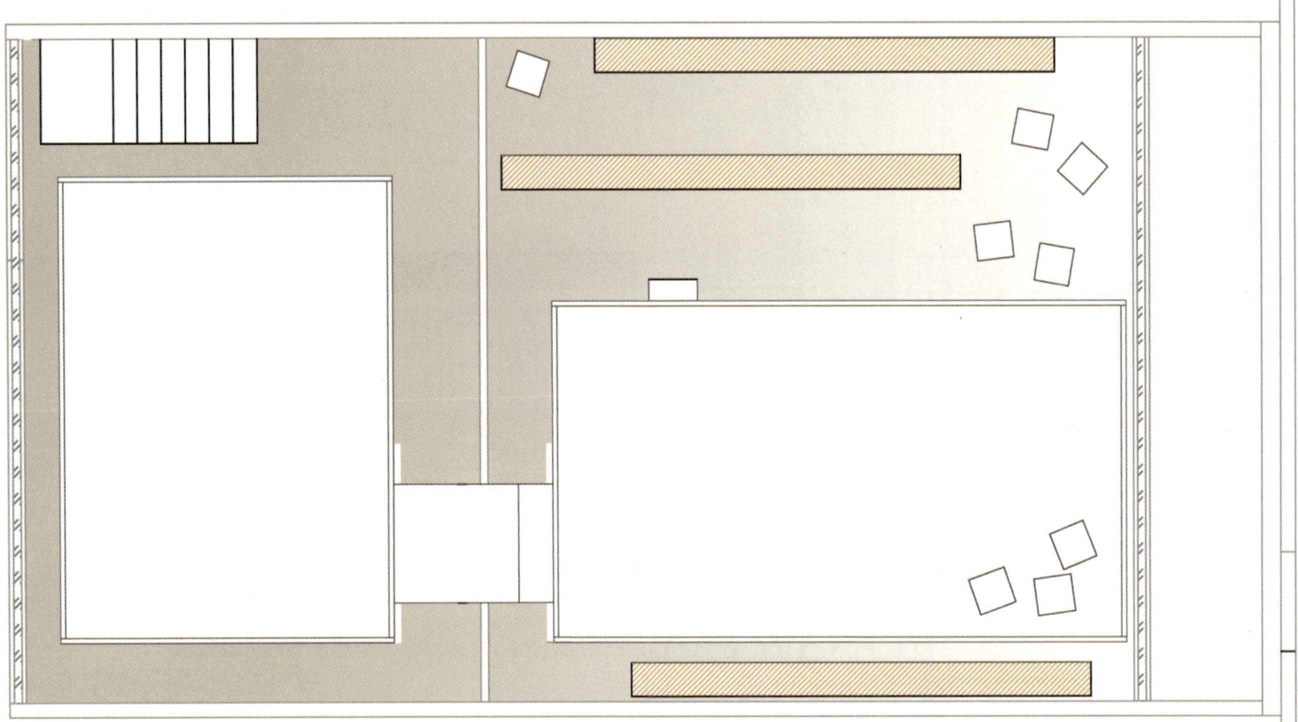

INTERIOR PLAN 2

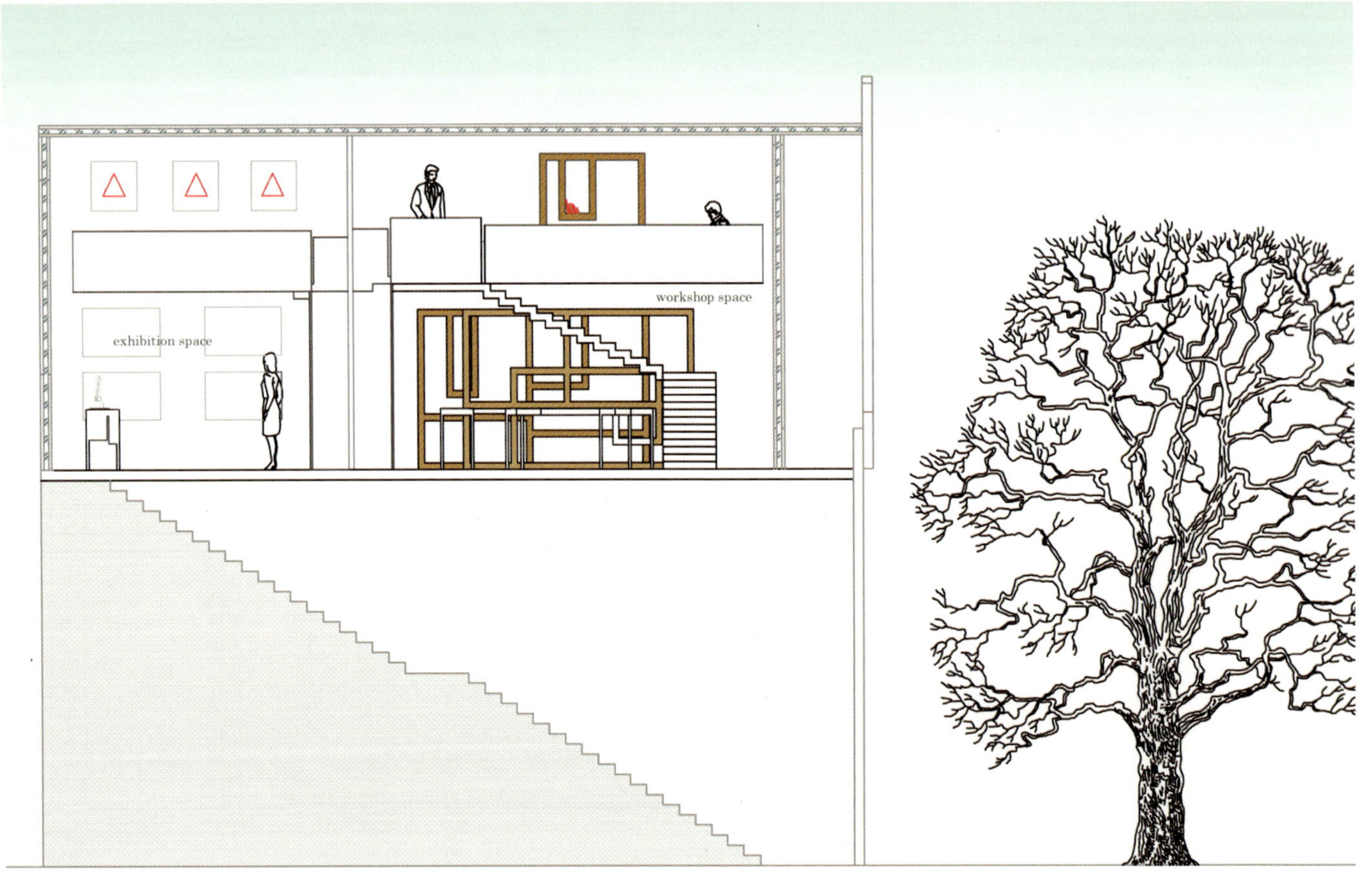

exhibition space

workshop space

ELEVATION

SPITALFIELDS

The Spitalfields site is a continuation of the Centre for Developing Creative Skills in Whitechapel. If the workshop and gallery is located there, here will be the organizational part offices and meeting rooms. The team will not only seek to encourage skills development within the space in Whitechapel, but through various open spaces located within Braham Park - Spitalfields
In terms of design, principles of simplicity and brightness are used here as well, based on the GAP concept.

SPITALFIELDS

Spitalfields 사이트는 Whitechapel의 예술공간의 지속이다. 워크샵과 갤러리가 Whitechapel에 있다면, Spitalfields에는 회의실과 사무소가 있을 것이다. 이곳의 팀은 기술 발전뿐이 아니라 Braham Park과 Spitalfields의 열린 공간에서도 창의적 활동을 격려할 것이다.
디자인 면에서는 Whitecapel과 마찬가지로 간단하고 밝은 요소들이 사용될 것이다.

Spitalfields concept sketch

Spitalfields concept front elevation

Braham Park diagnostic pictures

concept early sketch

back elevation

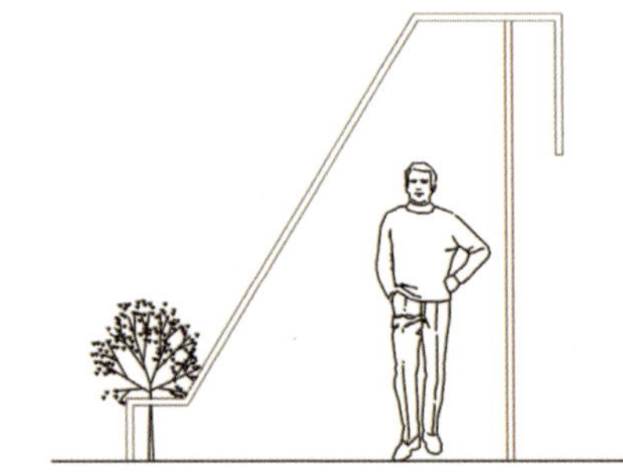

side elevation

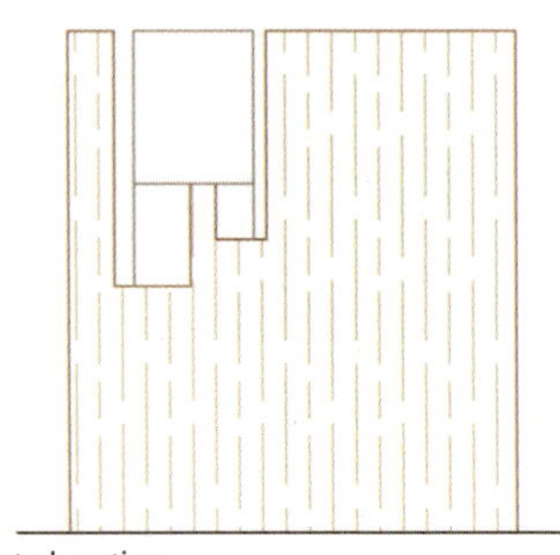

front elevation

Hidden between tall office buildings, The Braham Park site is a quiet green area. Initially a street, transformed into a discreet park it gives natural colours and textures to the grey surroundings being a calming place for a break from the City. The design for this site is purely conceptual: a wooden panel with "gaps" in it combined with a Corian structure will give the message of the link between traditional principles and modern new elements of the communities in Whitechapel area. The Corian structure "hides" a tree suggesting the hope communities have in new things, and that the valuable cultural principles won`t be lost no matter how old they are.

높은 회사 건물들 사이에 감춰진 Braham Park 사이트는 조용한 녹색의 공간이다. 원래는 도로였지만, 숨겨진 공원으로 변하면서 주변의 인공적인 회색 환경에 대비된 자연스러운 색깔과 질감을 보여주어 안정적이고 평화로운 휴식을 취할 수 있다. 이곳의 디자인은 순수하게 개념일 뿐이다.

'틈' 이 있는 나무 판자를 코리안 조형물과 합쳐 전통적인 원칙들과 새로운 요소의 관계를 보여주고 지역사회간의 관계를 표현한다. 코리안은 나무를 '감추고' 있는데, 이는 지역사회들이 새로운 것에 대해 가진 희망을 나타내며, 동시에 전통은 아무리 오래돼도 잊혀지지 않으리라는 것을 보여준다.

Grangewood Business Centre
LONDIS
CITY LEGAL PARTNERSHIP
EAST LONDON LAW CHAMBERS
IMRAN TRAVELS
Dry cleaners
TAILORING & ALTERATIONS
fresh

HIGH STREET ; in Seoul_Teheran-ro
가천대학교
Gachon University

GACHON UNIVERSITY
Department of Interior Architecture

2012 Junior ; Design Seminar

The 3rd Hidden Space Project
HIGH STREET

가천대학교 실내건축학과
2012 3학년 1학기 디자인 세미나 스튜디오 결과물

461-701
GACHON University Dept.of Interior Architecture
1342 Seongnamdaero, Sujeong-gu, Seongnam-si, Gyeonggi-do, korea
Tel + 82 31 750 5379

http://www.gachon.ac.kr/english

학과장 **김용성**
가천대학 실내건축학과
ysk1495@gachon.ac.kr

서 문

Hidden Space Project의 2012년 주제인 High Street를 바라보는 가천대학의 시선은 보다 문화적이고 사회적인 관점을 담고 있다고 할 수 있습니다. 이는 도시의 건축이 생산한 공간에 대한 해체와 숙고의 기회를 제공한 것입니다.
거대담론과 일반화된 단순개발의 차원을 너머 우리의 삶을 이루는 가장 일상적이고 기본적인 작은 편린들의 소중함을 '공간' 을 통해 끌어내고자 한 시도였다고 볼 수 있습니다. 고층 랜드마크와 하이테크 기반의 도시건축물들의 틈새에서 미래의 건축가들인 학생들이 바라보아야 할 지향점들은 과연 무엇인가에 대한 리서치이면서 동시에 우리 기성 건축가들에게도 자문해 볼 수 있는 의미 있는 과정이었음이 분명합니다.
High Street 의 숨겨진 공간에서 우리는 숨겨진 개념들을 찾아내게 됩니다. 결국 건축이 가지고 있는 창의적 조형성과 실현 가능한 기술 사이의 항상성, 그것의 주체는 늘 사람이라는 것입니다.

국제교류는 그것이 가지고 있는 장점과 교육적 성과, 그리고 미래적인 가치를 충분히 인식하고 있음에도 선뜻 시작되기 어려운 여러가지 제반 여건과 진행하는 사람들의 노력을 필요로 하는 쉽지 않은 교육활동입니다. 때문에 이번 프로젝트를 통한 한국과 영국의 학생들의 학문적 문화적 교류가 자랑스럽고 앞으로도 이러한 교류전시와 출판, 그리고 올해와 같이 다른 건축가들을 초대하여 런던에서 개최되는 심포지움 등 다양한 프로그램들을 통해 지속적인 발전이 있기를 기대합니다.

학기 중에 독일과 중국과의 교류전시로 인하여 많은 학생들이 이 프로젝트에 참여하지 못한 것이 아쉽지만 또다시 새로운 주제와 흥미로운 도전이 있을 것을 믿어 의심치 않습니다. 지원이 많지 않은 가운데서 끝까지 노력을 아끼지 않은 지도교수를 비롯한 참여자 모두에게 축하와 감사의 인사를 전합니다.

Head of Interior Architecture Department | Prof. Kim, Yong Sung
Department of Interior Architecture GACHON University
ysk1495@gachon.ac.kr

Preface

Gachon University's perspective of the 2012 HIDDEN SPACE project, the HIGH STREET, contains a more Social and Cultural view point. This has provided us with the opportunity to break down and think over the spaces of the city. This can be seen as an effort to overcome the mega-discourse and simple development and bring forth the value of small bits of everyday life through 'space'. The Urgan High Street project has proven to be a meaningful process that provided the students of today, the architects of tomorrow, an opportunity to research the perspective that needs to search for spaces between high-tech structures and landmark skyscrapers and the established architects to question themselves.

Within the hidden spaces of High Street we begin to find hidden concepts. in the end, the subject of the adaptation between the creative formativeness and realizable technology is always people.

I regret not being able to participate more due to the joint exhibitions with Germany and China, but I look forward to working with new subjects and students next year.
I am sending regards to all members of participants, tutor and students who did their best effort with not an exellent condition.

안성희
가천대학 실내건축과
sahn@hotmail.com

Lecturer in Interior Design Department | Ahn, Sung Hee

WHAT IS THE MISSING ELEMENT IN TEHERAN-RO?

인구 천만이 살고 있는 거대도시 서울에는 몇 군데의 high street이라 할 수 있는 도심의 중심축들이 있다. 그 중에서도 서울의 중심을 두 개로 나누는 한강의 남쪽에 위치한 강남은 서울의 금융과 쇼핑, 레지던시의 중심으로 80년대 이후 급속히 떠오르면서 부동산 투기와 과속성장의 아이콘이 되었다.

2012년의 Hidden Project 시리즈의 주제인 〈High Street〉의 대상지로 선택한 테헤란로는 이곳 강남에 위치하고 있다. 테헤란로 라는 이름은 1977년, 서울특별시와 이란의 수도인 테헤란의 자매결연을 기념하여 붙여진 것이다. 강남을 가로지르는 이 도로에는 국제무역센터와 인터컨티넨탈을 비롯한 호텔들, 여러 금융회사들, 포항제철을 비롯한 기업들, 그리고 최근 이주한 글로벌 기업인 삼성의 본사가 자리하고 있는 곳이다.

또한 1988년, 서울 올림픽 경기장이 지어진 곳이기도 하기에 2012년 런던 올림픽을 맞이하여 다시 되돌아 볼 기회를 주며 올림픽과 같은 큰 행사가 지나간 후 이러한 경기장과 건물들의 이용과 관리에 관해 돌아볼 수 있는 기회를 제공해주는 매우 흥미로운 사이트이기도 하다.

테헤란로의 사진을 찍어서 본다면 어느 각도에서 찍어도 그곳이 한국의 서울인지 모를 것이다. 30층이 넘는 현대식 건물과 영문간판들 사이에서 우리는 이곳에서 찾을 수 있는 아이덴티티는 무엇인지 질문하게 되는 것이다.

그렇다면 21세기 서울의 도심 high street 에서는 과연 무엇이 결여되어있는것 일까?

학생들은 그 길 끝에서 반대쪽까지 여러 번 걷고 조사하면서 그 질문에 대한 자기만의 시나리오를 만들어냈다.

거대한 인공적인 요소들로 가득 차있는 그곳에서 숨겨진 공간(Hidden space)를 찾아내어 이식 시키고자 하는 요소들은 크게 세 가지인데, 첫째, 보다 인간적인 공간 디자인, 둘째, 꿀벌이나 나비 등 자연과 이어지는 그린스페이스, 셋째, 한국의 과거와 이어지는 전통적인 요소들이라고 할 수 있다. 이 세 가지의 요소들은 일중독과 개발중독이 된 테헤란로의 거리에서 우리에게 숨 쉴 곳을 제공할 수 있는 새로운 층을 형성할 수 있는 가능성을 보여준다.

결과물이 가지고 있는 기술적인 미숙함 너머에 학생들이 저마다 노력한 과정의 흔적들과 함께 생각해 보아야 할 질문들을 발견해낸 것에 대해 자랑스럽게 생각하며 이제는 건축이 '일상의 삶' 과 만나지는 시대가 이미 왔다는 것에 대해 의심의 여지가 없다.

테헤란로의 화려한 건축물과 경기장 그리고 10차선 도로는 이제 이 프로젝트 안에서 작고 소중한 일상과 접점을 찾기 시작했다. 발전과 개발의 그늘에 가려서 보이지 않는 것, 미루어 두었던 것들 것 하나씩 열어서 양지로 끌어내는 이 시도에 함께 하게 되어 즐거웠다. 언제나 학생들과 함께하는 프로젝트 수업에서는 나 역시 참여자이다. 특정한 목표를 설정하고 이끈다기 보다 학생들이 발견한, 혹은 가고 싶은 방향을 함께 바라보고, 갈 수 있도록 도와주는 조력자인 것이다.

그럼에도 불구하고 마감에 쫓겨 많은 과제를 안겨주었던 것은 사실이다. 책에 포함되지는 못했으나 프로젝트에 참여했던 다른 학생들과 자기 작업 외에 익숙하지 않은 편집과 번역 등을 도와준 학생들과 연구원들에게 감사를 표한다.

HIGH STREET - SITE

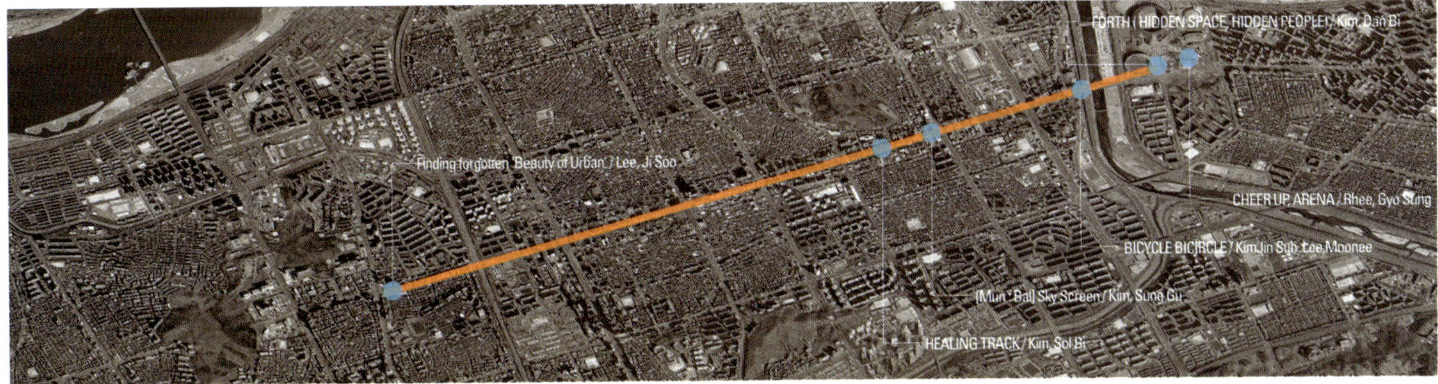

서울특별시 서초구 서초동 서초동–강남구 삼성동 삼성교에 걸치는 약 4km 10차선 도로변

10 lanes 4km from Seocho-dong to Samsung bridge, Seocho-dong, Seocho-gu, Seoul, Korea

Seoul, a city with a population of 1 million, has several central points that can be called high streets. Among them, KangNam is a place situated just south of the Han River, which separates Seoul city into two halves.

KangNam is the center of economy and shopping, as well as residency, which has rapidly gained attention since the 80's and become an icon of accelerated growth and land speculation.

Teheran-ro, the site chosen for the <High Street> project, is a street in KangNam. The name "Teheran-ro" comes from the capital of Iran, celebrating the 1977 MOU between Iran and Korea.

On this street that runs right through KangNam sits Chung-Ui World Trade Center, several hotels such as the Intercontinental Hotel, many banks, companies such as Posco, and the recently moved headquarter building of the world-class company Samsung.

Also, in 1988, the Seoul Olympic Stadium was built nearby, and as the 2012 London Olympics approach, it gives us a chance to look back at how these stadiums and buildings left should be cared and used after such large events.

No matter what angle you take a picture from, Teheran-ro will not look like Seoul, Korea. We begin to question what the identity is that we can find in the modern buildings over 30 stories tall, high-end fashion stores. Which element is missing in the 21st century urban high street?

We wish to find hidden spaces in the manmade jungle of tall buildings and high-end stores, and stress three special elements. Firstly, a more humanely scaled space. Second, a green space where we can connect with nature, such as bees or butterflies. Third and last, traditional elements that connect with Korea's past. These three elements give us the possibility of discovering a new layer that serves as a place to breathe in the workaholic, develop-obsessed mood of Teheran-ro.

Above the technical imperfections are the evidence of hardworking process of the students, and the questions that we should think about. Of this I am proud, and now there is no question that architecture is an influential part of everyday life.

The fancy buildings, the stadium, and 10 lane road of Teheran-ro are now coming in contact with the small, precious things within this project. The things that we could not see, we put off for later in the shadow of development are now being pulled into light, and I am glad to have been a part of this challenge. I am also a suppoter and participant in this kind of project which the students and I are working together. Instead of pushing the students towards a final goal, looking toward the goals that the students have found or wish to pursue.

I am appreciate to those researchers who helped to edit and translate for the book and students whose work were not a part of this years exhibition.

members

Design Seminar
안성희 Ahn, Sung Hee

門簾 [Mun : Bal] Sky Screen
김성구 Kim, Seoung Gu

CHEER UP, ARENA
이교성 Rhee, Gyo Sung

THE FORTH
김단비 Kim, Dan Bi

HEALING TRACK - URBAN GALLERY
김솔비 Kim, Sol Bi

BICIRCLE, BICYCLE
김진섭 Kim, Jin Sub | 이무늬 Lee, Moo Nee

도심…잊혀진 '美' 를 발견하다
Finding forgotten 'Beauty of The Urban'
이지수 Lee, Ji Soo

Kim, Seoung Gu / Interior Architecture Dept.
getgam825@hotmail.com

門簾 [Mun : Bal] Sky Screen

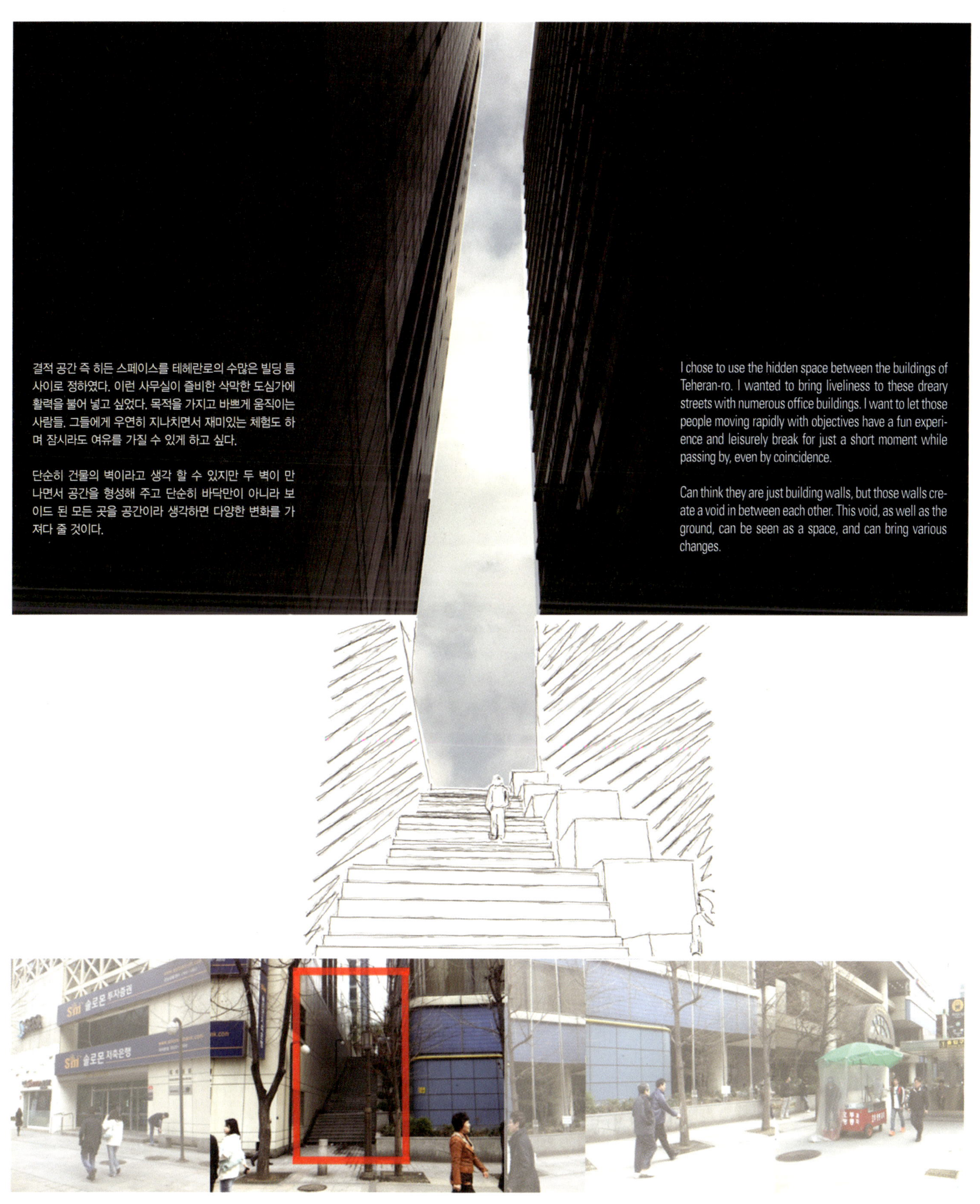

결적 공간 즉 히든 스페이스를 테헤란로의 수많은 빌딩 틈 사이로 정하였다. 이런 사무실이 즐비한 삭막한 도심가에 활력을 불어 넣고 싶었다. 목적을 가지고 바쁘게 움직이는 사람들. 그들에게 우연히 지나치면서 재미있는 체험도 하며 잠시라도 여유를 가질 수 있게 하고 싶다.

단순히 건물의 벽이라고 생각 할 수 있지만 두 벽이 만나면서 공간을 형성해 주고 단순히 바닥만이 아니라 보이드 된 모든 곳을 공간이라 생각하면 다양한 변화를 가져다 줄 것이다.

I chose to use the hidden space between the buildings of Teheran-ro. I wanted to bring liveliness to these dreary streets with numerous office buildings. I want to let those people moving rapidly with objectives have a fun experience and leisurely break for just a short moment while passing by, even by coincidence.

Can think they are just building walls, but those walls create a void in between each other. This void, as well as the ground, can be seen as a space, and can bring various changes.

CONCEPT

테헤란로에서 선택한 장소는 건물의 사이 작은 틈 공간이다. 도심의 높은 빌딩에 근무하는 직장인들은 휴식시간에 밖에 나와도 빛을 보기 어렵다. 좁은 틈이기에 빛(光)이 잘 들어오지 않기 때문에 빛에 대한 고민을 하였다. 좁은 틈에서 들어오는 빛과 예로부터 내려오던 한옥에서 볼 수 있는 문발(門簾)을 모티브로 삼았다. 이로부터 나온 것이 문창살(門窓–)이다. 문창살은 문이나 창에 살을 입혀 특유의 모양으로 빛을 받아들이는 재미있고 효과적인 한국의 전통 문양이라고 생각하였기 때문이다.

The place I chose is a small area between two buildings. It is a narrow passage that is rather dark, so I had to contemplate the lighting. I used the light coming through between the buildings and Mun-bal, a traditional Korean housing element, as the motif. Mun-chang-sal is the result. Mun-chang-sal is a time-honored Korean pattern that is made from putting multiple ribs on windows or doors, which is an interesting and effective way to receive light.

PROCESS

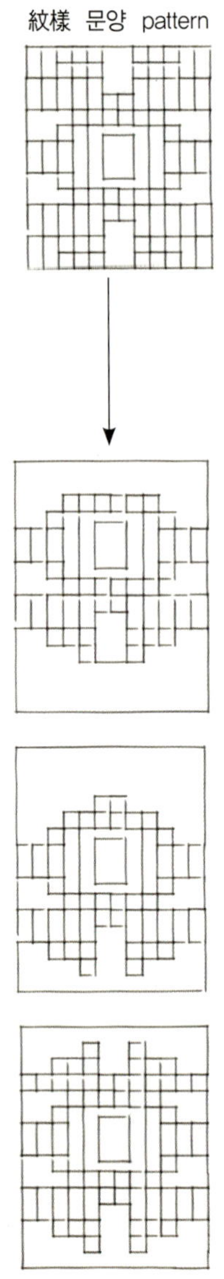

경복궁 전각의 문창살 패턴을 일정한 규격 안에서 재해석 해보았다. 그 패턴을 가지고 3가지 형태로 위해 나누어 보았다. 겹치면 재미있는 형태로 보인다.

I recomposed the ribs based on Gyung-bok Palace's doors and windows, based on a certain measurement. With those patterns I made 3 shapes. When they overlap, it should be interesting.

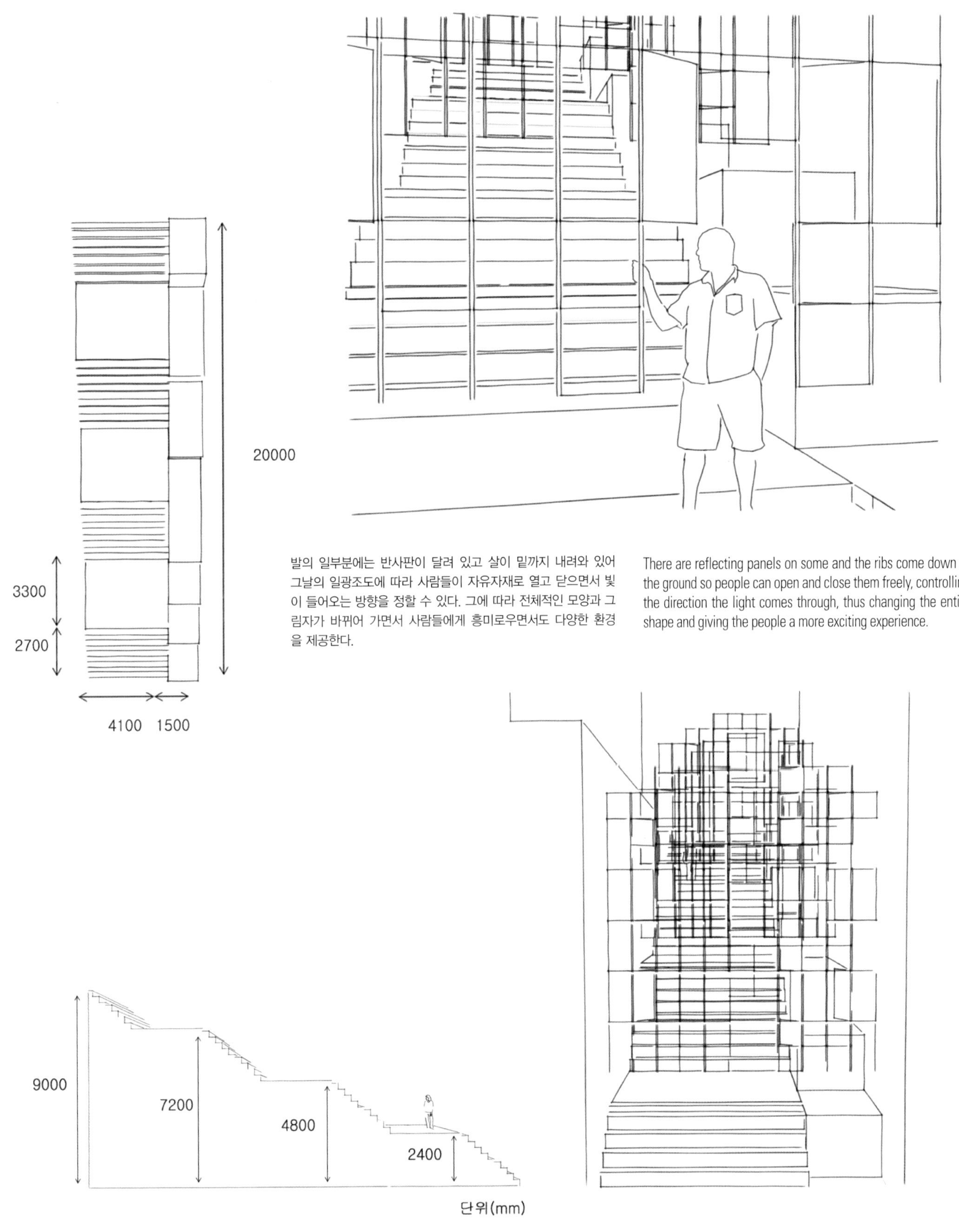

발의 일부분에는 반사판이 달려 있고 살이 밑까지 내려와 있어 그날의 일광조도에 따라 사람들이 자유자재로 열고 닫으면서 빛이 들어오는 방향을 정할 수 있다. 그에 따라 전체적인 모양과 그림자가 바뀌어 가면서 사람들에게 흥미로우면서도 다양한 환경을 제공한다.

There are reflecting panels on some and the ribs come down to the ground so people can open and close them freely, controlling the direction the light comes through, thus changing the entire shape and giving the people a more exciting experience.

3개의 레이어를 서로 다른색 혹은 채도를 가지도록 구분 하였고 약간 투명한 재질로 하였기 때문에 서로 겹치면 각각의 레이어가 합쳐진다는 느낌을 준다.

각 레이어의 색은 안으로 들어가면서 색이 진해지는데 이는 깊이감을 더 강조하기 위함이다. 또한 뚫려진 부분으로 하늘을 액자에 담아 보는 듯한 느낌을 가지게 한다. 각자 사용자가 밑에 달린 손잡이를 돌리면서 보여지는 하늘 부분도 자유자재로 변형해가는 움직이는 재미를 준다. 한국 특유의 문발이라는 컨셉으로 시작하여 단순히 딱딱한 구조물이 아닌 사용자가 변형할 수 있는 재미있는 구조를 만들어 보았다. 테헤란로의 복잡하고 바쁜 도심 빌딩 사이에서 잠깐 쉬어 가거나 재미있는 체험을 해보면 어떨까 한다.

I defined the three different layers, and because the material is translucent, the layers seem combined when overlaid.

Each layer's colour becomes darker as it advances in order to exaggerate the depth of the street. Overmore, the installation frames the sky with its openings. The viewer can have an enjoyable experience using the handle below and decide the part of sky that is visible. I used the uniquely Korean concept of Mun-bal to make not a hard, solid structure but a flexible, amusing one. How would it be to have a small experience or take a short break while rushing through the streets of Teheran-ro.

Rhee, Gyo Sung / Interior Architecture Dept.
luckyboy21c@naver.com

CHEER UP, ARENA

SITE

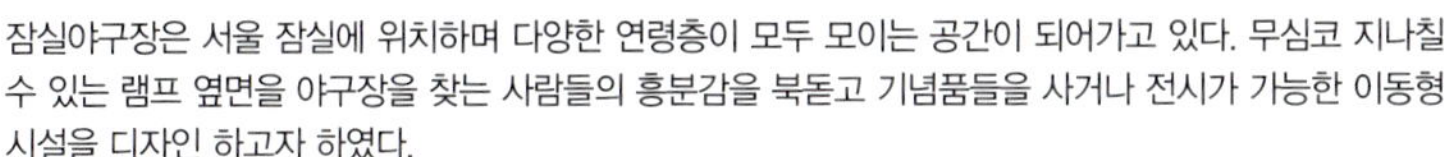

잠실야구장은 서울 잠실에 위치하며 다양한 연령층이 모두 모이는 공간이 되어가고 있다. 무심코 지나칠 수 있는 램프 옆면을 야구장을 찾는 사람들의 흥분감을 북돋고 기념품들을 사거나 전시가 가능한 이동형 시설을 디자인 하고자 하였다.

The Jamsil Baseball Stadium is becoming a place where people of all genders and ages come together. The ramp next to the stadium could be utilized as a space for a movable facility to excite the crowd and sell souvenirs. bicycle road, utilize the hidden place beneath the bridge as a shade for people to rest in. Also, we can provide the people that already use the bicycle road a resting area as well.

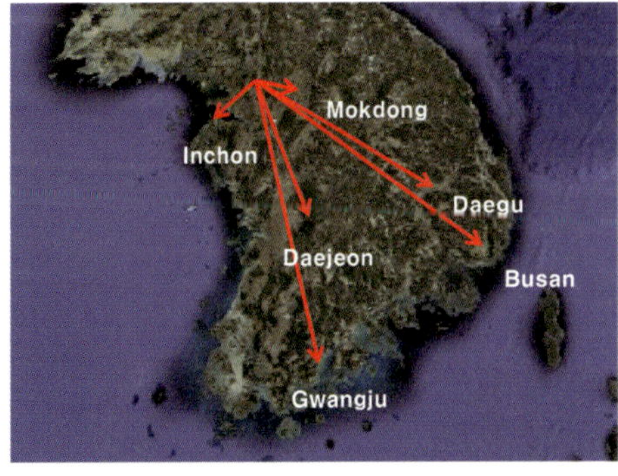

서울에서 가까운 인천부터 시작하여 부산까지 많은 원정 경기를 치루기 위하여 많은 이동이 필요한데, 함께하는 응원단과 선수단에게 안정감을 줄 수 있다.

Starting from Inchon near Seoul all the way to Busan, players have to play away games. This movable facility can give the players stability even at away games.

CONCEPT

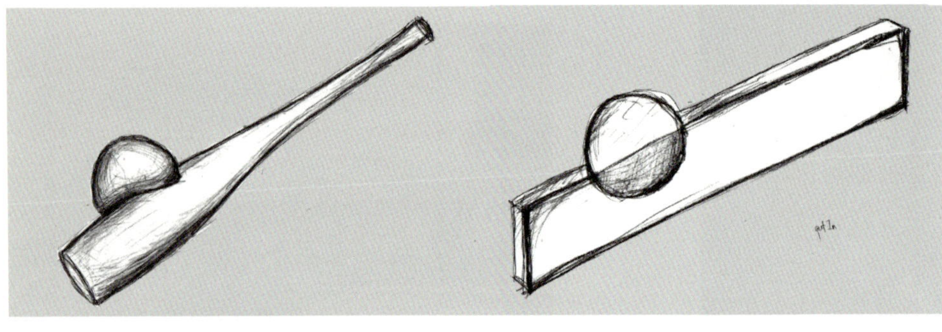

야구공이 배트에 맞는 순간 순간적으로 모양이 변형되는데서 아이디어를 얻었다. 하나의 매스가 다른 하나의 매스로 삽입되어지면서 램프 옆면에 고정이 된다.
옆면에 고정시키기 위하여 공중에 올리는 방법으로 일정 유닛의 프레임이 형성이 되고 그 프레임들이 조합되어 높이 조절이 된다.

The idea was envisioned from the baseball the moment it hits the bat. One mass is inserted into another mass, becoming fastened onto the ramp. In order to keep the facility on the ramp, frames of certain measurements will be produced and put together to match the height.

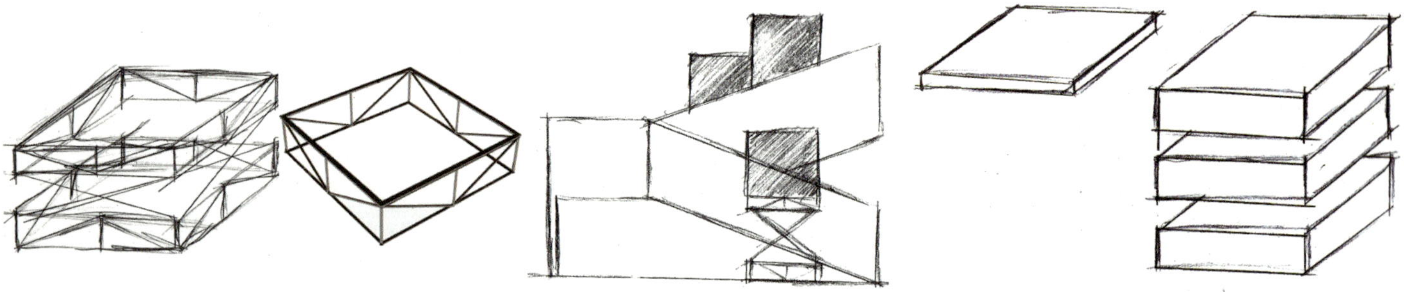

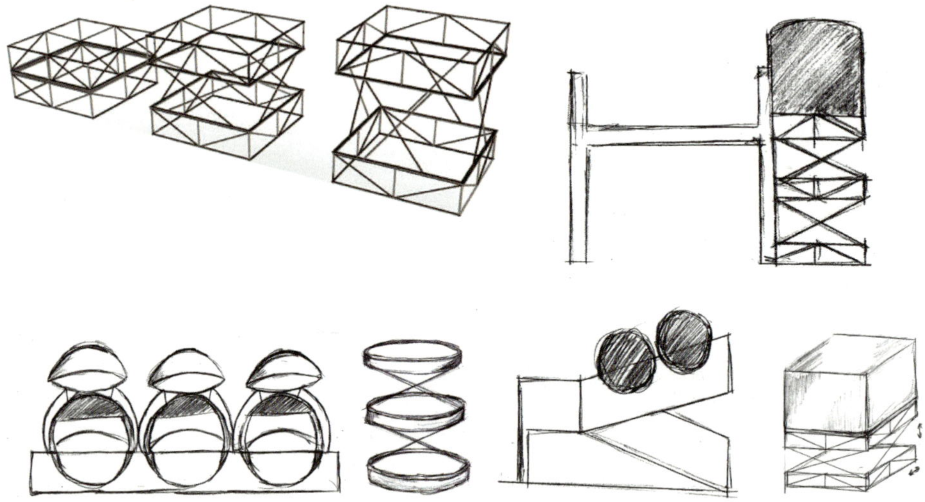

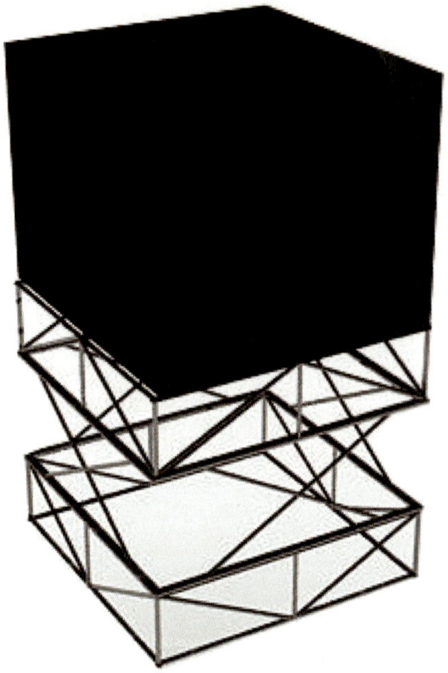

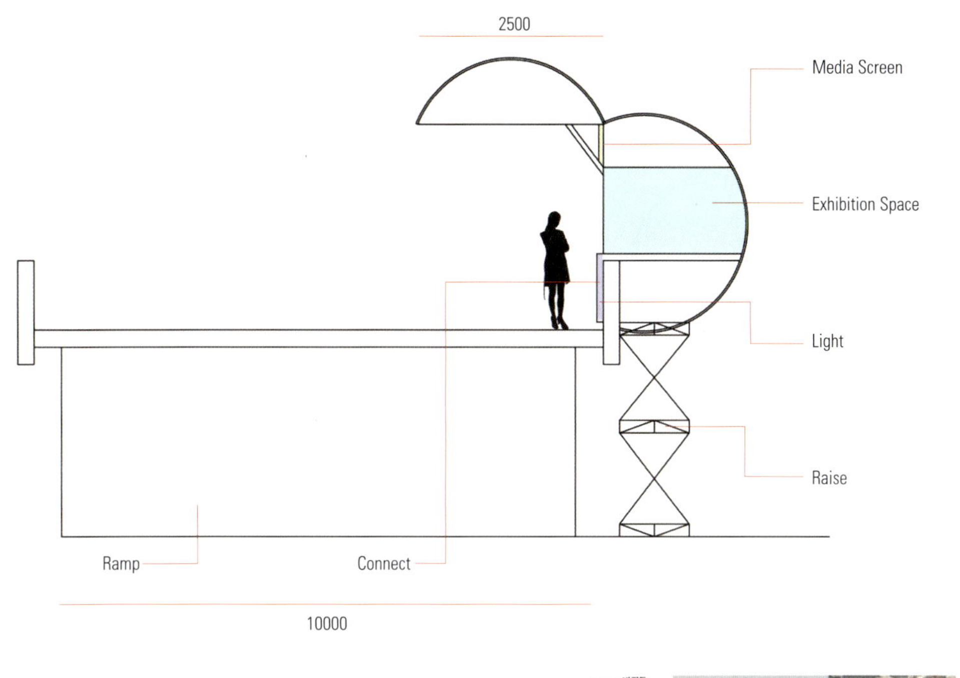

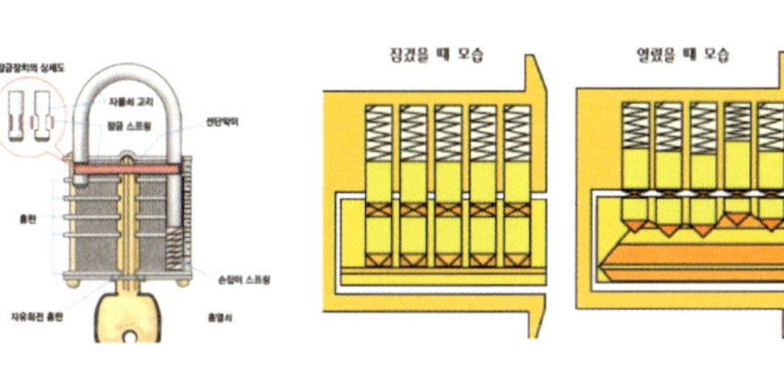

유압작기라는 요소를 통하여 설치물을 손쉽게 높은 곳으로 올리고 고정 후 뺄 수 있게 디자인 했고, 설치물을 램프 옆면에 고정시키기 위하여 자물쇠의 고정원리와 선박의 도크방식을 참고하여 디자인 하였다.

Using a mechanism called the oil-pressure jockey, we can easily lift the fixture up high. To fix the facility onto the ramp, I used the idea of a padlock and a boat dock.

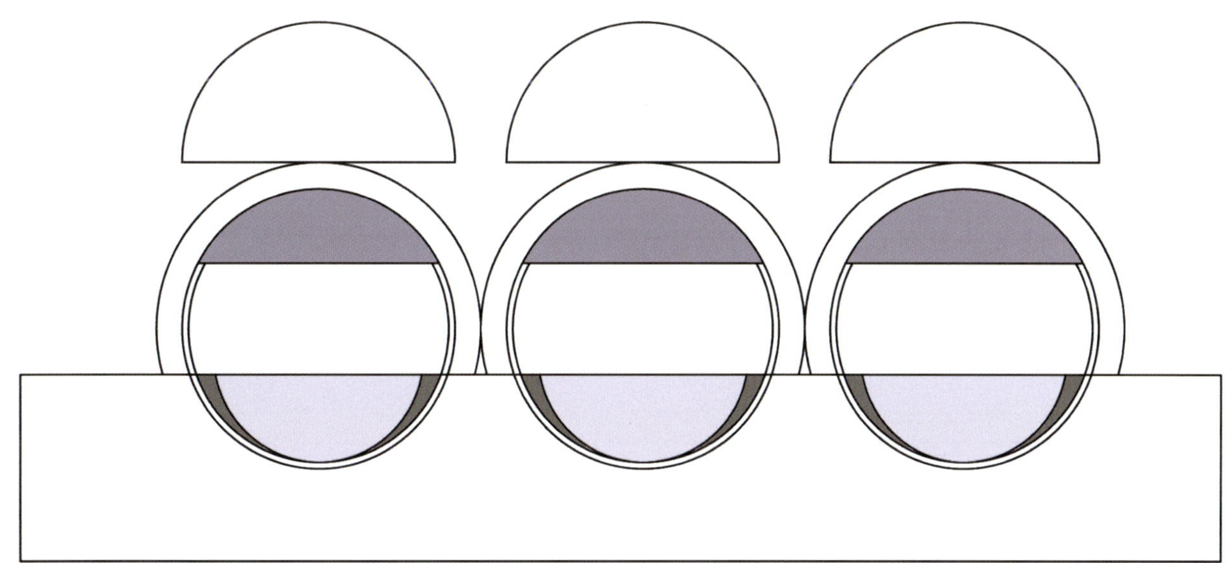

미디어 스크린을 통해 입장하는 관중들에게 다양한 영상을 통하여 야구장에 입장전에 기대감과 흥분을 더욱더 증폭 시켜 줄 수 있는 요소로서의 적용. 대한민국 야구장은 대부분 오래되어 경기가 끝나는 밤시간에 많이 어두운 문제가 있다. 이 문제를 해결하기 위하여 아랫면에 동선을 유도 할 수 있는 조명을 설치하여 퇴장하는 관중들에게 도움을 줄 요소를 덧붙였다.

The media screen shows entering guests various videos to excite them and make them look forward to the game even more. However, because many stadiums are old, the ramps are dark at night. The screens on the lower part of the facility also serve as a lighting component, assuring the safety of leaving guests.

DESIGN

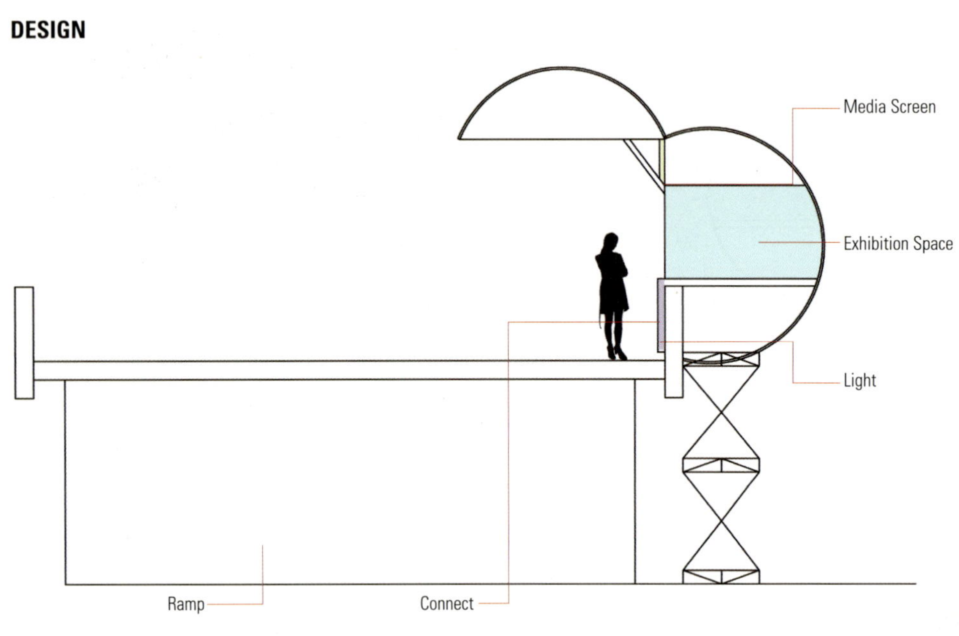

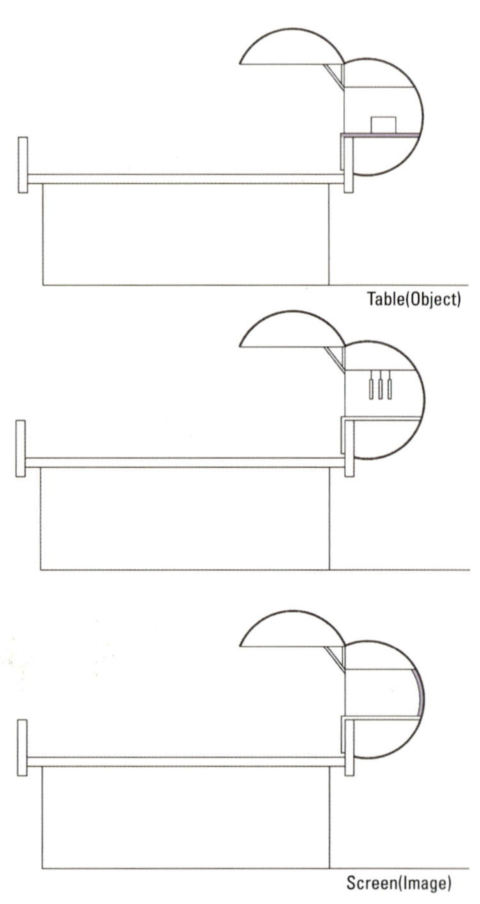

INSTALLATION PROCESS

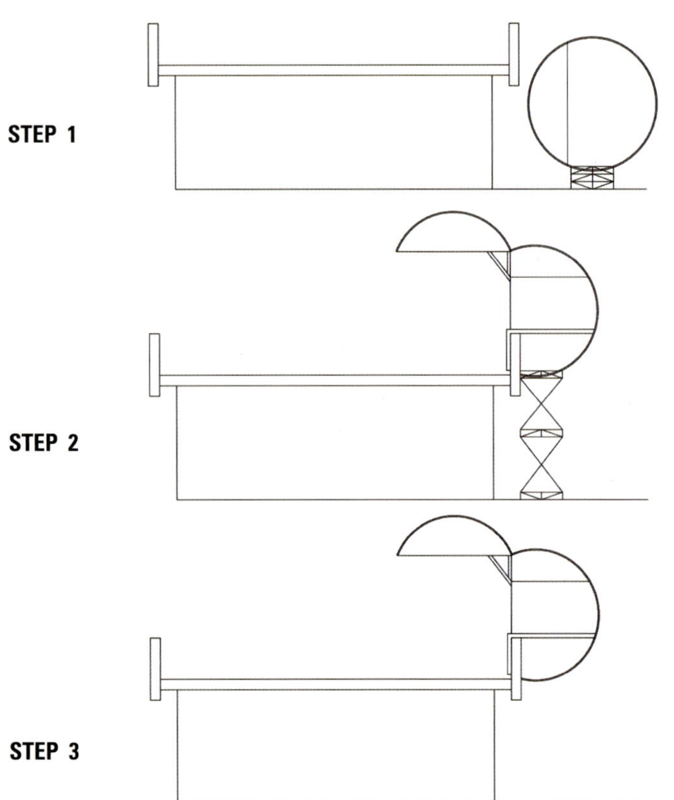

COLSE

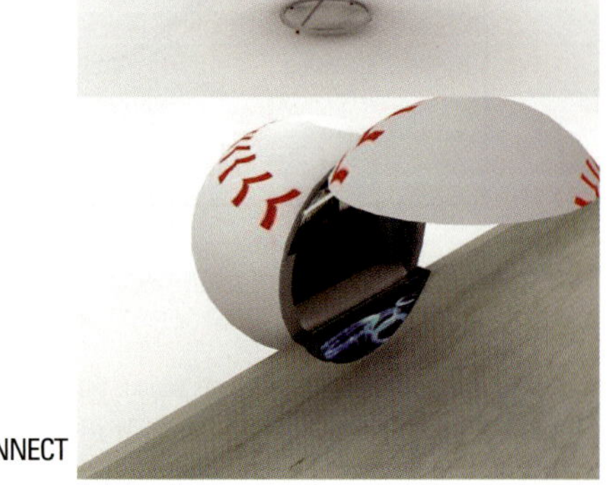

CONNECT

각각의 야구장으로 필요시마다 이동되어서 램프 옆면에 부착됨으로 간단하게 설치 철거가 가능하여 쉽게 이동 가능하여 지속적인 사용이 가능하게 디자인 되었다.

Designed to be easily attached and removed from ramps all around, and to be used continuously.

Kim, Dan Bi / Interior Architecture Dept.
edeyein@naver.com

THE FORTH

BACKGROUND

HIDDEN?

드러나지 않은 / 잊혀지고 있는 / 무언가에 의해 가려진
Not shown / Being forgotten / Concealed by something

현대사회에서 장애인들을 배려하는 공간디자인은 얼마나 나아졌는지를 보면 아직 장애인들은 공간디자인에서 마치 숨겨진 사람들이며 중요한 것은 스스로에 의해서가 아니라 그들과 다른 입장에 서서 그들을 바라보는 사람들에 의해 저절로 가려지게 된 사람들이라는 점이다. 종합운동장 뒤쪽 숨겨진 공간인 장애자들을 위한 창작공간을 더욱 확장시키고 보다 적극적인 공간으로 만들고 싶었다.

Look around to faces of the disabled, try to see how much considerd and have become better in modern society. Their faces are dark, their expressions solemn. The disabled are the hidden people of society. The important part is that they are not hiding themselves, but rather are automatically hidden by other people from dissimilar points of view.

서울특별시 송파구 잠실동 10 잠실 종합 운동장
JAMSIL Sports Complex, 10 Jam-sil dong, Song-Pa Gu, Seoul

잠실 종합 운동장은 1988년 서울 88올림픽 개최를 위해 오픈된 공간으로 2012년 현재는 여러 스포츠 종목의 시즌 경기가 있을 때마다 사용되고 있다. 매 종목의 시즌마다 수만명의 사람들이 다녀가는 이 곳에 사람들의 눈에 쉽게 보이지 않는 그러나 묵묵히 자신들의 색깔을 지켜나가는 Hidden Space가 있다.

DESIGN PROCESS

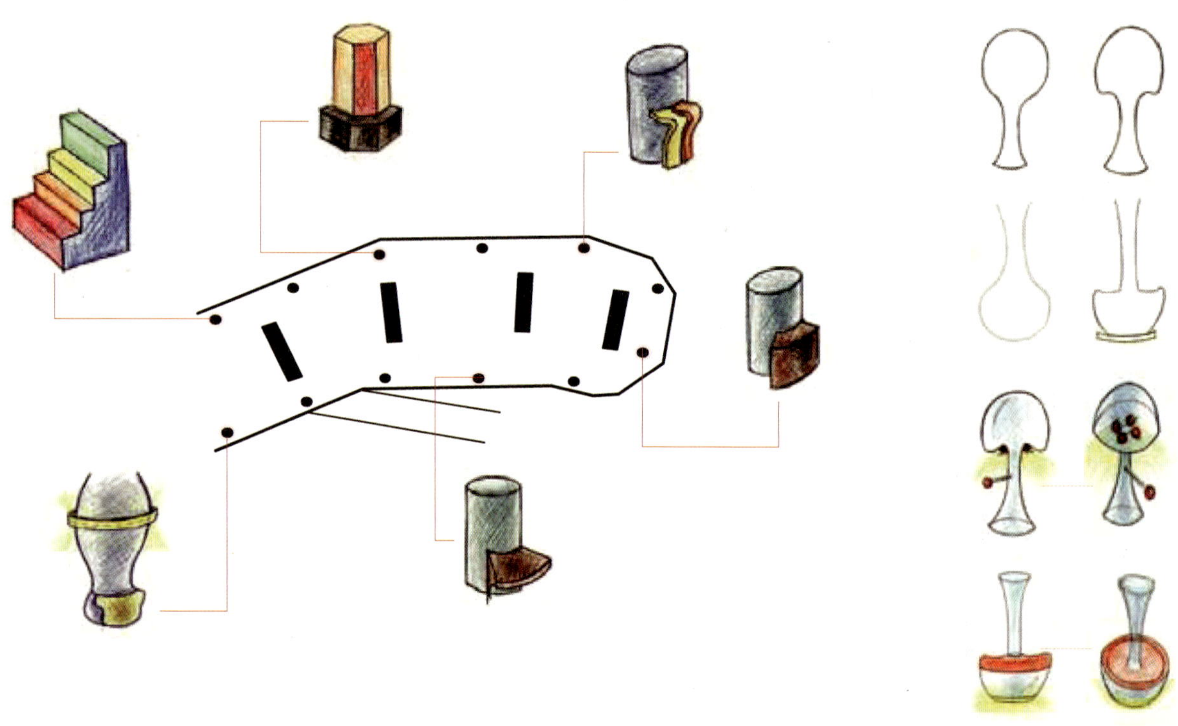

첫 번 째 디자인 포인트는 여러 개의 원기둥이다. 이 원기둥은 구조 전체를 받쳐주는 동시에 공간을 만들고 있는 역할도 하고 있다.
즉, 가장 기본적이면서도 중요한 역할을 하고 있다. 하지만 그 모양과 색은 너무 딱딱하고 건조하다. 여기에 변화를 주어 보다 시각적으로 즐겁게 사용할 방법을 찾으려 했다.

DESIGN

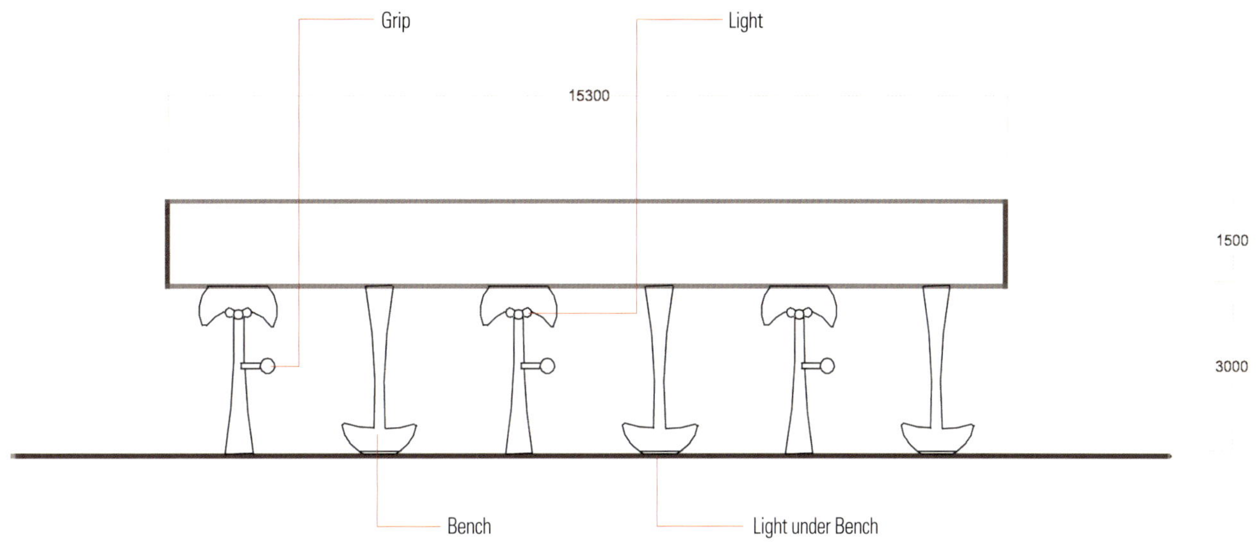

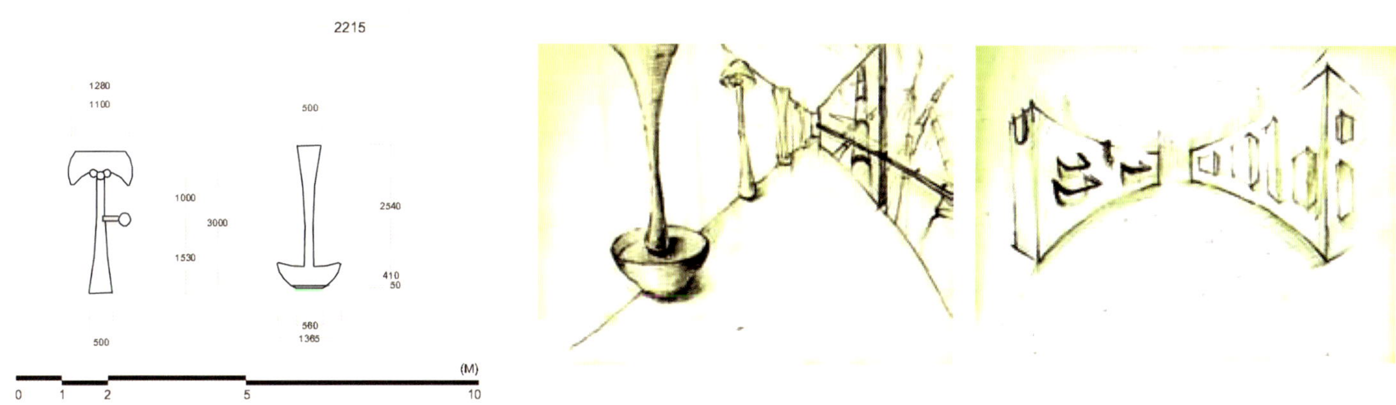

이 공간은 앞서 소개했던 '걷기 운동' 을 모티브로 한 기둥 디자인을 적용시킨 모습이다. 장애인들은 다른 사람의 도움 없이 쉽게 움직일 수 없다. 때문에 여행은 물론 산책 정도도 일상 생활에서 쉽지 않다. 그런 사람들을 위해 기둥의 도움을 얻으면서 원하는 풍경을 볼 수 있는 공간을 생각했다. 벽에 붙어있는 손잡이를 잡고서 벽을 따라 걸으면 그 동안 볼 수 없었던 아름다운 풍경들을 감상할 수 있다. 동시에 그 공간에 있는 것 같은 기분이 들게 한다.
반대쪽 기둥들은 걷기 연습을 할 수 있는 장치와 쉴 수 있는 장치를 동시에 나타낸다. 결과적으로 이 공간은 걸을 수 없는, 몸이 불편한 사람들이 바라는 다른 사람들처럼 평범한 활동을 할 수 있게 하는 공간이다. 더불어 생활에서 체육을 행할 수 있다는 장점도 있다.

다음 공간은 장애인들의 운동활동을 집중적으로 도와줄 수 있는 공간이다. 장애인들이 필요로 하는 지구력, 근력 등을 기를 수 있는 장애인 전용의 특별 운동 기구들이 설치되어 있다. 장애인을 더불어 비장애인들도 자유롭게 이용할 수 있으며 동시에 함께 어우러지고 장애인들에게 자존감을 주는 효과까지 얻을 수 있다.
장애인들을 위해 개별적으로 디자인된 공간과 기구들이 함께 어우러진 나만의 생활체육센터는 대한민국에서 많이 부족한 장애인 시설을 새로운 시점에서 부각시킬 수 있는 새로운 공간이다.

This space is following the 'walking exercise' motif, using single pole designs. The disabled cannot move freely without other peoples' help. Thus it is difficult to travel, even take a walk in everyday life. For those people, I have envisioned a place where the disabled can walk around taking in the scenery with help from the poles. Walking along the wall, holding on to the railing on it, they can see the surroundings that were harder to see before. At the same time, they can feel involved with the surroundings.
The poles on the other side offer a place to practice walking as well as a resting area. In the end, this space becomes a place where those who cannot walk, that wish to be like the average people, can act like common citizens. Also, it is a place where the disable can exercise daily walking.

The next space is one that focuses on helping the disabled get exercise.
It is a special place for the disabled where they can exercise their stamina and muscle strength with equipments throughout the space. Disabled people as well as able bodied people can exercise there, interacting with each other, giving disable people self regard.
A place that is designed for disabled people equipped with fitness tools for disabled peoples is a new space that can stand out in Korea which, in truth, lacks facilities for the disabled.

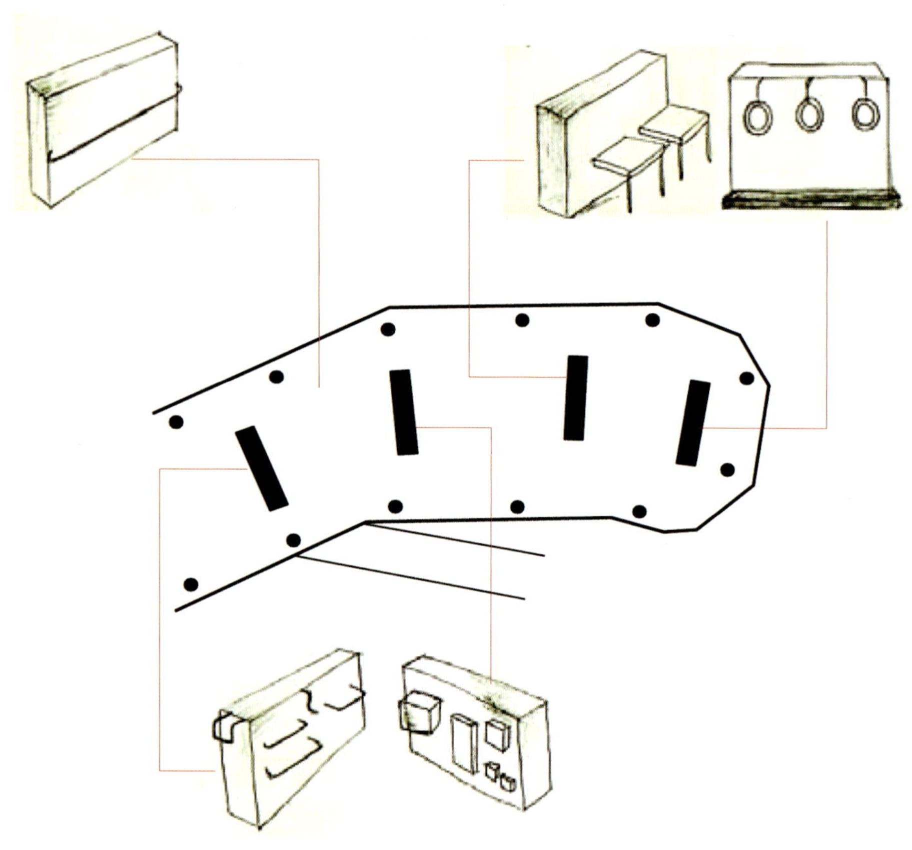

장애인에게 무엇보다 필수적인 요소는 꾸준한 운동이다. 하지만 장애인들의 실제 생활에서 그러한 활동이 적용되기는 상당히 어렵다. 이런 문제점을 해결하기 위해서 각각 장애인에 따라 반드시 필요로 하는 운동들을 적용시킨 새로운 개념의 기둥들을 디자인했다. 즉, 두 번째 디자인 포인트는 듬성 듬성 놓여있는 사각 기둥이다. 본래 기둥의 위치와 갯수는 유지하고 그 형태를 얼마나 더 효율적으로 사용할 수 있는지 고민했다.

The most important thing for disabled people is to exercise consistently. However, it is difficult for the disabled to encounter such activities in everyday life. To solve this problem, I have designed new poles that have the needed exercise equipment for the different types of disabilities. That is to say, the second design point is the square pillars placed sparsely in the space. While keeping the pillars placed as they are, it is key to more effectively use their form.

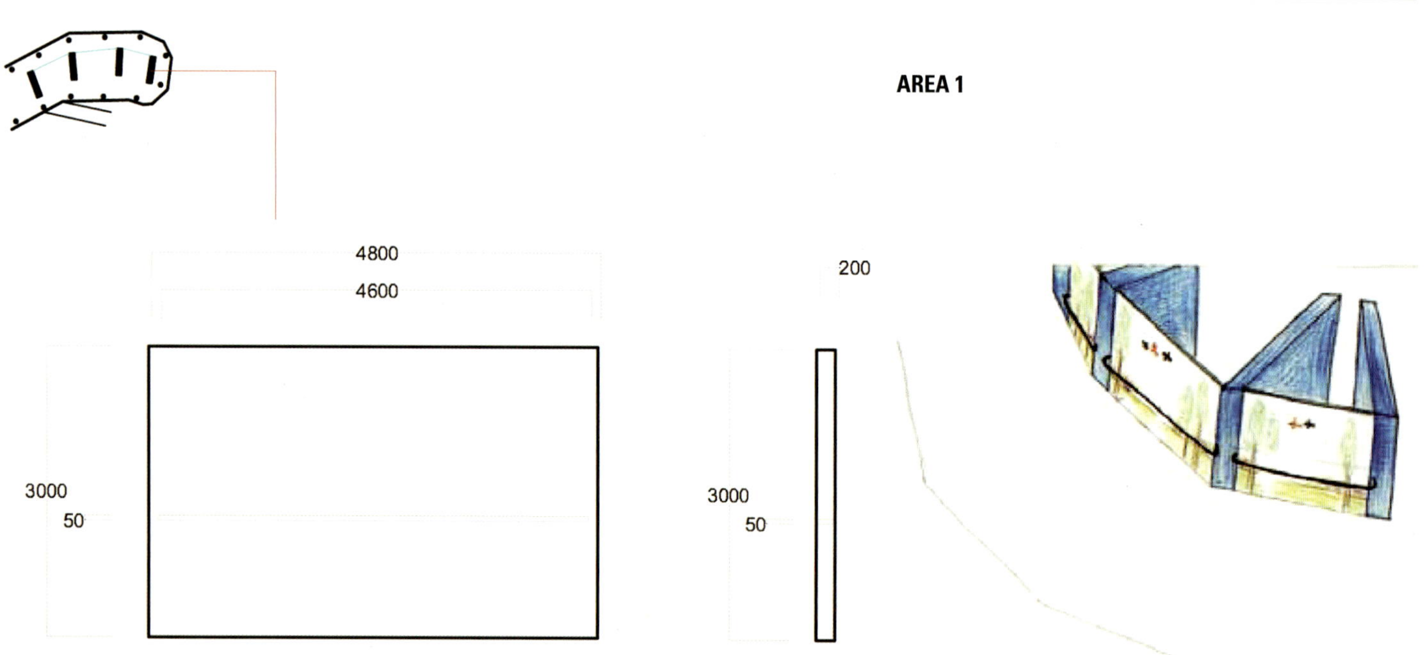

스스로의 걷기 운동이 어려운 장애인들을 위한 기둥 디자인이다. 기존에 있는 기둥에 가로로 별개의 기둥을 세워 마치 벽처럼 만들었다. 이 벽 내부에는 미디어 스크린 시스템이 내장 되어 있다. 따라서 장애인들이 벽을 따라 걷는 연습을 할 때 스크린에서 보여지는 영상을 보며 걸을 수 있다.

This is a pillar design for disabled people who have difficulty walking. I have placed horizontal panels in between the already existing rectangular pillars to look like walls. Within these panels are media screen systems. The disabled people can practice walking while looking at the video playing on the screen.

AREA 2

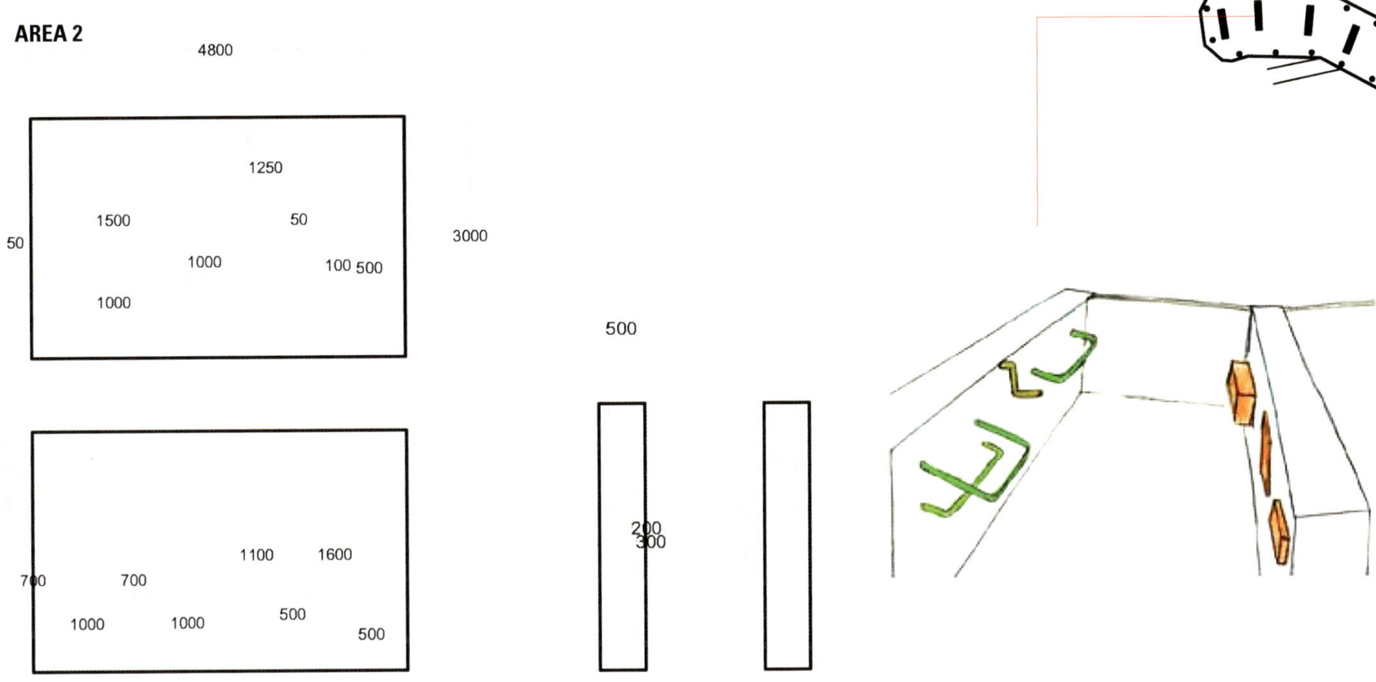

장애인들의 근력과 활동력을 높여주는 밀고 당기기 운동 장치이다. 장애인에게 직접적인 신체적 자극을 주기 때문에 장애인에게 부족한 균형감, 지구력 등을 기를 수 있다.

These are tools to help disabled people gain muscle strength. Because these tools come in direct contact with people, they help increase stamina and balance.

AREA 3

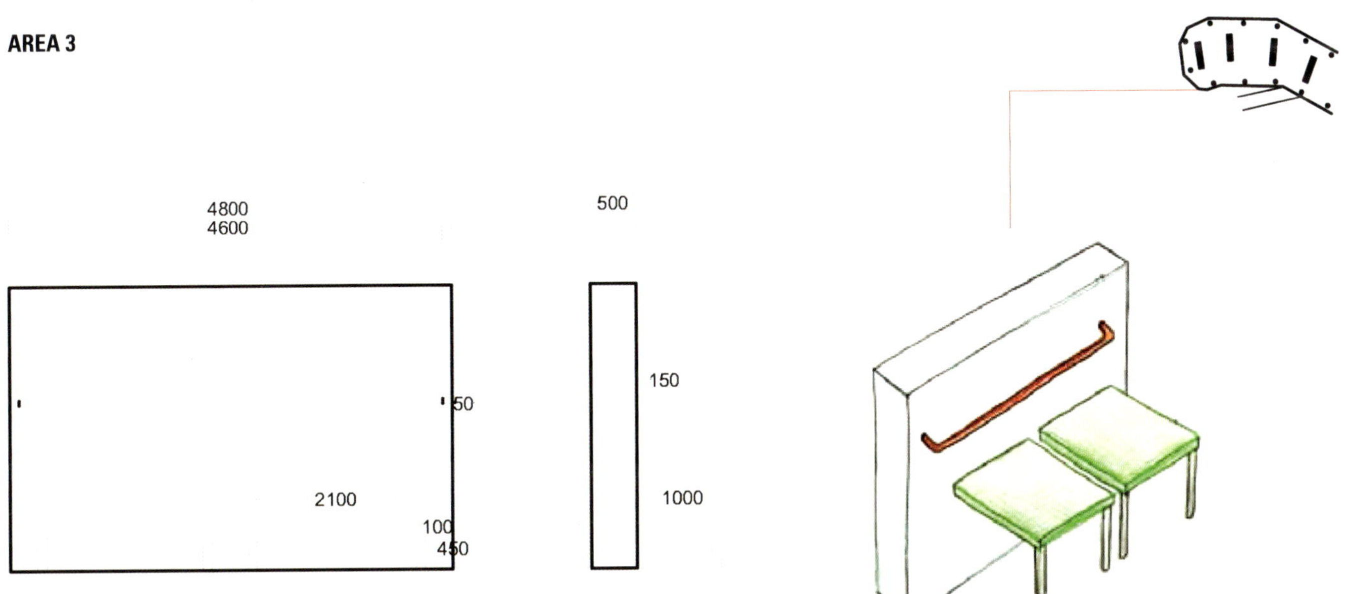

체조 운동은 장애인에게 많이 이롭다. 스트레칭이나 간단한 체조운동을 일상 생활에서 자연스럽게 해준다면 신체적으로 많은도움이 될 것이다. 또한 그림처럼 평균대 운동을 한다면 균형감이부족한 청각장애우들에게 역시 많은 도움이 될 수 있다.

Gymnastics are very helpful for disabled people. Stretching and simple gymnastics in everyday life can help the body in many ways. Also, the balance beam can help the hearing-impaired who lack balance more than others.

Kim, Sol Bi / Interior Architecture Dept.
soldori11@gmail.com

HEALING TRACK

URBAN GALLERY

BACKGROUND

건물과 건물 사이의 경계공간에서 보이는 틈에 주목함으로써 공간과 공간 사이에 존재하는 '틈새적' 시각이 단순히 주변적이고 우연적인 것이 아니라 새로운 의미의 공간으로 생각해보았다.
틈이라는 빈 공간은 쓸모없는 공간 즉,부정적으로만 여겨지고 있지만, 공간에 있어 틈을 적극적으로 고려하고 활용하는 방법 및 가능성을 제시하고자 한다.

I focused on the gaps between buildings, not just vicinities or coincidences but as new spaces. Gaps are regarded as voids that are useless and therefore have a negative representation. However, I wish to illustrate a way to consider gaps in space and utilize them.

SITE ANALYSIS

서울특별시 강남구 삼성동 테헤란로

테헤란로에서 거리를 유유자적하게 걷는다는 개념은 아직 성립되지 않는다. 보행자와 방문객을 위한 다각도의 배려가 부족하다

TEHERAN-RO, SAMSUNG DONG, GANGNAM-GU, SEOUL

The idea of idly walking the street of Teheran-ro is not yet established. There is not enough consideration for passer-by's and visitors.

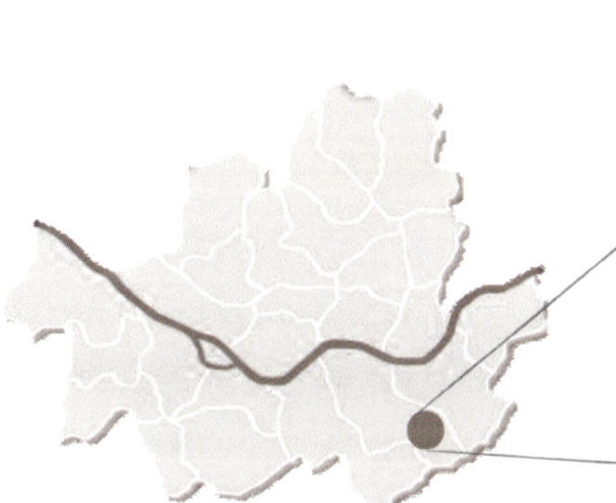

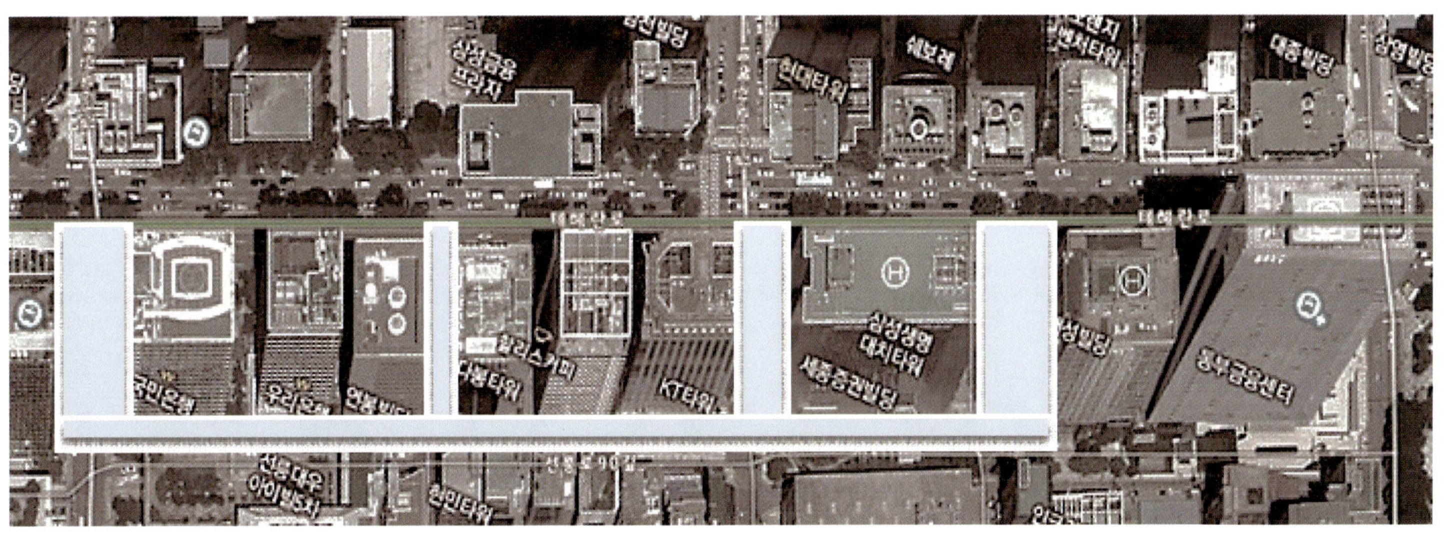

CONCEPT

ㄱㄴㄷㄹ
ㅁㅂㅅㅇ
ㅈㅊㅋㅌ
ㅍㅎ

한글

훈민정음 기본 자모음의 형태소는 극도로 간결한 최소화의 형태를 추구하고 있다.
훈민정음의 자형을 이루는 기본 도형들이 있는데 이 삼각형, 원, 사각형, 점, 직선, 사선의 형태를 이용하면 한글의 어떤 형태도 만들어 낼 수 있다.

HANGUL

Hangeul, rooting from Hun-min-jeong-eum, uses Ja-mo, and maintains minimalist form.
The consonants of Hun-min-jeong-eum are based on a few simple forms: the square, the circle, the triangle, the dot, the straight line, and the diagonal line. Using these forms, we can create any new form of Hangeul.

DESIGN CONCEPT

골목길

문화공간

도시갤러리

PROCESS

한글의 자음과 모음을 정사각형, 원형, 삼각형모듈시스템을 적용함으로써 독특한 형태를 가지고 있는 한글의 배경을 정리하여 주며, 엄격한 질서를 통해 명확성과 함께 기능적, 조형적 의미를 더해 문자를 배치할 수 있는 표준 단위를 만들 수 있다.

Using square, circular, and triangular module systems for Hangeul vowels and consonants establishes an idea for the background of the distinctive Hangeul alphabet. Through strict rules comes precision along with functional and sculptural aesthetic, as well as measurements that the letters can be organized in.

UNIT STUDY

특히 정사각형 모듈의 질서적 표현과 각이 없어 안정감을 주는 원형 모듈, 독특한 형태의 조형성을 보인 삼각형모듈은 한글을 의사 전달 수단 뿐 만 아니라 시각 요소의 역할로써 충분히 활용되어 그래픽 디자인과 제품에 응용함으로써 한글 글자 자체가 갖는 기능성과 조형성에 대한 가치를 알 수 있다.

The square modules' organization, the circular modules' stability, and the triangular modules' peculiar modeling can be used in graphic and product designs not only to communicate literally, but also the aesthetic value of Hangeul.

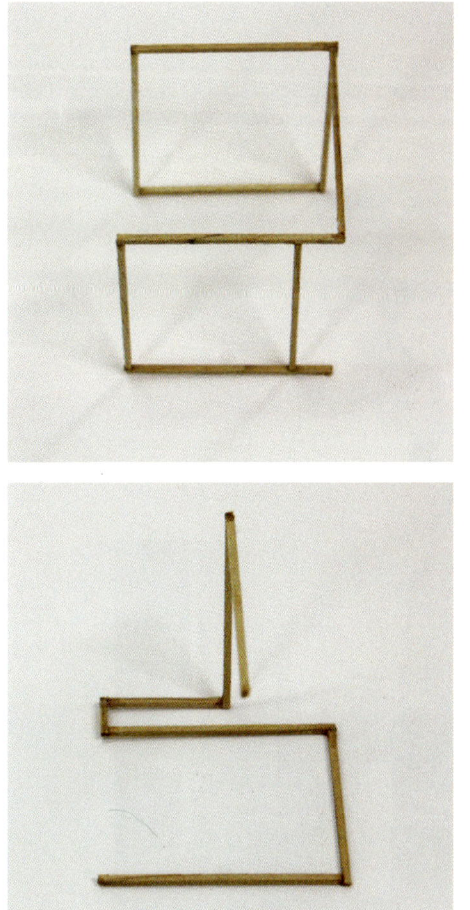

DESIGN

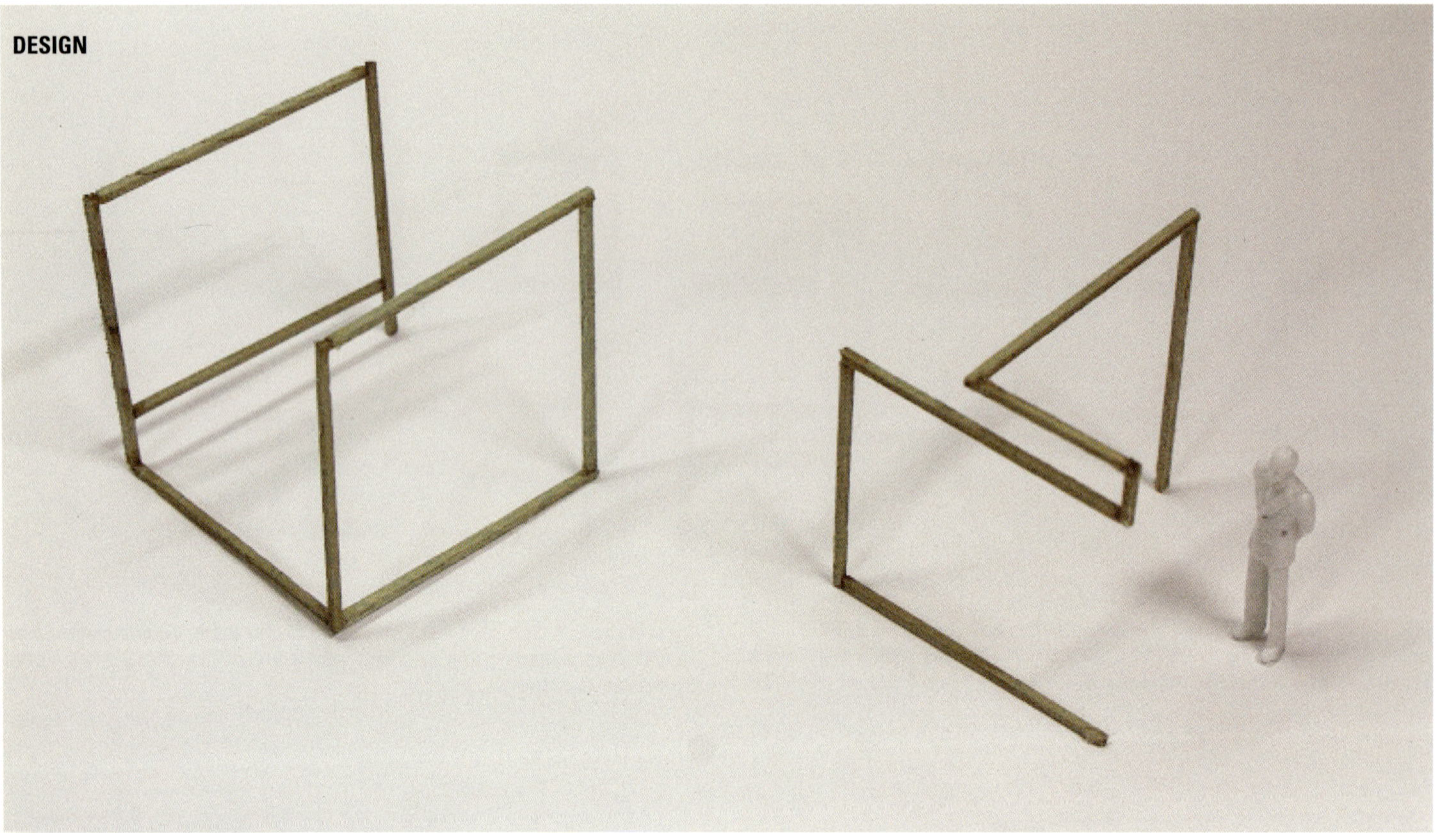

한글 활자에 사용된 규칙적인 수평, 수직의 그리드 시스템을 통해 구조적 체계를 살펴보았다. 한글디자인에서 모듈시스템은 한글 글자체를 규칙화시킴으로써 효과적인 활자제작을 가능하게 하며, 일정한 단위의 형태로 공간을 나누어 시각적으로 정리되어 보이게 하고 균형을 잡아 준다.

The right angles and parallel lines used in Hangeul create a grid that was observed to find an underlying system. Hangeul design's module system gives an organization that enables an efficient production of fonts, and creates balance through dividing the letters into planes of equal measure.

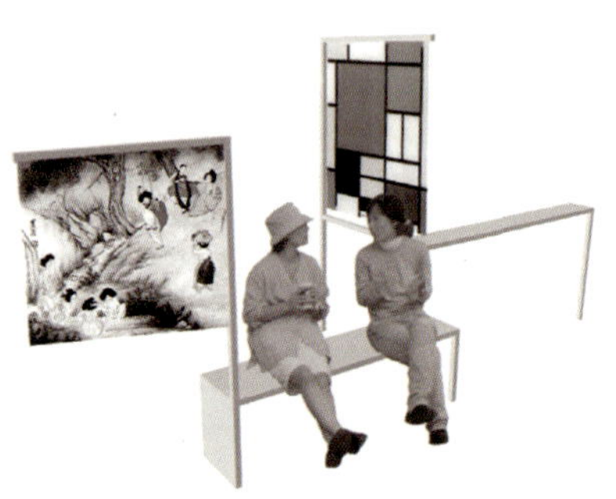

전시중심형 공공 갤러리는 미술을 가깝게 접할 수 있는 기회를 제공하고, 시민들의 커뮤니티 형성을 목적으로 하는 일종의 공공 아트 프로젝트이다. 프레임 하나가 유닛이 되어 테헤란로 전체가 갤러리가 되도록 하였다.

The public gallery centered on exhibiting provides easy-access art, and is a project that focuses on creating a community for the citizens. One frame becomes a unit and Teheran-ro itself becomes a gallery.

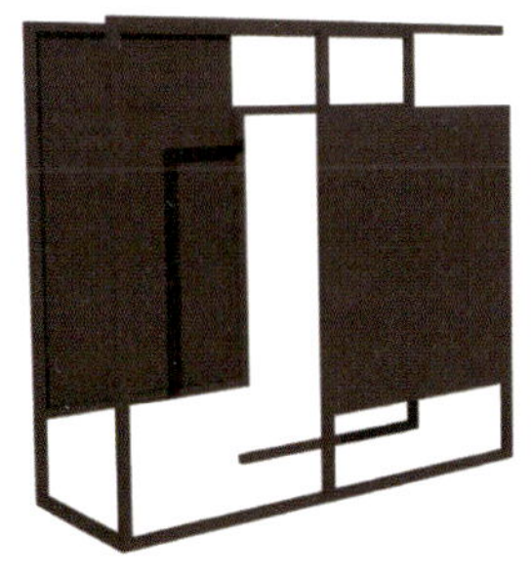

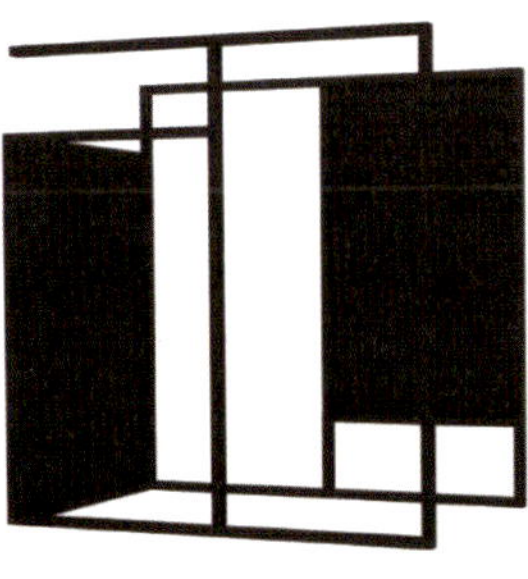

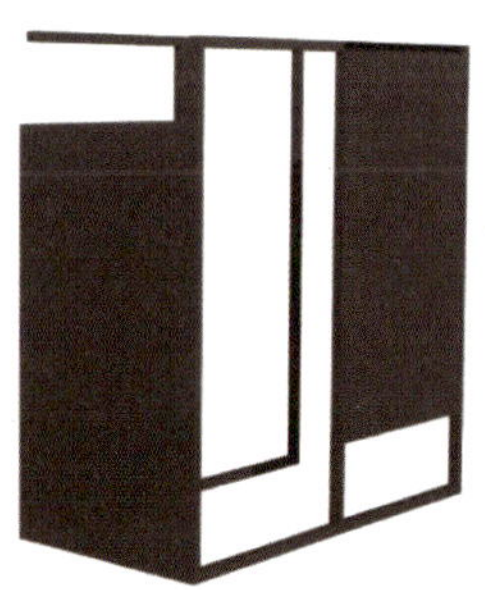

Kim, Jin Sub | **Lee, Moo Nee** / Interior Architecture Dept.
suvu@nate.com | muni130@nate.com

BICIRCLE, BICYCLE

BACKGROUND INTETNTION

테헤란로에는 한강의 지류인 탄천이 있다. 그러나 직장인의 이용도는 낮다. 자전거를 타고 출근을 하는 테헤란로의 많은 직장인들을 위해 매일 강을 볼 수 있는 곳에 위치한 자전거 주차장에 주차를 할 수 있도록 하고, 본래 자전거 도로에 인접한 사이트이 특징을 이용해 다리 하부공간 그늘진 부분을 Hidden Space로 찾아내 활용하는 것을 컨셉으로 한다. 또한 기존의 자전거 도로 사용자들에게도 휴식을 취할수 있는 가벼운 휴식공간도 함께 제공한다.

Teheran-ro crosses paths with Tancheon, a branch of the Han River. However, the usage of the riverside by office people is low. For workers commute by bicycle every day, I wish to make a parking space for bicycles where people can see the river, and using the fact that the site is close to the already existing bicycle road, utilize the hidden place beneath the bridge as a shade for people to rest in. Also, we can provide the people that already use the bicycle road a resting area as well.

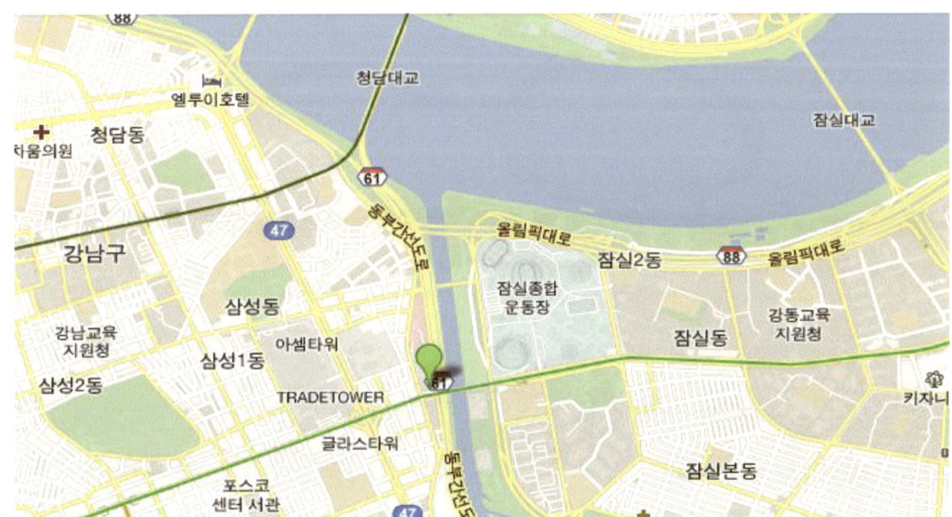

SITE

Samsung bridge, Samsung dong Gangnamgu Seoul, South Korea.

테헤란로와 잠실종합운동장을 잇는 삼성교 탄천을 가로지르고 있으며, 다리 밑으로는 탄천을 따라 자전거 도로가 있다.

CONCEPT

Circle, 원

기존의 사이트에서 연상되는 형태로 원들은 각각의 Zone들을 형성하며 곡선을 따라 이동하는 사용자들의 동선도 만들어진다.

The circle, envisioned from the site that is already there, creates separate zones, along with a curved path that cyclers can follow.ell.

SCENARIO

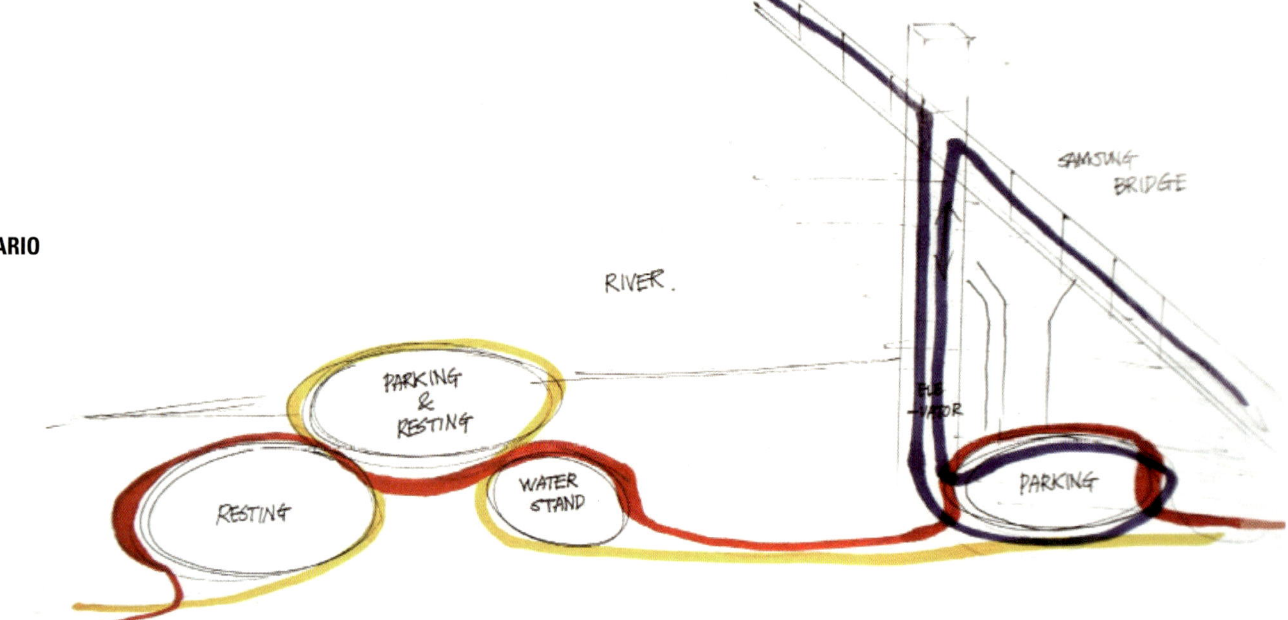

Red	original rout
Blue	elevator > parking > elevator
Yellow	parking & resting place > water stand > bike road
	parking Ⓟ > water stand > resting place

User A (Blue Line)

자전거로 출근하는직장인들을 위한 자전거 보관소

Parking area for those who commute by bicycle.

User B (Yeollow Line)

탄천자전거 도로 사용자들이 사용할 수 있는 자전거 휴게소

Parking and resting area for those who use the Tancheon bicycle road.

DESIGN

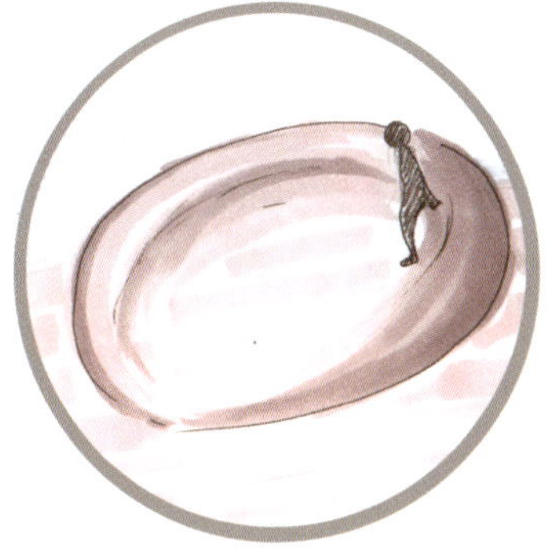

Resting Place

resting place는 bicircle의 컨셉에 맞도록 돔형태의 의자가 둘러 싸여있어 파킹을 마친 사용자 혹은 보행자들이 쉴 수있는 공간이다.

Surrounded by dome-shaped chair where passer-by's and parkers can rest.

Details

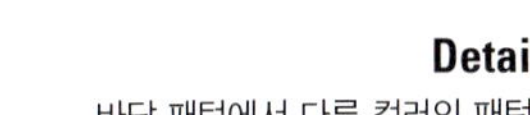

바닥 패턴에서 다른 컬러의 패턴이 팝업되는 형태로 플랫한 바닥에서 팝업되는 바닥은 사용자의 동선을 유도장치이다.

Different colored patterns pop out from the floor, the risen pattern indicating the path.

Temporary Resting Place

자전거를 타고 잠깐 앉아 쉬고싶을때 자전거는 자전거 거치대에 잠시 세워두고 벤치에 앉아 쉴 수 있다.

A place for those who want a short break on their bicycles.

Parking Place

자전거 주차장은 다리 밑의 공간을 활용하여 설치되어 있으므로 비를 피할수도있으며, 자전거를 밀어 올려 잠그는 형식이다.

The area under the bridge is utilized, affording shelter, and a locking system for parked bicycles.

Resting & Parking place
자전거는 강가에 잠시 파킹을 하고 바닥패턴이 팝업되어 형성된 벤치에 앉아 쉴 수있다.

Colour User Rout
평면상에서 circle이라는 컨셉을 적용하여 각 영역을 원으로 지정하였다. 또한 각 영역과 영역사이를 사용자들이 오고갈 수 있도록 색을 이용하여 동선을 유도한다.

Bike parking place

엘리베이터를 타고 나와 자전거를 주차시키는 공간이다.
테헤란로에 위치한 직장으로 출퇴근하는 직장인들은 이곳까지 올 때 출퇴근 시간의 붐비는 교통수단 대신 편한 복장으로 와서 자전거를 주차하고 출근할 수 있다.

Resting & Water stand

주차 후 쉬거나 손을 씻거나 물을 마실 수있는 식수대이다.
간단한 식수대를 설치해서 자전거를 타고 탄천을 오가는 자전거족들이 이용할 수 있는 휴식공간이다.
자연을 감상할 수 있도록 지붕없는 오픈스페이스로 기획하였다.

Elevator

삼성교에서 자전거 주차장을 연결하는 Bicycle 엘리베이터이다.
삼성교와 테헤란로의 레벨과 탄천의 레벨이 다르고
현재는 자전거가 내려갈 수 있는 환경이 조성되어있지 않기에
자전거와 사람이 같이 이용할 수 있고 장소를 찾는 이를 위한
상징적인 아이콘의 개념으로 설치하고자 한다.

Lee, Ji Soo / Interior Architecture Dept.
Cdside@nate.com

도심… 잊혀진 '美' 를 발견하다

Finding forgotten 'Beauty of The Urban'

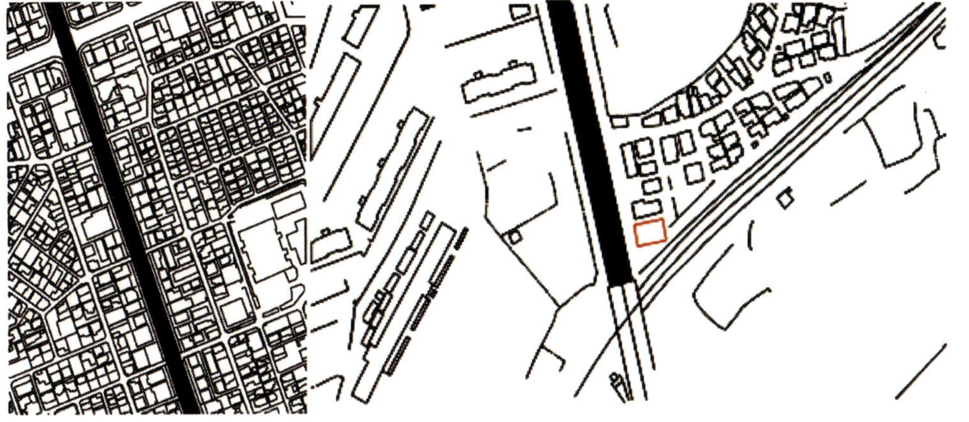

DESIGN MOTIVE

문양(紋樣) Traditional Door Frame

문양은 한마디로 미적 감각을 불러 일으키기 위해 점이나 선, 색채를 도형과 같이 형상화한 것. 언어나 문자와 마찬가지로 사용 주체인 민족과 그 민족이 처한 역사적 배경에 따라 고유한 형태. 문양은 단순히 장식적인 기능에 국한되지 않고 인간 본연의 기원과 욕구를 다분히 종교적 성격을 띠면서 담아내고 있다.

서울특별시 강남구 역삼동의 강남역 사거리에서 삼성동 삼성교의 구간에 이르는 도로로, 가로 길이는 4km, 너비는 50m이다. 강남지역을 동서로 가로지르는 왕복10차선 간선도로로, 국제금융과 무역이 활발하고 도시설계지구에 해당한다. 주요 통과지역은 역삼동 · 대치동 · 삼성동이다.

Teheran-ro is a street that crosses through Gangnam, one of the busiest regions of Seoul. It is four kilometers long and 50 metres wide, consisting of 10 lanes altogether.

1.Site의 거리 : Site의 총 길이는 99m로 가로의 폭은 좁고 길게 뻗어 있는 직사각형의 형태로 있어서 공간을 표현할 때 옆에 있는 산책로와 같은 동선의 형태로 비교적 조화로운 동선을 보인다고 생각한다.

The streets of the site the total length of the site is 99 meters, shaped like a narrow, long rectangle. It is harmonious with the walking path that next to it when illustrating space.

2.Site의 면적 : 비교적 좁은 폭의 길게 형성된 공간을 보아 앞서 말한 듯 공원과 비슷한 동선을 형성하기 때문에 시작과 끝을 명료하게 나눌 수 있다고 생각한다.

The area of the site as mentioned, the street shows a similar path with the park next to it, so it should be easy to mark the start and finish.

3.Site의 주변상황 : 도로와 밀접한 곳에 위치하고 있기 때문에 소음은 직접적으로 전달되며, 소음을 줄이고자 출입구를 꽃담으로 공간으로 형성하여 주변의 단점을 보안하고, 집중할 수 있도록 계획한다.

The surroundings of the site because it is near a road, sounds directly affect the site, so I plan to make flower walls (Ccot-dam – traditional Korean walls) at the entrance and exit, making a closed space and letting people focus on what is inside.

CONCEPT

Traditional fence of KOREA

꽃담

예로부터 우리민족은 자연과 더불어 사는 것을 신의 섭리라고 여겨왔고, 자르고 다듬고 파헤치기 보다는 자연의 생활 속에서 아름다움과 행복을 느끼며, 더불어 가정의 건강과 풍요로움을 기원하였다. 그리고 그 생활 속에서 아름다움을 표현하는 방식의 한가지로서, 전통주택 공간의 담장에 상징적인 무늬를 형상화하여 표현하였던 꽃담을 들 수가 있다.

전통문양의 의미적 의미로 중앙은 우주를 의미 곧 중심공간으로 죽을 이루는 공간으로 표현

중심을 축으로 전통 꽃담을 표현하여, 담장 이지만 소통적인 담장의 개념으로 공간을 표현

문양의 변화와 식물의 조화 문양의 형태에 따라 시각적 효과를 연출

담장을 묘사할 부분이며 공간의 시조이자 끝의 개념으로 출입구를 표현

SCENARIO

1. Entrance & Exhibition

Site의 대지 분석 결과 도로면과 밀접 – 실내의 구성. 실내 진입의 유도와 자연스러움을 표현하기 위하여 내 외부를 전체적으로 차단하지 않고, 중립의 성격을 띄기 위해 계획. 시각 / 촉각 / 청각의 다양한 체험을 계획.

The site is close to the road, so we need an interior space. Create a more neutral space by not completely isolating the interior from the exterior, but rather naturally lead the people inside. Experience in sight, touch, sound.

2. The inside & outside

Site의 주변 거주자들의 참여를 불러 일으킬 수 있도록 사진전 및 다양한 행사를 행할 수 있는 공간. 실내와 밖의 중간공간으로 자연스런 동선으로 연결을 유도.

Create a space that will call to nearby tenants as well, opening photography shows and diverse events.
A neutral space in between interior and exterior, naturally connecting the two spaces.

3. Park

Open된 공간으로 사람들의 휴식처가 되는 공간. 전통문양을 활용한 조형물과 식물, 그리고 물을 이용하여 도심 속 녹지공간을 연출. Site옆 산책로와도 연결이 되는 공간.

An open space where people can rest. Sculptures and plants that use traditional patterns, along with water, representing a green area within the city. A place that connects to the walkway next to the site.

DESIGN

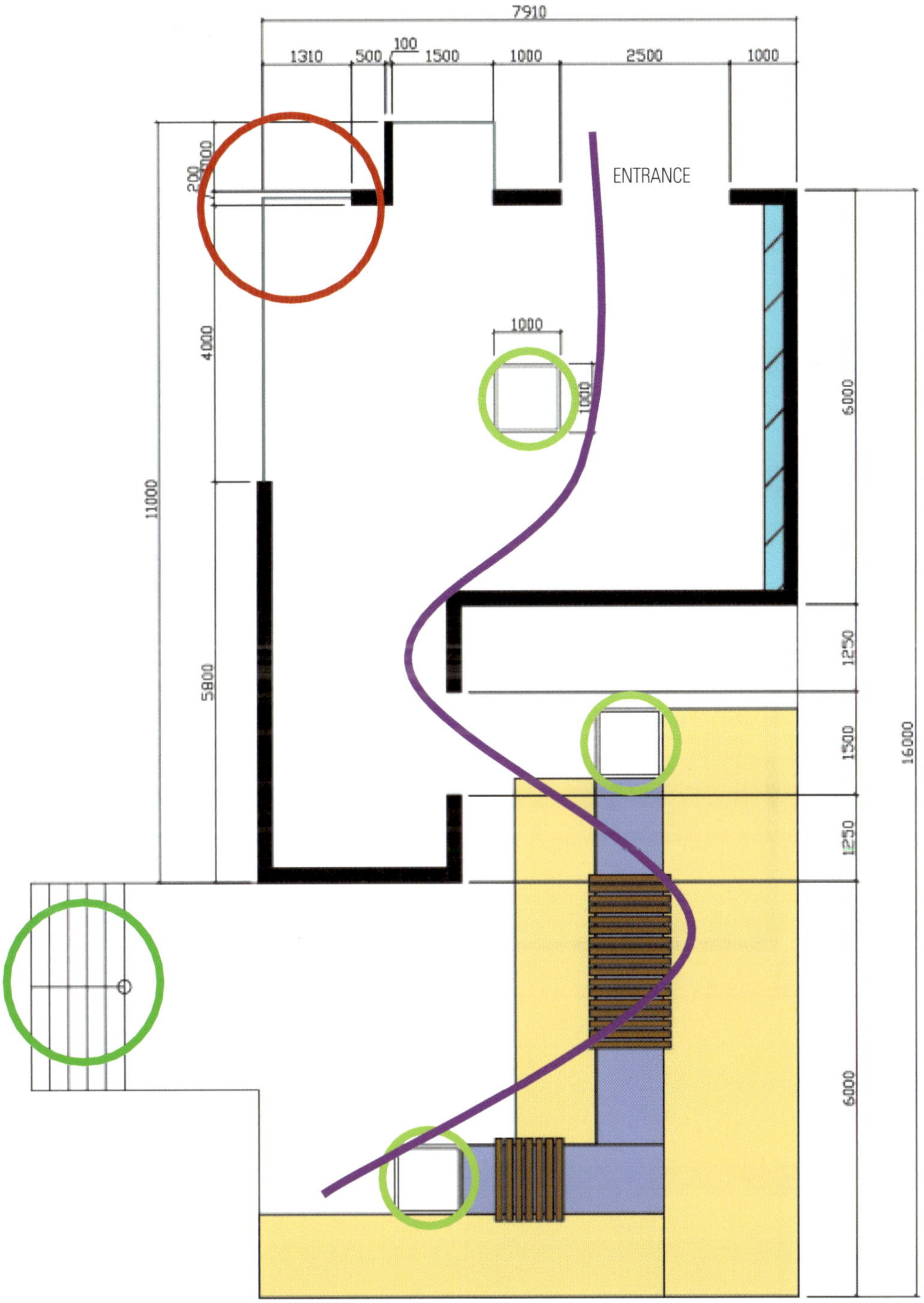

7910
1310 500 100 1500 1000 2500 1000
ENTRANCE
200 1000 4000 11000 5800
1000 1000
6000 1250 1500 1250 6000 16000

외부와의 단절을 줄이기 위해 유리와 루버를 이용하여 내외부의 중립적인 성격을 띄게 유도

To decrease the division between interior and exterior, use glass and louvers, create a neutral feeling.

우리나라 전통 문양 "완" 자를 이용하여 조형물 설치. 조형물안에 우리나라를 대표하는 식물을 넣어 사람들에게 전통미와 자연을 함께 느낄 수 있도록 계획.

Use traditional Korean pattern "wan" to install sculptures. Put plants that represent Korea to show traditionalism as well as feel nature.

Site와 인접한 옆 산책로와의 연결을 유도.

Try to connect with walkway next to site.

ELEVATION

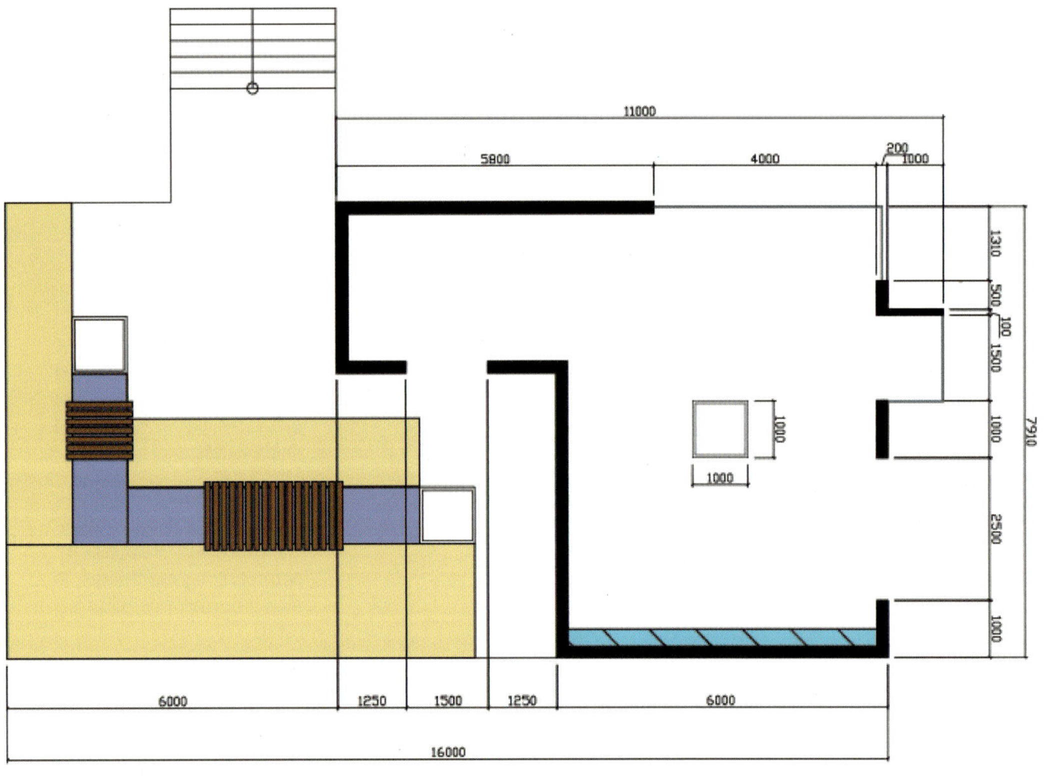

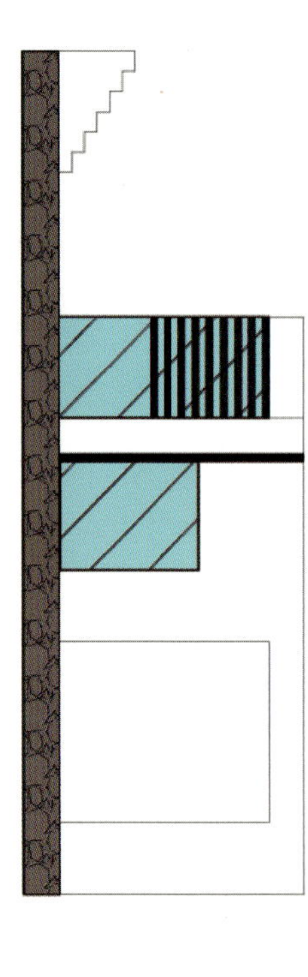

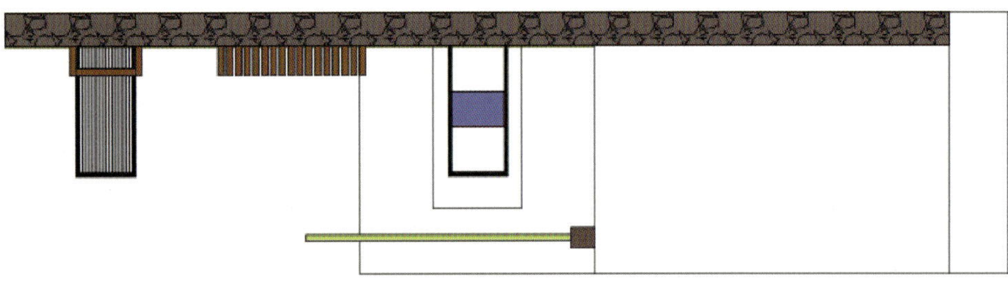

Scal 1 : 300

FINAL MODEL

HIGH STREET ; in Flyovers
DONGYANG MIRAE
UNIVERSITY
DONGYANG MIRAE University
interior design

DONGYANG MIRAE UNIVERSITY
Department of Interior Design
2011 sophomore design studio
The 3rd Hidden Space Project
HIGH STREET
동양 미래대학 실내디자인과
2011 2학년 2학기 설계스튜디오 결과물
152-714
DONGYANG MIRAE University Dept.of Interior Design
62-160, gocheok-dong, guro-gu, seoul, korea
phone + 82 2 2610 1962
fax + 82 2 2610 1960
http://interior.dongyang.ac.kr

학부장 **김홍기 교수**
동양미래대학 디자인학부 실내디자인과

서 문

도시는 문화의 심장이자 문명의 탄생지입니다.
도시가 한권의 두터운 책이라 한다면
건축은 문장을 구성하는 단어라 할 수 있습니다.
단어와 단어가 연결되어 문장이 구성되는 것처럼
건축과 건축이 연결되어 거리가 형성되고,
거리와 거리가 연결되어 도시가 됩니다.
도시는 다양한 삶의 무대입니다
수많은 기능들이 충돌하고 있습니다.
그 속에는 버려진 공간,
숨어있는 공간들이 있습니다.
이러한 틈새 공간을 어떻게 활용할 것인지에 대한
아이디어 탐침이 2012년도
국제교류 프로그램의 주제입니다.

본 프로젝트는 디자인을 통해
도시의 품격과 삶의 품격을 높이고자 하는
예비 디자이너들의 꿈이 담겨있습니다.
디자인으로 인간의 삶을 행복하게 하고
디자인으로 환경을 바꿀 수 있다는 꿈과 열정이
작품 속에 축적되어 있습니다.
치열한 자신과의 싸움을 통해
작품을 완성해 낸 학생들의 몸짓에 감사드립니다.
실험성과 창조성의 결집을 통해
실내디자인 교육의 새로운 버전을 만들어가는
[2012 히든스페이스 국제교류전]을
견인해 온 지도 교수의 노고에 감사드립니다.

Head of Interior Design Department | Prof. Kim, Hong Ki
Department of Interior Design DONGYANG MIRAE University
hkkim@dongyang.ac.kr

Foreword

The city is the heart of culture and the birthplace of society.
If a city were a thick book,
architecture would be words that make up sentences.
As words combined create sentences,
architecture combined create streets,
and streets combined created cities.
The city is a stage for various lives.
Countless functions clash.
Inside are abandoned spaces, Hidden spaces.
The probing and pondering for ideas of how to use
these gaps in space is part of
the 2012 International Exchange program.

This project shows the dream of the young designers,
and how through design,
the city's class and lifestyle can be improved.
The work includes the designers' visions
and passions of making people's lives more abundant,
and changing the environment.
I thank the students who fought with themselves
to complete their projects.
Through the combination of experimentalism
and creativity comes Interior Design,
and I thank the advisors who helped in making
this '2012 International Exchange Exhebition'.

2학년 설계담당 **박영태 교수**
동양미래대학 디자인학부 실내디자인과

responsible for sophomores | Prof. Park , Young Tae
Department of Interior Design DONGYANG MIRAE University
robotomy@naver.com

Hidden Space project - Episode 3 "HIGH STREET- FLYOVER"

올해로 3번째로 진행된 HIDDEN SPACE 프로젝트는 "HIGH STREET" 라는 주제로 진행되었다. 다소 포괄적인 성격의 이번 주제는 우리 스튜디오 내부적으로 시작단계에서 많은 고민을 하게 하였다.
사실 전 세계의 어느 나라보다 재개발과 재생에 대한 역동적인 시도가 진행되고 있는 한국과 서울의 모습들 속에서 "숨겨진 공간" 이라는 교류전 주제의 중심을 잡기위해서는 좀 더 구체적인 통찰력과 집중력이 요구되었다. 역사성과 시대성들을 단서로 접근할 수도 있었겠지만, 함께 지도하는 교수님들과의 논의 결과 도시공간 안에서 "가장 방치된 공간" 이 어느 곳일까?라는 질문으로 시작되었다. 무엇보다도 공공성의 영역이라는 두꺼운 보호막으로 인해 결코 쉽게 접근되기는 어렵지만 항상 우리 옆에 공존하고 있는 실재적 공간들에 대해서 토론하였으며, 결국 런던측에서 제공한 자료, 정보교환을 종합, 내부적 직관에 의해 "FLYOVER" 라는 공간적 모색으로 주제가 좁혀지게 되었다.
기반시설물들이 가지는 도시환경 속 존재적 정의에 대해서 우선 고민하였는데, 그 본질적 합목적성 – 적절히 비어있고, 또 묵묵히 구축되어 있어야 할 다양한 FLYOVER들 – 을 위협하는 다소 과장된 우리들의 공간적 욕망이 투영되지 않을까하는 점이 진행되는 내내 가장 큰 고민이었다.

지면상 실리지는 못하였지만, 이번 프로젝트에서는 기존에는 거의 비어있는, 또 비어있어야 자연스러운 공공의 여백을 과연 어떤 자세로 채워나가야 할 것인가에 대한 물음과, 그 방법적 최선을 위해 시나리오를 시작으로 만화식 스토리보드 작성에 집중하였다. 진행되었던 스토리보드 중에는 쪽방촌 사람들을 위해서 주거공간을 제안하겠다고 한 학생들도 있었다. 지도교수로서 공공의 공간에 특히 주거라는 사적공간의 흡수는 복합적인 문제가 있을 수 있으며, 그 문제해결은 감당이 어려울 것이라고 조언하였다. 하지만 최근 서울시에서는 쪽방촌 리모델링을 위해 이러한 교차로(Flyover) 하부에 콘테이너 박스에 임시주거를 진행 중이다. 많은 생각을 하게하는 부분이다. 공공영역에 대한 해석과 그 적용, 그리고 구성원들의 가치와 판단이 그 사회현실의 반영과의 관계에서 어떠한 양상으로 경계를 해체할 수 있는 지 비록 임시주거이기는 하지만 지켜 볼만한 부분이라는 생각이 들었으며, 또 다행스럽게도 이런 현상으로 이번 프로젝트가 가지는 실효성과 구체성의 측면도 우리 모두 체험하게 될 것이라는 생각도 들었다.

해마다 느끼는 부분으로, 2학년 학생들이기에 공간적 전략전술과 그 표현의 숙련도에서의 아쉬움이 남기는 하지만, 상당수를 보편적 수준에 도달시키기 위해 각자 많은 노력들을 하였다. 여기서 정해진 시간에 이러한 교육성과를 내기 위해 가장 안타까운 부분은 학생들 각자의 잠재적 가능성들의 보존과 향상이 이 체계내에서 얼마나 보상될 수 있을까 하는 부분이었다.
벌써 3번째 HIDDEN SPACE 프로젝트에 함께해 준 김석영 교수님과 처음이지만 많은 정성을 쏟아내신 박진호 교수님, 그리고 묵묵히 각자 역할로 화이팅을 보여준 학생들 모두에게 감사하게 생각한다. 또 공간적, 언어적 제약으로 충분한 교감은 부족했지만, 각자의 프로젝트 진행내용을 블로그와 워크샵 자료들의 왕래로 채워 준 런던 메트로폴리탄 대학과 가천대학교 구성원들 모두에게도 깊은 감사의 마음을 전하고 싶다.

콘테이너 박스 활용으로 진행중인 Flyover하부 임시주거단지안 – 2012 서울시
Container box temporary house plan below the Flyover - 2012 Seoul City

The third HIDDEN SPACE project progressed with the subject of "HIGH STREET". This subject, with its broad nature, made us worry from the beginning.
In order to establish a central idea such as the "hidden space" in Seoul, a city that is making more dynamic attempts for redevelopment and regeneration, we needed a more specific focus and insight. We could use tradition and period as a base, but after much discussion, the professors' question of "Where is the most 'neglected place'?" was chosen as the starting point. We discussed the places that were more difficult to access wrapped in the word "publicity", but always existed around us. By combining the information from London and the data we exchanged, we were able to narrow down the search to a "FLYOVER" space.
First we thought about the meaning the existing infrastructure gave to the city. The fundamental purpose – the properly empty and silently established FLYOVERs – that was being threatened by the desire for space might be projected in our project.

It is not on paper, but this project focuses on what is or should be empty but we are filling in, and how we are going to do so. We made a comic-like storyboard from the beginning scenario in order to find the best way. Included in the storyboards was a plan to provide housing for people in Jjok-bang Chon(dosshouse town). As the advisor, I told the student that a personal space, especially like a house, could bring along multiple problems, and those problems would be hard to solve. However, recently, Seoul has started a Jjok-bang Chon remodeling plan, and has moved people to temporary housing in container boxes. This makes me think. Although it is only temporary housing, it is a result of putting the interpretation and application such interpretations of public territory together with the values and decisions of the community members, and it questions how those elements will work together to bring down the border between the analyses and the members' values. Also, thankfully, this plan lets us experience the effectiveness and specificity sides of the HIDDEN SPACE project.

Each year, I feel that because these are sophomores, they have limits to their skills in spatial planning and expressiveness. However, they all work hard to bring up the work quality to common standards. But to overcome the average, it is important to keep in mind how the preservation and challenging of the students' potential will be rewarded.
I wish to thank Professor Kim Seok-young, who has already been part of three HIDDEN SPACE projects, and Professor Park Jin-ho, who was a first-timer but showed much devotion nonetheless, as well as the students who fought on silently. I also thank the London Metropolitan University who, even though there were some constraints in conversation due to language barriers, kept communicating with us, providing data and information through their blog and workshop.

1st day_05, september ORIENTATION | Hyosung Building 5F AM 11:00

Prof. Kim, Seok Yeong

Prof. Park, Jin Ho

Prof. Park, Young Tae

1ST day_02 september - MID-TERM CRITIQUE | AM 10:00

63th day_07, november MID-TERM CRITIQUE | Event Hall PM 14:00 28 WORKS

2011 AUTUMN DESIGN STUDIO PROJECT
HIGH STREET 2012

103th day_17, december EXHIBITION

EVENT HALL AM 11:00 23 WOR

with OPENING PARTY & AWARDS CEREMONY

2012 HIDDEN SPACE PROJECT "HIGH STREET-FLYOVER" MEMBERS

A community cluster
이세미 Lee, Sae Mi | 박수진 Park, Su Jin

The Gathering
김소연 Kim, So Yeon | 장선영 Jang, Sun Young

Maze in Barley Field
김현숙 Kim, Hyun Sook | 장아영 Jang, A Young

Urban Oasis
최정호 Choi, Jeong Ho | 박성우 Park, Seong Woo

BIBS and BOBS
송지혜 Song, Ji Hye | 변유진 Byeon, Yoo Jin

Tale of Tales
김성국 Kim, Seong Kuk | 박종남 Park, Jong Nam

Stumble on microdot
양승철 Yang, Seung Cheol | 송정은 Song, Jeong Eun

Enjoy in the cat's cradle
박현진 Park, Hyun Jin | 손소윤 Shon, So Yoon | 정우순 Jeong, Woo Soon

Off-line Network
현혜현 Hyun, Hye Yeon | 전주희 Jeon, Ju Hee

TA-DA
박혜진 Park, Hye Jin | 김미성 Kim, Mi Seong | 김혜진 Kim, Hye Jin

Language to Language
이유림 Lee, Yoo Lim | 강담희 Kang, Dam Hee

Hearts are connected with hearts
전상배 Jeon, Sang Bae | 최준호 Choi, Jun Ho

But a significant change
김동수 Kim, Dong Soo | 전영우 Jeon, Young Woo

See Saw Seem
이진영 Lee, Jin Young | 강명구 Kang, Myoung Koo

ECO HERE : eco in boxes
이윤선 Lee, Yun Seon | 이혜리 Lee, Hye Ree

The Return
이광수 Lee, Kwang Soo | 방준수 Bang, Jun Soo

間加樓
임소영 Lim, So young | 이샛별 Lee, Saet Byeol

I am a City Farmer.
최지혜 Choi, Ji Hye | 오화인 O, Hwa In

The Starry Night
이경은 Lee, Kyoung Eun | 정가희 Jeong, Ka Hee

倉洞-下里
김남희 Kim, Nam Hee | 김혜현 Kim, Hye Hyeon | 이수진 Lee, Soo Jin

Recycle
신예지 Shin, Ye Ji | 이미애 Lee, Mi Ae

Studio_A
박영태 Park, Young Tae

Studio_B
김석영 Kim, Seok Yeong

Studio_C
박진호 Park, Jin Ho

우리가 최선을 다해야 하는 이유는 사람들을 감동시키기 위해서가 아니다.
최선을 다할 때만이 자신이 즐겁게 일할 수 있기 때문이다.

[앤드류 매튜스]

A community cluster

귀속감, 심리적 안정감, 다양한 사회적 행태를 포함한 한 송이의 결합체
Attachment,psychological stability, including a variety social behavior a cluster combinations.

자연에 가깝게 조성하는 자연친화적 조경공간은 환경 디자인으로서의 뜻이 담겨있다.
심신 수련 및 친목 도모와 레크레이션 효과를 제공 하며,주변의 경관을 더욱 풍요롭고 윤택하게 한다.

The nature-friendly landscaped area close to the composition of the natural environment as a means design is embedded.Promote physical and mental training, and social effects, and provide recreation and scenery around makes more prosperous.

SITE 서울특별시 관악구 신대방동 Gwanak-gu, Seoul Sindaebang

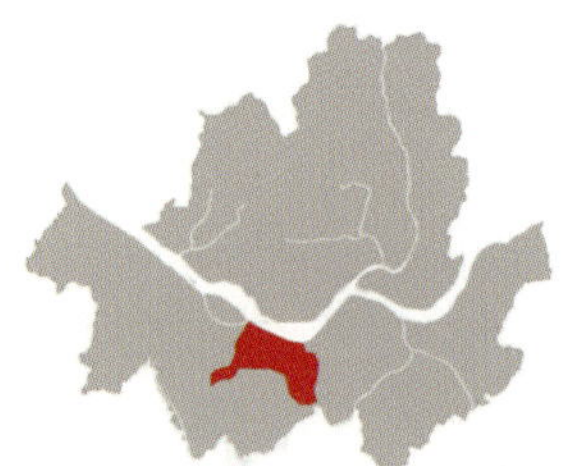

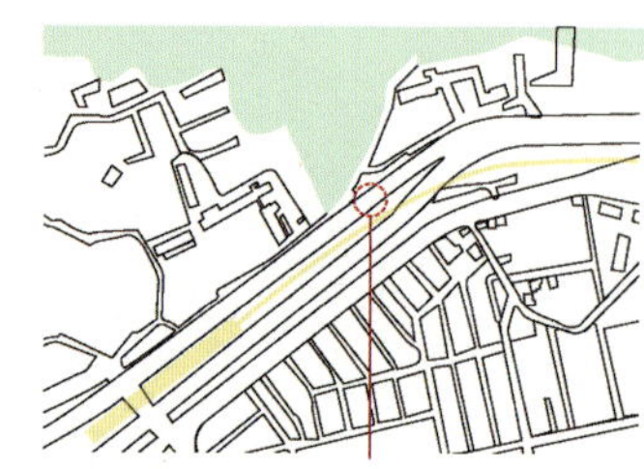

SITE ANALYSIS

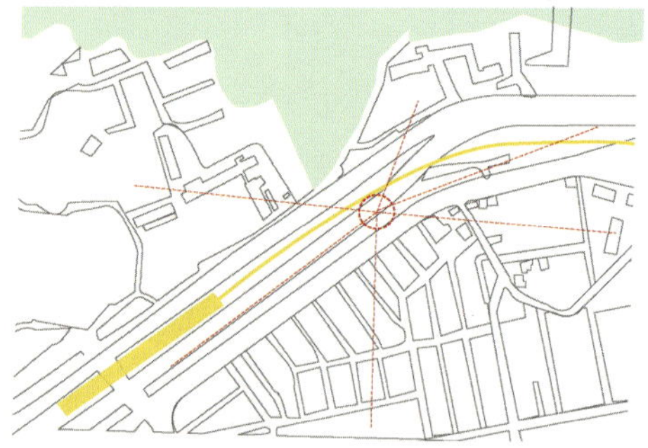

상가밀집 지역 Downtown area

Guro digital danji station Sindaebang Station

많은 상가가 밀집한 신대방역 구로디지털단지가 인접해 있다. 환승센터인 구로디지털단지 역으로 인해 유동인구가 많다.

Guro Digital complex at Sindaebang station adjoin it. there are many floating population due to Guro Digital complex is transit transfer centers.

주택가 밀집 지역 Residential area

Southern Disabled Welfare Center

사이트 주변에 주택가가 밀집해 있어다양한 연령층이 있다. 남부 장애인 종합 복지관이 있다.

There are various age group because around site adjoins residential area. there is South welfare center for the disabled.

SITE VIEW

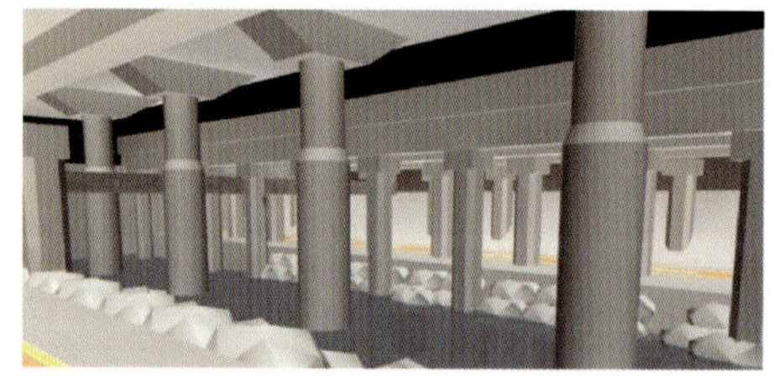

두 갈래의 봉천로에 하나의 지하철 2호선 고가가 교차하고 있다. 레벨이 다른 세 개의 고가는 이 사이트가 가지고 있는 가장 큰 특징이다.
Two parts of Bongcheon street cross overpass of seoul metro Line 2. overpass has three level are important feature.

BACKGROUND

사이트는 서울시 동작구에 위치해있는데 동작구는 복지동작이라는 슬로건을 가지고 있다. 하지만 장애인복지시설에 치중이 되어있어 노령화사회로 진입하고 있는 동작구의 노인복지 시설은 부족한 현실이다. 사회적으로 세대 간의 갈등이 심화되고 ,인간소외 현상이 문제가 되고 있다. 또, 급격하게 일어난 도시화는 삭막한 도시경관를 만들어냈다. 이러한 사회 전반의 미시적, 거시적인 문제점을 해결 하기 위해서 어떤 공간을 필요로 할까?

Site is in Dong Jak Gu, Seoul City. Dong Jak Gu has slogan that is welfare Dong Jak. but because most of welfare center in DongJak is welfare center for the disabled, Dong Jak Gu is joining aging society lacks welfare center for the senior citizen. socially a generation gap is deeping, phenomenon of human alienation becomes bads. and rapid urbanization cause desolate cityscape. be like this, to solve microscopic or macroscopic problem of the society at large, we need what place?

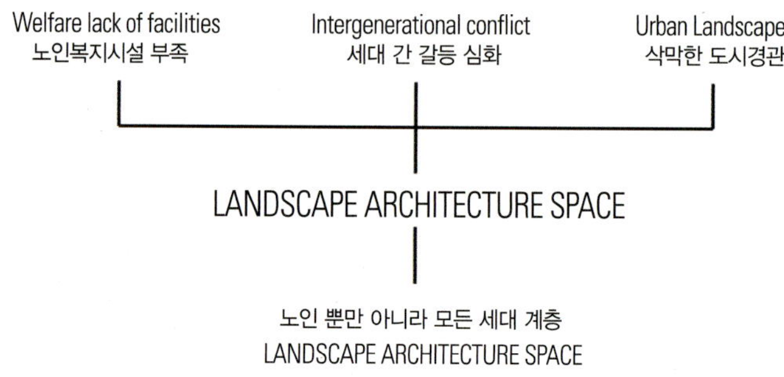

SOLUTION

이 공간은 모든 세대 계층의 갈등과 인간 소외 문제를 해결 하기 위함에 목표를 두고있다. 어린아이들에게는 학습능력을 키워주고 노인들은 취미생활을 할 수 있는 곳이다. 사이트 근처에는 장애인 복지시설들이 많기 때문에 장애인들의 학습 효과 향상에도 도움을 준다. 오픈스페이스는 방문자의 다양한 활동이 이루어지는 공간이고 휴식과 심리적 안정감을 제공하는 역할을 한다. 이 공간을 통해 도시민들에게 보다 더 질 높은 인간행태의욕구를 충족 시킨다.

this place has purpose is resolve generational conflict, phenomenon of human alienation. this is place children can improve learning ability in, senior citizen can enjoy hobby. because there are many welfare center for the disabled around site, disabled people can improve learning ability. Openspace is place visitor has various activity in, Openspace is place provides rest and psychological stability. for citizen, this place fill desire of human behavior.

DESIGN CONCEPT

STRATA

생성시기가 다른 단층 , 시간적연속성, 공간의 불연속성, 수직적 공간진화
When to create a different fault, temporal continuity and discontinuity of the space, vertical space, the evolution

각각의 단층을 세대계층 으로 봤을때, 지층은 현재의 사회 문제점을 보여주듯 서로 섞이지 않고 수직적으로 공간진화하고 있다. 이 각각의 단층에 사이트의 특징을 반영하여 단층이 수직적 공간진화가 아니라 혼합되며 서로 섞여 각 세대 층이 상호작용 하며 교감 할수 있는 공간을 구성하고자 한다.

getting each floor are generation class, uppermost floor is not mixing like social problem. reflecting feature of site in floor, each floor is mixed, not vertical evolution, each floor is will be place that each generation can interact and share each other's feeling. I will construct place like this.

DESIGN MOTIVE

VERBENA

꽃 색이 풍부, 여러 송이의 작은 꽃이 무리를 지어 피는 모습이 특징.
하나의 줄기에서 여러 송이의 꽃이 핀다.
꽃말-화합

color of flower is abundant. that many flowers blossom is featur.e
many flower blossom at one stem.
Flowers - Harmony

모티브인 버베나의 특징은 하나의 줄기에서 여러 송이의 꽃이 핀다는 것이다. 이러한 줄기의 특징을 하나의 기준면에 부여했다.모든 공간은 이 기준면으로부터 기능적으로 분화된다. 이 기준면은 여러 세대층을 하나로 묶어주는 역할을 하여 다양한 세대 층의 방문자의 화합의 장소가 된다.

many flower blossom at one stem. feature of VERBENA is motive is many flower blossom at one stem. I invest this feature of stem with one base level. every place is functionally divided from this base level. this base level harmonize generation.

KEYWORDS

LEVER
LAYER
ISOLATION & RELATION
HYDROPHILIC

사이트의 세 갈래로 갈라진 도로의 레벨 차를 이용
지층의 각각의 층을 이용하여 다양한 지층의 속성을 공간에 부여
모든 공간이 하나의 기준면으로 연결되어 있지만 기능적으로 각각 분화
수변공간으로 본래의 생태를 존중하고 인위적인 요소를 최대한 배제하여 친수성을 높임

I use 3 difference level of site road.
Each layer of strata by using the various strata of the space given to the properties of
To all the space is connected to a single plane, but each functional differentiation
Respect to the waterfront and the ecology of native artifacts as much as possible to increase the hydrophilic

SECTION - LONGITUDINAL VIEW

SECTION - CROSS SECTIONAL VIEW

(B-boy)1970(Kool DJ Herc)break-boys, B-boys(B-girl)

DETAIL VIEW

입구에서 본 OPEN SPACE
The open space at the entrance

조경공간으로 올라가는 길에서 본 VIEW
The view to the landscaped area on the road dating

온실 내부에서 본 VIEW
View from inside the green house

퍼블릭 가든 내부에서 본 VIEW
View from inside Public garden

오픈 스페이스에서 휴게공간으로 들어가는 VIEW
View from the resting place going into the open space.

SECTION

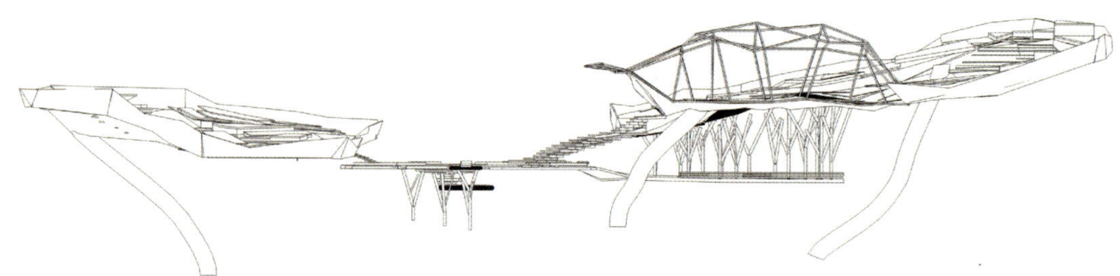

SPACE PROGRAM

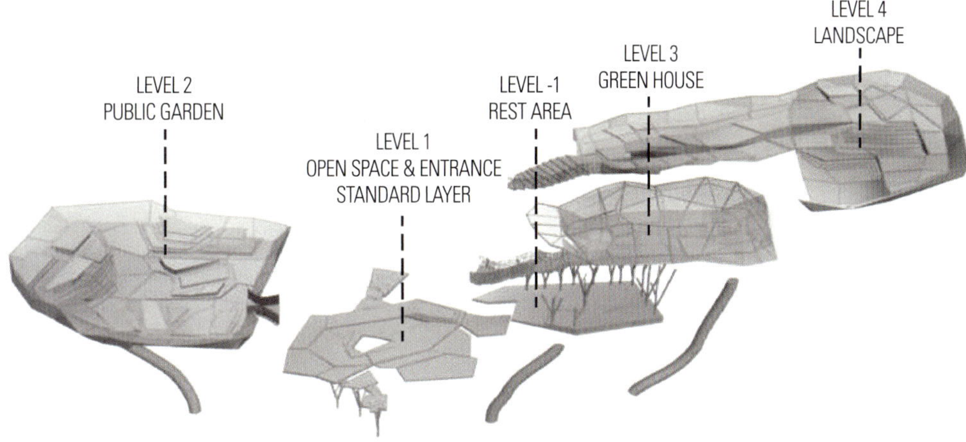

LEVEL -1 REST AREA

조경공간으로 올라가는 길에서 본 VIEW
The view to the landscaped area on the road dating

다양한 세대 계층이 섞여 상호 작용할 수 있는 곳을 만들기 위해 기준면에 휴게공간을 만들었다. 레크레이션을 할 수 있다.
Will be harmony place for visitor of various generation. to construct interacting place for various generation, rest place is made in base level. can take recreation.

LEVEL 1ENATRANCE (STANDARD LAYER)

좁은 입구로 방문자의 동선을 유도 하여 오픈스페이스로 연결된다. 기준면에는 한 줄기에서 여러 송이로 분화되는 버베나의 특징을 반영했다. 이 기준면은 모든 공간을 여러개의 면으로 기능적으로 분화시킨다.
A narrow enterance induce visitor, open space lead from narrow enterance. base level is reflected by feature of VERBENA is many flower blossom at one stem. every place is functionally divided from this base level.

LEVER 2. PUBLIC GARDEN

공공화단으로 방문자가 자신이 키우고 싶은 식물을 가져와 심을 수 있는 곳이다. 노인들은 취미 활동을 할 수 있고, 어린 아이들은 자신이 키우는 식물을 보면서 학습 효과를 향상시킬 수 있다.
Public flower garden is place visitor can plant in. senior citizen can enjoy hobby, children can improve learning activity seeing that they plant.

LEVEL 3 GREEN HOUSE

온실은 우리나라에서 쉽게 기르기 어려운 채소나 식물을 키우는 곳이다. 온실이라는 특성 때문에 유일하게 밀폐되어 있는공간이다. 온실의 외피는 투명 태양광 집열판으로 되어 있어 이 곳에서 쓰이는 전기를 절약할 수 있다.
Greenhouse is place that we can plant thing that is difficult to plant in korea. this place is sealed uniquely. because cover of greenhouse is composed of solar module, electricity can be saved.

LEVEL 4 LANDSCAPE

조경공간은 다양한 종류의 식물을 심어 방문자가 감상 할 수 있고, 산책을 즐길 수 있는 공간이다. 간단한 운동을 할 수 있고, 온실에서 재배 되는 식물을 보고, 만지고, 먹어보며 직접 경험 할 수 있다.
Visitor view landscape space is planted, can walk in view landscape.A simple exercise can, and the plants are grown in the greenhouse to see, touch, eat watching can be experienced directly.

CIRCULATION

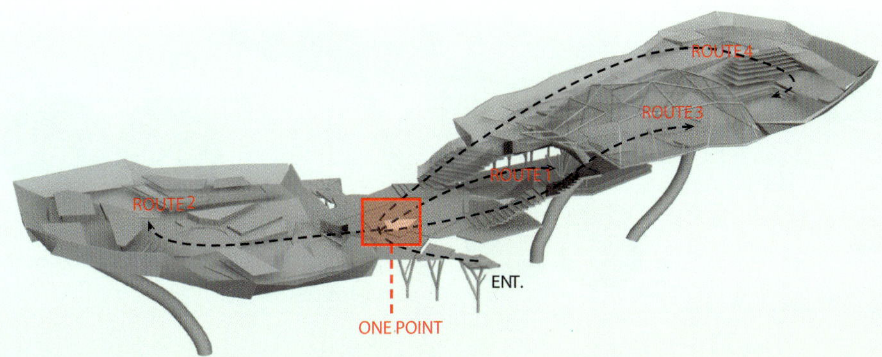

MASS PROCESS

지층형성
Formation of strata

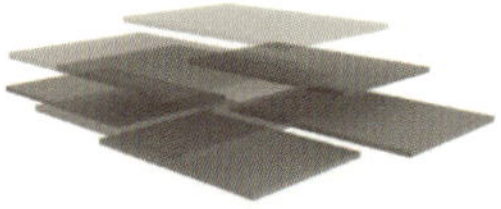

SITE의 특징인 LEVEL을 이용하여 단층 분리 & 흩어짐
SITE LEVEL by the characteristic separating and scattering tomography

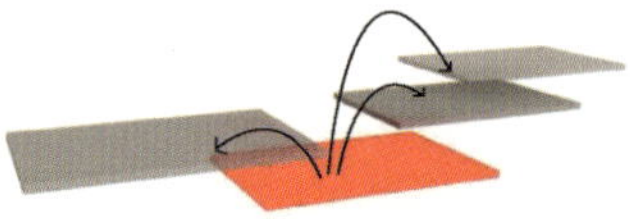

줄기 반영하여 하나의 기준면에서 분화
One basis to reflect the differentiation in terms of stem

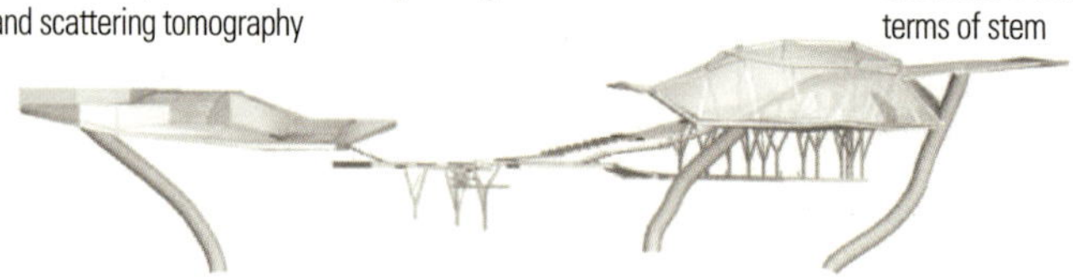

DESIGN PROCESS 1

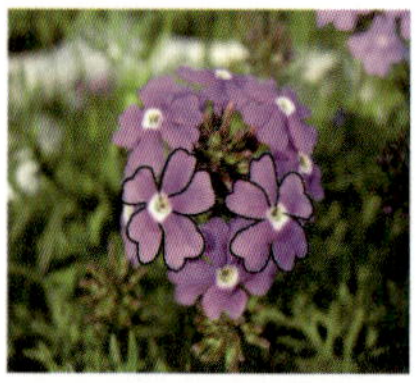

버베나의 꽃잎
The petals of VERBENA

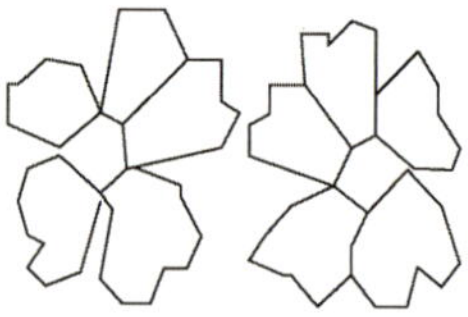

버베나의 꽃잎의 단순화
Verbena petals simplify

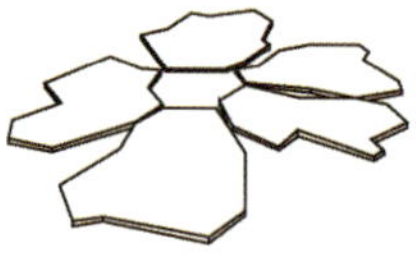

꽃잎에 여러 겹의 레이어를 중첩
Multiple layers of petals overlapping layers

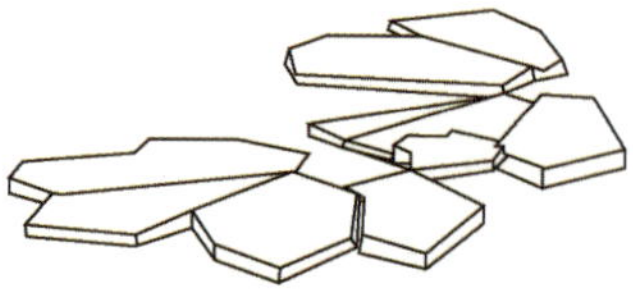

버베나 꽃잎에 LEVEL 차를 반영
Verbena petals reflect LEVEL differences.

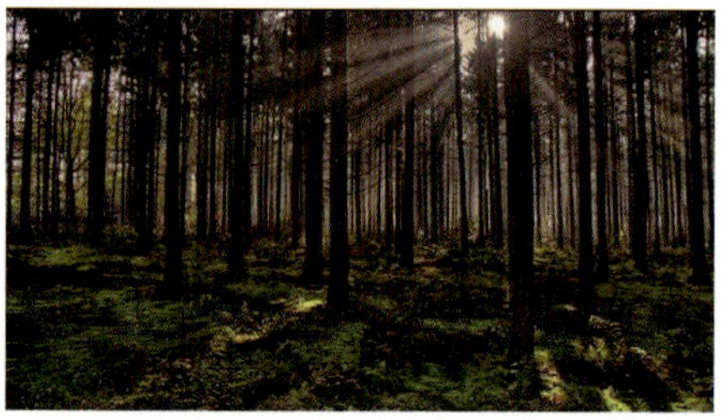

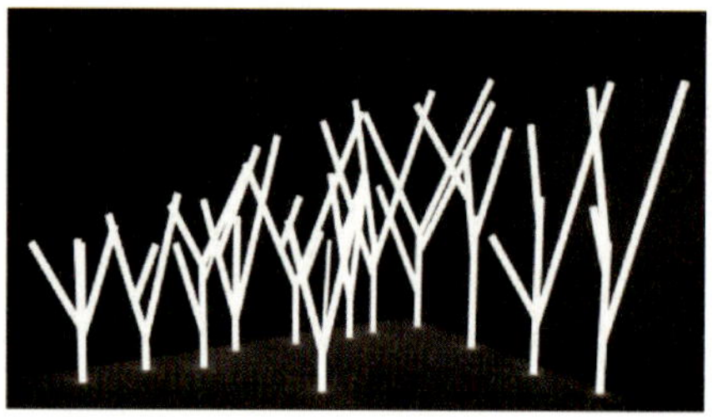

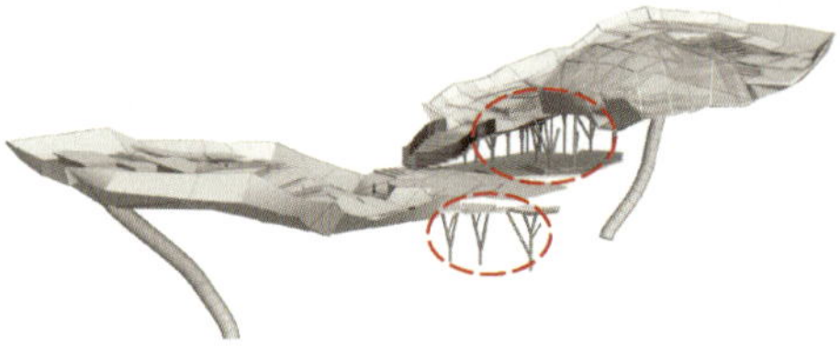

원래의 자연과 최대한 비슷하게 하기위해 인위적인 요소를 최소화한다. 공간에 배치된 조명을 나무와 같은 형태로 만들어 멀리서 봤을 때, 이 공간 주변에 있는 나무와 중첩 되는 효과를 낸다.
To be similar natural, artificial thing is minimized. arranged light of tree shape is overlapped with around tree.

PLANTS PROGRAM

홍콩야자 & 행운목
Hong Kong, coconut & haengunmok

잔디 & 돌미나리
Lawn & dolminari
오염된 늪지를 정화하여 깨끗한 물을 흘려보낸다.
Flow of clean water to purify polluted swamps sent.

스파티필름
Seupatipilrum
햇빛없이 잘크는 다년초로 수경재배도 가능
Hydroponic growing without sunlight as possible perennial jalkeuneun

칼라페페 & 아펠란드라 & 야자수 & 산세베리아
Pepe & ahpelrandeura & palm trees & sans collar Siberia

담쟁이덩굴
Ivy

은행나무 & 포플러
Ginkgo & Poplar
오염된 공기를 전화한다.
To purify the polluted air.

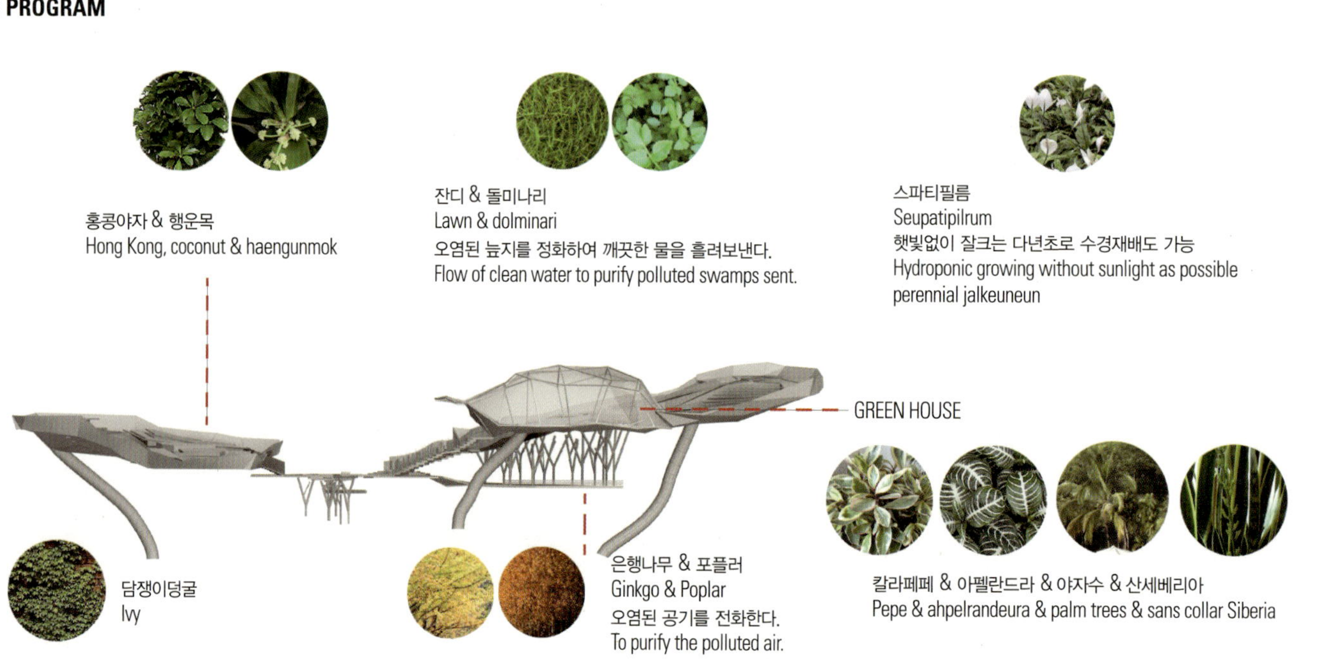

FLOOR PLAN

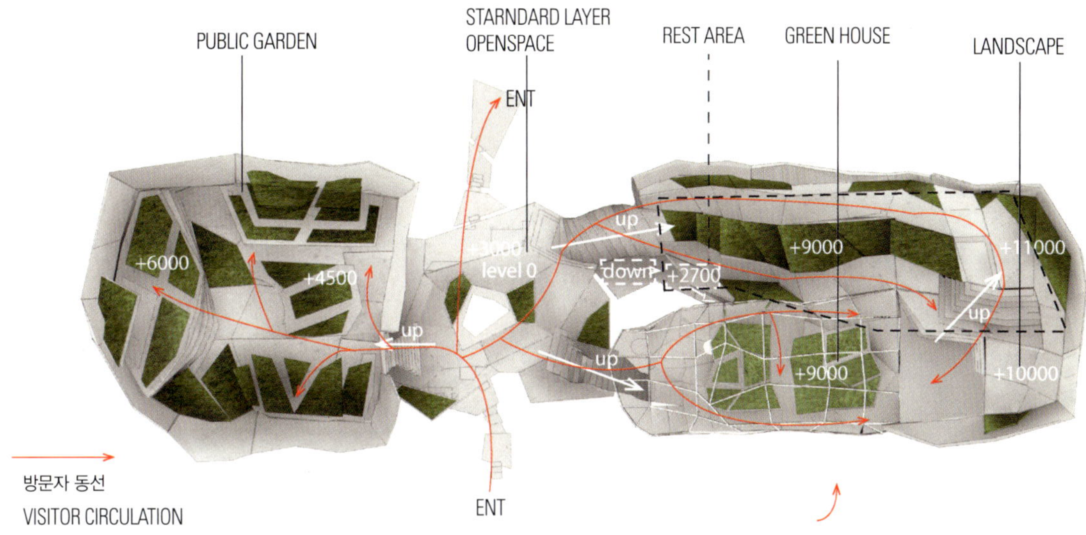

사용자들의 연령별 공간 분할
SPACE DIVISION OF USERS BY AGE

식물별 공간 분할
SPACE DIVISION OF PLANTS SPECIES

용도별 공간 분할
Divided by application area

사용자들의 연령별 공간 분할
SPACE DIVISION OF USERS BY AGE

ACCESSIBILITY

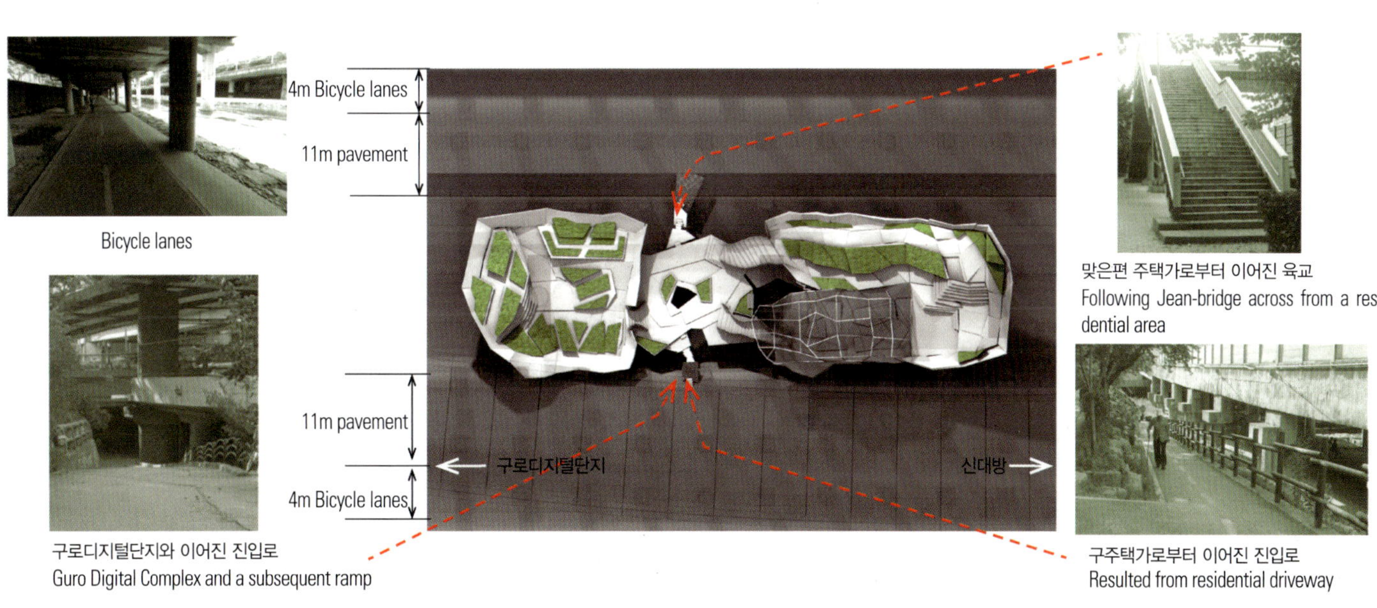

Bicycle lanes

맞은편 주택가로부터 이어진 육교
Following Jean-bridge across from a residential area

구로디지털단지와 이어진 진입로
Guro Digital Complex and a subsequent ramp

구주택가로부터 이어진 진입로
Resulted from residential driveway

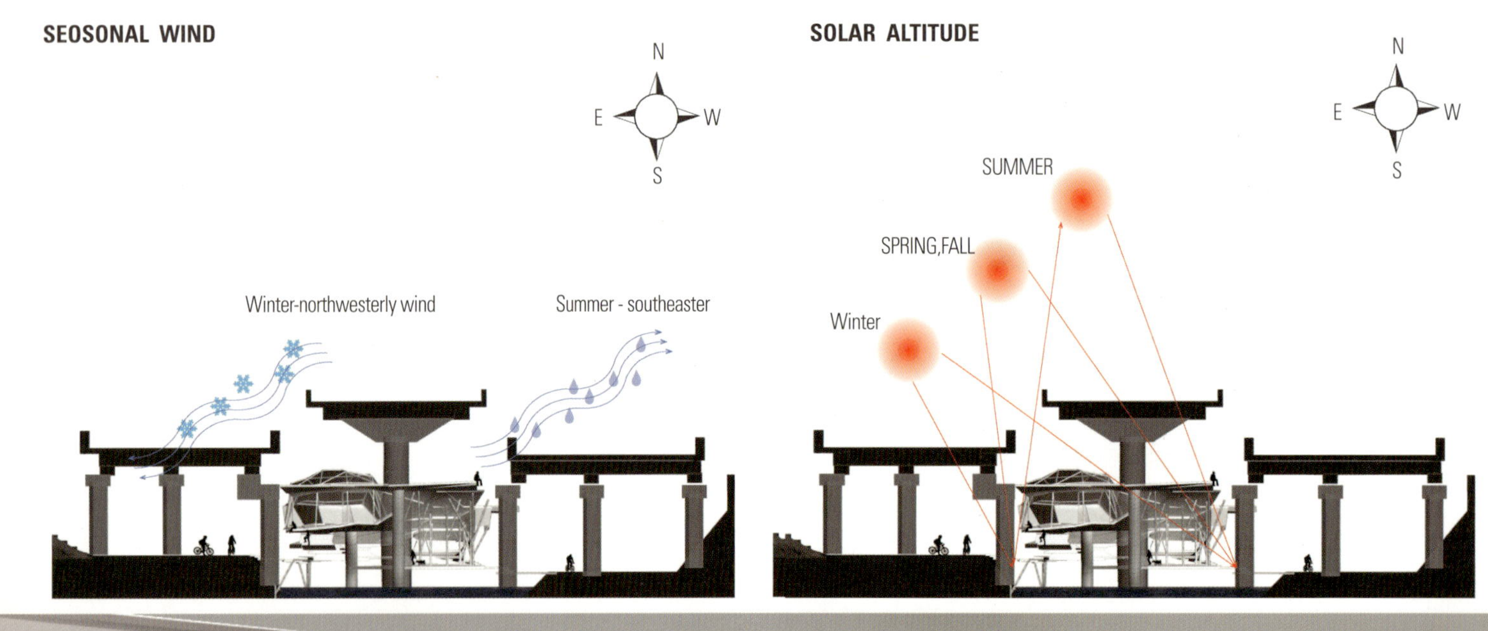
SEOSONAL WIND
N
E
W
S
Winter-northwesterly wind
Summer - southeaster
SOLAR ALTITUDE
N
E
W
S
SUMMER
SPRING,FALL
Winter

IRRIGATION SYSTEM

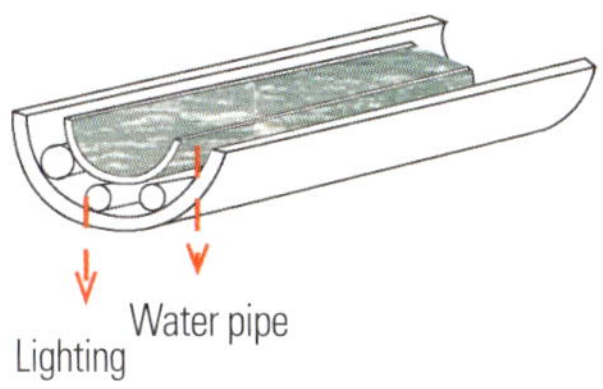

WATER ROAD

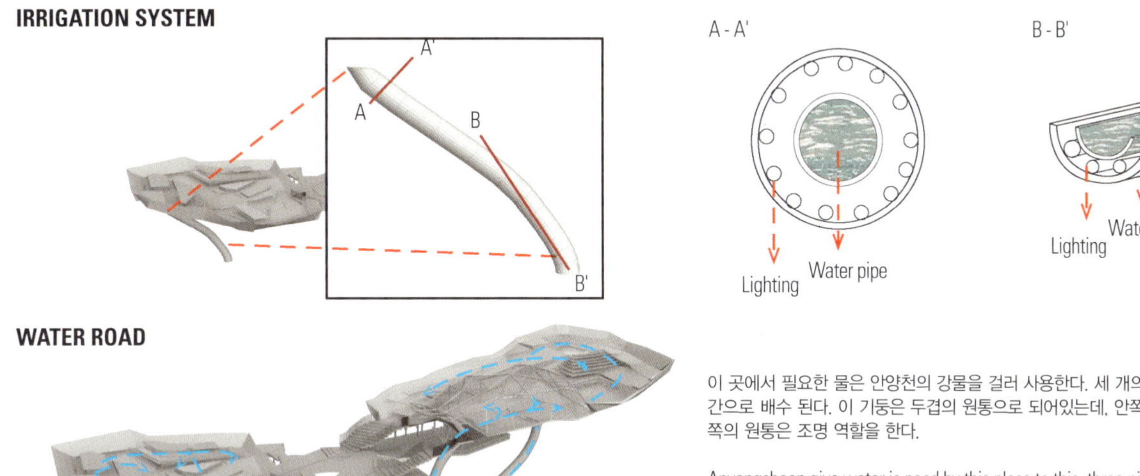

이 곳에서 필요한 물은 안양천의 강물을 걸러 사용한다. 세 개의 기둥으로 물을 끌어 올리고, 그 물은 각 공간으로 배수 된다. 이 기둥은 두겹의 원통으로 되어있는데, 안쪽의 원통은 배수를 하는 용도로 쓰이고 바깥쪽의 원통은 조명 역할을 한다.

Anyangcheon give water is need by this place to this. three pillars raise water, underdrian to each place. this pillar is compose of two layers of cylinder, inside thing plays role supply water, outside thing plays role light.

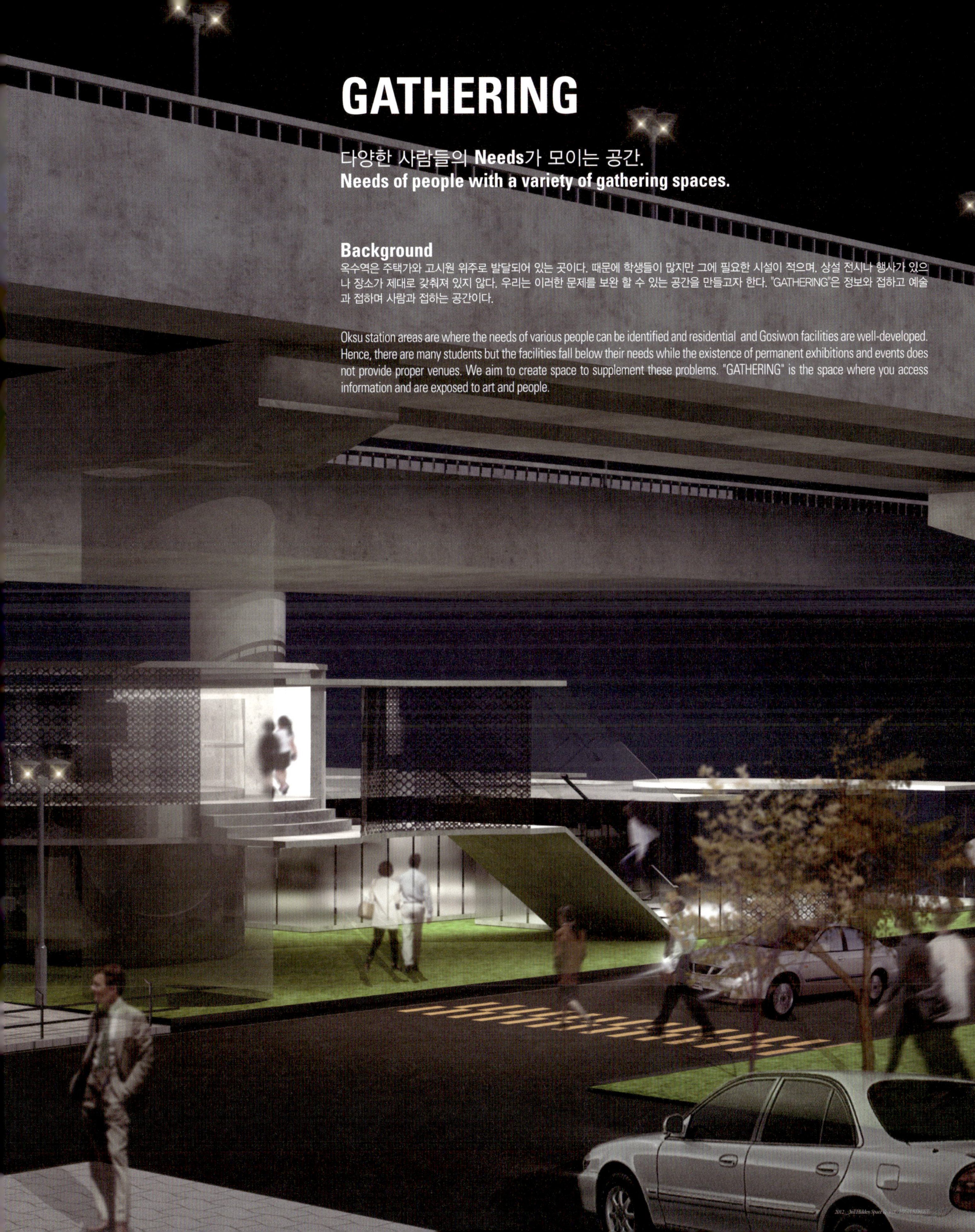

GATHERING

다양한 사람들의 **Needs**가 모이는 공간.

Needs of people with a variety of gathering spaces.

Background

옥수역은 주택가와 고시원 위주로 발달되어 있는 곳이다. 때문에 학생들이 많지만 그에 필요한 시설이 적으며, 상설 전시나 행사가 있으나 장소가 제대로 갖춰져 있지 않다. 우리는 이러한 문제를 보완 할 수 있는 공간을 만들고자 한다. "GATHERING"은 정보와 접하고 예술과 접하며 사람과 접하는 공간이다.

Oksu station areas are where the needs of various people can be identified and residential and Gosiwon facilities are well-developed. Hence, there are many students but the facilities fall below their needs while the existence of permanent exhibitions and events does not provide proper venues. We aim to create space to supplement these problems. "GATHERING" is the space where you access information and are exposed to art and people.

PROPOSAL

이 곳에 정보 제공을 위한 공간과 상설 전시 시 장소제공을 할 공간을 제안한다. 책을 빌리러 다른 지역까지 다니던 학생들이나 바쁜 회사원들은 필요에 따라 이 곳에 찾아와 정보를 공급받을 수 있다. 무료한 노인분들이나 어린아이들은 이곳에서 지루하지 않게 머무를 수 있다. 이곳은 모든 주민을 위한 소통과 만남, 정보 나눔의 공간이다.

The suggestion is the provision of space for information provision and permanent exhibitions. Busy students or salaried people who used to go to other areas to borrow books come to this place for information if necessary. Bored seniors or children can enjoy being in the space. The space is for communication, gathering, and information provision among all residents.

TARGET

주 타겟은 전문 자료를 찾는 바쁜 직장인이나 학생들이지만, 모든 주민들을 대상으로 한다.
Although it is provided mainly for busy salaried people or students who want professional materials but aims to help all residents.

USER NEEDS

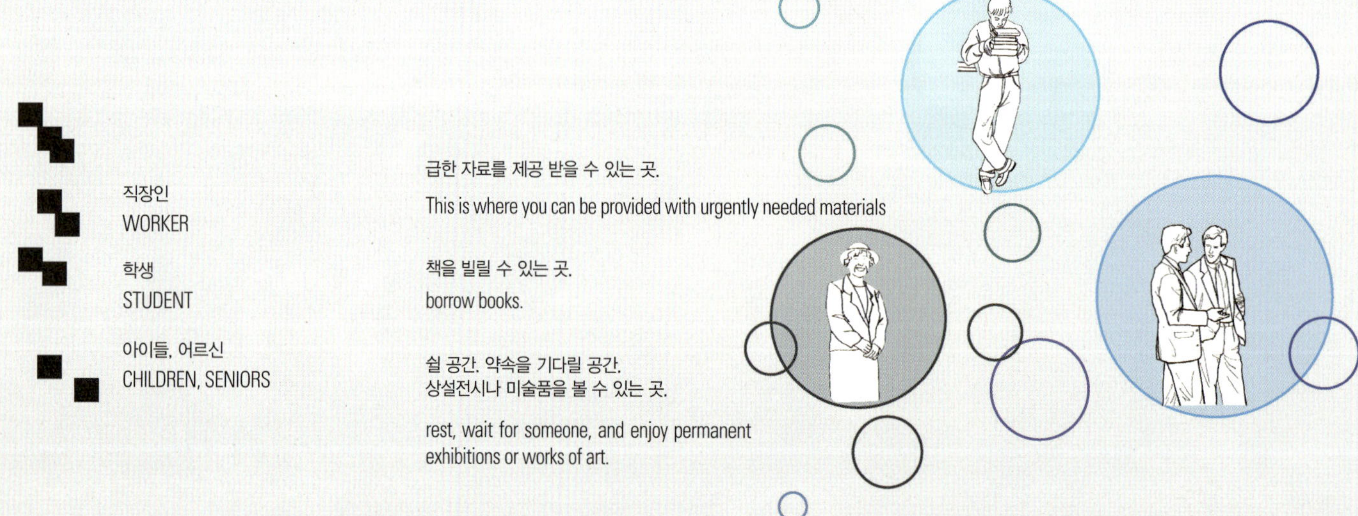

직장인
WORKER

학생
STUDENT

아이들, 어르신
CHILDREN, SENIORS

급한 자료를 제공 받을 수 있는 곳.
This is where you can be provided with urgently needed materials

책을 빌릴 수 있는 곳.
borrow books.

쉴 공간. 약속을 기다릴 공간.
상설전시나 미술품을 볼 수 있는 곳.

rest, wait for someone, and enjoy permanent exhibitions or works of art.

도서관은 주로 직장인들과 학생들이 급히 자료를 찾을 때 사용 할 수 있는 무인 도서관으로 계획하였다. site가 역에 근접해 있기 때문에 출퇴근 시간에 편히 이용 할 수 있을 것으로 예상하며, 무료 대출이 가능하다. 갤러리의 경우 상설 전시 때마다 공간을 빌려주는 용도이기 때문에 내용이 수시로 바뀌므로 약속을 기다리는 사람들이나 무료하신 어르신들에게 시간을 보낼 수 있는 공간을 제공한다.

As for the library, it is designed to be manless so that salaried people and students can use the facility when they find it urgent. As the site is near the station, anyone can use it conveniently while commuting and you can borrow books for free. As for the gallery, it will be provided for permanent exhibitions and the contents will change frequently. So it provides good space for people waiting for someone or seniors who live in boredom.

SITE

위치: 서울 성동구 옥수동 옥수 고가차도 밑.
접근성: 삼면에서 도보로 접근가능.
옥수역 7번 출구 앞.

주변정보
주변에 교육시설과, 아파트, 고시원이 많아 조용한 편이며, 진입하기 편한 곳에 위치해 있다.

Location: Under the Oksu Overpass, Oksu-dong, Seongdong-gu, Seoul, Republic of Korea
Accessibility: accessible on foot from three directions, in front of Exit # 7, Oksu Station

Information on the nearby areas: there are many educational facilities, apartment houses, and Gosiwons, so it is quiet. The access is convenient.

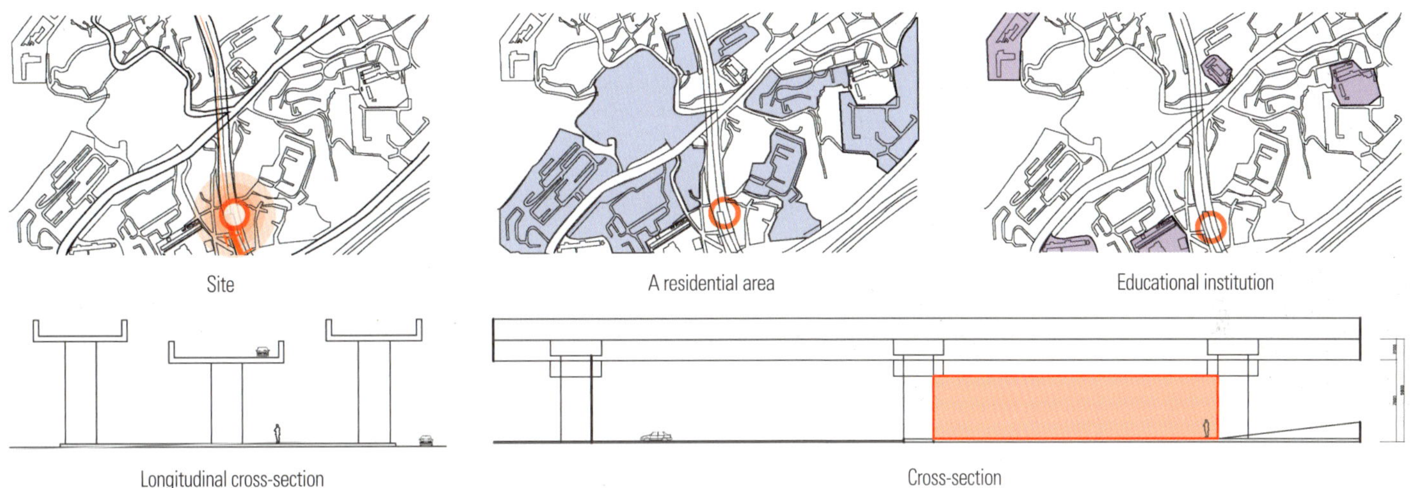

Site

A residential area

Educational institution

Longitudinal cross-section

Cross-section

좌우로 횡단보도가 위치해 있으며 삼면으로 차가 다니지만, 주변에 아파트가 많아 서행하므로 진입함에 있어서 안전성 문제는 없다.
There are crossroads on both sides. Cars pass by in three directions but the speed is low as there are many apartment houses. So there is no problem with safety in terms of entrance.

CONCEPT

What is mosaic?

1.〈미술〉 여러 가지 빛깔의 돌이나, 유리, 금속, 조개껍데기, 타일 따위를 조각조각 붙여서 무늬나 회화를 만드는 기법.' 짜 맞추기' 로 순화 ≒조각무늬 그림.
2.사진이나 화면 따위에서, 특정 부위를 가리기 위하여 그 부위만을 잘 안보이게 처리 하는 일. 또는 그렇게 처리된 상태. 각각의 개성& 어울림– 각각의 서로 다른 물질들이 만나 하나의 모자이크를 이루듯, 이 지역의 다양한 사람들이 어울릴 수 있는 공간을 만들고자 함.

1.<Art> It is a technique to make patterns or draw paintings by attaching patches and pieces of various-colored stones, glasses, metals, shells and tiles.
Refinement through 'match-up' glyptics.
2. The technique is to hide a certain area on a picture or a screen and treat it not to be seen properly or involves such a status. Each personality and seemliness - just like different substances come together to make a mosaic, the aim is to produce space where various people from this are can come together and enjoy being with each other.

Design Keywords & Story

픽셀–이미지를 이루는 가장 작은 단위. 평면을 픽셀로 나누어 형태의 기반을 잡음. 조각–픽셀이 모여서 조각 단위를 이룸. 각각의 조각들로 메스를 재 배치함.
형태–규칙적인 형태의 모자이크와, 불규칙적인 형태의 모자이크 중, 불규칙적인 형태를 적용.
투명 & 불투명–가시성을 서로 다르게 하여 모자이크의 또 다른 느낌을 나타냄.

픽셀–이미지를 이루는 가장 작은 단위. 평면을 픽셀로 나누어 형태의 기반을 잡음. 조각–픽셀이 모여서 조각 단위를 이룸. 각각의 조각들로 메스를 재 배치함.
형태–규칙적인 형태의 모자이크와, 불규칙적인 형태의 모자이크 중, 불규칙적인 형태를 적용.
투명 & 불투명–가시성을 서로 다르게 하여 모자이크의 또 다른 느낌을 나타냄.

PROPOSAL 01

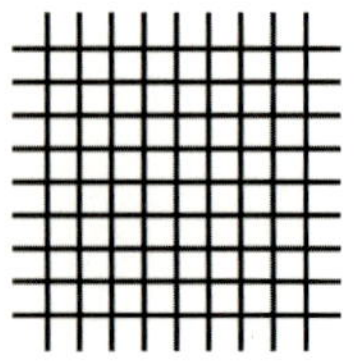

규칙적
Regularly

불규칙적
Irregular

선으로 인해 생기는 면
Area formed by lines

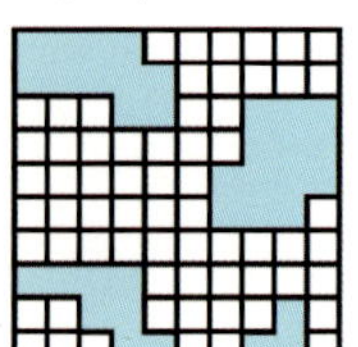

불투명성
Opaqueness

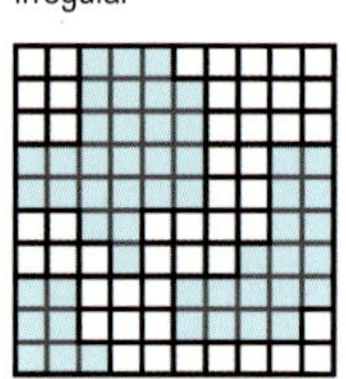

투명성
Transparency

레이어 중첩, 호기심 유발
Layers overlap, curiosoty induced

형태의 다양성
Diversity of form

크기의 다양성
Diversity of size

배열의 다양성
Diversity of array

PROPOSAL 02

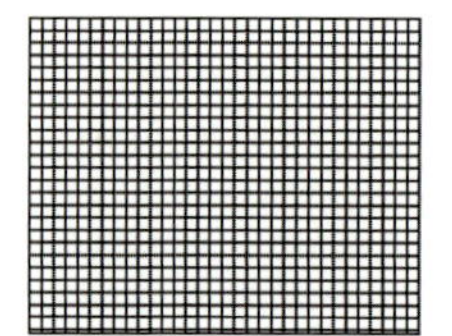

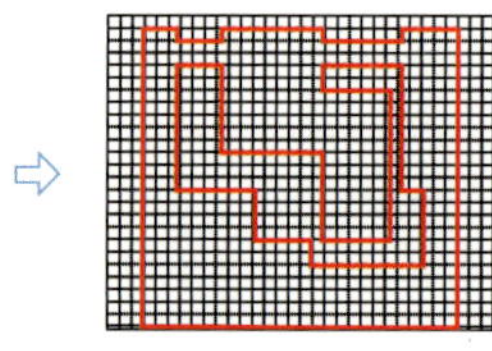

규칙적인 격자 안에서 불규칙적인 매스를 형성.
Irregular mases are generated within regular grids.

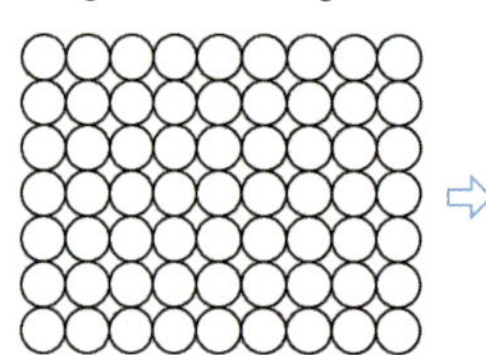

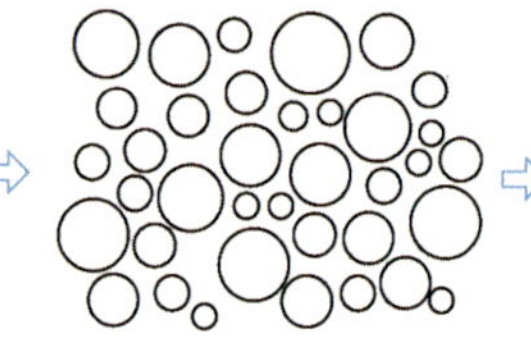

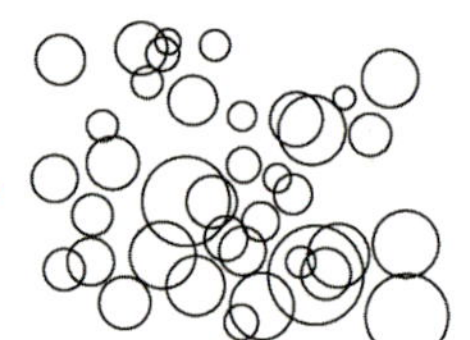

다양한 크기, 불규칙적 배치, 중첩을 적용.
Various sizes, irregular arrangement and overlaps are applied.

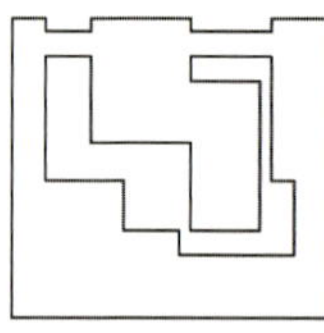

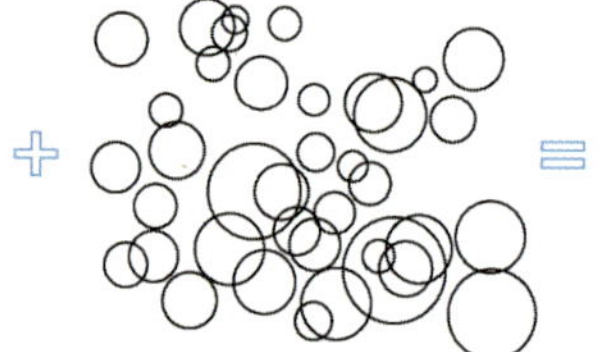

서로 다른 형태의 도형을 집합시켜 매스 완성.
Differently formed figures are connected to complete masses.

FLOOR PLAN

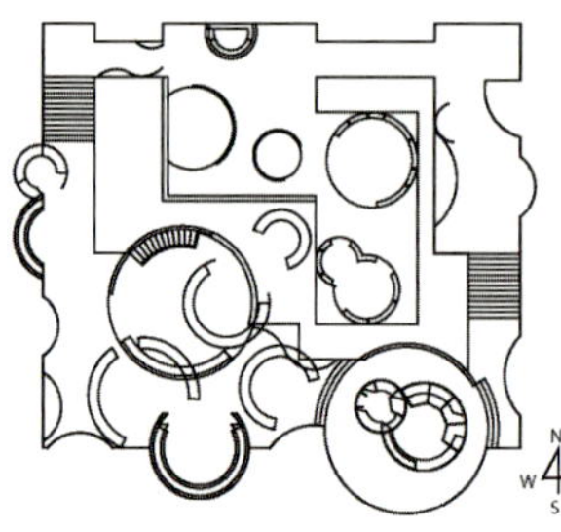

1 FLOOR

1층– 무인 시스템의 도서관과 휴식을 취할 수 있는 필로티 공간.
First floor - Pilotis for an unmanned library and rest.

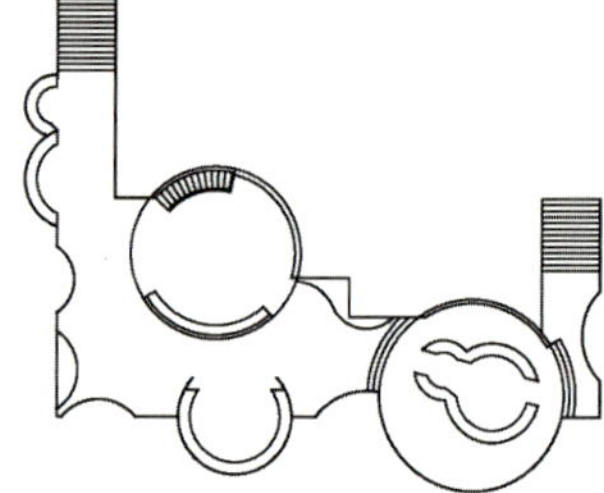

2 FLOOR

2층– 상설전시를 위한 갤러리 공간.
Second floor - a gallery for permanent exhibitions.

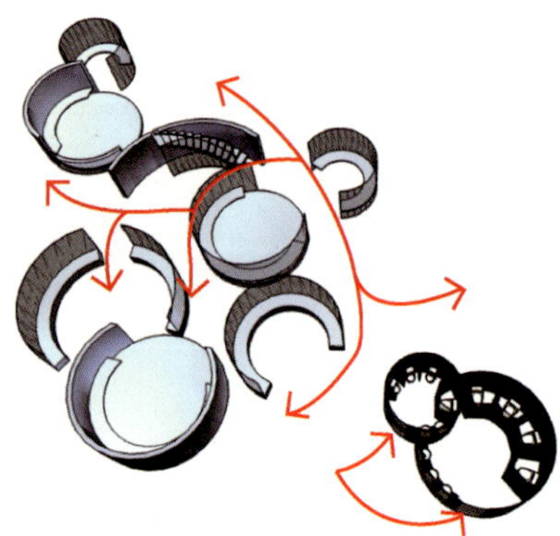

Pilotis movement line.

필로티 공간의 자유로운 동선 선택.

Pilotis movement lines. Unrestricted movement lines in Pilotis space to be chosen.

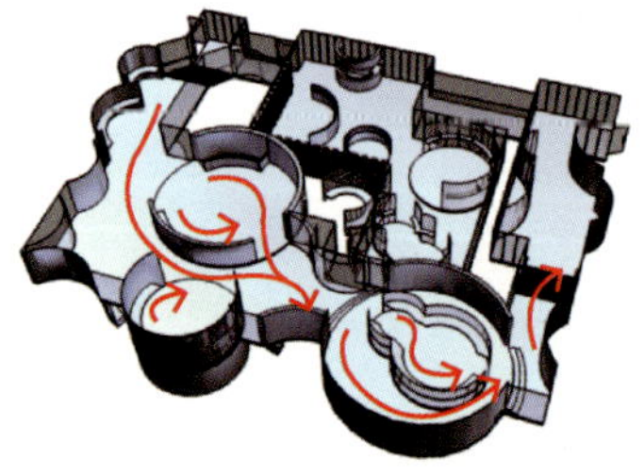

Gallery traffic line.

원의 도입으로 인한 동선 유도.

Inducement of movement lines based on the introduction of circles.

PROCESS 1

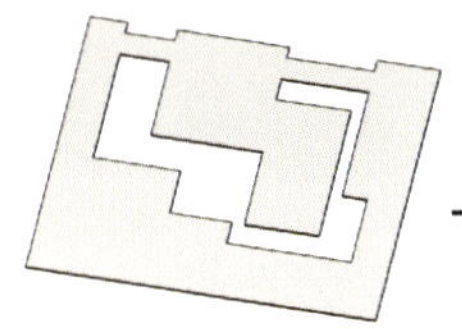

규칙적 격자에서 불규칙적 매스 형성
Generation of irregular masses from regular grids.

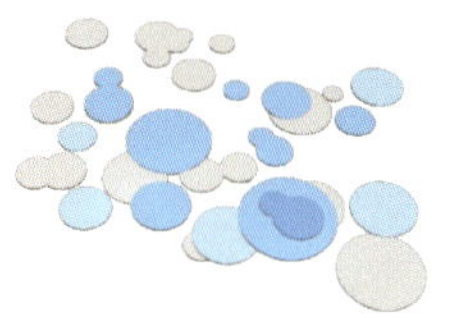

다른 형태 도입.
Introduction of other forms

다양한 크기의 원형에 중첩 적용.
Overlaps applied to circles of various sizes.

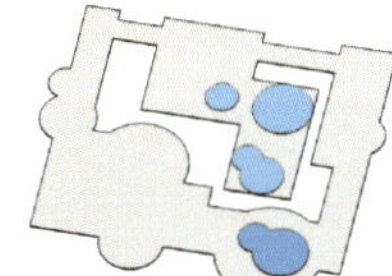

기존 매스에 추가.
Addition to the existing mass.

규칙적 형태와 불규칙적 배열 적용.
Application of regular forms and irregular arrangements.

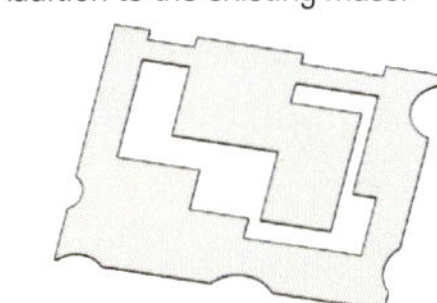

기존 매스에서 제외.
Exclusion from the existing mass.

PROCESS 2

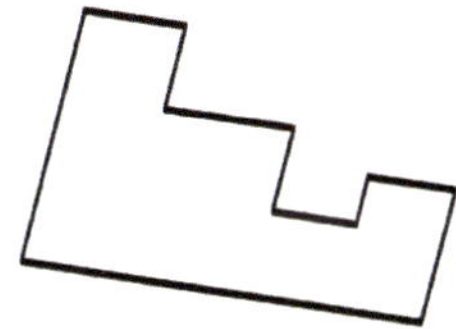

2층 매스 아래에 있는 필로티 공간.
Pilotis space under the second floor mass.

규칙적 형태와 불규칙적 배열 도입, 중첩 적용.
Introduction of regular forms and irregular arrangements as well as application of overlaps.

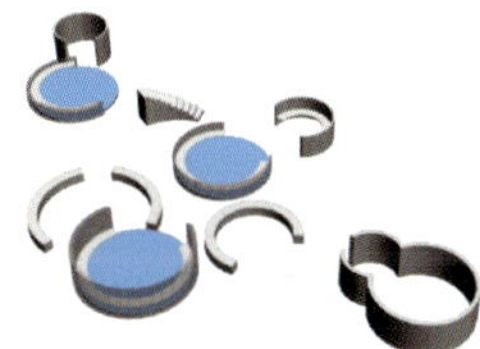

선과 면으로 이루어진 서루 다른 형태의 필로티 공간 생성.
Generation of variously formed Pilotis space composed of lines and planes.

PROCESS 1 + 2

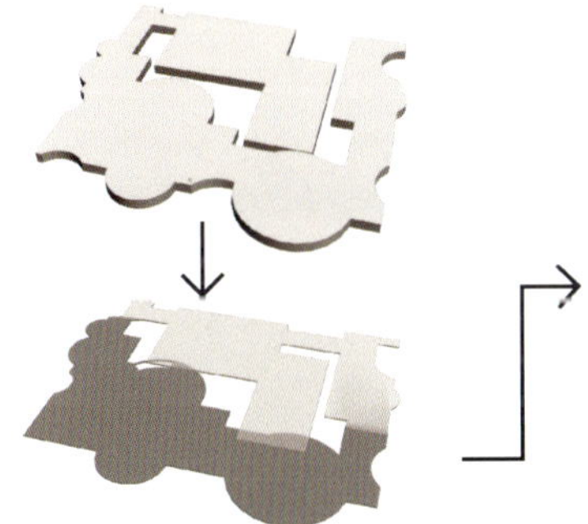

1,2층 세분화.
Segmentation of the first and the second floors.

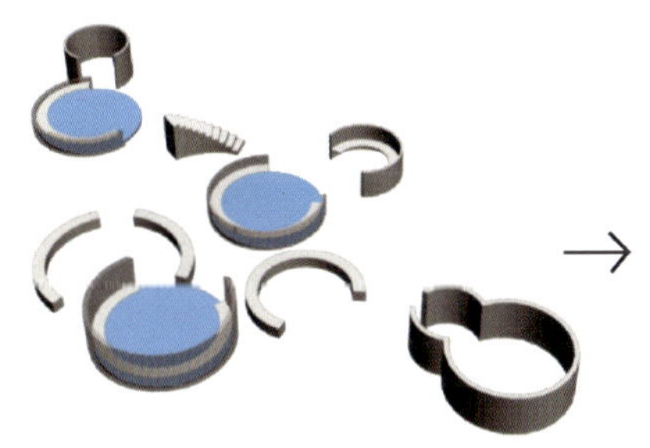

필로티 공간.
Pilotis space.

공간 형성
Space generation.

ELEVATION

NORTH ELEVATION

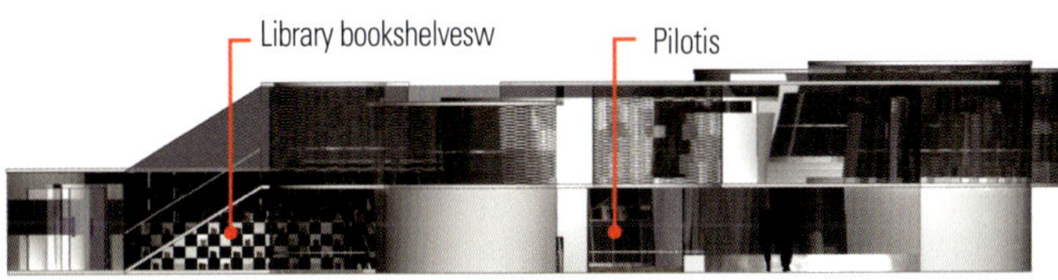

WEST ELEVATION

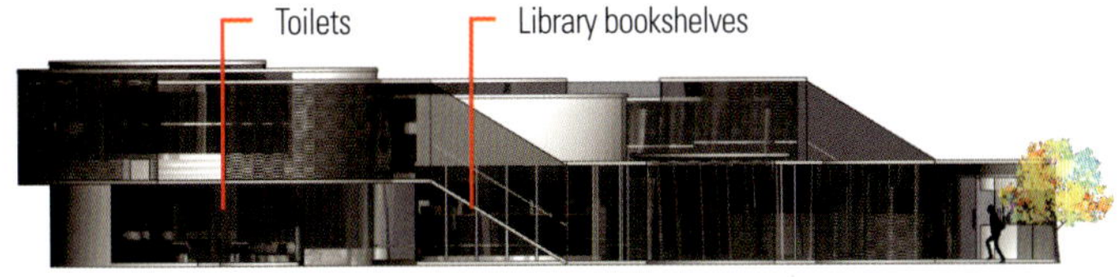

SOUTH ELEVATION

EAST ELEVATION

SPACE COMPOSITION

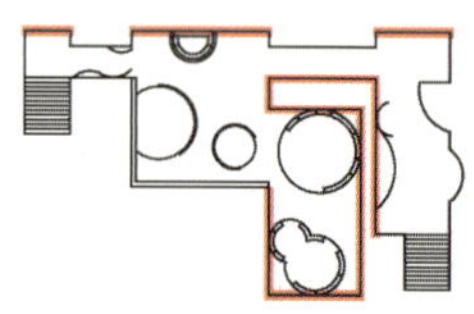

Transparent and opaque glass walls.

투명성과 불투명성을 적용, 비교적 덜 답답한 느낌의 모자이크를 나타내려 함.

The application of transparency and opaqueness is to present less boring mosaics.

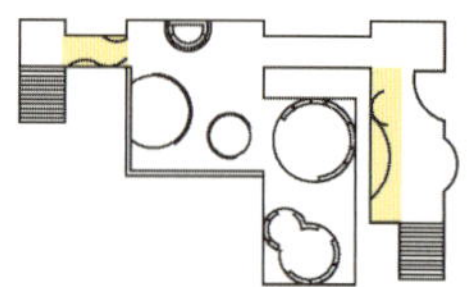

Dual glasses and layer overlaps.

이중 유리, 레이어 중첩. 레이어 중첩으로 인해 외부로부터 내부의 상황에 대한 호기심을 유발하도록 함.

Overlapped layers are to induce curiosity of the exterior about the interior situation.

Toilets

역과, 도로, 외부 필로티와 가까운 곳에 위치해 사용이 편리하며, 불투명의 유리 소재로 이루어져 있음.

Located in areas near from stations, roads, and external Pilotis, usage is convenient and the composition is of opaque glass materials.

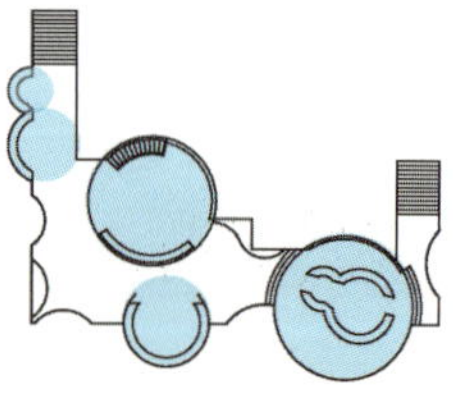

A Scuttle

주로 도서관 위쪽의 채광창이 많으며, 최대한 햇살을 받을 수 있도록 많은 천창과 측창들을 설치.

Mostly, there are many skylights in upper areas of the library and many scuttles and side wall windows for the sunlight to enter as much as possible.

Library bookshelves.

외부와 내부의 소통을 위한 개념. 책을 꽂는 위치에 따른 벽의 시각적 형태 변화.

A concept for communication between the exterior and the interior.
Visual changes of the walls based on the locations to place books.

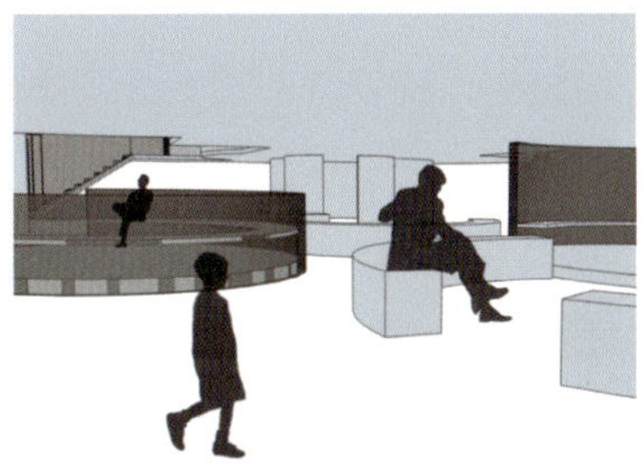

Relaxation

필로티 공간에 투명성과 불투명성을 적용. 투명 유리벽과 반투명 유리벽으로 인한 개인적 공간과 오픈된 공간 모두 제공.

Application of transparency and opaqueness to Pilotis space. Both types of individual and open space are provided through transparent and semi-transparent glass walls.

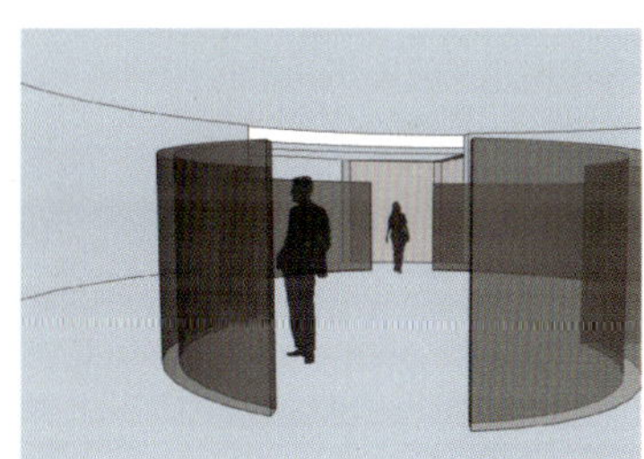

Gallery inside

원들의 중첩으로 인한 내부 공간 세분화.

Segmentation of the interior space through overlapped circles.

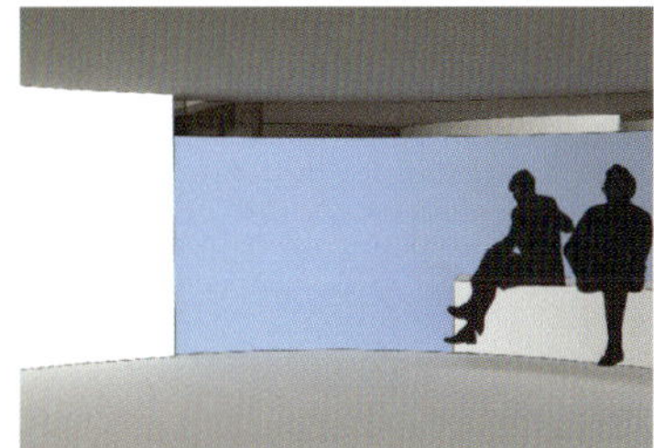

Visual Screening room

용도 변화에 따른 벽의 움직임으로 인한 시각의 다양화.

Visual diversification through the movement of walls according to changes in purposes of use.

1st Floor _ library

1층은 도서관으로 사용하고 필로티 공간은 개인적이면서도 공공장소로 사용할 수 있는 휴식공간을 제공.

The first floor is the library. Pilori space that can be used as personal and public places provide relaxation.

2nd Floor _ Gallery

2층은 갤러리공간이다. 벽체가 공간을 분할하고 동선을 유도한다.

The second floor is the gallery. Wall to divide the space and leads traffic line.

Part of a 1floor library space

1층 도서관의 부분 이미지이다. 세분화된 공간과 불투명성을 보여주고 있다.
Library on the first floor. This image shows divided space and opacoty.

Hearts are connected with hearts

마음과 마음을 잇다

사회적 이슈가 되고 있는 상처입은 유기견, 그리고 외로운 노인. 우리는 이들이 한 공간에서 만나게 됨으로써 서로에게 활력을 줄 수 있는 공간을 제안한다.
In social issues and wounded yugigyeon, and lonely elderly. We meet at one place they can give vitality to each other in becoming space is proposed.

PROPOSAL

공간으로서 버려진 Flyover, 충동적인 관심으로 입양되어 쉽게 버려지는 유기견, 그리고 우리나라 뿐만 아닌 전 국가에서 심화되는 노인증가, 이러한 사회적 문제를 이슈화 하여, 버려진 공간으로 하여금 강제 퇴거 되어진 사람과 동물이 만남을 통해 서로에게 위안이 되며, 단순한 카페나 센터가 아닌 사회적 약자를 위한 공간을 제안한다.

A spatially abandoned flyover, stray dogs purchased impulsively and abandoned so easily, and the increasing population of old people which is a common phenomenon found both in Korea and every other country. By understanding these social problems as issues, the abandoned space can be a place where forcefully displaced people and animals meet together to give solace to each other. This is a suggestion of space for the socially vulnerable rather than the existing cafes or centers.

TARGET

미국처럼 반려동물 문화가 확실하게 정착된 나라에서도 유기견이 많이 발생하고 있다. 경제화로 인한 경제적, 금전적 문제, 그리고 충동적인 입양과 그 후의 무책임한 행태로 인한 급증하는 유기견. 경제 사회로 변화하는 사회의 특징 및 문제점인 저출산과 독신주의들의 급증으로 인한 인구 고령화와 경제적 형편 및 경쟁사회에서 뒤처지고, 계속되서 증가되는 독거노인. 점점 반려동물이라는 사회적 관심으로 인해 넓어지는 애견에 관한 직업군 등. 이러한 주체들을 서로 엮어주고 서로의 문제점을 해결해줄 수 있는 공간을 제시한다.

Even in advanced countries such as the US where the responsible pet ownership is established, there are many stray dogs. The number of stray dogs rapidly increases due to economic and financial problems coming from capitalization, impulsive purchase and irresponsible behavior afterwards. The number of old people living alone keeps increasing due to low birth rates and an increase of celibacy, which are changing social and economic characteristics and problems. They tend to fall grow older and fall below the economic standards in the competitive society. The occupational area becomes larger and larger as the social interest in companion animals increases. The aim is to suggest space that can link these subjects and solve problems in each subject.

USER NEEDS

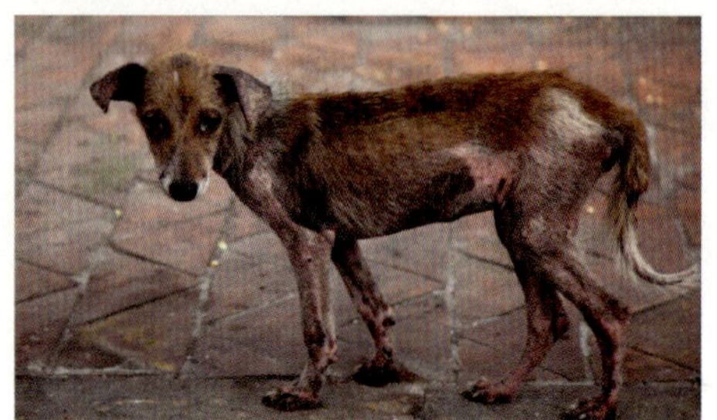

Abandoned dogs
–방황하고 쫓겨다니는 떠돌이 생활에서 탈피하는 공간.
–상처나 피로한 몸을 치료하고 회복하는 공간.
– Space to stop the wandering and being driven away.
– Space to heal and recover from scars or tiredness.

Senior citizen
–친구(노인, 애견)와 소통 혹은 휴식하는 공간.
–새 가족(애견)을 입양 할 수 있는 기회 제공.
– Space to communicate with friends (old people, dogs) or take a rest.
– Provision of an opportunity to adopt a new family member (a dog).

Administrator
–애견에 관한 직업군을 희망하는 시민들에게 경험 제공.

–Provision of experiences to citizens who wish to engage in works related to dogs.

SITE ANALYSIS

정의
서울특별시 구로구 구로동과 고척동을 연결하는 다리.
분석
구로구는 밀집거주지역으로 인구가 많고 주택, 아파트단지가 밀집해 있으며, 다른 도시와 서울을 잇는 경인로와 남부순환로 등 도시를 가로 지르는 도로시설 때문에 유동인구가 많다. 서울 외곽에 위치하고, 사회 계층의 중층 서민들이 주로 주거하는 지역으로써 그에 따른 경제적 이유로 유기견이 많이 발견되고, 독거노인 또한 많은 편에 속한다. 구로구를 가르지르는 안양천은 독거노인들이 자주 이용하는 산책로로 많이 이용되고 있다.

Definition
The bridge connecting Guro-dong and Gocheok-dong, Guro-gu, Seoul.
Analysis
Guro-gu is highly concentrated and has a lot of residential facilities and apartment houses. The flowing population is large with the road facilities crossing the city such as the Gyeonginro and the Nambusunhwanno connecting Seoul with other cities. Located in the outskirts of Seoul and inhabited by middle to lower classes, the economic conditions lead to more stray dogs and old people living alone. The Anyangcheon Stream crossing Guro-gu is a promenade favored by old people living alone.

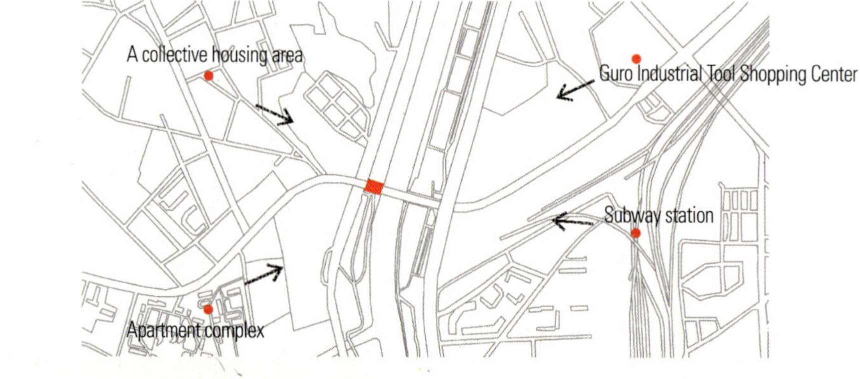

MOTIVE

유기견과 노인의 소통
Communication between stray dogs and old people

떠나간 것을 위로하다
Consoling what has been lost

소통
1. 막히지 아니하고 잘 통함.
2. 뜻이 서로 통하여 오해가 없음.

위로
따뜻한 말이나 행동으로 괴로움을 덜어 주거나 슬픔을 달래 줌.

사회적 흐름으로 유사한 이유로 상처받은 주체들에게 서로 소통하고 위로하는 공간을 만들고자 했다.

Communication
1.On the same wavelength without standoff.
2.Well-understanding each other and there is no misunderstanding.

Consoling
Alleviating pain or sadness with warm messages or behaviors.

The aims was to provide space where the subjects hurt from similar reasons in the social trends can communicate with and console each other.

CONCEPT & KEY WORDS

연결
고리라는 단어에서 기본적으로 가장 먼저 떠오르는 사슬이라는 형태에서 고리와 고리의 연결이라는 성질을 사용한다.

Link
The form of a chain, which is a word that the term loop reminds you of firstly, the characteristic of connection between and among loops will be applied.

폐곡선
폐곡선은 점에서 시작해서 시작한 점으로 다시 돌아오는 성질을 가지고 있다.

Looped curve
A looped curve will start from a point and come back to the original point.

Void & Solid
고리는 입면적으로 끝없는 면에 불과하다. 하지만 평면으로 봤을때는 공간이 구현된다.

Void & Solid
A loop is only a three-dimensionally endless plane. But it will generate space when seen two-dimensionally.

고리(명사)
1.긴 쇠붙이나 줄, 끈 따위를 구부리고 양 끝을 맞붙여 둥글거나 모나게 만든 물건.
2.어떤 조직이나 현상을 서로 연관되게 하는 하나하나의 구성 부분 또는 그 이음매를 비유적으로 이르는 말.
-노인과 유기견의 서로 연관되는 공통점인 외로움을 어떠한 구성 또는 이음매체를 통해 서로 이어지게 만든다.

Loop (noun)
1.A square or round object as long metals, strings or cords are bent or their both ends are attached to each other.
2.A noun metaphorically referring to each component or medium that connects a certain organization or a phenomenon.
-Loneliness, common to and connecting old people and stray dogs, will be communicated through a certain component or a medium.

연결
1.긴 쇠붙이나 줄, 끈 따위를 구부리고 양 끝을 맞붙여 둥글거나 모나게 만든 물건.
2.어떤 조직이나 현상을 서로 연관되게 하는 하나하나의 구성 부분 또는 그 이음매를 비유적으로 이르는 말.
폐곡선
곡선의 하나. 그 위를 한 점이 한 방향으로 움직였을 때, 다시 제 위치에 돌아오는 곡선. 양끝이 일치하고 있는 연속된 호
VOID
(커다란) 빈 공간, 공동; 공허감
SOLID
견고한, 속이 꽉 찬, (속이) 빈 데가 없는

Connection
1.A square or round object as long metals, strings or cords are bent or their both ends are attached to each other.
2.A noun metaphorically referring to each component or medium that connects a certain organization or a phenomenon.
Looped curve
One of curves. When one point moves toward a certain direction on a curve, the point will come back to the original location. An arch whose both ends coincide with each other and continued.
VOID
(Large) empty space, hole; a feeling of emptiness
SOLID
Fortified, full, (interior) without emptiness

DESIGN PROCESS

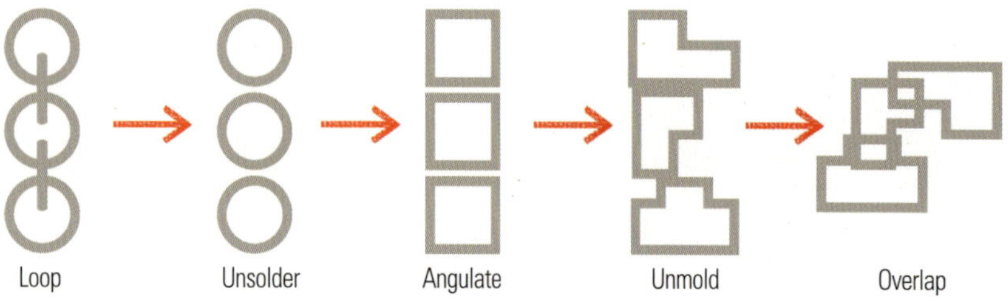

Loop → Unsolder → Angulate → Unmold → Overlap

고리를 분리시키고 평면화한 폐곡선을 곡선에서 직선으로 형태 변형시킨 ㄴ자, ㄱ자, T자의 다각형은 평면에서 고리의 형태를 보여주고, 그것을 중첩, 겹침을 통해 고리의 연결을 보여준다.

ㄴ-shaped, ㄱ-shaped, and T-shaped polygons generated from looped curves transformed to straight lines based on the separation and complanation of loops will show the forms of loops on the plane. The overlap or offset will show the connection of loops.

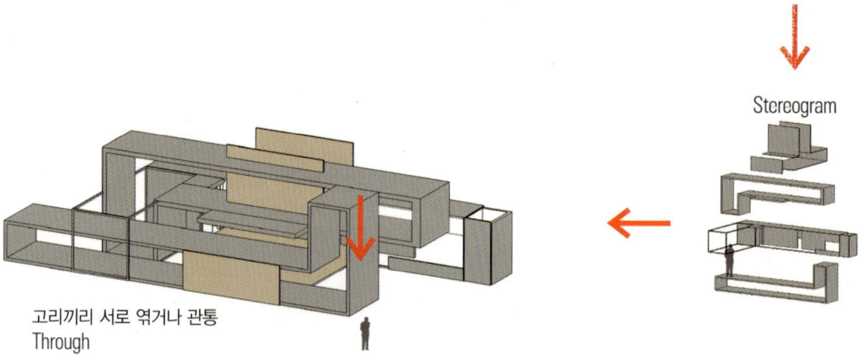

Stereogram

고리끼리 서로 엮거나 관통
Through

그렇게 생성된 평면을 입체화시켜 공간을 생성한다. 그러한 공간들을 입체적으로 겹치거나 관통시켜서 디자인적 요소를 완성하고, 이것을 확장하여 짜임새는 있지만, 단조롭고 일관적인 디자인의 사용으로 다소 무료하고 지루한 공간에, 무채색 톤을 사용하여 통일적 느낌안에서 절제된 포인트를 살려준다.

Thus generated plane will become three-dimensional to create space. The space will be three-dimensionally overlapped or penetrated to complete an element of design. The expansion of it will provide well-structured but monotonous and consistent design, which will be supplemented with temperate points within the feeling of unity generated from achromatic colors and tones in the rather boring and dull space.

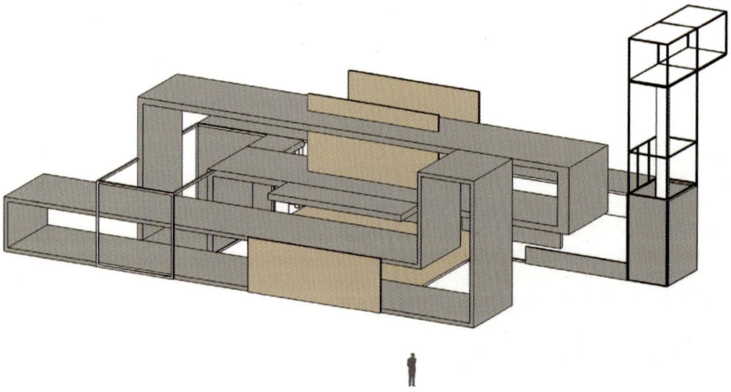

공간에 필요한 부분을 확장
Expand

Facade

공간들이 중첩되어 생기는 속이 꽉 찬 공간은 폐쇄된 마음을 표현하고자 하였고, 그러한 상처 받은 주체에게 최소한의 사생활 보호를 실현시킨다. 그와 반대로 각 공간의 빈 공간은 서로의 교감을 할 수 있는 공간을 구현해준다. 폐곡선에서의 공간끼리 교차되는 빈 공간과 꽉 찬 공간은 소통과 차단이라는 이중적 의미를 모두 담고 있다.

The space with its interior filled with overlaps of different types of space is meant to express a closed set of mind. But it realizes the minimum protection of privacy of the injured subjects. On the contrary, the empty area of each space realizes space where they can sympathize with each other. The empty space and full space where different types of space come across on the looped curve provide a dual meaning of communication and cutoff.

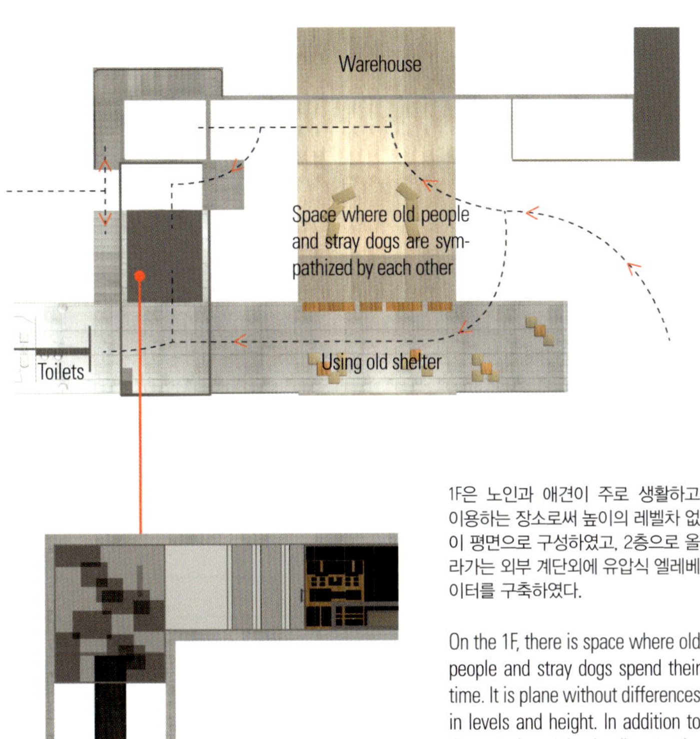

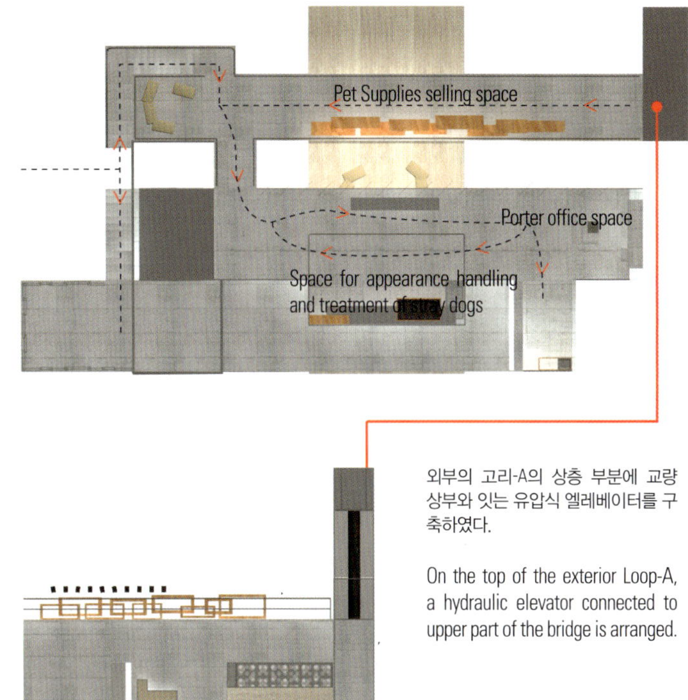

1F은 노인과 애견이 주로 생활하고 이용하는 장소로써 높이의 레벨차 없이 평면으로 구성하였고, 2층으로 올라가는 외부 계단외에 유압식 엘레베이터를 구축하였다.

On the 1F, there is space where old people and stray dogs spend their time. It is plane without differences in levels and height. In addition to the exterior stairs leading to the second floor, a hydraulic elevator is arranged.

외부의 고리-A의 상층 부분에 교량 상부와 잇는 유압식 엘레베이터를 구축하였다.

On the top of the exterior Loop-A, a hydraulic elevator connected to upper part of the bridge is arranged.

FLOOR PLAN

1 1F 창고
Warehouse

2 1F 유기견 생활공간
Stray dogs living space

3 1F 노인이 이용하는 쉼터
Using old shelter

4 1F 유기견과 노인이 교감하는 공간
Space where old people and stray dogs are sympathized by each other

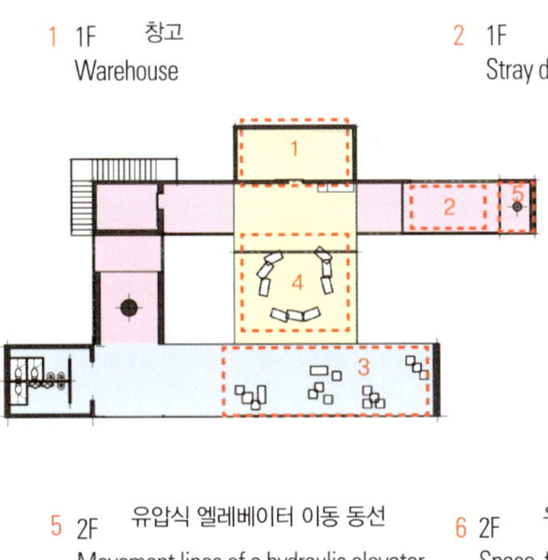

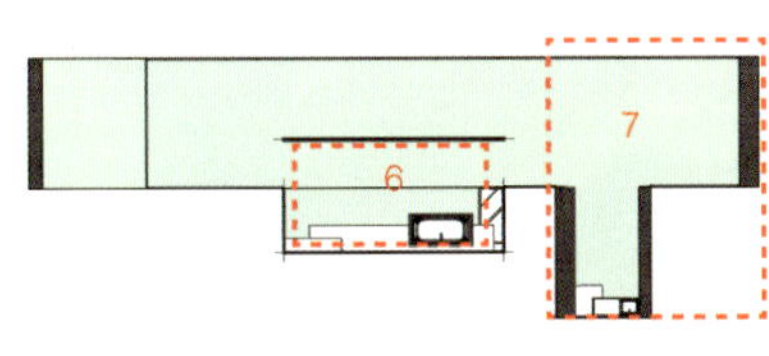

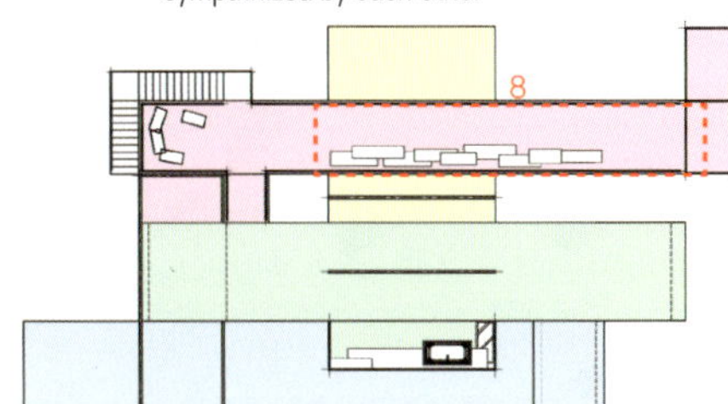

5 2F 유압식 엘레베이터 이동 동선
Movement lines of a hydraulic elevator

6 2F 유기견의 미용 및 치료 공간
Space for appearance handling and treatment of stray dogs

7 2F 관리인 사무공간
Porter office space

8 2F 애견 용품 판매 공간
Pet Supplies selling space

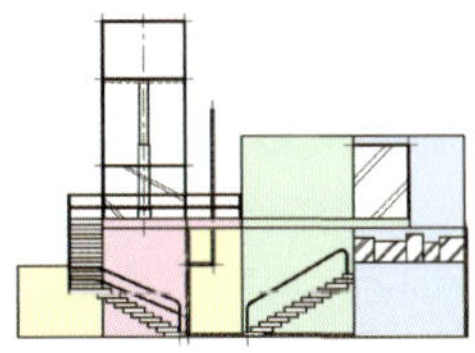

2층으로 연결되는 계단
That is connected to the second floor staircase

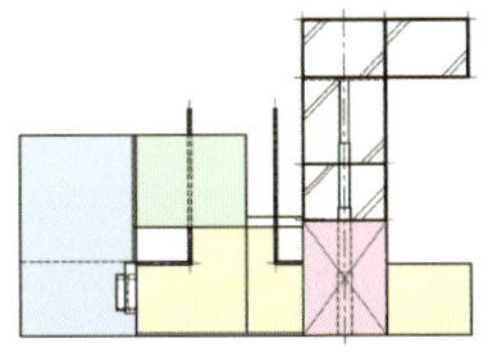

다리로 연결되는 엘레베이터
Elevator connected to the bridge

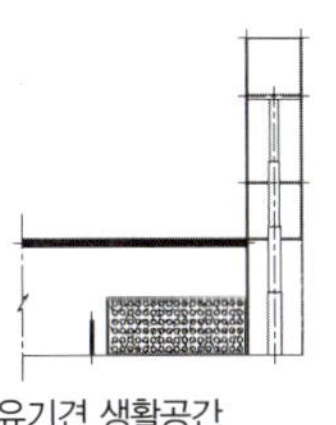

유기견 생활공간
Stray dogs living space

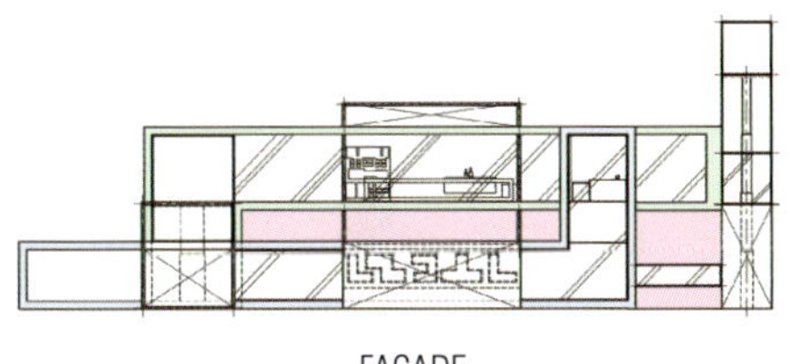

FACADE

SPACE PROCESS

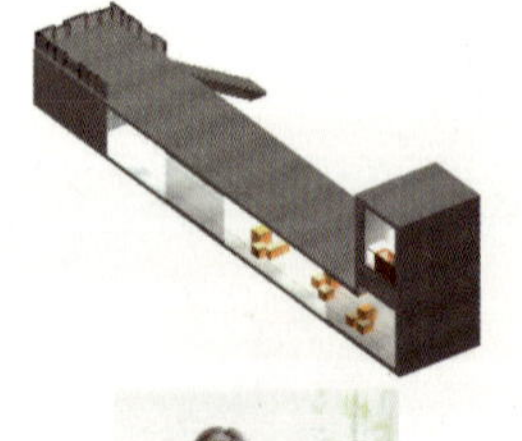

Loop - A

주로 노인을 위한 공간
카페 및 쉼터, 화장실

Space used mainly for old people
Cafes and shelter, toilets

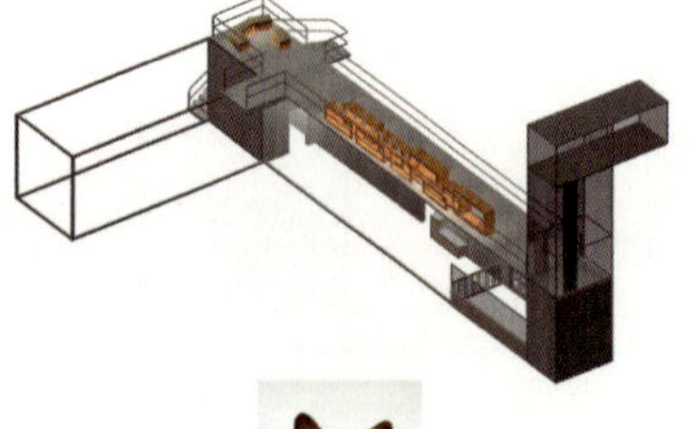

Loop - B

유기견을 위한 공간
유기견 보관, 유압식 엘레베이터

Space for stray dogs
Stray dogs protected, hydraulic elevator

Loop - C

A와 B를 연결시켜
노인과 유기견이 만나는 공간

A and B are connected
Space where old people encounter stray dogs

Loop - D

관리자를 위한 공간
사무실

Administrator for Space Office

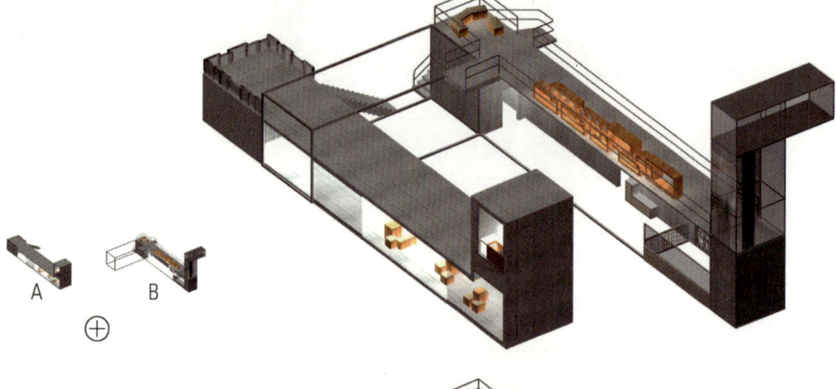

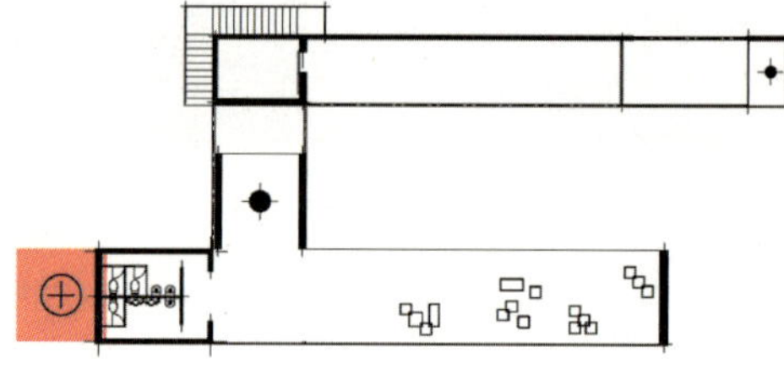

고리-A와 고리-B를 연결 시킨다.
Loop-A and Loop-B are connected.

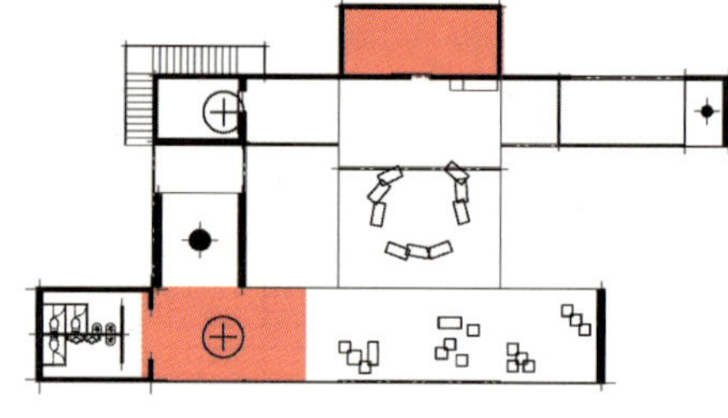

고리-C가 고리-A와 고리-B를 관통하여 연결 시켜주는 역할을 한다. 또한 유기견과 노인을 연결 시켜주는 공간을 구현.

Loop-C penetrates into and connects Loop-A and Loop-B. Also space connecting stray dogs and old people is materialized.

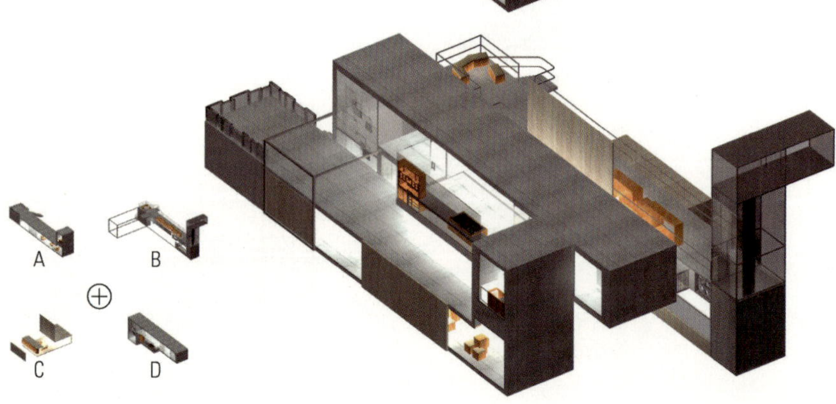

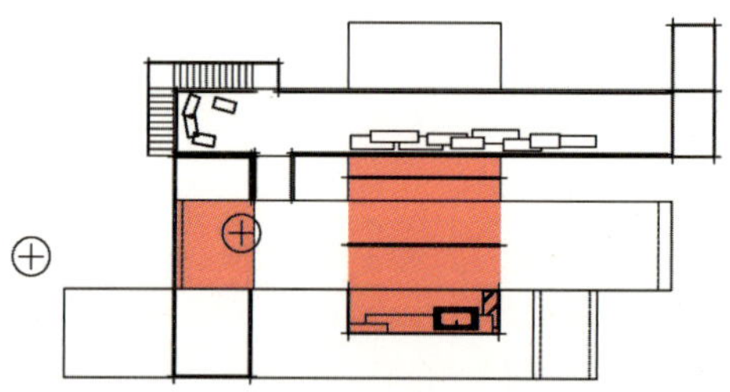

고리-D는 고리-A, 고리-B 그리고 고리-C를 모두 연결 시켜준다. 그리고 관리자 공간을 형성한다.

Loop-D connects Loop-A, Loop-B, and Loop-C altogether while generating space for administrators.

DETAILED VIEW

1F

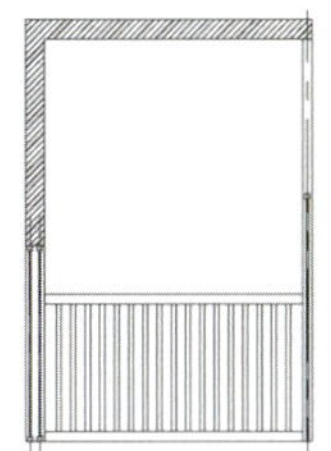

유기견들이 회복하고 생활하는 공간

Space where stray dogs recover and live

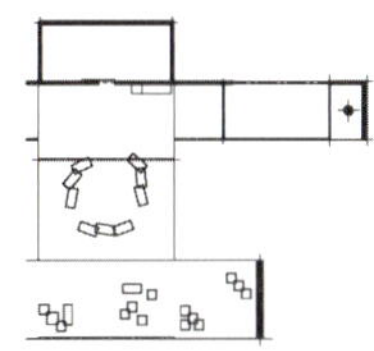

입구에서의 한 장면

A scene from the entrance

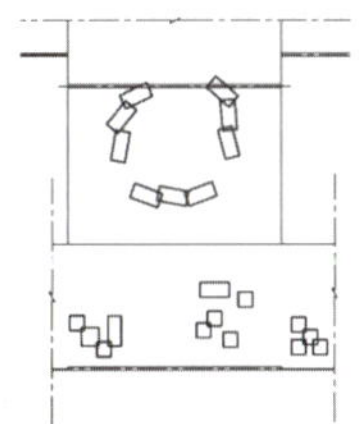

유기견과 노인이 만나서 놀거나 서로 소통하는 공간

Space where stray dogs and old people met to play or communicate with each other

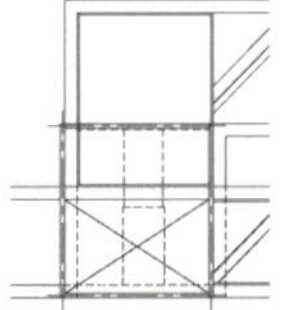

1층과 2층을 잇는 유압 엘레베이터와 노인들이 쉴 수 있는 쉼터

Hydraulic elevators moving to and from the first and the second floors and a resting area for old people

2F

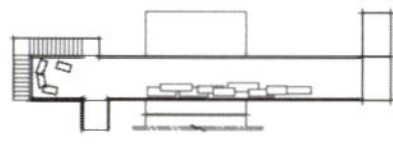

노인들이 제작하는 애견용품을 판매하고, 유기견을 위한 캠페인 광고를 하는 2층 외부공간

Space on the second floor for appearance handling and treatment of stray dogs

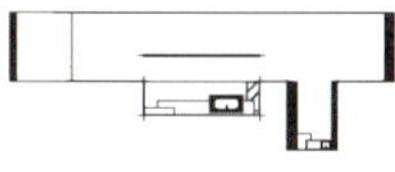

2층에 위치한 유기견의 미용 및 치료를 위한 공간

Space on the second floor for appearance handling and treatment of stray dogs

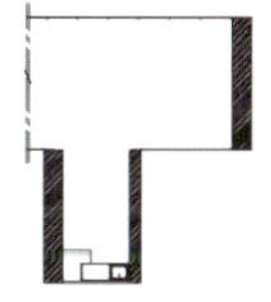

유기견 상태관리 및 개인 사무실로 쓰이는 2층 사무공간

An office on the second floor to manage the conditions of stray dogs and provide space for personal administrative affairs

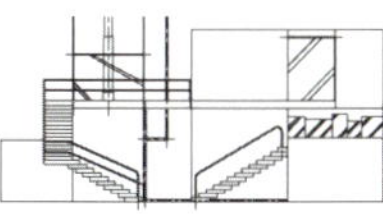

건물 측면에 있는 외부 계단

Exterior stairs on the sideways of a building

SPACE PROCESS

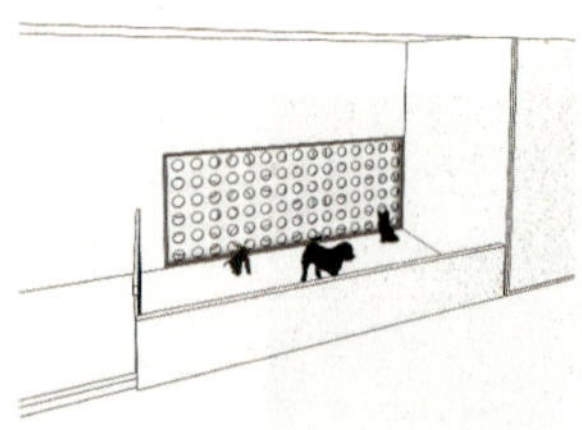

자연 환기설비
각각 다른 크기 구멍을 뚫은 두 개의 유리를 겹쳐 고리가 서로 엮이는 모양을 구현하고, 구멍으로 냄새가 배출되는 것을 이용

Natural Ventilation Equipment
Two pieces of glass with two differently-sized holes will be overlapped to materialize a shape of loops connected to each other while the holes can emit smells.

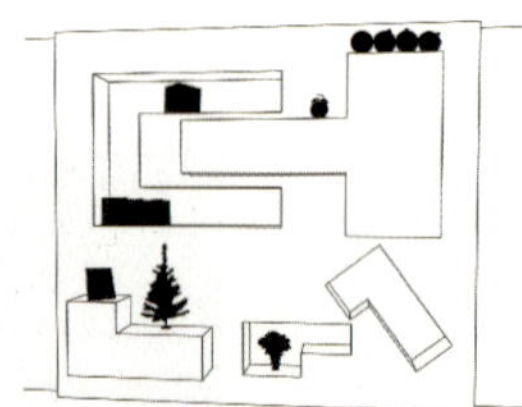

수납장
벽 두께의 레벨 차를 주어 벽 자체의 수납공간을 구현

Storage closet
Levels of the wall thickness will differ to materialize storage of the walls themselves.

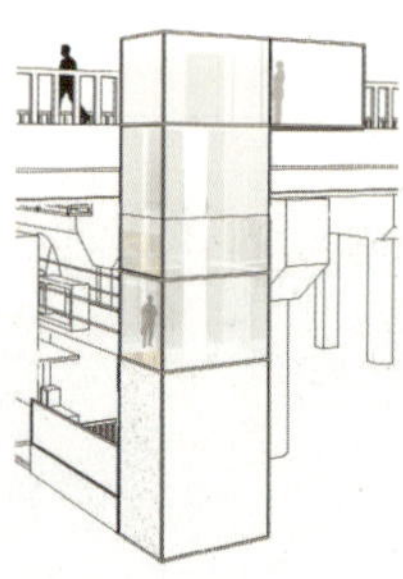

엘레베이터
1F -> 2F 그리고 2F -> 다리로 연결되는 엘레베이터를 구축하여, 불편한 다리의 노인 혹은 위급시 유기견을 이동시키는데 어려움이 없게 하였다.

Elevator
Building an elevator covering the movement from 1F to 2F and from 2F to the bridge, there will be no problem with moving an old person with disabled legs or a stray dog in an emergency.

SECTION

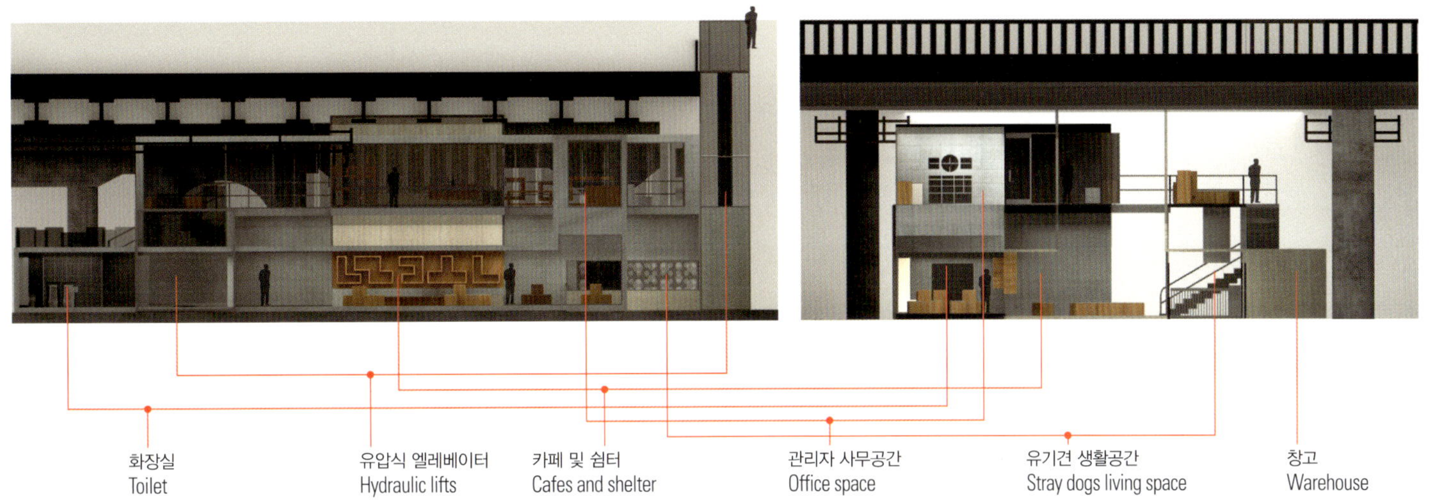
화장실
Toilet
유압식 엘레베이터
Hydraulic lifts
카페 및 쉼터
Cafes and shelter
관리자 사무공간
Office space
유기견 생활공간
Stray dogs living space
창고
Warehouse

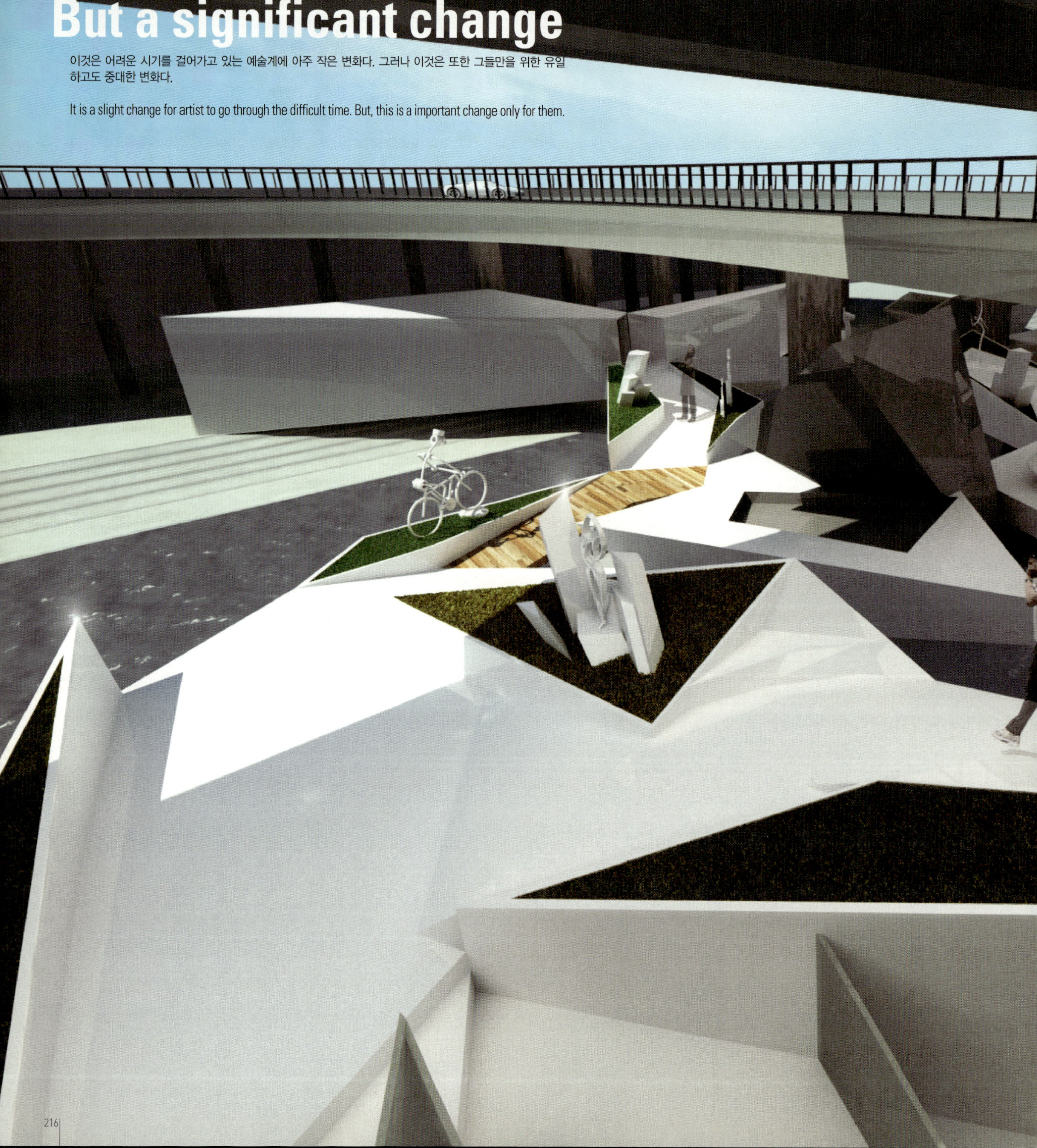

But a significant change

이것은 어려운 시기를 걸어가고 있는 예술계에 아주 작은 변화다. 그러나 이것은 또한 그들만을 위한 유일하고도 중대한 변화다.

It is a slight change for artist to go through the difficult time. But, this is a important change only for them.

PROPOSAL

경제 수준의 성장과는 대조적으로 여타 선진국들과 달리 우리나라의 예술계는 여전히 침체기를 벗어나지 못하고 있다. 성장한 경제 수준과 비례하지 못하고 예술가들은 높은 물가와 임대료로 인해 여전히 힘든 시기를 보내고 있다. 우리는 이러한 예술가들에게 일정기간마다 공모전을 기획하여 경쟁하며 상생하는 공간을 제공함과 동시에 선발된 인들에게 작업공간을 제공한다. 또한 바쁜 현대인들에게 문화적 역량을 향상시킬 수 있는 공간이다.

In contrast to level of economic growth, unlike other developed country, Korea's artist world does not getting out stagnation.
Still, Artists who cannot be proportional to developed economic level are going thorugh the difficult time because of the high price and rent. Each of these artists for a period of time, we plan competition and the competitive coexistence while providing a space to provide working space to the chosen people.
Also, This space is improve to cultural competency of busy modern people. the place.

생활수준의 향상으로 인해 단순한 경제 활동으로 부의 축적만을 생각하던 사람들은 문화, 예술 산업으로 시선을 달리 하고 있다. 이러한 현재의 트랜드에 맞춰 미래의 문화, 예술계를 이끌어갈 예술가들의 창작역량강화를 위한 지원 프로그램을 개설하고 신진 예술가 발굴하는 공간을 형성한다.

Improvement of living standard, people the accumulation of wealth as a simple economic activity has changing to culture and art of industry. A busy modern people to improve the cultural competency is the place. Improvement of living standards due to the accumulation of wealth as a simple economic activity, people who think only of culture, art, industry attention has been otherwise. The future according to current trends in culture, creative artists who will lead the support for capacity-building program to identify open space and to form budding artists.

USER NEEDS

ARTISTS

창작공간, 휴식공간, 전시공간

– 개인 창작공간을 제공하는 차원을 넘어 예술가들 간의 다양한 예술장르를 융합, 실험하는 공간.
– 정보를 공유하고 도움을 주면서 자신을 성찰하고 발전할 수 있는 계기를 만들어 주는 공간.

문화향유 공간

– 바쁘게 돌아가는 사회 속에서 문화적으로 소외된 일반 서민들에게 예술적 기질을 발견할 수 있는 기회를 주는 공간.

ARTISTS

Creative space, resting space, exhibit space

- Over the individual creative space, among the artists, this space unite various art genres and test.
- Sharing information and giving assistance to reflect on their own to create an opportunity to develop a space.

Enjoyment of cultural space

- Busy, culturally marginalized in society to the general populace the opportunity to discover the artistic temperament that space.

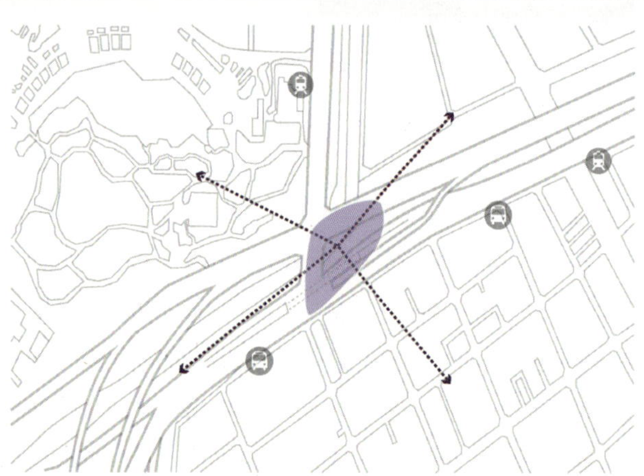

STADIUM PARKLAND

RESIDENTIAL HAN-RIVER

주소: 서울시 마포구 성산동
지역지구: 홍제천 입체교차로
주변시설: 마포구청역 / 월드컵경기장 / 주거단지 / 한강시민공원
진입로: 마포구청역 과 월드컵경기장역, 한강 둔치 방면 3곳으로부터 이어지는 길이 형성 되어 있어 접근성이 용이하다.

Address: Dong, Mapo-dong
Regional District: hongjecheon Cloverleaf
Surrounding facilities: Mapo-gu Office / World Cup Stadium / residential complexes / Hangang Park
Ramp: Mapo-gu Office and worldcup Stadium station, leading from the river where Dunchi Area 3 is the formation of a Load is easy accessible.

UPLIFT
UP(위로) + LIFT (끌어올리다)

UPLIFT는 해수면 밑이나 지하에서 드러나 있지 않았던 지각이 어떤 현상에 의해 지표면으로 드러나는 현상을 의미한다. 재능있는 예술가들을 세상 밖으로 유도한다는 공간의 특징을 UPLIFT만의 진보적이고 진취적인 수직적, 사선적 구도와 새로운 지각의 형태 등을 사용하여 디자인으로 표현하였다.

UPLIFT is that under the sea or not exposed in the basement of the crust is exposed to the surface by some phenomenon refers to symptoms. Induction into the world of talented artists that spatial characteristics of the UPLIFT own liberal and progressive vertical, diagonal line and new forms of crust and representation of the design was used.

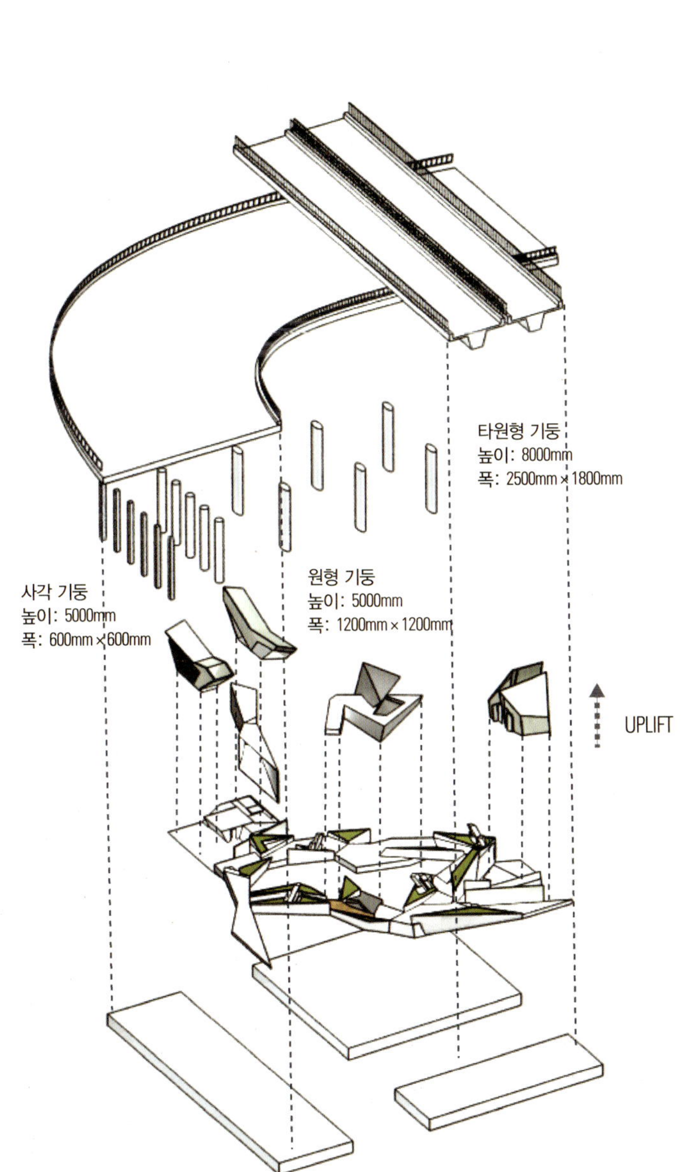

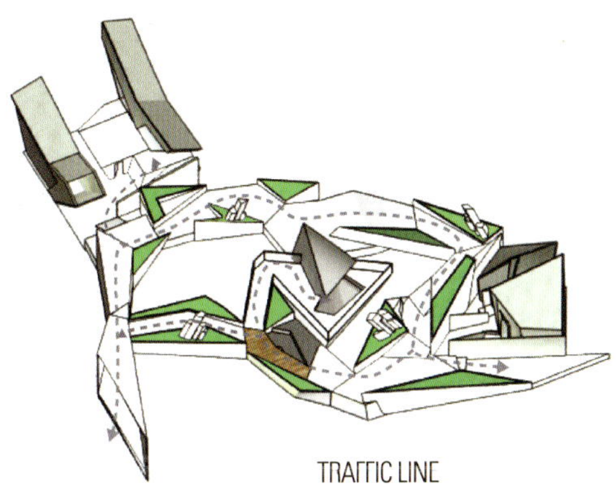

TRAFFIC LINE

– 3개의 공간을 연결시켜주는 이동통로적 역할은 보행자들에게 기존 교각 하부공간의 연결통로의 부재란 문제점을 해결해 주고 있다.
– 공간의 곳곳의 배치된 갈림길은 보행자들에게 다양한 동선을 선택할 기회를 준다.

- Path connected three spatial is solved problem of the existing piers of the underlying space to pedestrian.
- Fork of a path gives a chance to choice various of circulations.

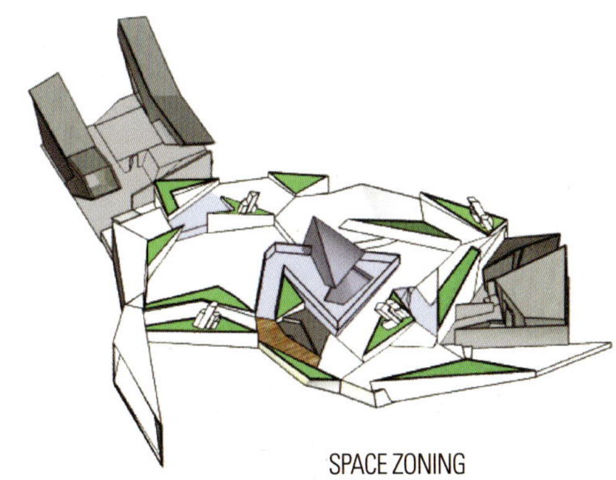

SPACE ZONING

– 총 4개의 스튜디오와 3개의 크고 작은 휴식공간으로 구성되어 있다.
– 자연스럽게 이동하며 전시를 구경할 수 있게 이동 통로 곳곳에 전시대가 배치되어 있다.

- It is 4 of studios and 3 of large and small break space.
- Move smoothly and be able to witness the exhibition moving walkways throughout the exhibition are arranged board.

SPACE PROCESS

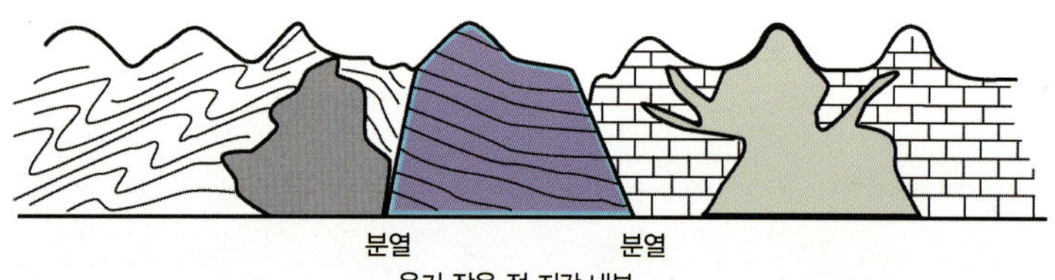

융기 작용 전 지각 내부

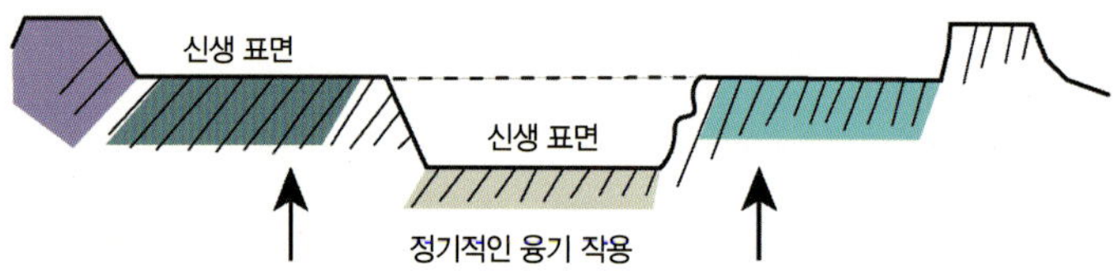

정기적인 융기 작용

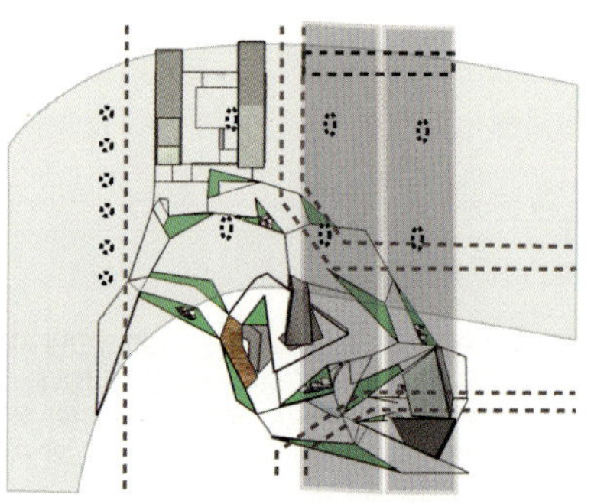

– UPLIFT 지표면의 형상을 대지에 적용
– 세 개의 공간을 이어주는 통로적인 역할
– 전체적 선을 사선구도로 통일하여 상승감과 속도감을 가진다.
– 교량의 교각의 위치를 고려한 동선을 사용 해당 FLYOVER의 특징을 보여주는 디자인을 하였다.
– 공간의 확장과 수축을 반복하여 동선의 지루함을 최소화

– UPLIFT ground's form apply to site.
– Passage of three roles to bridge the space.
– all of the line united diagonal composition have pace and upward.
– We designed that FLYOVER's characteristic, circulation considering position of bridge pier.
– We minimize flatness of circulation repeating expand and contract of space.

– 지면의 Mass 상승과 그에 비례하여 아래로 내려간 침강 현상이 나타나며 지표면의 Active한 굴곡이 나타난다.
– 미증유의 자연적 힘에 의하여 매스의 분열이 일어나 융기의 상승현상이 일어나며 사선적 형태로 솟아오른다.

– Activity line is made by ground level's mass uplift and downlift.
– By the natural forces of unparalleled mass bump up the divisions in the form of a rising phenomenon occurs diagonal form.

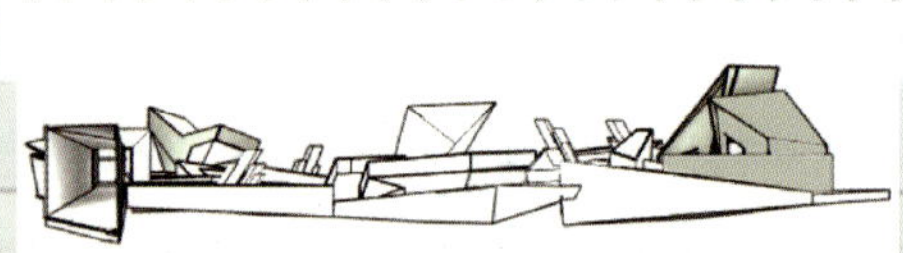

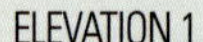

ELEVATION 1

ELEVATION 2

ELEVATION 3

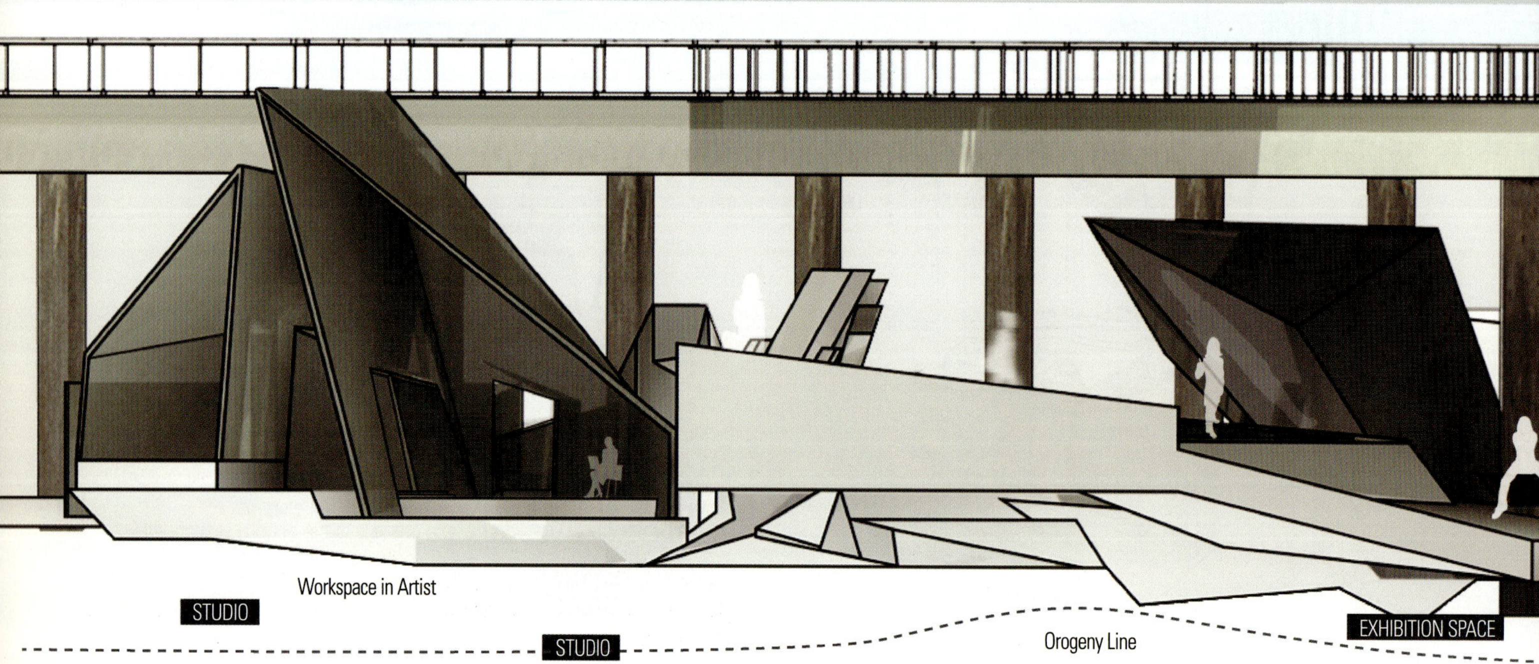

MASS PROCESS

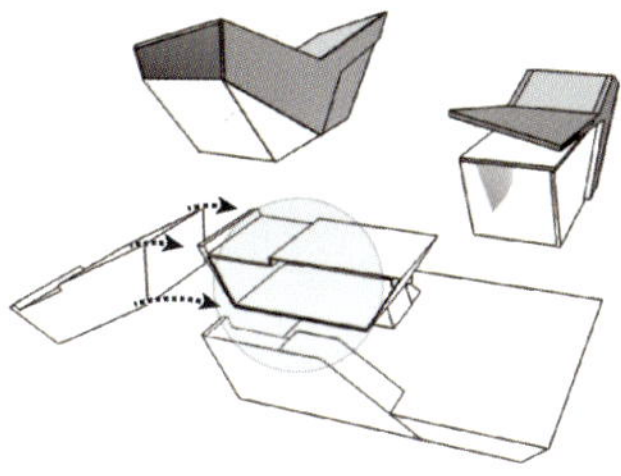

지각의 충돌로 인해 개입된 MASS가 하나의 새로운 공간으로 형성. 밀도차이에 따른 MASS의 관계성으로 인한 형태. 상승하고 침투하는 지각의 특성을 이용한 사선적 형태의 MASS 디자인

Mass of crust of a conflict involving one of the new space is formed. The difference in density according to the form of mass due to the relationship of the castle. To penetrate the crust to rise and with the characteristics of the biomass in the form of diagonal design.

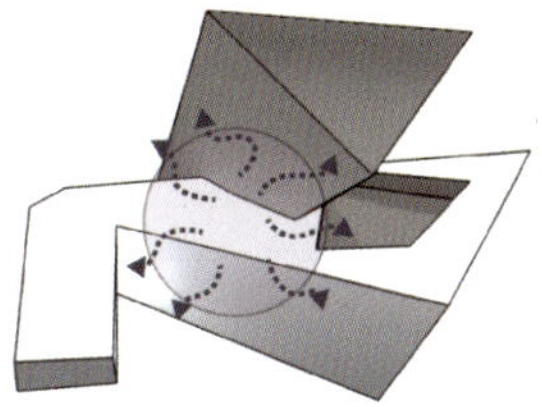

두개의 강력한 지각판의 힘겨루기로 인해 방향이 전환되고 다른 지각판에도 영향을 주는 힘의 파장의 중심이 되는 MASS.

Two powerful crust switch the direction of plate due to a trial of strength, and other forces that affect the crust of the wave-length plate in the central mass.

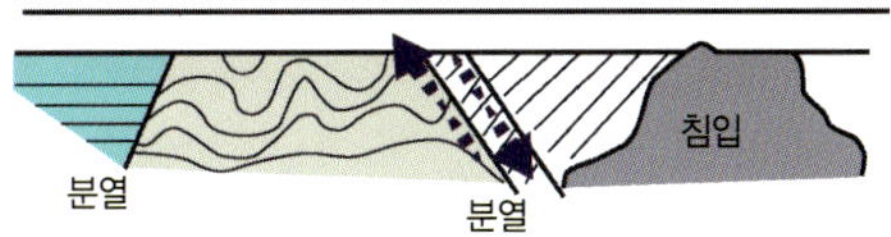

융기 작용전 지면의 내부 모습이다. 퇴적층이 형성되 있으며 유기적인 모습과 사선의 모습이 서로 공존하고 있다.

Formed by sediment and organic shape and appearance of diagonal can coexist with each other.

융기 현상이 나오기전 조산운동으로 인한 지면 매스의 분열 및 분할이 이루어진다.

Division and split of the ground mass that symptoms caused by orogeny before uplifted This is ground inside before uplifted.

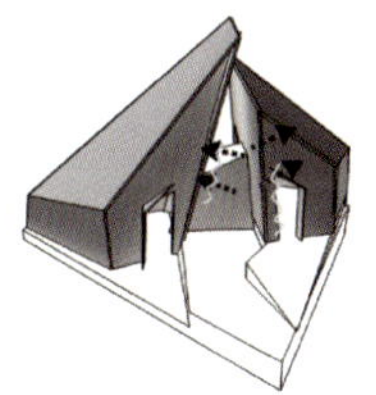

조산운동으로 수직적이면서 동일한 방향으로 상승하던 MASS가 새로이 개입된 MASS와 충돌하여 양쪽이 두개의 MASS로 관통 양 방향에서 진입이 가능한 공동 화장실이 형성.

Orogeny in the same direction as the vertical rise, yet the mass was in conflict with the newly-engaged by the mass to the mass penetration of the two sides, two-way shared bathroom on the entry form possible.

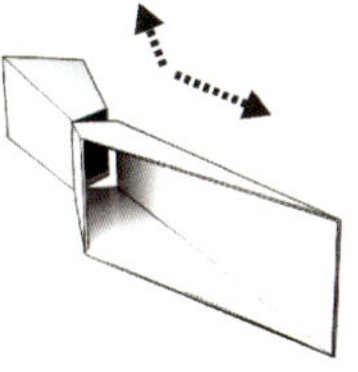

서로 다른 방향으로 점진적으로 작용하던 힘의 영향으로 하나의 MASS가 두개로 쪼개지개 되고 그 틈으로 열린 공간과 확장되는 공간으로 나누어 지게 됨. 쪼개진 틈을 갈림길로 사용.

Gradually in different directions under the influence of the power that worked with one of two broken mass and its extension through the cracks in the open space and the space becomes also be divided into. Crossroads is used to split a break.

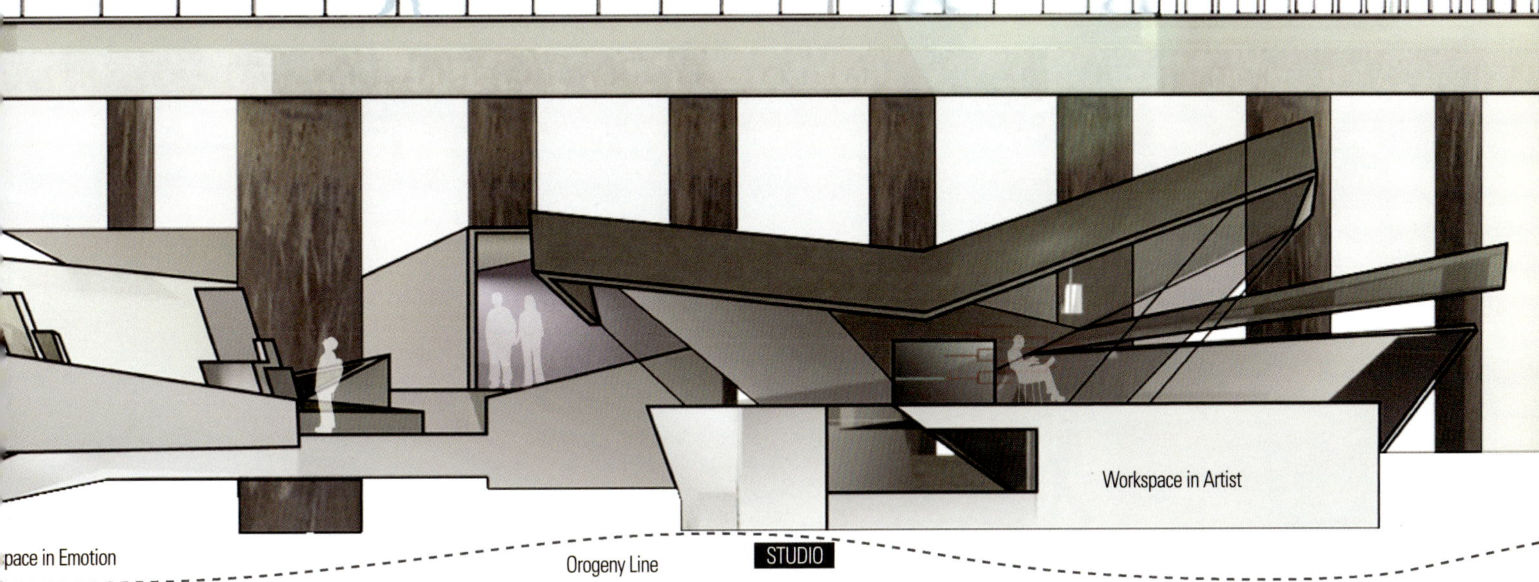

SPACE FUNCTION

진입로

입구가 사선으로 나있어 동선이 자연스럽게 유도되었고 천정고의 차이로 인해 호기심을 유발하였다.

Naturally, entrance designed a diagonal It is got a diagonal entrance and ceiling, and the copper-induced a natural curiosity was due to differences.

▼

갈림길

인식하지 못하는사이 좁아진 통로는 확장되는 왼쪽 길과 개방된 오른쪽 길로 쪼개져 있다.

Unknowingly, the narrow passageway broken as extended left way and opened right way.

▼

갈림길 우측 동선

교량의 곡선적 형태와 사선으로 길게 나있는 길은 마치 빨려 들어가는 듯한 속도감으로 동선을 유도한다.

Curved shape of pier and long diagonal of road and seem to be sucked into predrilled like the way copper is induced by fast-paced.

▶

중앙 통로 갈림길

두 교각 사이 하천 위를 지나는 길로 탁트인 시야로 해방감을 느낄수 있고 자연적인 재료인 나무와 콘크리트, 잔디로 인해 대지가 가지고 있는 자연적인 느낌을 받을 수 있다.

The road passing through two bridges over rivers, unobstructed view can feel liberating to the natural material of wood and concrete, due to the grass with the earth can give a natural feel.

벤치

의도되지 않은 절제된 형태의 의자는 공간과의 조화를 이루고 가구 역시 공간의 개념을 일관되게 통일시킨다.

In moderate form of unintended chair furniture blends well with the spatial concept of the space makes a consistent unification.

▼

전시대

단층의 형태를 단순화시킨 전시대와 파편적 형태의 전시대를 UPLIFT 다양한 이미지를 표현하고 있고 다양한 크기와 형태의 작품을 전시할 수 있다.

Simplified form and broken form of flat board are presented a variety of images and sizes of work may exhibit.

▶

오르막길

교각으로 인해 시선이 차단되는 오르막길은 다음공간에 대한 호기심을 유발하고 사선적인 형태의 MASS와 유기적인 동선으로 이루어진 공간은 다양성을 가지게 된다.

An uphill road blocked gaze by bridge cause curiosity about next space. The mass of diagonal form and space made by flexible circulation will be have varieties.

휴게공간 진입로

두번째 갈림길에서 언덕으로 올라와 있는 휴게공간은 무거운 MASS 두개로 이루어져 있다. 속도감 있는 공간에서 안정감과 무게감을 실어준다

Come up the hill at the fork in the second rest area is composed of two heavy mass. In the area of speed stability and gives weight to pick up.

▼

휴게공간

휴식공간을 덮치는 듯한 형태의 MASS와 교량상판으로 인해 외부와의 시선을 차단시켜 준다.

Like covering the retreat of the Mass and the bridge deck due to the form with the outside world will shut eye.

▼

STUDIO 2

수직으로 솟구치듯 강렬하게 표현된 스튜디오는 진보적이면서 예술가들의 상승욕구를 자극 시킨다.

Studio expressed strongly is progressive and stimulus of artist desire.

▶

내리막길

UPLIFT현상을 축소시켜 놓은 듯한 절제된 형태의 전시대와 구불구불하게 나있는 내리막길은 기존 전시공간에서 오는 지루한 동선에서 탈피하여 호기심을 자극하는 전시공간으로 받아들여진다.

The form restrained such as UPLIFT reduced and a winding downhill escape boring circulations. coming from the existing exhibition space by moving away from intriguing exhibition space will be accepted.

▶

STUDIO

교각에서 노출되어 전망이 좋은 한강방면으로 배치되어 있다. 세련된 사선구도와 Cantilever를 사용한 두개의 MASS는 세련되고 날렵한 형태를 취하고 있다. Solid한 면과 Void한면을 사용하여 Privacy를 중요시하는 공간과 좋은 전망을 필요로하는 공간으로 구분하였다.

Exposed at the piers are arranged in the direction of the Han River with great views. Two mass composed cantilever and stylish diagonal composition is a stylish and sleek. Use a face on one side Solid Void Privacy-conscious and a good view of the space delimited by the space was needed.

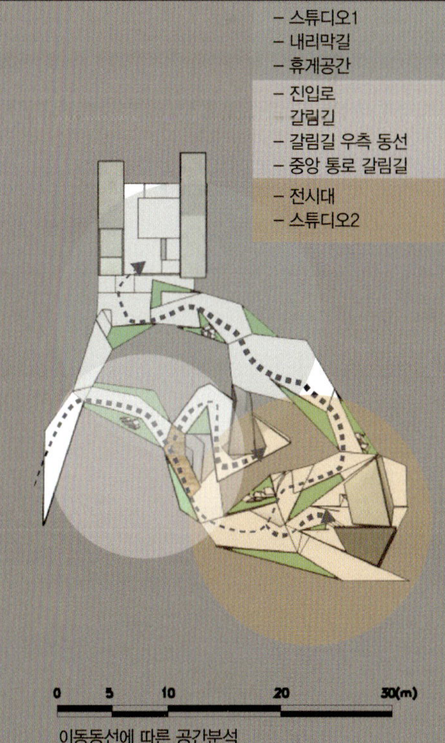

이동동선에 따른 공간분석

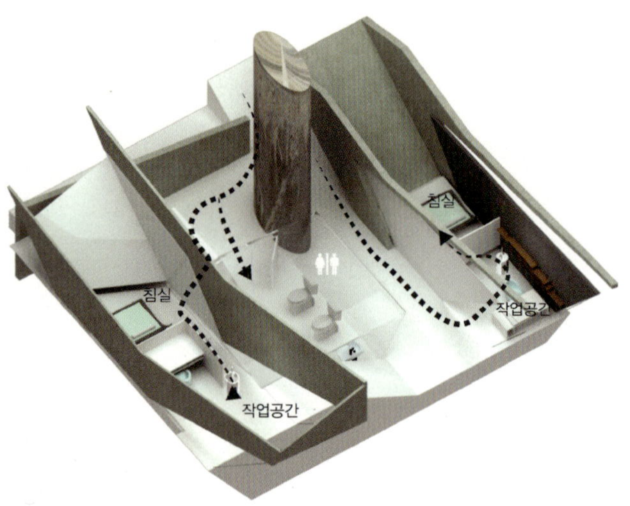

STUDIO 1 **AXONOMETRIC**

STUDIO 1

- 레벨차를 가지는 바닥면에 두개의 스튜디오 MASS가 위치.
- 작업공간과 침실로 구성되어 있고, STUDIO 중간지점에 공동화장실이 배치되어 있다.
- Solid한 면과 Void한 면으로 영역이 나누어져 있어 Privacy가 요구되는 공간을 제공하였다.

- Two different heights in the bottom position of the studio MASS.
- Work area and bedroom, and, STUDIO bathroom are arranged on the middle joint.
- Solid Void one side and a side region is divided Privacy has provided the required space.

STUDIO 2

- 두개의 거대한 MASS가 충돌하기 전 침입한 밀도가 높은 MASS로 인해 양쪽으로 침투한 공간이 형성되었다. 양쪽에서 출입이 가능하여 공동화장실로 사용된다.
- 비교적 넓은 작업공간과 1개의 침실로 구성되어 있다.

- Two before the conflict broke into a huge MASS MASS due to high density were formed on both sides invaded the space. Access from both sides can be used as a shared bathroom.
- A relatively large working space and consists of one bedroom.

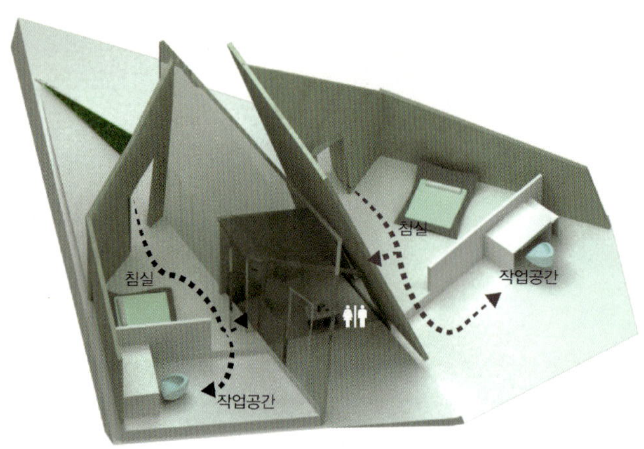

STUDIO 2 **AXONOMETRIC**

교량 상판을 사용하여 태양 빛을 차단해주고 교량과 사선으로 뻗어 있는 MASS가 시각적인 속도감을 부여한다.
Deck using the sun is blocked. Bridge and mass gives a visual sense of speed.

Maze in Barley

도시시설물 중에서도 상당부분을 차지하고 있고, 도시문제를 야기하고 있다고 생각되는 곳, '고가도로' 하부공간에 초점을 두고 프로젝트를 진행한다. 도시의 고가도로는 단지 원활한 교통흐름을 의도한 인공물에 불과했다. 그 고가도로의 밑은 어둡고, 불쾌하고, 소음과 먼지가 가득한 공간이 된 채 방치되어 왔다. 하지만 도시를 하나의 생활공간으로 가정해 본다면, 도로를 포함한 도시공간은 교류와 의사소통을 시각적으로 드러내는 매개공간의 기회를 가진 색다른 장소라고 볼 수 있다.
다리 밑이든, 어디서든, 다양하고, 풍요로운 인간의 행위는 다양한 수단을 통해서 의도할 수 있다.

This project will be carried out with an emphasis on space below 'overpasses' which are supposed to take significant part of urban facilities and cause urban problems. Urban overpasses used to be simple artificial structures intended for smooth traffic and transportation. The lower space under overpasses have been neglected as that is dark, unpleasant and filled with noises and dusts. However, based on an assumption that a city is space for living, urban space including roads can be described as unique space that provides opportunities for media visually presenting communication and exchanges. Anywhere, be it space under a bridge or others, various and rich human activities can be intended via diverse means.
Mityideun legs, anywhere, diverse and rich human behavior can be uidohal through various means.

SITE

도림 고가도로 : 영등포구 도림동 162-94일대
Dorim high road : 162-94, Youngdeungpo dorimdong a

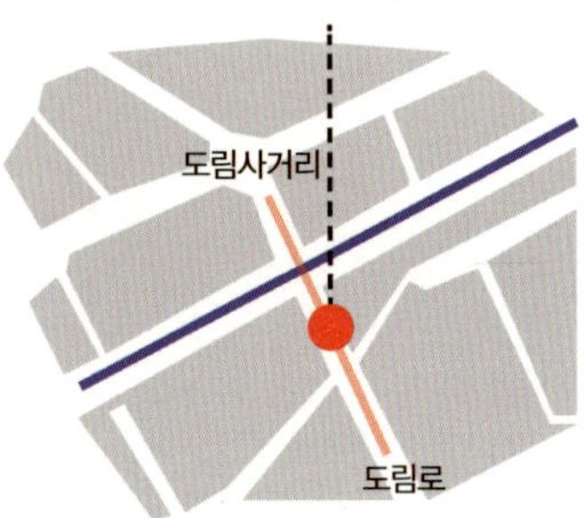

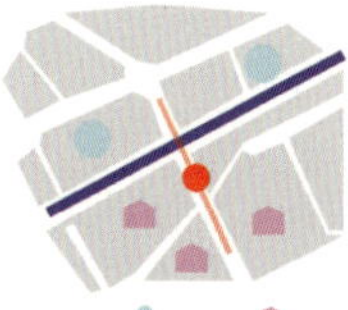

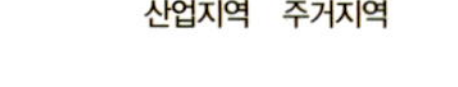

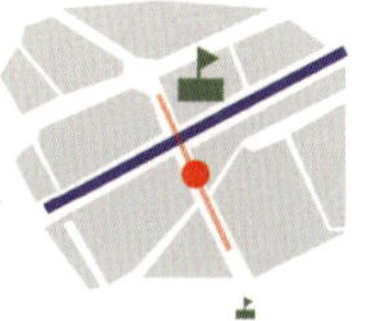
초등학교

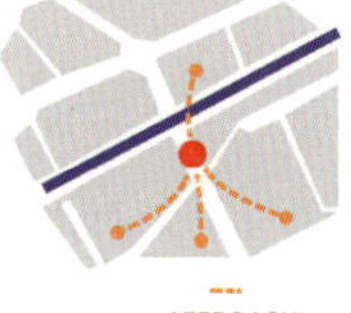
APPROACH

- 위치 : 영등포구 도림동 162-94번지 일대
- 용도지역 : 준 공업지역, 일반주거지역
- 인구 : 19,383명 (남: 9,894명 / 여: 9,489명)
  총 9,156세대 – 2010.10.31 기준
- 면 적 : 0.69㎢
- 경인선 남쪽에 위치한 도림동은 주택가로서 비교적 조용한 지역

- Location: Yeongdeungpo-gu, a 162-94 dorimdong
- Usage Area: given industrial area for residential
- Population: 19,383 people.
  The total 9156 generation
- Area: 0.69 ㎢
- As a residential area located to the south dorimdong gyeongin-seon A relatively quiet area

SITE ANALYSIS

도림고가도로 밑 부분은 기존의 4차선도로 아래에 위치해 있다. 차도나 인도가 연정되어 있지 않아 막다른 골목의 느낌을 주고 있고, 주변에는 주택들이 밀집되어 있다. 도림 고가도로를 지나면 초등학교가 있어서 어린이들의 왕래가 잦고, 이 고가도로 밑 하부 공간은 넓고 높아서 예전부터 교육시설과 놀이시설을 갖춘 어린이들을 위한 공간으로 사용되어 왔었으나 지금은 그 기능을 잃었다.

The space under the Dorim Overpass is located below the existing four-land road. Roadways or pavements are not extention, which gives an impression of a dead-end while residential houses are concentrated in the vicinity. There is an elementary school if you pass by the Dorim Overpass, involving frequent comings and goings of children while space below the Overpass is wide and high. So this space has been a place for children covering children's facilities for education and entertainment for a long time. However, this function has been lost by now.

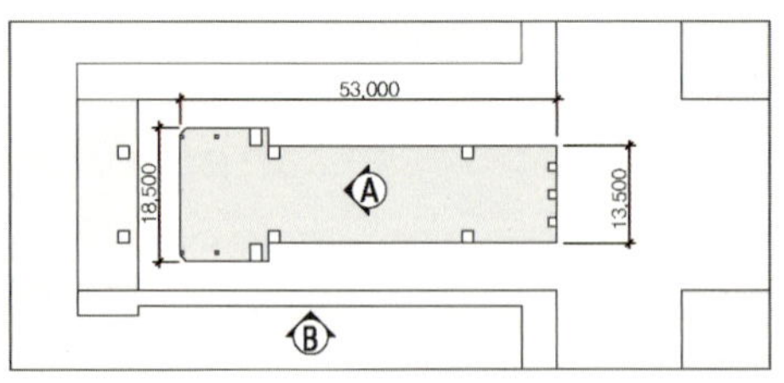

FLOOR PLAN

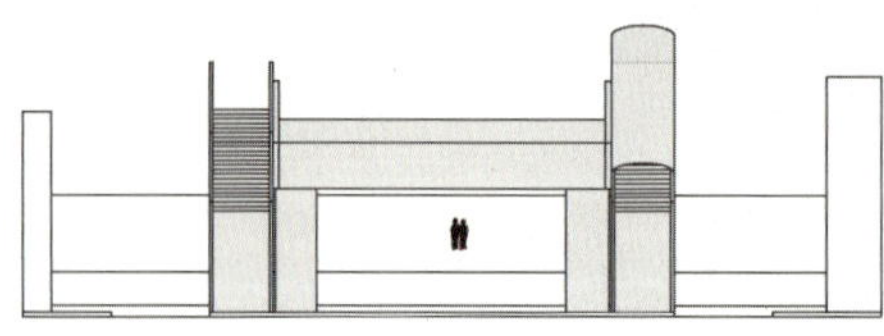
ELEVATION - A

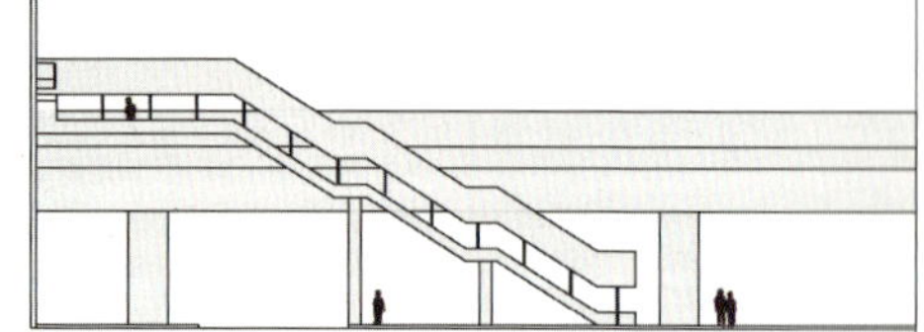
ELEVATION - B

SITE PROBEM

초등학교를 가기위해서는 도림고가도로를 지나쳐야하기 때문에 아이들의 왕래가 잦다. 그 곳은 광장처럼 넓고, 차도 별로 다니지 않기 때문에 비교적 안전한 공간임에도 무용지물로 사용되고 있다.

You will need to pass by the overpass to reach this elementary school, which involves a lot of comings and goings of children. Although this place is as wide as a plaza and there are few cars passing by, which makes it a relatively safe area, it has become useless.

TARGET

초등학생
Elementary school

PLAYING AREA

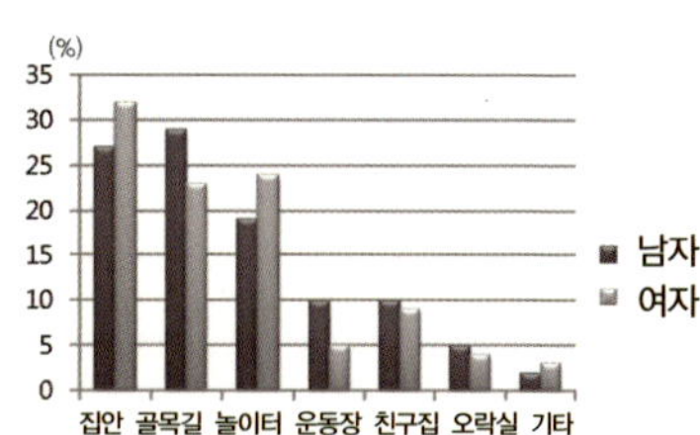

CONCEPT

學遊

學遊[학유] : "배울학" + "놀유" [놀이를 통한 어린이의 학습]
學遊[hakyu]: "learn" + "play" [Children's learning through play]

GOAL

디지털화되고 규격화 되어있는 도시에 살고 있는 아이들에게 아날로그적이고 유기적인 옛것과 자연의 교육을 도입시킨다. 아이들의 호기심을 자극하여, 체험과 놀이를 할 수 있게 하고 체험과 놀이 안에 학습을 넣어 놀이를 통한 재미있는 학습을 체험하게 한다.

The aim of this project includes the introduction of analogue, organic, old and natural education to children living in digitalized and standardized cities. This is to stimulate children's curiosity, enable their experiences and play, include them in learning, and help them to find learning exciting.

MOTIVE

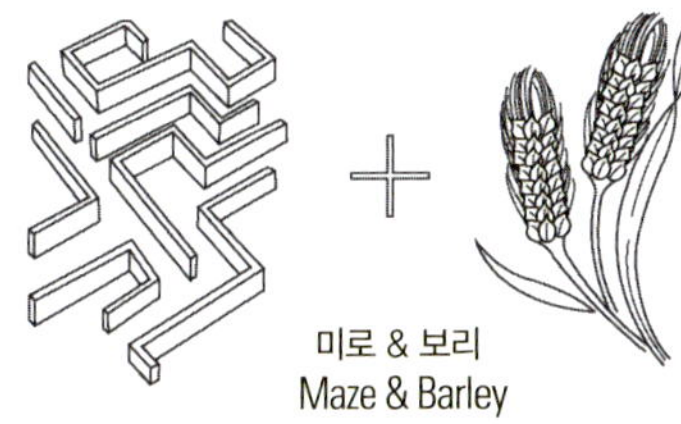

옛날에 이곳은 보리밭 이었다. 보리와 같은 곡물들을 아이들에게 알려주고자 한다. 그리고 이곳의 아이들에게 흥미를 끌 장치로 미로를 선택 하였다.

In the past, this area provided fields of barley. So the aim includes the introduction of grains such as barley to children. Also mazes were chosen as the media to help children to get interested.

SHAPE

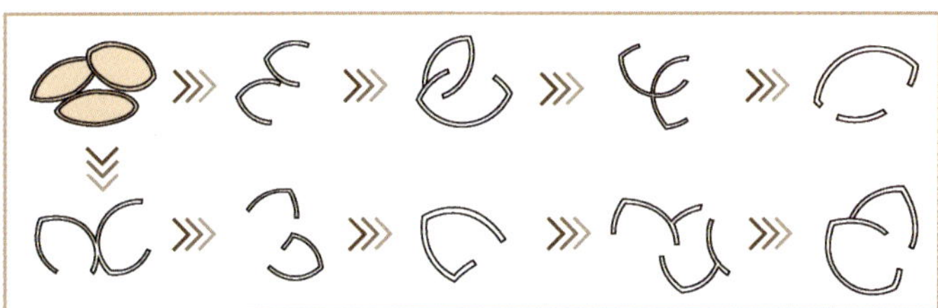

FUNCTION

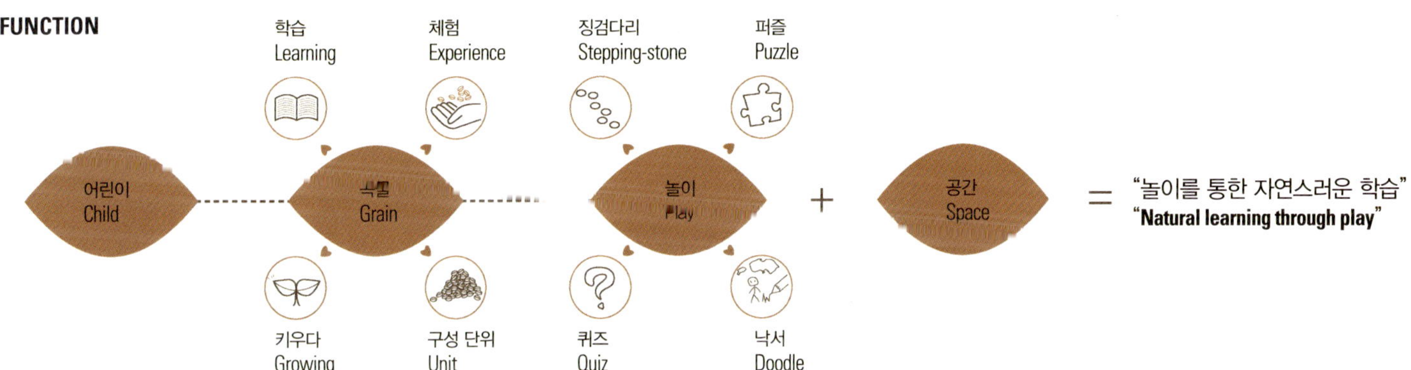

FUNTION PROCESS

▶ 보리뿌리기 Barley seeding

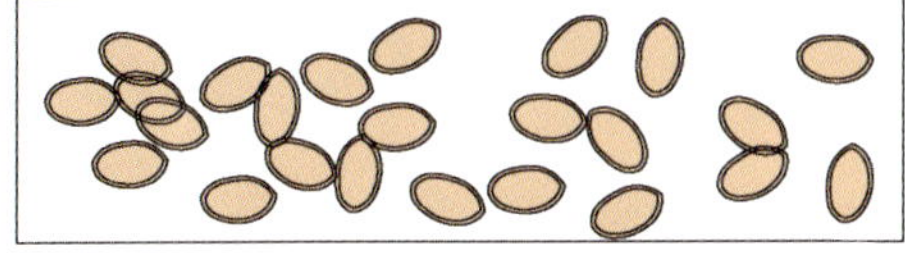

보리를 심는 방법을 적용해 뿌린다. 비규칙적으로 자리잡게 된다.

The method of barley planting will be applied. The arrangement will be established irregularly.

▶ 구역(기능)나누기 Section (feature)breaks

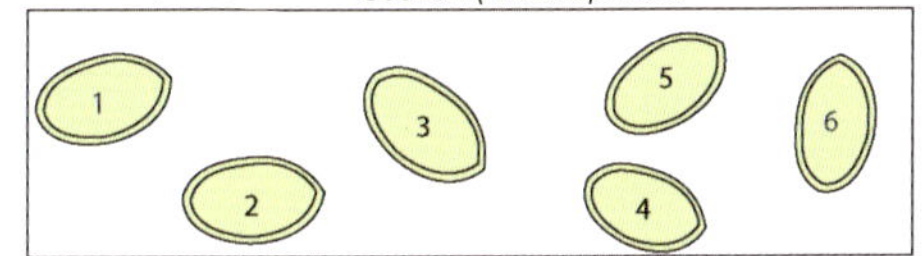

간격과 위치를 잡고, 총 6개의 구역으로 나눴다.

The intervals and locations were decided and the area was divided into six zones.

▶ 필요에 따른 형태 Forms to meet the needs

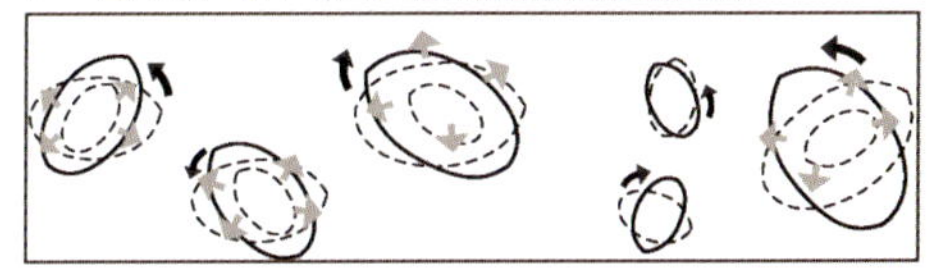

기능과 필요에 따라 형태를 잡기위해 크기조절을 하고 회전을 시켰다.

In order to set up forms according to functions and needs, the sizes were adjusted and rotation took place.

▶

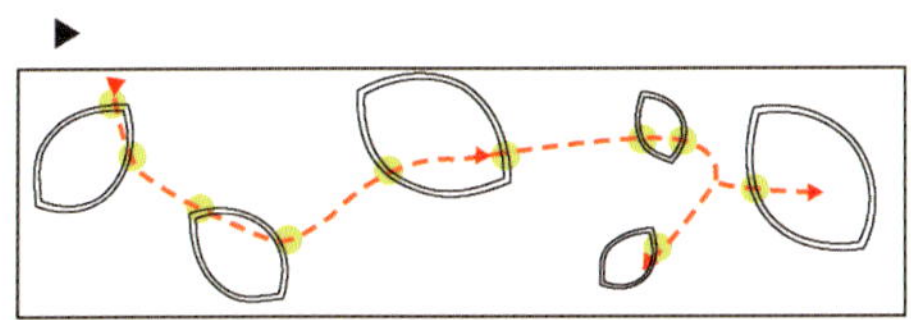

동선을 분석하고 출입구 위치를 선정하였다.

Movement lines were analyzed while the entrance location was selected.

▶

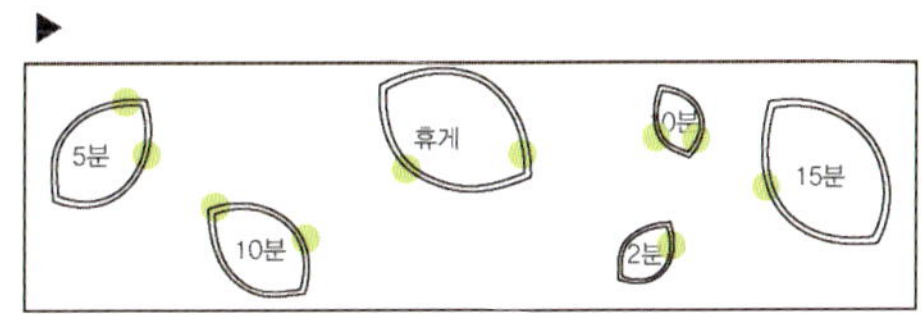

각 프로그램별로 소요되는 시간과 충당할 수 있는 인원수를 분석해 보았다.

The number of people and length of time necessary for each program were analyzed.

▶

기능을 담지 않은 보리알 모양의 유닛들을 배치하였다. 이것들은 사용자의 의도에 따라 사용된다.

Functionless barleycorn-shaped units were arranged, which will be used according to a user's intent.

SPACE COMPOSITION

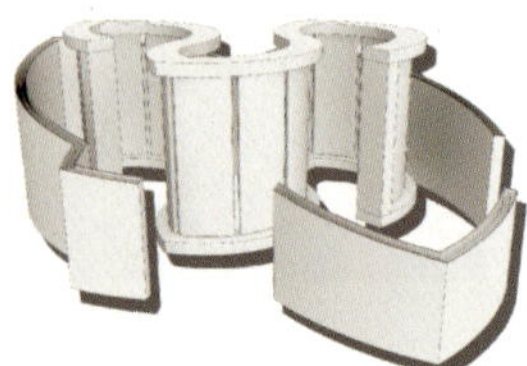

곡물패널
앞면에는 씨앗이, 뒷면에는 다 자란 식물이 그려져 있다. 아이들은 패널을 밀고 돌릴 수 있다.

Grain pannel
You can see seeds on the front while a fully-grown plant is placed on the rear. Children can push and turn the panels.

징검다리
각각 유닛에 문제가 제시되어 있다. 틀린답의 디딤판은 밑으로 들어간다.

Stepping-stone
You can see the maze game installed under the overpass at one glance. Problems are presented in each unit of stepping-stones.

광장
중심이 되는곳에 설치되어 여럿이 모일 수 있고, 시야를 다양한 곳으로 분산시킬 수 있다.

Plaza
This is installed in the center so many people can gather while the view can be dispersed to various directions.

전망대
고가밑에 설치된 미로게임을 한눈에 볼 수 있다.

Observatory
You can see the maze game installed under the overpass at one glance.

LEFT VIEW

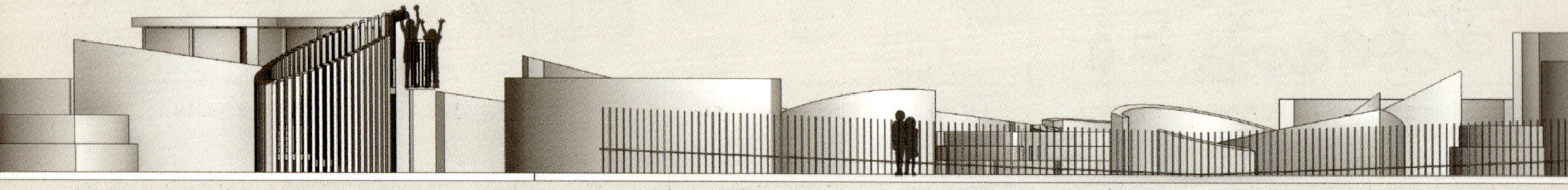

RIGHT VIEW

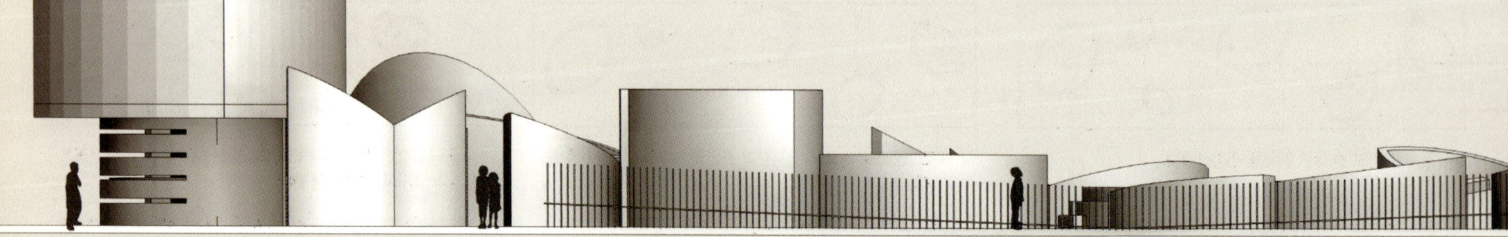

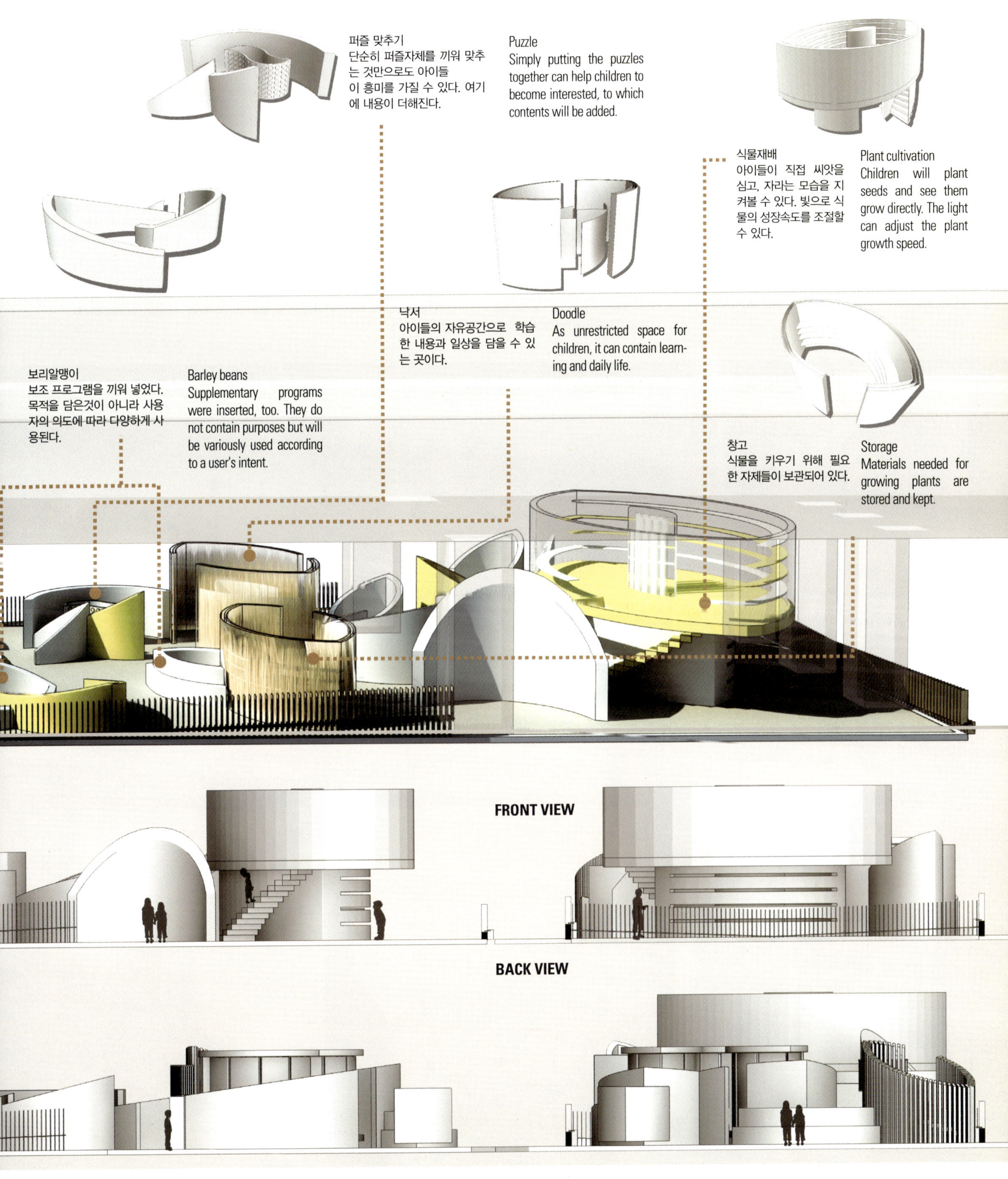
퍼즐 맞추기
단순히 퍼즐자체를 끼워 맞추는 것만으로도 아이들이 흥미를 가질 수 있다. 여기에 내용이 더해진다.
Puzzle
Simply putting the puzzles together can help children to become interested, to which contents will be added.
식물재배
아이들이 직접 씨앗을 심고, 자라는 모습을 지켜볼 수 있다. 빛으로 식물의 성장속도를 조절할 수 있다.
Plant cultivation
Children will plant seeds and see them grow directly. The light can adjust the plant growth speed.
낙서
아이들의 자유공간으로 학습한 내용과 일상을 담을 수 있는 곳이다.
Doodle
As unrestricted space for children, it can contain learning and daily life.
보리알맹이
보조 프로그램을 끼워 넣었다. 목적을 담은것이 아니라 사용자의 의도에 따라 다양하게 사용된다.
Barley beans
Supplementary programs were inserted, too. They do not contain purposes but will be variously used according to a user's intent.
창고
식물을 키우기 위해 필요한 자제들이 보관되어 있다.
Storage
Materials needed for growing plants are stored and kept.
FRONT VIEW
BACK VIEW

SPACE PROGRAM

SORT

Learning + Road

앞면에는 씨앗형태의 곡물과, 뒷면에는 다자란 곡물의 그림이 있음.

Children's turning panels will bring about changes in the roads (movement lines).

OX QUIZ

Stepping-stone + Unit

옳은 답이 적힌 징검다리 일경우 안전한 디딤역할. 올바르지 않은 징검다리로 올라서면 불안정함.

You can cross by using stepping boards on which information on grains is provided. If you use a wrong stepping board, the stepping-stones will be directed downwards.

FREE

User + Purpose

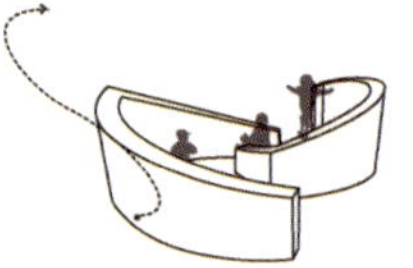

다양한 형태의 유닛들이 배치되어 있어서, 사용자의 의도에 따라 여러가지로 사용된다.

By directly inserting the block bearing the answer into the wall bearing the question, children can learn while playing.

INTERACTIVE SPACE VIEW

Puzzle
+
Quiz

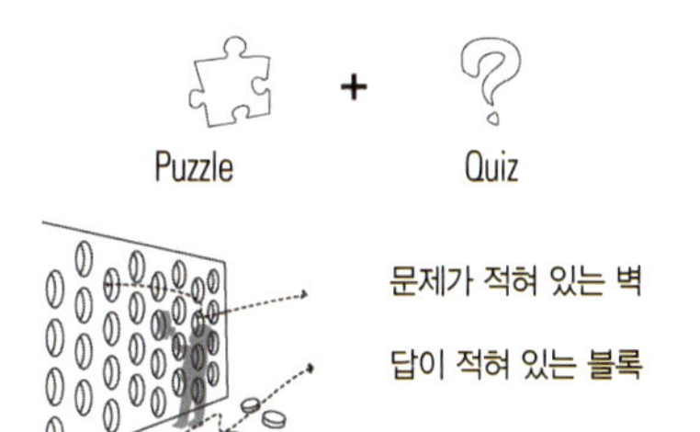

문제가 적혀 있는 벽
답이 적혀 있는 블록

In this space, children will write down what they have learned while playing on the glass wall or just scribble as learning involving play.

Doodle
+
Learning

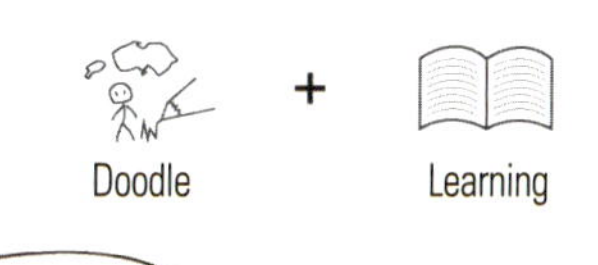

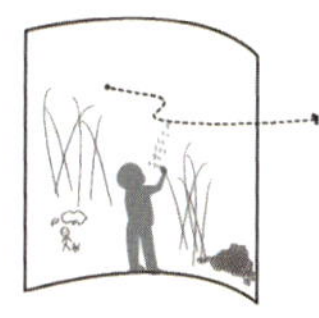

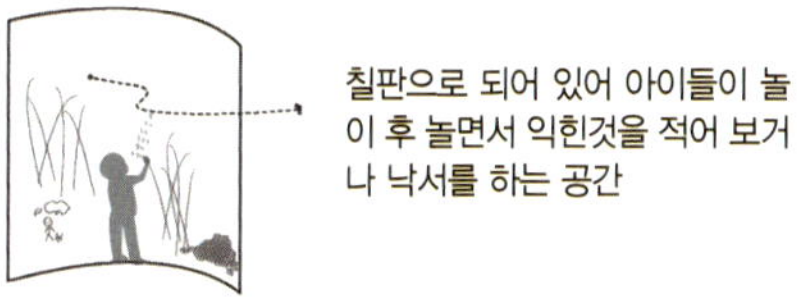

칠판으로 되어 있어 아이들이 놀이 후 놀면서 익힌것을 적어 보거나 낙서를 하는 공간

One child will be provided with one flowerpot so that certain space to tend a plant can be provided.

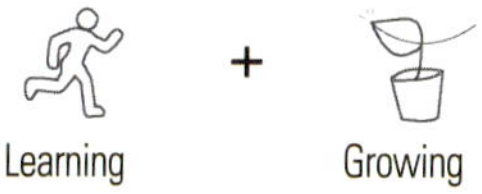

Learning
+
Growing

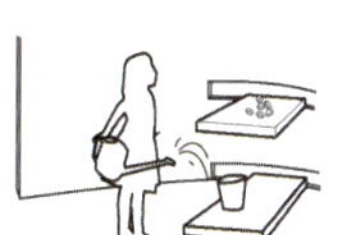

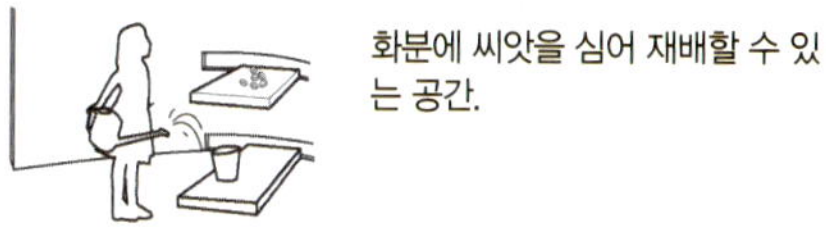

화분에 씨앗을 심어 재배할 수 있는 공간.

The surrounding walls can bear various characteristics based on users' intents while they play or rest by the functionless walls in the middle of moving from space to space.

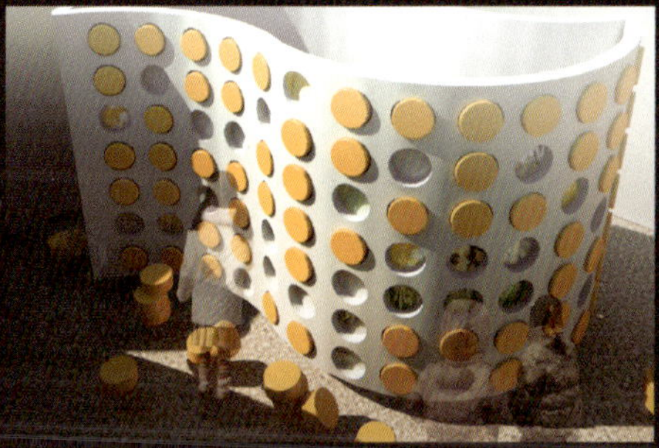

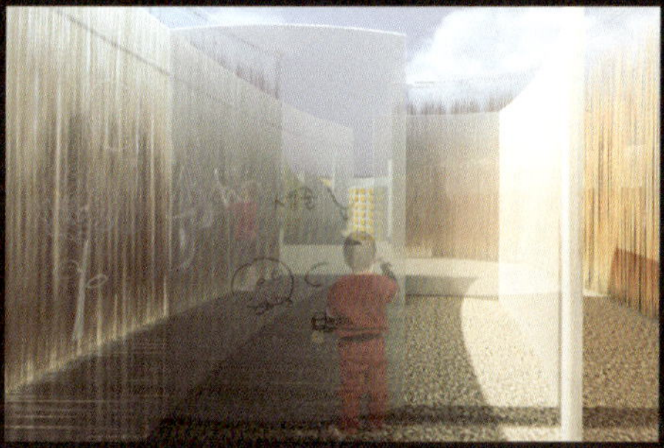

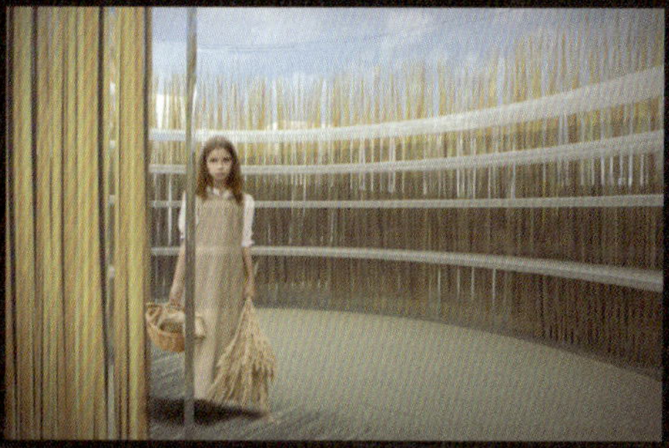

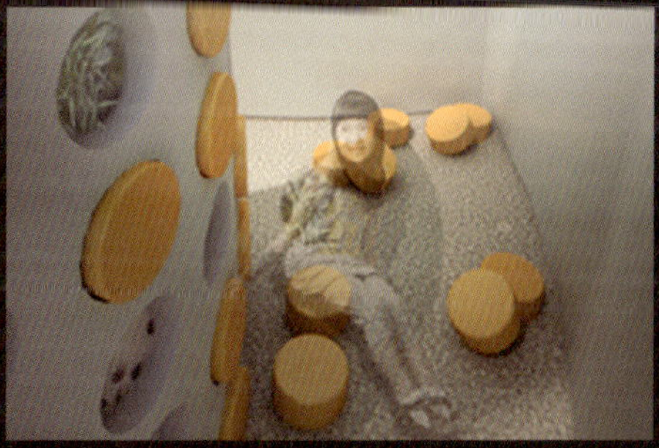

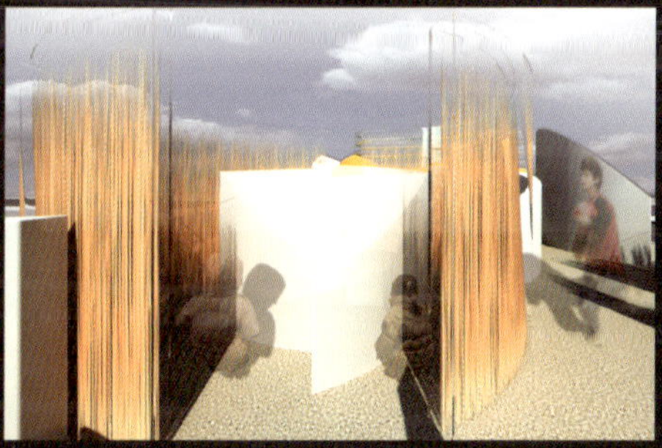

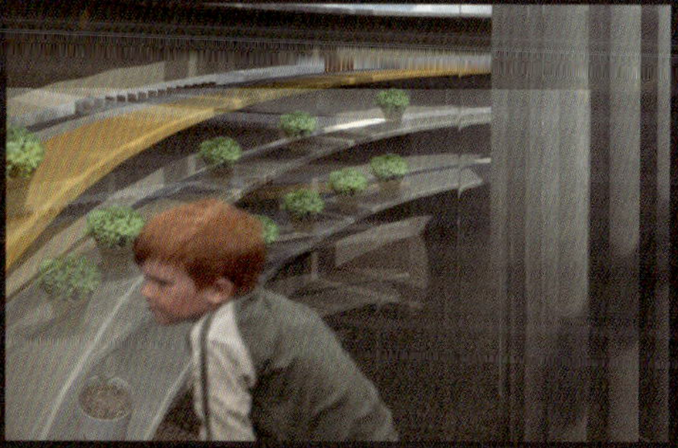

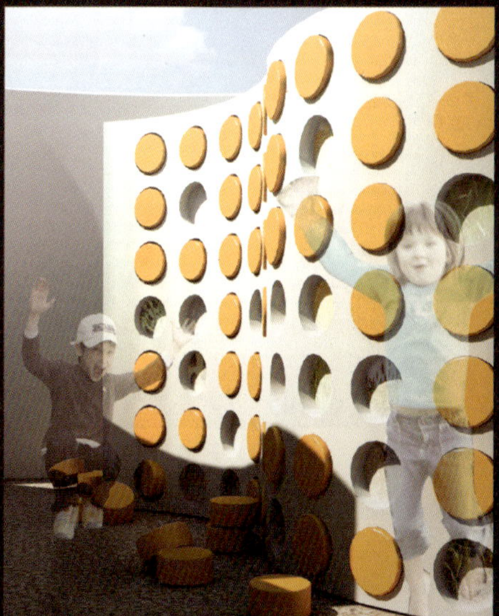

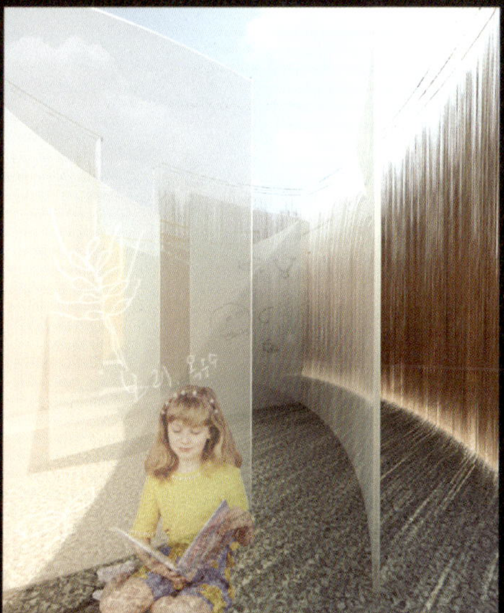

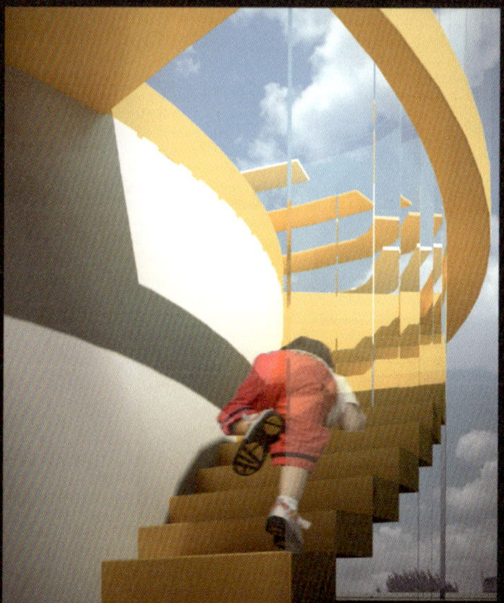

영상앞에서 Wii를 이용해 도리깨질을 간접적으로 체험을 할 수 있는 공간이 있다.
There is space where you can indirectly experience flailing by using Wii in front of images.

특정기능이 없는 공간에서 사용자의 행위가 이뤄지면 기능이 있는 공간으로 성격을 변하게 한다.
In space without specific functions, if a user's behavior takes place, the space will change to have certain functions.

식물을 키우기 위해 필요한 자제들이 선반 위에 보관되어있다.
plant growing are kept on the shelves.

미로처럼 다양한 형태의 벽들이 존재한다.

There are walls that are as variously formed as mazes.

정면에서는 보이지만 측면에서는 각제가 중첩되어 보이지 않는 공간이다.

While the facade can be seen, the sideways will be overlapped and can't be seen in this space.

사용자가 직접 들어오는 출입문을 만들어 낸다.

A user will create an entrance on his/her own.

SEE SAW SEEM

과거, 현재, 미래를 보다.

BACKGROUND

교통체증을 줄이기 위해 세워진 FLYOVER. 우리는 FLYOVER 하부 공간이 방치되어 있거나 낭비되고있는 것을 주변에서 손쉽게 볼 수 있다 이에 우리는 FLYOVER 밑 유휴공간을 경관적으로 개선하여 주민들에게 돌려주고, 나아가 그 지역주민들이 활용할 수 있으며, 지역활성화에 영향을 미칠 수 있는 방안을 찾아보려 한다.

There are lots of overpasses to reduce a traffic jam. However, we can easily see the under space of overpasses is being wasted. So, we try to find out the way to improve it with a grand view for residents. It could be utilized by residents, and effect to invigorate an area.

PROPOSAL

과거 맑은 물이 흐르던 안양천은 산업화로 인한 생태계 파괴로 지난 50년간 생태 복원 프로그램을 시행하고 있었다. 이에 현재는 생태하천으로 변모하여 철새 보호구역으로 지정되었으며 매년 안양천으로 찾아오는 철새의 종 수는 증가하고 있다. 이에 우리는 시 자체 철새 관련 프로그램과 연관하여 이 지역을 찾아오는 사람들에게 철새의 모습을 보여주며 매년 변화해 나가는 안양천의 자연을 관측하고 볼 수 있는 공간을 만들고자 한다.

Anyang-steam was clean before, but it was polluted by industrialization. Finally The Ecological Restoration Program has been existing for 50 years. Presently, Anyang -steam which is changed to Ecological steam, has been designated as a migratory bird sanctuary, and the number of species of migratory birds, stop in Anyang-steam, has increased annually. So, with this program, we want to make a space which can show a migratory bird sanctuary, and see a process of changes of Anyang-steam.

OBSERVATORY

관측

휴식

전시

SITE ANALYSIS

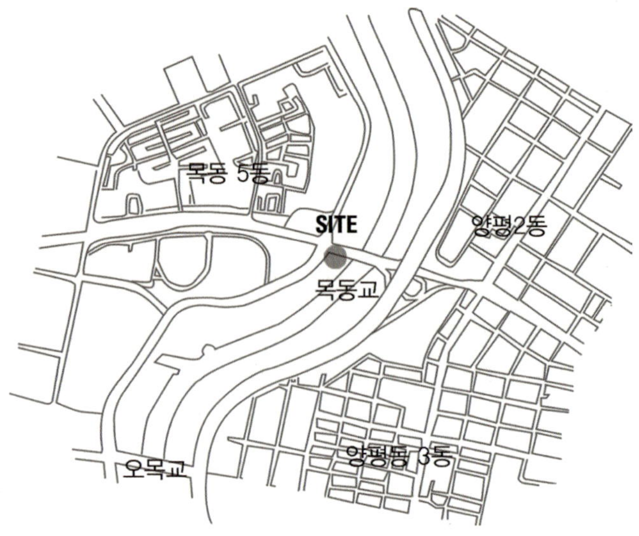

서울 양천구 목동 목동교

목동교
양청구와 영등포구를 가르는 안양천을 잇는 다리

목동
전형적인 농촌지역으로, 작은 마을 단위로 공동체를 이루고 살았으나 1970년대부터 대규모 주택단지로 바뀌기 시작하였다. 1983년 목동지구 신시가지 개발 계획이 발표되었고, 1988년에 신시가지가 완성되었다. 목동종합경기장, 서울국제우체국, 서울세관출장소, 기독교방송국, SBS, 현대백화점 목동점, 파리공원 등이 있다.

The Mocdong Bridge, Moc-dong, Yangcheon-gu, Seoul

The Mocdong Bridge
It cut through Anyang-steam between Yangcheon-Gu and Yeongdeungpo-gu

Mocdong
This typically was a agricultural area. This area consisted of small village communities. Since 1970's, it has been changed to a large scale of a collective housing area. In 1983, a project of Moc-dong New Town was announced, and it was completed in 1988. Mok-dong Stadium, Seoul International Post Office, Seoul Custom station, Christian Broadcasting System, SBS, Hyundai-department store, Paris Park etc,, are built in that time.

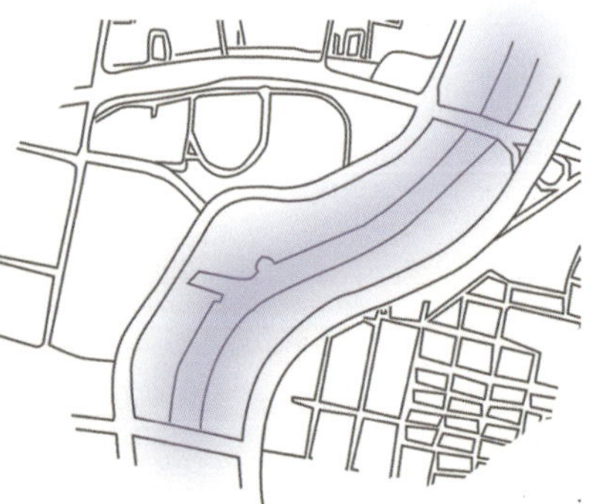

안양천

안양시 도심을 중앙으로 관류하여 광명, 서울시를 거쳐 한강에 유입되는 하천
면적 : 286
하천연장 : 32.5

Anyang-steam
A steam flows into Han-river, and the center of Anyang town through Gwangmyeong and Seoul city.
Area : 286
Length : 32.5

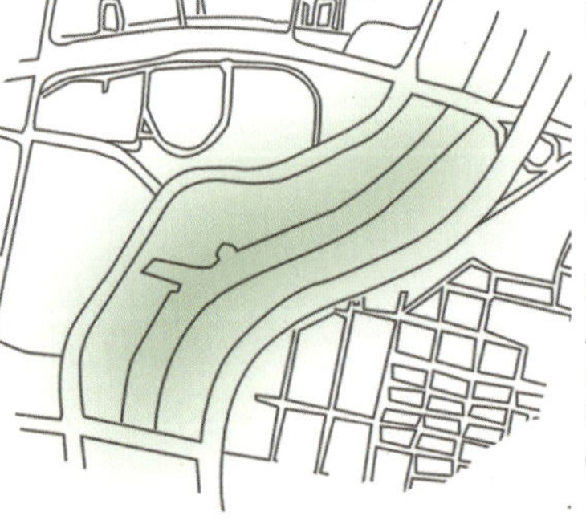

철새보호구역

서울시는 안양천 하류 지역을 서울시에서 3번째 철새보호구역 으로 지정했고 자연생태 학습 공간으로 적극 활용해 나간다는 계획이다.

Seoul City has designated the lower Anyang-steam area as third migratory bird sanctuary in Seoul, and the spcace has efficiently used for a natural ecology study.

DESIGN CONCEPT

둥지를 틀다

빈사상태의 하천에서 생태하천으로 변모한 안양천 이는 자연으로의 완벽한 회귀가 아닌 시작이라는 것을 공간의 변화로 보여주고자 하며, 둥지를 틀다는 진행형 컨셉으로 지어지고 있는 둥지가 가지고 있는 변화의 속성과 미완성적인 구성을 공간에 담아보고자 한다.

Building a nest

We want to show that Anyang-steam, changed from the moribund steam to ecological steam, is the beginning, not completely return to the nature by changing spaces. We try to express attributes of a change and incomplete constitutions into the space.

KEYWORDS

HANG	나무나 지붕위에 매달려있는 둥지의 특성을 반영	Reflecting a trait of a nest, is hanging from trees or roofs
IRREGULAR	불규칙 적인 패턴의 변화	Change of patterns irregularly
PILE	나뭇가지를 쌓아올린 둥지의 구조적 특성 반영	Reflecting a structural trait of a nest, is piled with branches
CURVE	자연에서 가져온 유기적 곡선 반영	Reflecting organic curves from nature

APPROACH

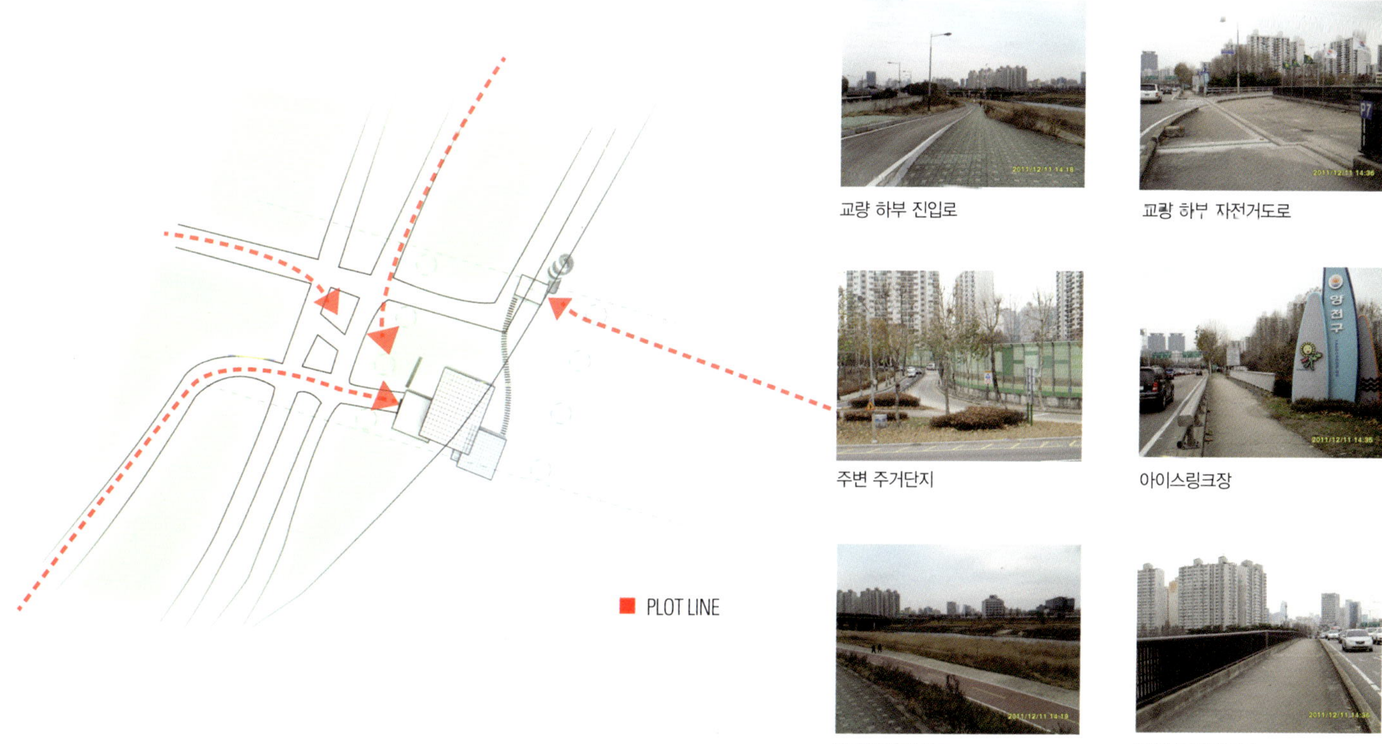

교량 하부 진입로

교량 하부 자전거도로

주변 주거단지

아이스링크장

주변 자전거도로

교량 상부 진입로

MASS PROCESS

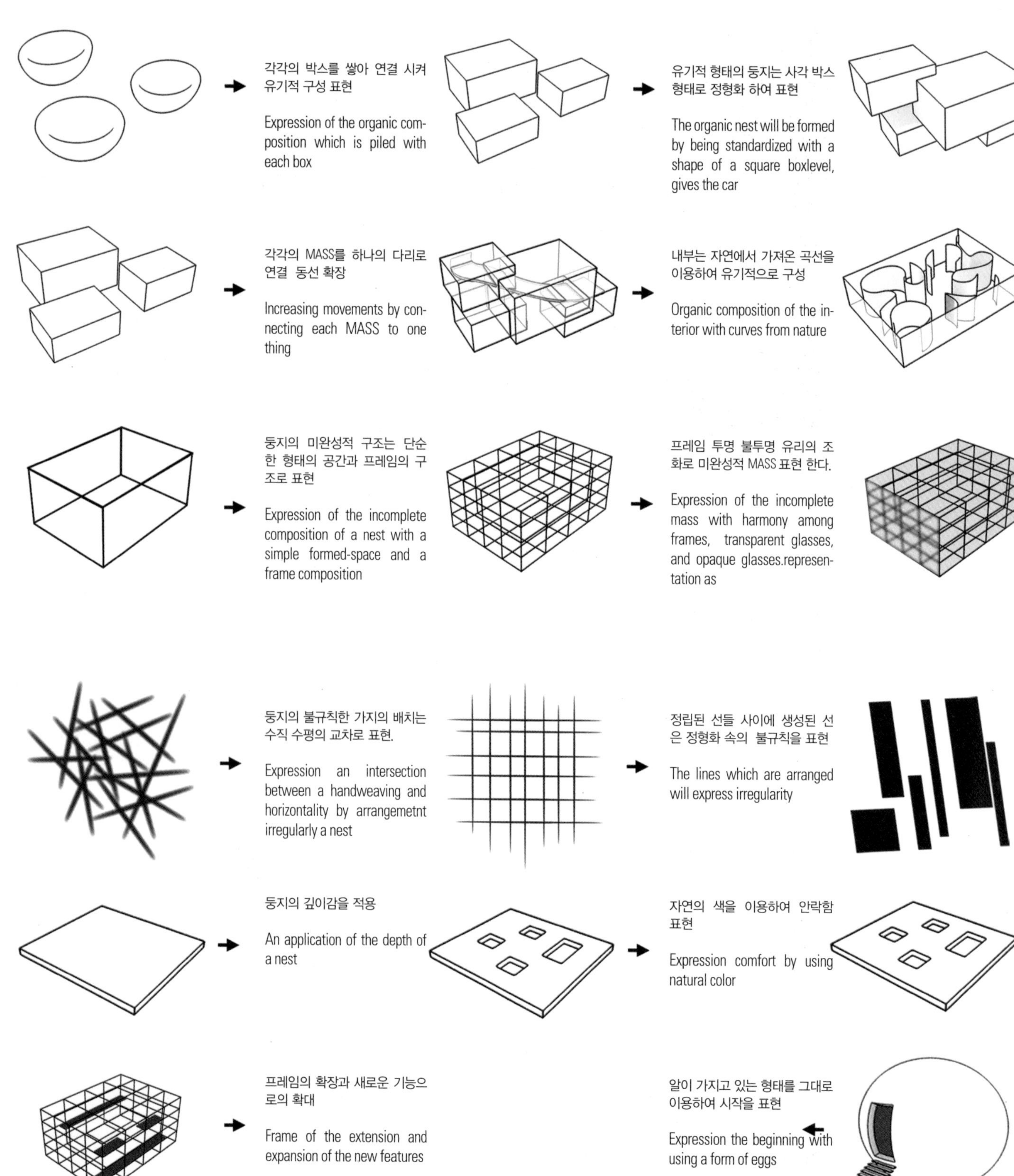

01

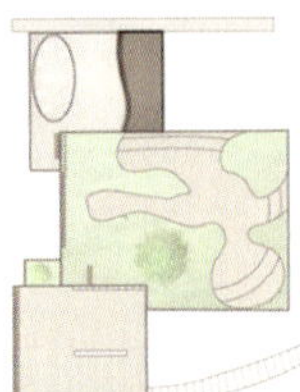

입구 부분은 아무것도 존재하지 않은 좁은 통로로 하여 과거 아무것도 존재하지 않은 안양천을 되새기게 하며 허전함을 전해준다.

The entrance part will be a narrow passageway with nothing, and it will make people recall the empty Anyang-steam in the past through

03

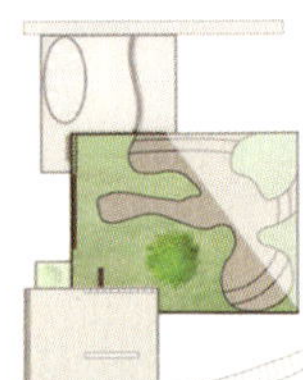

유기적인 벽면과 그 위에 존재하는 철새들은 서로 자연스럽게 존재한다.

The wall, above which there is organic to each other migratory birds, there is naturally

02

입구 부분은 아무것도 존재하지 않은 좁은 통로로 하여 과거 아무것도 존재하지 않은 안양천을 되새기게 하며 허전함을 전해준다.

The entrance part will be a narrow passageway with nothing, and it will make people recall the empty Anyang-steam in the past through

04

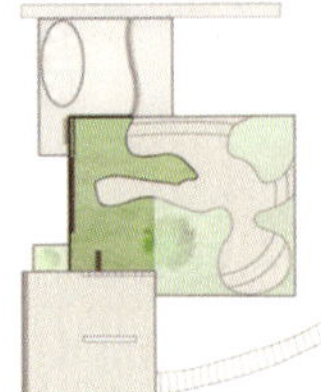

전시 관람과 휴식이 자연스럽게 연결되도록 되어 있으며 외부의 프레임은 내부로 들어와 가구를 형성하게 한다.

The view of exhibition and the space of rest are naturally connected, and outside frame will be furniture inside.
Making small nests in a next will make a comfortable space to let people take a rest.

MASS PROCESS

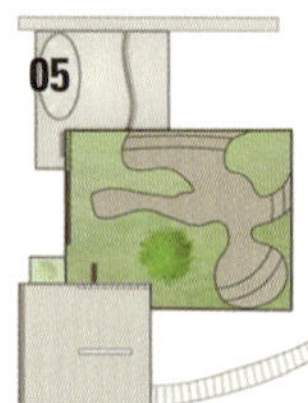

05

예측할수 없는 유기적 형태의 공간과 통로는 기하학적인 외부와는 대조를 이루며 다양하고 복잡한 동선을 만들어 숲속 길을 걷는 듯 하게 해준다.

Unpredictable organic space and passageway will contrast to geometric external by making various and complex movements like as people are walking in a forest.

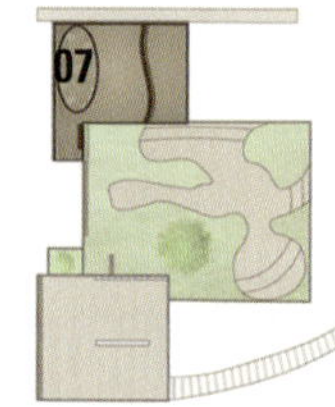

07

전시공간을 지나고 데크에 들어서면 외부의 자연과 서로 소통하며 휴식을 취할 수 있다.

If you enter deck through the exhibition space, people can take a rest by communicating with nature.

06

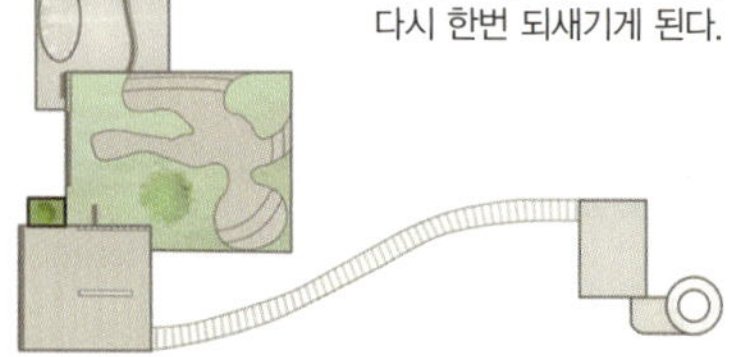

자연을 직접적으로 느낄수 있는 실내 정원이 존재하여 자연의 소중함을 다시 한번 되새기게 된다.

08

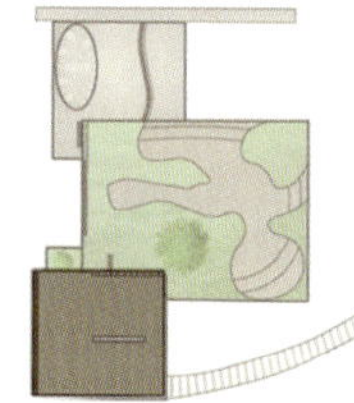

전시 관람과 휴식이 자연스럽게 연결되도록 되어 있으며 외부의 프레임은 내부로 들어와 가구를 형성하게 한다.

The view of exhibition and the space of rest are naturally connected, and outside frame will be furniture inside.
Making small nests in a next will make a comfortable space to let people take a rest.

ELEVATION

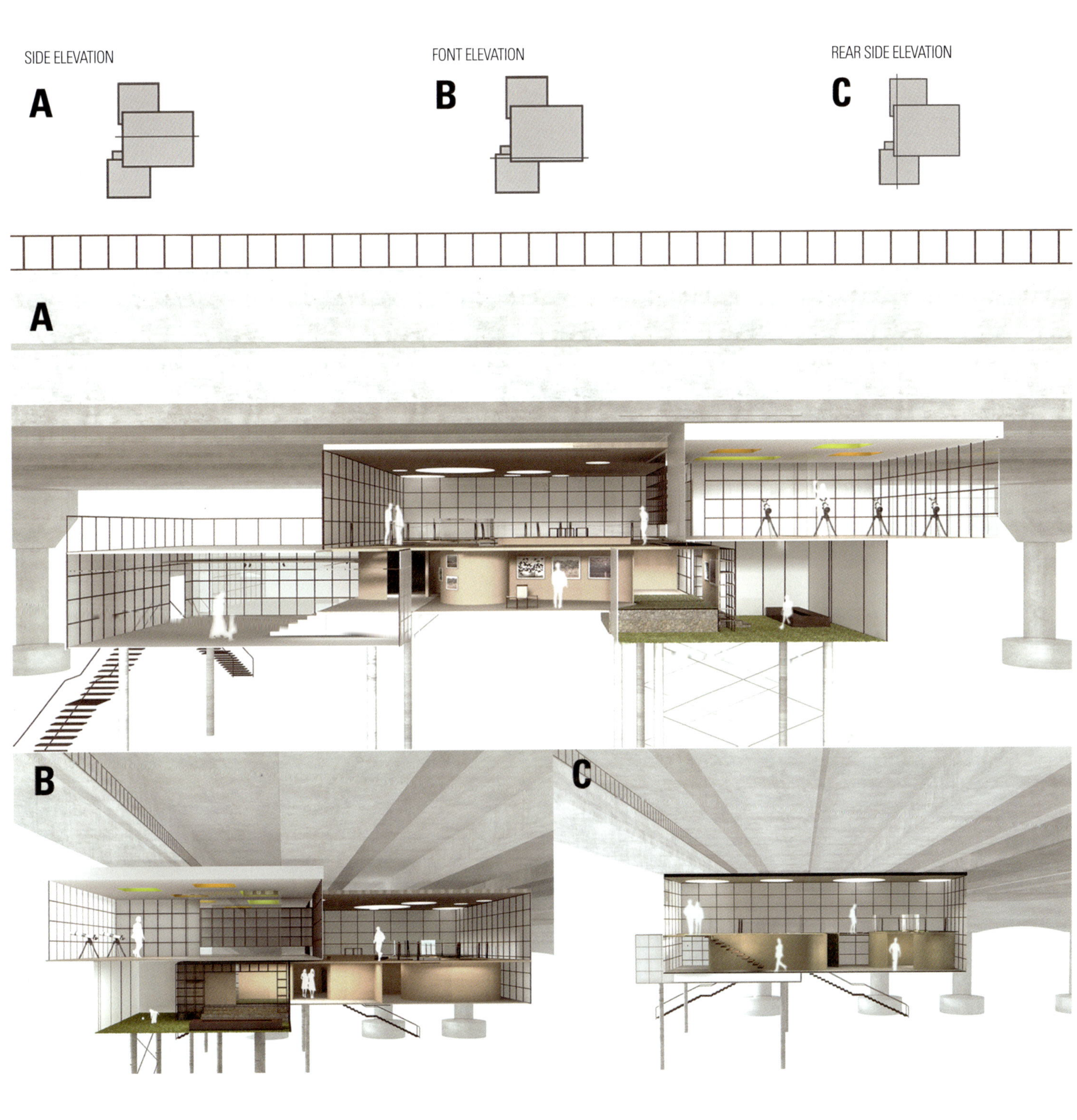

쌓아놓은 듯한 기하학적 매스들이 유기적으로 연결되어있음을 알 수 있다.
The geometrical space connect with other space flexibility.

예측할 수 없는 공간 각각의 공간을 연결 하여 주는 다리는 다양한 뷰를 이용자들에게 제공한다.
외부에서 뻗어져 나오는 프레임은 새로운 공간이 되고 가구가 되어 내부와 소통한다. 이러한 요소들은 한데 어우러져 자연적 공간이 되고, 도시의 사람들이 간접적으로 자연을 느낄 수 있는 공간이 된다.

unexpectable space and each space whichis connected, provide various views The frame will be a new space from outside, furniture inside.
Those characteristic things, could indirectly make people feel the nature.

전망 공간은 사면을 유리로 하여 다양한 뷰를 제공해 주며 철새를 관측 하며 철새뿐만 아니라 안양천의 자연을 보고 교류할수 있는 공간이다.

Views of the slopes of glass area gives you a variety of views, as well as migratory birds, migratory birds, and Anyang and the nature of observations reported is the place where you can exchange.

첫번째 전망대와 연결되어있는 전망공간으로 더 넓은 뷰를 확보해주는 공간.

As the space connect with first overview, It is space that guarantee more board space.

URBAN OASIS
자전거이용객들은 도심 속 오아시스를 갈망한다.
Bicycle users are wanting "URBAN OASIS"

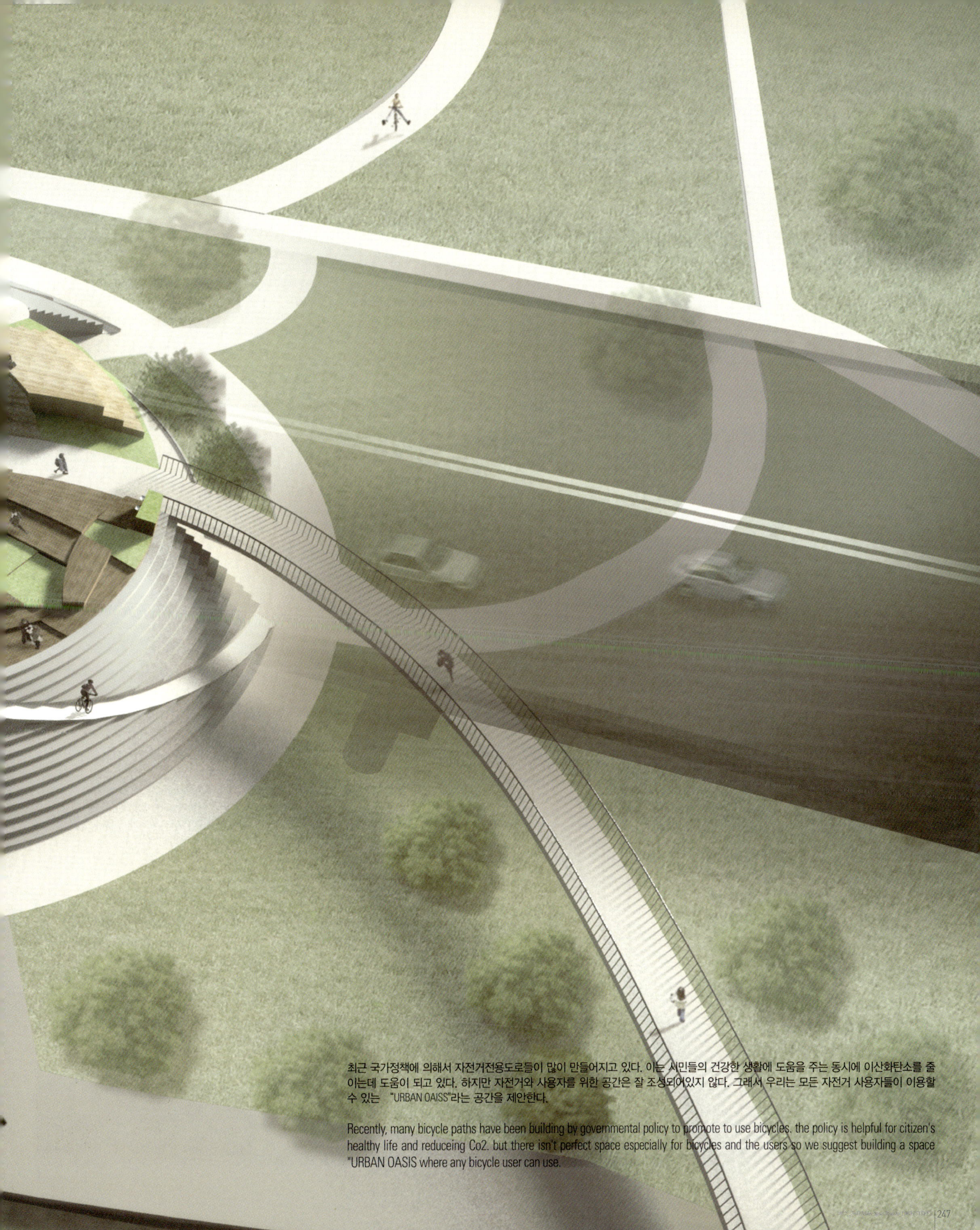

최근 국가정책에 의해서 자전거전용도로들이 많이 만들어지고 있다. 이는 시민들의 건강한 생활에 도움을 주는 동시에 이산화탄소를 줄이는데 도움이 되고 있다. 하지만 자전거와 사용자를 위한 공간은 잘 조성되어있지 않다. 그래서 우리는 모든 자전거 사용자들이 이용할 수 있는 "URBAN OAISS"라는 공간을 제안한다.

Recently, many bicycle paths have been building by governmental policy to promote to use bicycles. the policy is helpful for citizen's healthy life and reduceing Co2. but there isn't perfect space especially for bicycles and the users so we suggest building a space "URBAN OASIS where any bicycle user can use.

SITE

안양천은 총 34.8M로 안양시와 서울시 경계부터 한강 합류지점까지 국가하천으로 지정되어있으며 광명시와 서울 금천구 구로구 영등포를지나 성산대교 서쪽에서 한강으로 흘러든다. 신정교는 안양천에 위치해 있으며, 양천구 신정로와 영등포구 문래동을 이어주는 다리로 총 길이 635M, 폭은 26M이며, 신정교 주변에는 다양한 운동시설을 갖추고 있다. 또한 안양천과 도림천이 만나는 지점에 위치해 있어 사람들이 몰리는 지점이기도 하다. 주변에 주로 아파트 단지로 조성되어 있어 가족단위의 구성원이나 자전거이용객, 실제 자전거동호회 모임 장소로 활용 되고 있다.

Anyang and Seoul Metropolitan Government to a total of 34.8M from the boundary to the confluence of the Han River is designated as national and Gasan, Guro, Seoul, Youngdeungpo Gwangmyeong and west through the Sacred Mountain Bridge in Han River heulreodeunda.

Sinjeonggyo(Bridge) near by Anyang(stream) is connecting Sinjeong-dong, Yangcheon-gu with Mullae-dong, Youngdeungpo. total 635m width 26m various athletic facilities are occupied also It's the point people gather Anyang(stream) Dorimcheon(Stream) meeting point It is using gathering point for bicyle users, familys, because there are a lot of apartments complex.

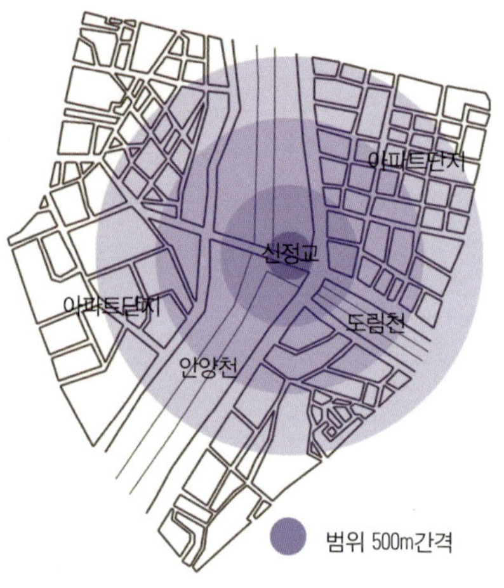

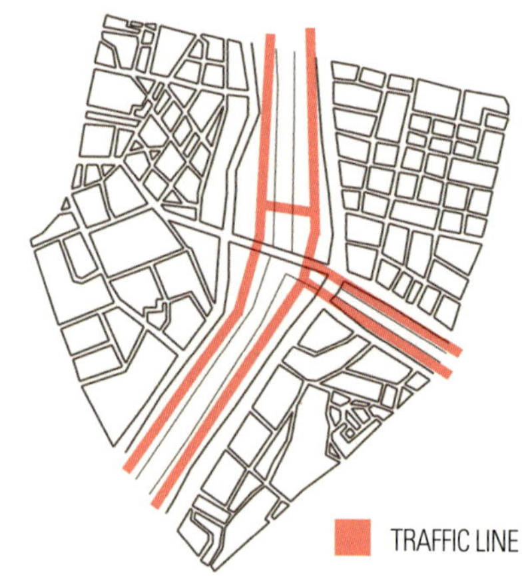

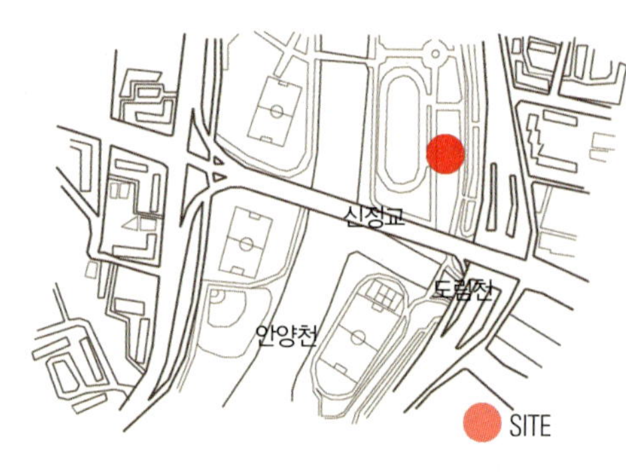

TARGET

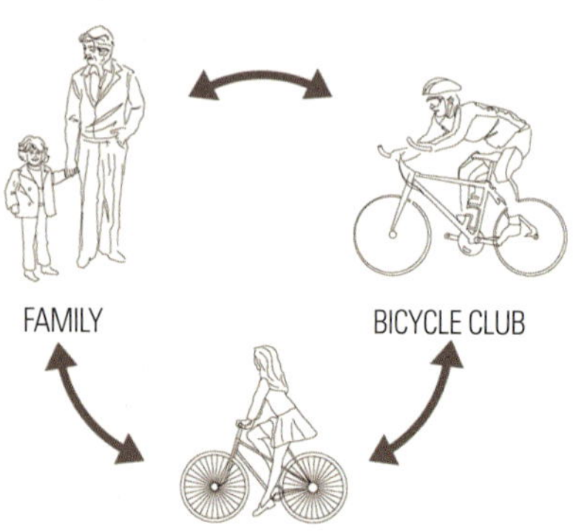

최근 자전거이용객들이 많아지면서 소수의 자전거 동호회만이 존재하였던 것이 점차 확대 되고 있으며, 서울시에서만 보더라도 한강을 중심으로 각각 하천들을 포함하여 자전거도로가 깔려있어 마음만 먹으면 자전거 하나로도 어디든 갈 수 있다. 하지만 정작 자전거를 위한 공간은 극소수이다. 특히 신정교 하부공간은 실제 자전거동호회의 모임장소로 사용이 되고 있으며, 자연스럽게 자전거용품을 판매하는 불법노점까지 생겨났다. 우리는 주 TARGET인 자전거 이용객의 불편을 보완 해줄 수 있는 공간을 제시한다.

Recently, bicycle user are increasing. only small bicycle clubs became big. we can go anywhere by using bicycle path. however, there is lest space for bicycles espeacially Sinjeonggyo(Bridge) clear head is using as gathering point. natually illegal stores for bicycle are increaing we suggest a space which supplement any inconvenience of bicyble users.

USE

DESIGN CONCEPT

WATER-COURSE

물이 한데 모여 개천이나 강으로 흘러가는 줄기

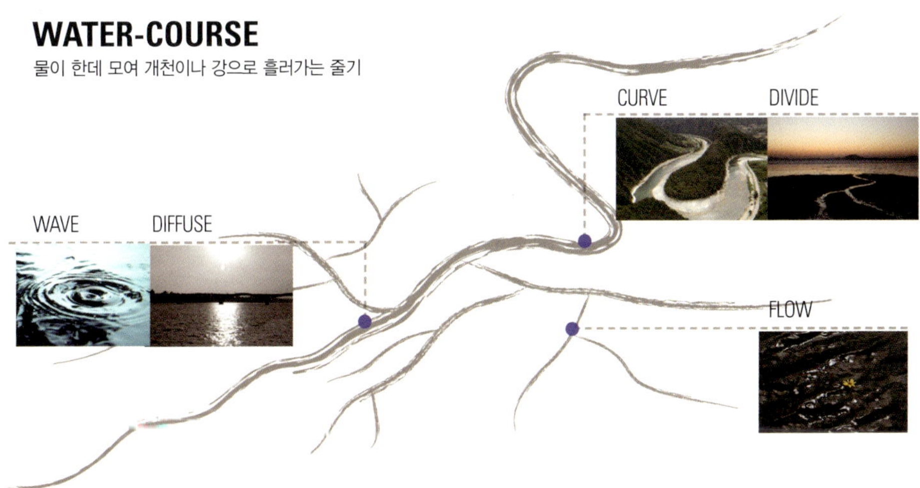

고유가시대 복잡한 도시에 지친 사람들은 자유를 갈망한다. 때문에 자전거동호회나 자전거이용객들이 늘어나는 추세이다. 우리는 이러한 사람들을 위해 좀 더 자유롭고 일상의 돌파구이자 활력소가 되어줄 수 있는 하천의 물줄기를 CONCEPT하여 함께 동화하는 공간을 만들어보고자 한다.

people who got a lot of stress from urban life want a freedom so bicycle clubs and usrs are increasing. we will make any space which is helpful for giving refreshment to these people our structure concept is stream pattern.

DESIGN

전체적인 흐름을 통한 공간의 변화 순환과정을 구체적인 사례로 제시
– 틀에서 벋어난 주변의 조건을 활용한 공간 제시

- illustrate space changing and the cycling through total proceess
- illustrat space using a unusual condition

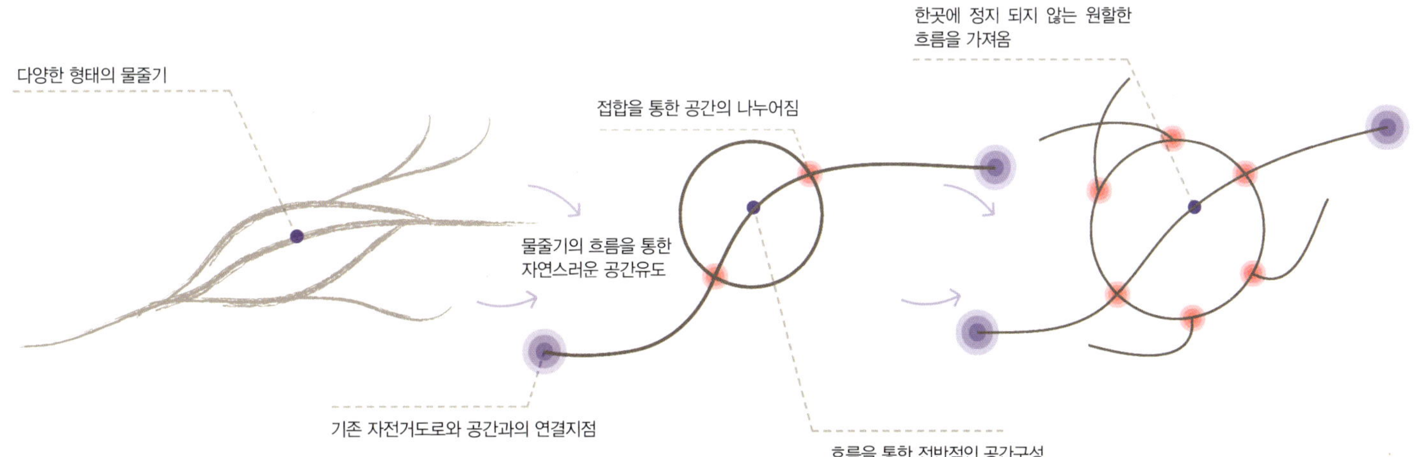

PROPOSAL 1

유기적인 선과 나누어짐

CURVE

일정한형태가 아닌 자연 그대로의 모습
The nature shape not artificail

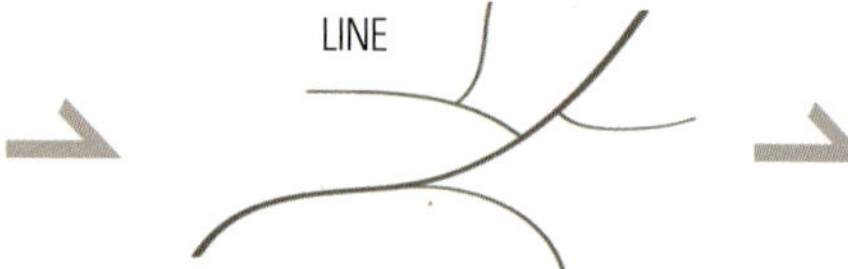

하나의 자연적 물줄기의 선 추출
Extract nature stream line

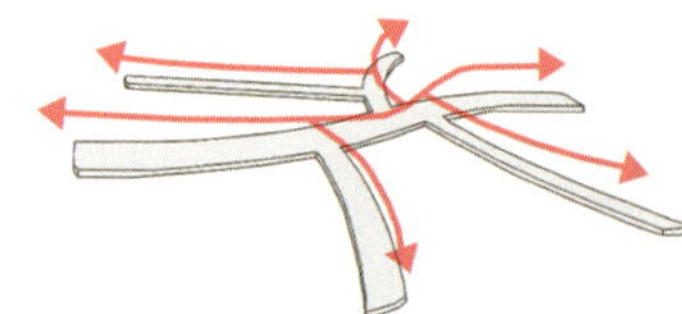

물줄기 형태에 의한 하나의 동선형성
Form moving line by the line

DIVIDE

불규칙한 선의 방향성
Direction temper by irregular line

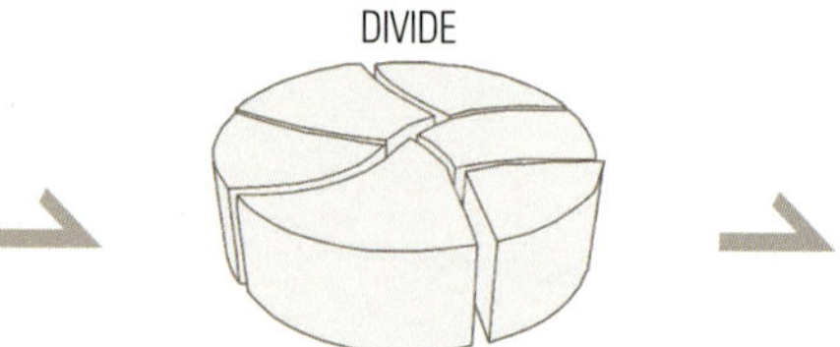

불규칙한 선에 의한 각각의 객체간의 나뉨 현상 발생
Objects cracking by irreglar line

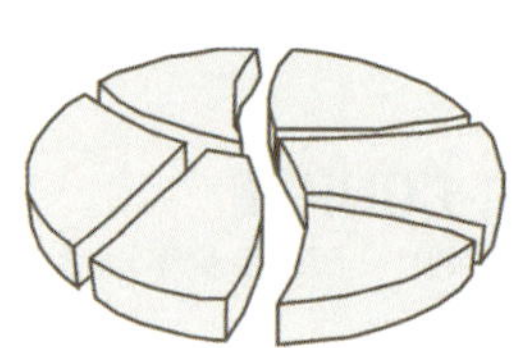

선의 나누어짐으로 각각의 공간형성
Forming each space by cracking

PROPOSAL 2

유기적인 선과 나누어짐

WAVE

규칙적인 파동의 패턴
The patter of regular wave

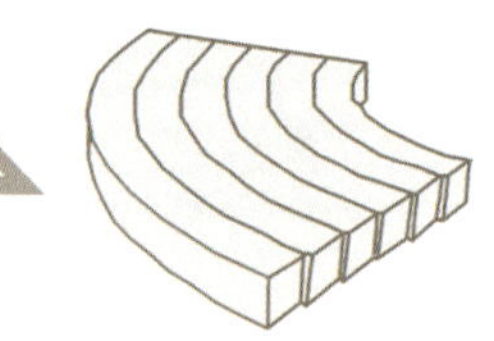

패턴의 구체화
Patter's materialization

물의 속성인 파동
Wave water's temper

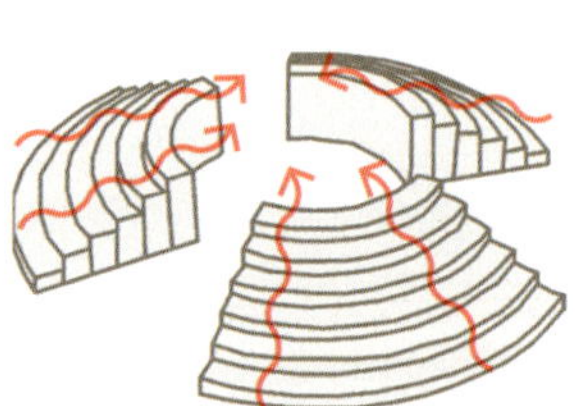

한 곳을 중심으로 퍼져나감
Spreading from the center one point

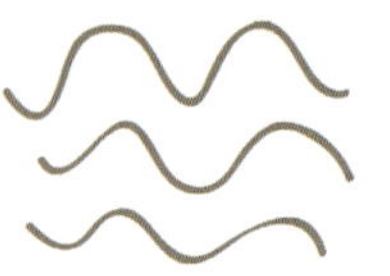

파동의 높낮이
Wave's undulating

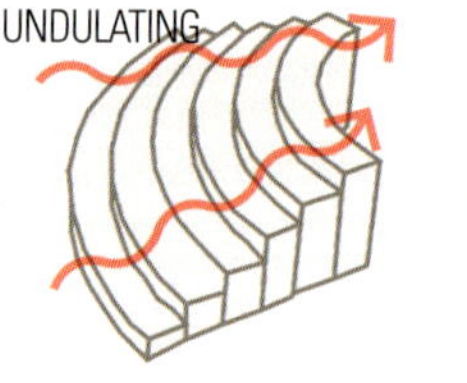

패턴의 높낮이를 이용한 단 형성
Formation using the undulation

PROPOSAL 3

물의 흐름 통한 공간유도 및 순환

FLOW

수로를 따라 한방향으로 흐르는 물
Water along a waterway

수로에 다른 수로로 인한 흐름 유도
Induce other waterway

흐름의 유도를 통한 기존의 자전거도로에 공간으로 유도하는 새로운 길의 생성
Flow through the induction of a bicycle on the road to the existing road space to encourage a new generation

SPACE STORY 1

1 Bike Lane

공간의 시작은 자전거 도로에서 부터 시작된다.
The first point begin from the bicyle path.

4 Bilcycle Rest Area

SLOPE는 2층 공간과 연결이 되어 있으며, 2층 공간의 전체는 휴식공간으로 이루어져 있다.
connected to the second floor space, and the whole of the second floor space is made up of a resting place.

2 Bike Lane And Slope

SLOPE를 통하여 자전거 노도와 공간과의 지연스러운 연결선을 나타낸다
Indicate natural connectivity analysis through slope between bicycle paths and spaces.

5 Bicycle Rack

높낮이가 다른 객체로 이루어진 공간으로 객체 일부분에 자전거 거치대가 설치되어 있어, 이용객들이 자전거를 세워두고 휴식을 취할 수 있다.
Consisting of pitch to another object in space, part of the object are equipped with bicycle holder, passengers will put the bike can take a break.

3 Splope

SLOPE는 하나의 자전거 통행로로 평탄한 경사가 있어 아이들도 자전거로 이동이 용이하다.
One of the bike lanes have a flat slope, go to the kids bike is easy.

6 Drinking water

하나의 객체에 음수대가 설치 되어 있어, 따로 물을 준비하지 않아도 물을 마실 수 있다.
Bicycle holder for one object in the same way as a negative consideration is the installation, failure to prepare drinking water is drinking water.

7 Surface and Deuk

객체의 일부분을 파동의 형상를 적용시켜 곡면의 형태가 생성되는데, 곡면 상부에 DEUK가 생성되어 하나의 휴식공간이 생성된다.

Part of the object by applying the shape of the wave in the shape of the surface but it was not, on the top surface is ganotyige deuk there is a resting place.

10 Complex Space

계단 슬로프를 통해 1층 공간으로 이동할 수 있으며, 1층 공간은 자전거를 위한 편의 시설들로 구성되어 있다.

Slope of the stairs can move through the first floor space and a floor space consists of facilities for bicycles are.

8 Slab's altitude gaps by cracking

하나의 면이 여러 방향으로 분할되며, 객체의 하나 하나에 높낮이를 달리 주어 앉아서 쉴 수 있는 벤치 역활을 하며, 공간 전체에 재미를 부여한다.

Shaping Formation and Nonlinear Analysis of Arch Shaped Space Structure with Multi-directional Joint. and fun in space will give the whole.

11 Bicycle Pump

1층 공간 곳곳에 설치되어 있는 공기 주입기를 통해 필요로 할 시에 간편하게 바람을 주입할 수 있다.

Installed throughout the first floor space needed by the air injector to inject the air is easy.

9 Step Rest Area

계단과 SLOPE의 일체형으로 계단에 걸터 앉아 휴식을 취하며, 동시에 자전거 통행이 가능하다. 이에 자전거의 원할한 이동이 가능하다.

Got to the stairs, slopes can relax and sit on the stairs, and at the same time is to allow for bicycle commuting. This bike can be moved smoothly.

12 Vlaintenance

다른 목적지로 이동하기 전 1층에 위치한 정비 공간에 들려 자신을 자전거를 살펴본 후 이동 하는 것도 하나의 방법이다.

Before moving to other destinations on the ground floor, step on the maintenance area, and then saw them move their bikes are necessary.

PLAN

배치도에 확인 할 수 있듯 물줄기라는 컨셉을 통하여 한 곳에 머무르지 않는 기존의 자전거도로와의 원활한 흐름을 볼 수있다.

as you see in plot plan, through the stream concept connecting existing bicycle and new one.

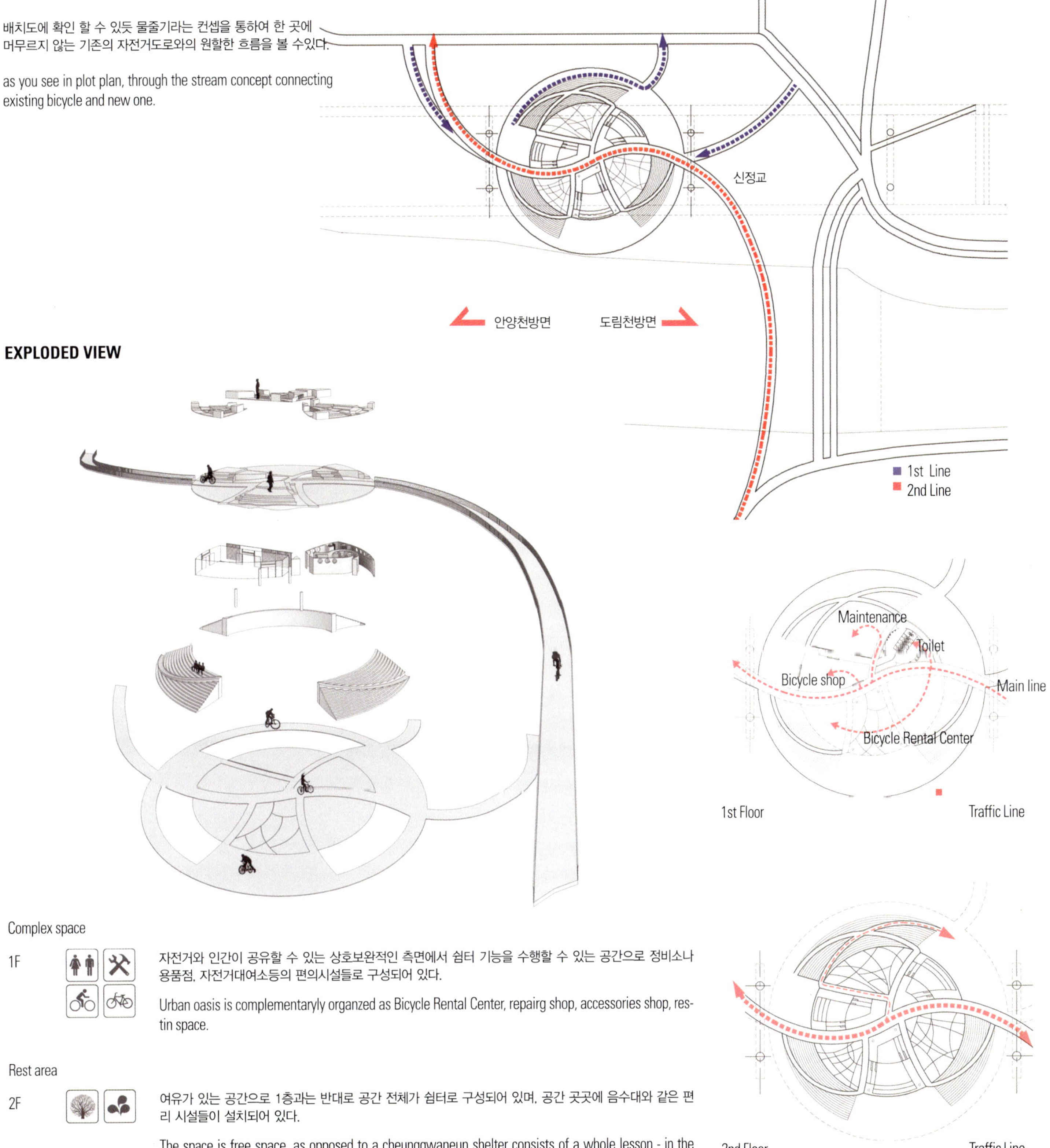

Complex space

1F

자전거와 인간이 공유할 수 있는 상호보완적인 측면에서 쉼터 기능을 수행할 수 있는 공간으로 정비소나 용품점, 자전거대여소등의 편의시설들로 구성되어 있다.

Urban oasis is complementaryly organzed as Bicycle Rental Center, repairg shop, accessories shop, restin space.

Rest area

2F

여유가 있는 공간으로 1층과는 반대로 공간 전체가 쉼터로 구성되어 있으며, 공간 곳곳에 음수대와 같은 편리 시설들이 설치되어 있다.

The space is free space, as opposed to a cheunggwaneun shelter consists of a whole lesson - in the negative space around the case of such facilities are installed.

SLOPE

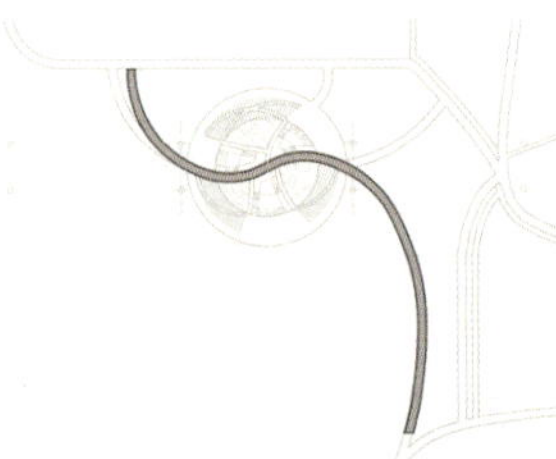

SLOPE를 통하여 기존의 자전거도로와의 연결로 공간의 자연스러운 흐름을 보여주며, SLOPE의 흐름으로 전체 공간이 생성된다. 이 SLOPE는 공간의 핵심이라 할 수 있다.

Slope of the bike path through the existing connection to the natural flow of the show space, Slope due to the flow of the entire space will be created. Slope of the two key requirements of the space can be Sora.

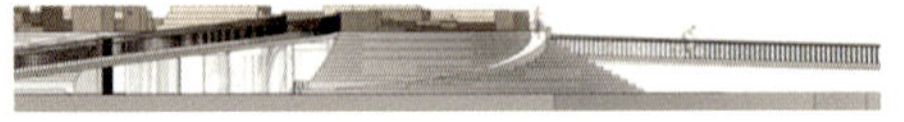

SLOPE

BICYCLE SHOP

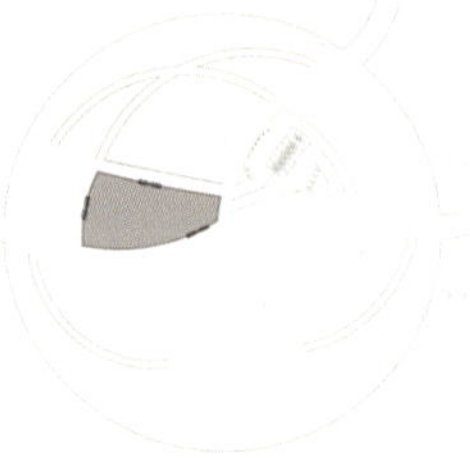

자전거 용품점으로 갑작스러운 고장에 대처하여, 사전에 미리 정비 할 수 있도록 자전거부품 및 기타 용품들을 판매한다.

The bike shop sells many bike parts and equipments tuning up the bike before sudden breakdown.

BICYCLE SHOP

SLOPE

자전거 대여소가 조성되어 있어 자전거를 준비하지 않은 일반 시민들도 간단히 자전거를 대여하여 즐길 수 있다.

I was not far-sighted enough to think of that.

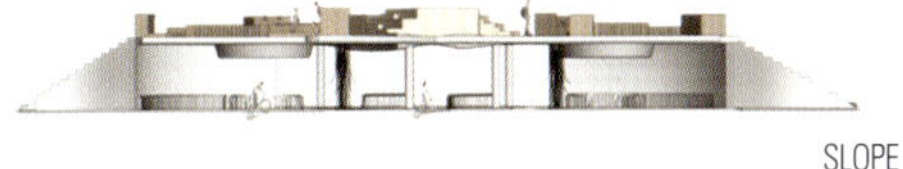

SLOPE

BICYCLE SHOP

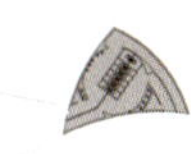

전체공간은 자연적인 곡선을 띄고 있으며, 이 곳 역시 동선의 흐름을 적용되었다. 일반 딱딱한 매스 형식이 아닌 자연적인 곡선을 통해 공간의 재미와 시퀀스를 보여주며, 벽면에 다양한 형태의 타공으로 인한 자연채광을 통해 공간의 쾌적함을 보여준다.

All-round space is organic curved surface this is applied from moving stream it indicates not only the natural light space by perforation but also space sequcence through organic surface not mass system.

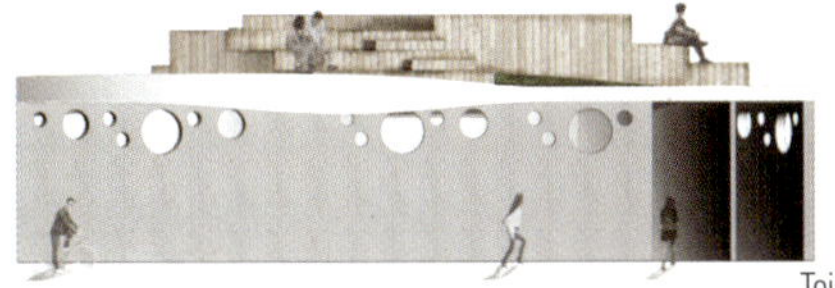

Toilet

ECO HERE : eco in boxes

재활용 예술을 통해 그 작품을 판매하는 마켓공간.자신의 작품으로 공간을 만드는 창작공간. 전시공간 그 공간속에 우리의 자연이 담긴다

Recycling through art that works selling market space. Create a space that his work as a creative space, the exhibition space is a space which contains our nature in

PROPOSAL

친환경 바람이 대세를 넘어서 필수로 자리매김하고 있는 2012년 요즘 환경문제가 대두되고 있다 이렇게 생활 폐기물을 이용한 재활용품을 주제로 공간을 계획하고자 한다
일반시민 등이 재사용이 가능한 생활용품 등을 직접 교환하고 또한 재활용품을 이용하여 만든 작품도 교환하고 팔 수 있는 새로운 마켓을 만들려고 한다 이렇게 벌어들인 수익금은 기부를 하여 뜻 깊은 의미를 만들고 더불어 재활용을 이용한 체험공간, 동시에 만남과 교류를 할 수 있는 공간을 더한다 자신의 손이 닿았던 재활용품을 의미 있게 교환하므로써 자원재활용 생활문화 확산과 자연에 대해 다시한번 생각해보고자 한다.

Eco-friendly wind on solidifying as required beyond these environmental problems in 2012 and is thus subject to the living space planning with the waste and recyclables should you wish to Ordinary citizens, such as reusable household goods, such as this and also the direct exchange of recyclables to create works that can be sold and exchanged for new markets to make a donation so your proceeds by means of deep meaning and experience of space, using with recycling
At the same time, we can throw his hands touched, adds to a meaningful exchange of recyclables, recycled life, culture and the natural resources so spread once again, I would like to think about.

RECYCLE ART

재활용 예술을 통해 그 작품을 판매하는 마켓공간, 자신의 작품으로 공간을 채울 수 있는 창작공간, 그 작품을 전시할 수 있는 전시공간, 갤러리공간을 이루고 그 공간속에서 우리의 자연에 관한 생각을 일깨우고 한번더 생각할 기회를 제공한다. 또 서로간의 커뮤니케이션을 통한 서로간의 작품을보고 느끼고, 함께 생각하며 감성을 공유한다

Recycling through art that works selling market space, as his own work, creative space that can be filled out a space for you to be able to display that works in the exhibition space, a gallery space in the midst of that space to fulfilling our thinking about the nature and gives you the opportunity to once again think. Also with each other to see the works between each other through communication among, and will share with the sensitivity to feel

TARGET

Think about nature, environment, and anyone can do, people who think nature increases the environment better. If anyone is interested in the nature of recycling, we both, young children can participate in all

자연,환경에 관한 생각은 누구나 할 수 있고, 자연을 생각하는 사람이 많아질수록 환경은 더 나아질 것이다. 재활용, 우리 자연에 관심이있는 사람이면 남녀노소, 어린아이 모두 참여 할 수 있다

USER NEEDS

재활용 예술 / 창작공간 / 자연주의

- 자연에 대한 생각을 서로 공유하며 더욱더 자연주의적 경향이 강해진다
- 유대관계 형성 , 많은 사람들과의 교류 기회 (타인과의 커뮤니케이션)
- 개인 창작공간을 가지며 재활용 예술을 활성화 시킬 수 있다

USER NEEDS

RECYCLE ART / CREATIVE SPACE / NATURALISM

- Natural to think about each other, share, and the stronger further naturalistic tendency
- Relationship formation, there are a lot of people's exchange opportunities (with communications)
- Personal creative space and can activate the recycling art

INSTALLATION ART

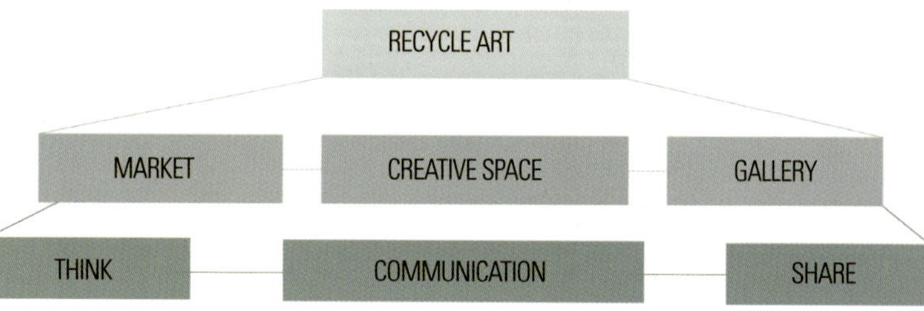

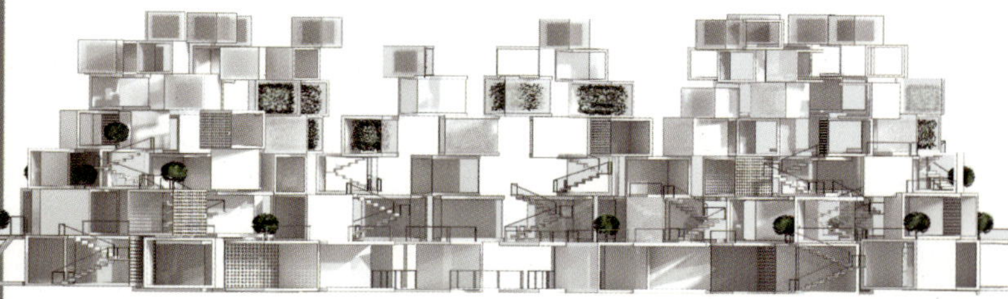

SITE ANALYSIS

양재천은 도심지 아파트 밀집지역에 버려져 있는 하천 공간을 정비하여 주민들의 휴식공간으로 제공하고, 맑은 물속에서 물고기가 살 수 있고 도시개발사업으로 훼손된 하천을 자연에 가까운 하천으로 복원시켜 자연생태계를 회복시켰다. 현대인들의 다양한 레크리에이션 욕구를 충족시키고, 옛 강의 모습을 회상토록 하여 시민의 정서 함양을 도모하였다.

Yangjaecheon is a downtown apartment areas abandoned in the rest of the residents of river space into space, and can buy fish in clear water, urban development business as close to the rivers that are undermining the natural ecosystem recovery by restoring the natural rivers. People to meet a variety of recreational needs, an old River called by the emotion of citizens on the amount of substrates downloads this week.

양재천로
Yangjaecheon stream

주거공간의 밀집
The dense residential space

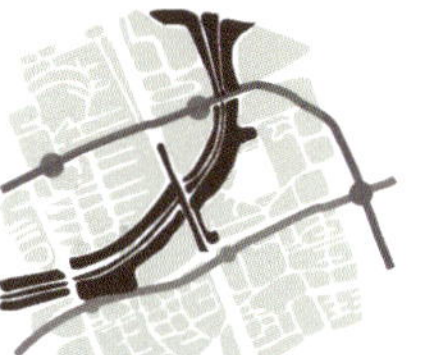

지하철 3호선, 분당선
Metro line 3, bundang line

DESIGN CONCEPT

그라데이션이라는 컨셉에 박스라는 개념을 입혀 박스를 통해 자연에 관한 변화를 보여주도록 계획하였다.크게 그라데이션의 단계적 변화를 박스라는 오브제를 통해 표현하고, 그 속에서 자연에 관한 생각, 관념을 변화시킨다.

The concept of the box in the gradient box, called the concept through the changing natural regarding clothed show, said the plan .Step change in the gradient greatly to express through boxes of my, and regarding the thoughts, notions in a natural turn

GRADATION

BOX

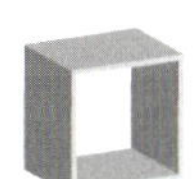

NATURE / CHANGE

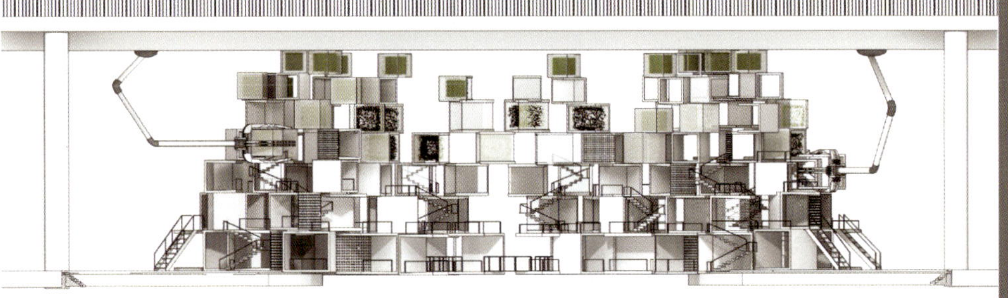

GRADATION

그라데이션이란 그래픽에서 사용되는 기법으로 어두운데서 밝은 값으로, 큰 모양에서 작은 모양으로, 껄끄러운 결에서 매끈한 결로, 또는 한 색상에서 다른 색상으로 등과 같이 점진적이며 매끄럽게, 단계적으로 변해 가는 것을 말한다

As a technique used in the gradient graphic Iran raises from the big, bright values in the form, in the shape of a small, sleek, and color condensation or gritty lodge in different colors, such as the gradual and smooth, step by step, says it has become

TARGET

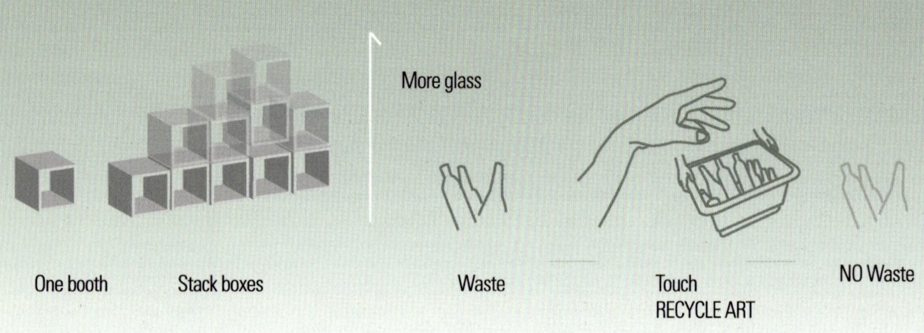

쓰레기에서 우리의 손이 닿아 새로운 재활용 작품으로 탄생하게 되면, 그것은 더이상 쓰레기가 아니다 생명이 없던 쓰레기에서 우리손에 의해 생명을 가진 작품으로 태어나게 되면 쓰레기는 더이상 존재하지 않고 소멸 되는것이다 우리는 쓰레기에서 작품이 되고 생명이 없던 것에서 생명을 갖게되는 점차 사라지거나 생겨나는 특징을 잡아 컨셉을 정하였다

Garbage in our hands to reach the new recycling work, the birth, it is no longer in our hands is not garbage by garbage is not in the life with no trash the work of the reincarnated do not exist, would we trash is destroyed, and the work on the life in life was gradually dismissed or made to grab aconcept arising features

BOX

RECYCLE ART

MARKET

GALLERY

CREATIVE SPACE

NATURE **VOID/GLASS +TREE**

BOX SHAPE

3 SIDE 3 SIDE 1 SIDE

WASTE SOILD / BOX

GRADATION

BOX SHAPE

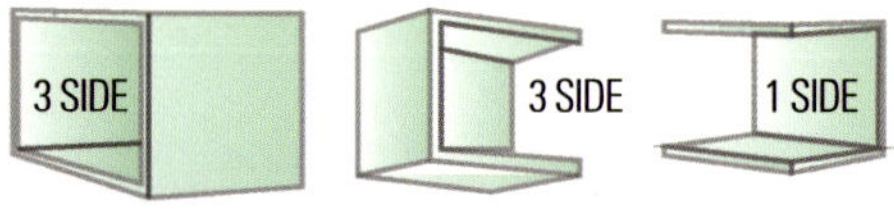

GLASS ANOTHER FACE

NEW SHAPE

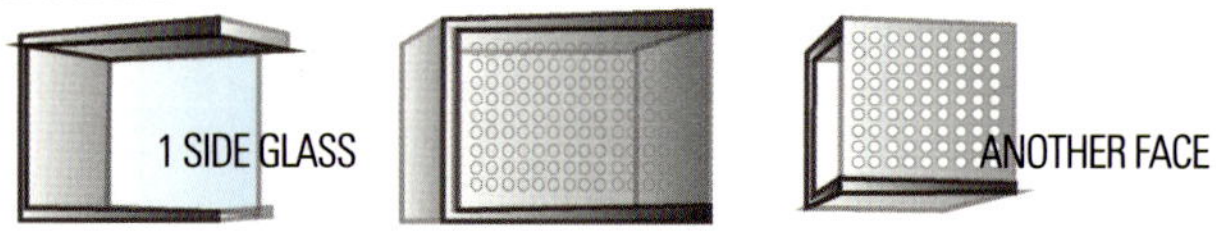

1 SIDE GLASS ANOTHER FACE

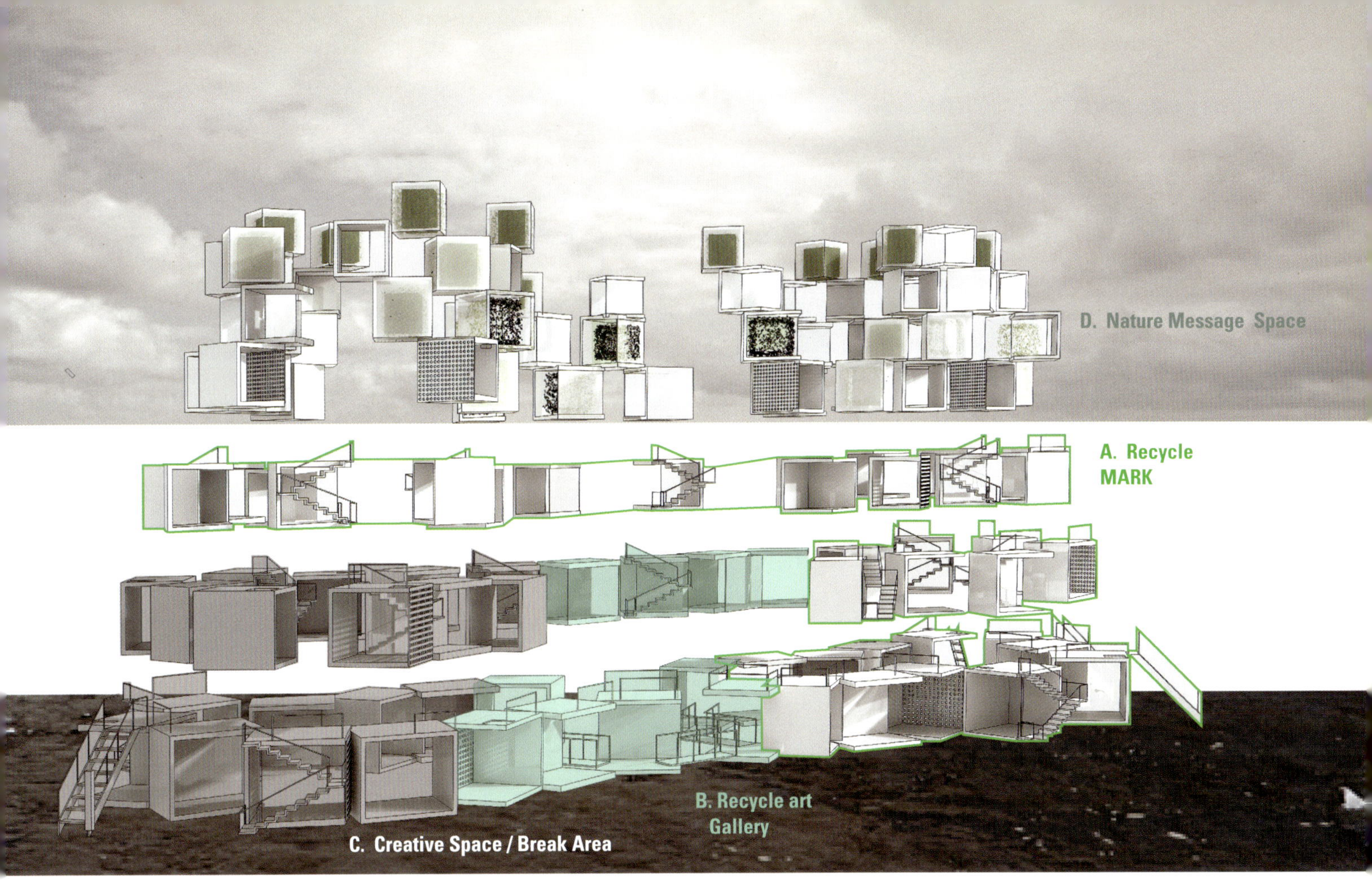

D. Nature Message Space

A. Recycle MARK

B. Recycle art Gallery

C. Creative Space / Break Area

SPECIAL ROUTE

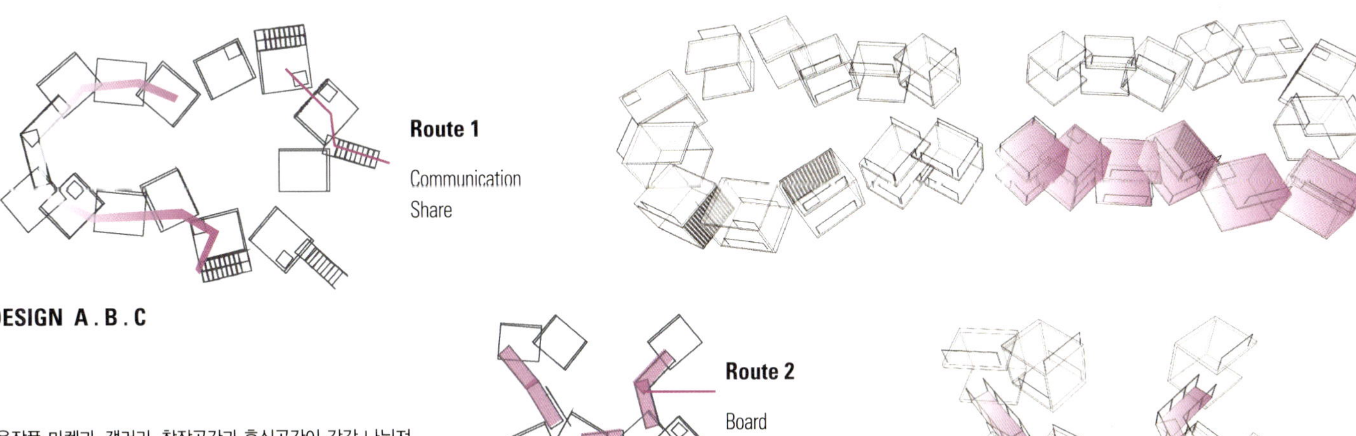

DESIGN A . B . C

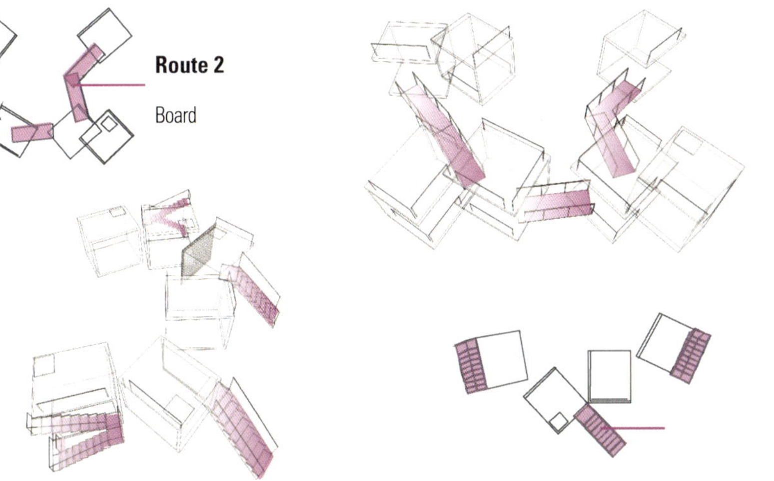

재활용작품 마켓과, 갤러리, 창작공간과 휴식공간이 각각 나눠져 있으며 박스에 의해 공간이 형성된다. 박스들 사이에서 특별한 루트가 생겨나는데, 박스들의 면모양 에 의해서 , 판자에 의해, 계단에 의해 각각 형성된다. 사람들은 이 특별한 루트를 통해 재미를 느끼며, 서로간의 커뮤니케이션과 감정적 교류를 할 수 있다.

Recycling markets, galleries, creative space and relaxation space works, each divided by a space in the box. Having a special root that arises among the box, on the appearance of boxes, shingles is formed by a staircase by, respectively. People walked through this communication and emotional exchanges between each other, fun.

A. RECYCLE ART MARKET

자신의 재활용작품을 팔 수 있는 시장이다. 하나의 부스안에서 자신만의 매장을 열 수 있다 그 공간들은 서로 독립된 개체로 보이지만 사실은 한곳도 빠짐없이 이어져있다. 박스와 박스 사이에서 형성되는 이동 동선은 서로 간의 예술을 공유할 수 있고, 자연에 대한 생각 또한 함께 나눌 수 있다

You can sell their recycled work market. One can open their own stores within the booth space as each other, but the fact that an independent waking one place. Move the box and the box is formed between copper can share with each other between art, nature, think also about the share

RECYCLE ART MARKET —— **1 FLOOR** —— **2 FLOOR** —— **3 FLOOR**

Recycle Market   Route

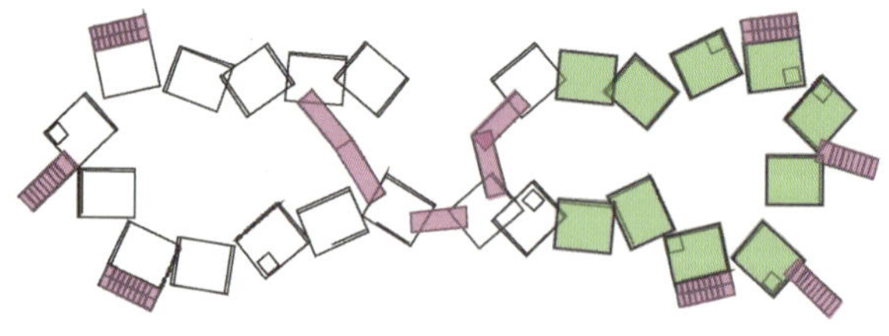

1 FLOOR

Furniture/appliances ZONE

ROUTE 1 . 2 . 3

2 FLOOR

Reform clothing Zone

ROUTE 1, 3

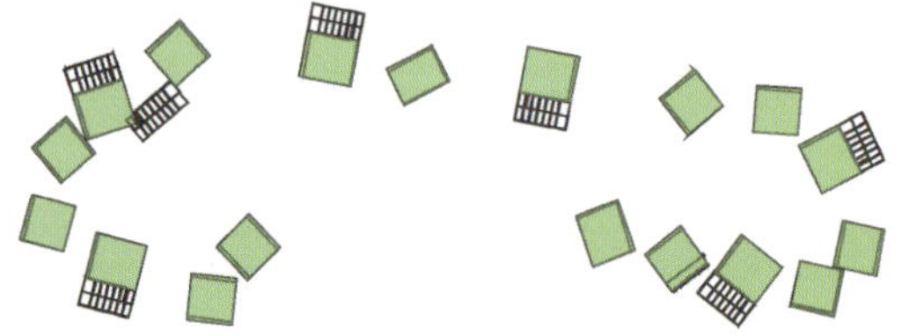

3 FLOOR

Props/goods ZONE

ROUTE 1, 3

A. RECYCLE ART MARKET

재활용을 이용해 만든 작품을 전시하는 공간이다. 전시공간은 시민의 소통공간으로서도 의미를 가지고 있다. 시민들은 이곳에서 폐품을 이용해 그들에게 질 높은 문화 향유 기회를 부여한다. 특히 예술– 재활용 –사람 을 잇는 문화공간으로 서로간의 소통과 문화를 활성화 시킨다.

Made using recycled work space. The exhibition space have the meaning as civil communication space. Citizens here and ask them to scavenge given the opportunity to enjoy a high quality culture. In particular, the art-recycling-linking people to each other as a cultural space that communication and culture.

B. RECYCLE ART GALLERY

—— 1 FLOOR —— 2 FLOOR

Recycle art Gallery Box

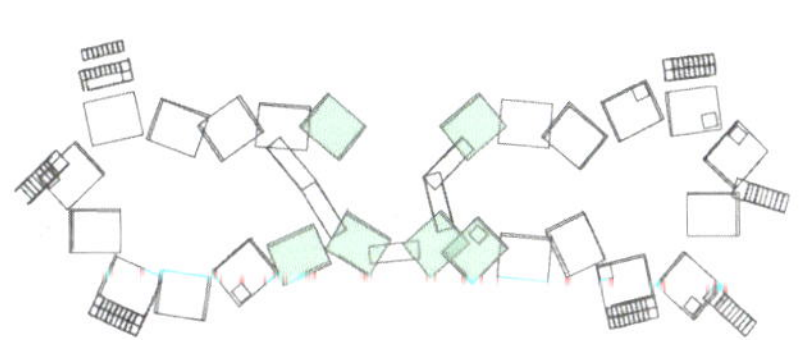

Creative Space / Break Area

—— 1 FLOOR —— 2 FLOOR

창작공간과 휴식공간으로 사용되는 이 공간은 시민들이 함께 공간을 이루어 나가는 공간이다. 소통의 공간으로도 중요하지만, 가장 중요한 시민이 참여를 실천할 수 있는 공간이다.

Creative spaces and relax with this space is used space made out of the citizens. Communication space is important, but, the most important citizen participation can live up to.

C. Creative Space / Break Area

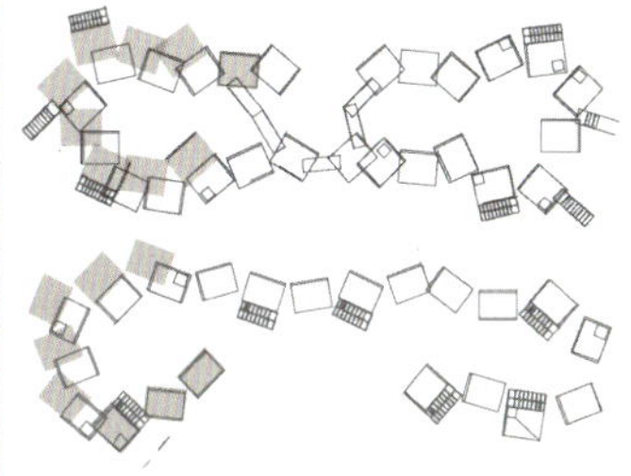

D. Nature Message Space

점층적으로 쌓여있는 박스들은 전체적으로 그라데이션을 보여주며, 아래에서 이루어지는 재활용아트는 우리의 환경을 점점더 자연으로 만들어준다 . 그것을 솔리드한 그냥 박스에서부터 위로 갈 수록 보이드한 유리의 면을 많게 함으로써 자연화 되가고있는 메시지를 전달한다. 글라스로 만들어진 유리들은 그 안에 자연의 녹색을 담게되고, 유리들은 점점 위로갈수록 투명도가 생긴다.

The points accumulated at the bottom of the box are on the whole gradient and, increasingly, our recycled art takes place as the natural environment. It's more solid and just go up from the box, and if a lot of glass Boyd at the message that by nature. Glass made of glass are in natural green wall, glass stems and less transparency over their increasingly.

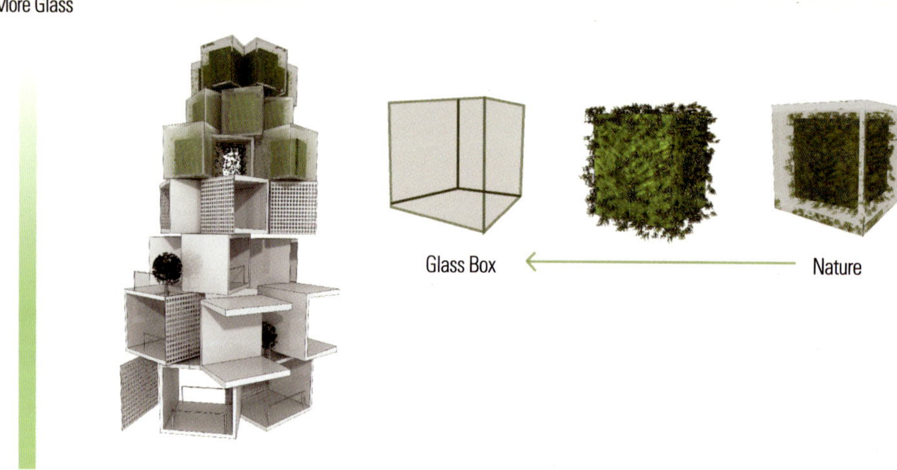

E. Robot Arms

로봇팔은 우리의 손을 대신하는 상징적인 존재로써 행동한다. 로봇팔이 우리가 가꾼 자연을 상징하는 박스를 옮기며 우리에게 자연에 대한 생각을 다시한번 상기시키고, 새로운 메시지를 심어 준다. 우리는 로봇팔을 통해 우리 스스로가 자연을 위해 중요한 행위를 하고 있는것 같은 감정을 느끼게 된다. 그것을 발판으로 더욱더 강한 자연주의적 생각을 갖게 만든다

Robot arm as a symbolic presence to our hands instead of Act. Robot arm symbolizes the nature we have dreamed of the box thinking about nature, moves us to remind you once again, the new message to indoctrinate. We through our own robotic arm is an important act for nature and the emotions it evokes. further makes a strong naturalistic idea

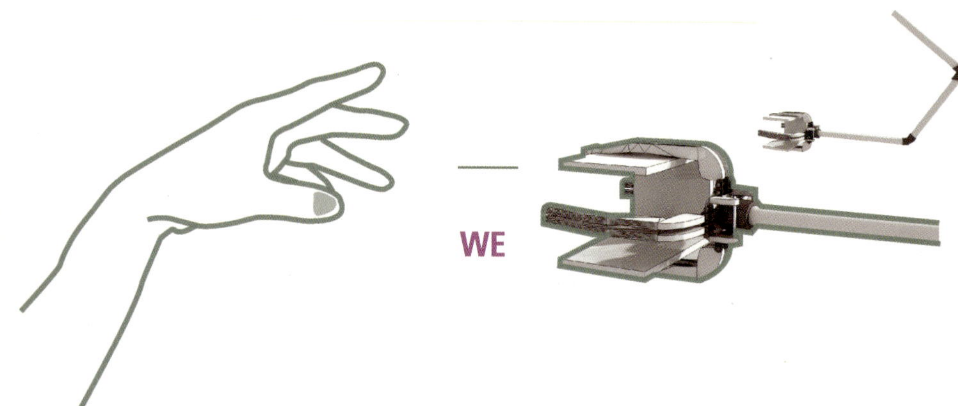

BIBS and BOBS

70년대와 80년대의 향수와 낭만을 느낄 수 있는 문화공간

BACKGROUND

랩과 힙합으로 대변되는 90년대 문화에 밀려 동화되지 못하고 갑자기 사라진 문화, 하지만 현재 다시 주목받고 있는 문화가 있다. 바로 7080문화이다. 7080세대라고 하면 40~50대의 중장년층을 말한다.

그들은 암울한 사회적 정치적 분위기 속에서 낭만 속으로 도피하여 청년기를 보냈다. 그런 그들에게 7080문화는 아련하고 향수를 불러일으키는 것이다. 현재 다시 주목받고 있는 그 시절 낭만을 재현하는 공간을 만들고자 한다.

'Culture of 70's and 80's' was once faded away by waves o f 90's Hip-Hop and rap. But, these days, culture of 70's and 80' revived again. People who enjoyed the culture are mostly at the age of 40 to 50.

They escaped to romance of 70's and 80's during their adolescence from politically and socially gloom mood of 1970s to 1980s. This romance, arouse people's nostalgia nowadays. This project is trying to reproduce cultural place of 70's and 80's.

TARGET

Main Target

–1970~1980년대에 청년기를 보낸 40대–60대 남녀

Sub Target

–7080문화를 경험해 보고 싶은 20~30대 남녀

현재 40~50대 중장년층은 자식들 뒷바라지에 힘든 나날을 보내고 있다. 힘들고 지친 일상에서 벗어나 그들의 청년기에 느꼈던 낭만으로 자녀들과 함께 되돌아가보는 즐거움을 느낄 수 있는 공간을 만들고자 한다.

-Ages at 40 to 50 male and female who spent their adolescence in 1970 to 1980.

-Ages at 20 to 30 male and female who want to experience 70's and 80's culture.

These days, those people who spend their adolescence in 1970 to 1980 are living tough days by supporting their offspring. This project aimed to make place for relaxation that they can communicate with their sons and daughters by using the culture of 70s and 80s.

SITE ANALYSIS

서울특별시 양천구 신정동 오금교

Ohgeumgyo, Sinjeong-dong, Yangcheon-gu, Seoul

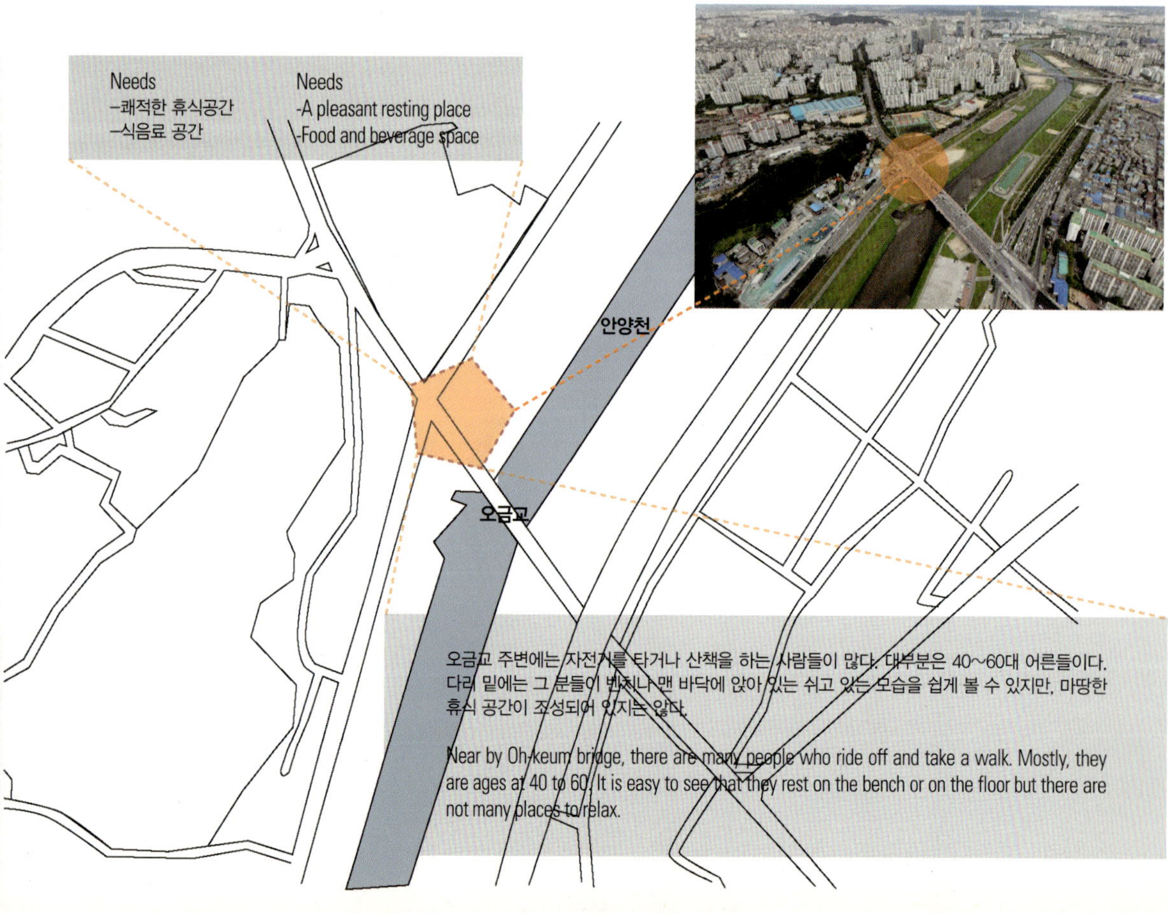

Needs
–쾌적한 휴식공간
–식음료 공간

Needs
-A pleasant resting place
-Food and beverage space

안양천

오금교

오금교 주변에는 자전거를 타거나 산책을 하는 사람들이 많다. 대부분은 40~60대 어른들이다. 다리 밑에는 그 분들이 벤치나 맨 바닥에 앉아 있는 쉬고 있는 모습을 쉽게 볼 수 있지만, 마땅한 휴식 공간이 조성되어 있지는 않다.

Near by Oh-keum bridge, there are many people who ride off and take a walk. Mostly, they are ages at 40 to 60. It is easy to see that they rest on the bench or on the floor but there are not many places to relax.

AREA ANALYSIS

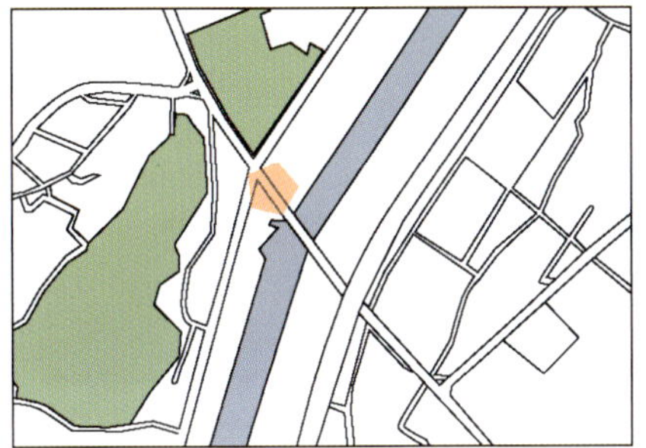

공원
-양천구 어린이 교통공원
-양천구 해누리 체육공원
-갈산공원

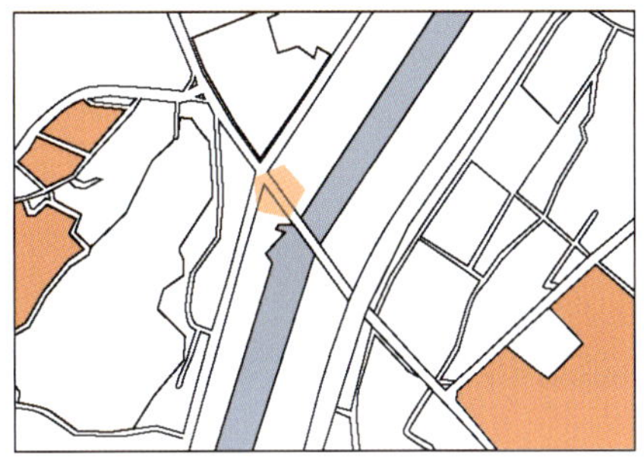

아파트 단지
-삼성레미안
-목동 2차 우성아파트
-라이프 미성아파트
-대림 E 편한세상아파트

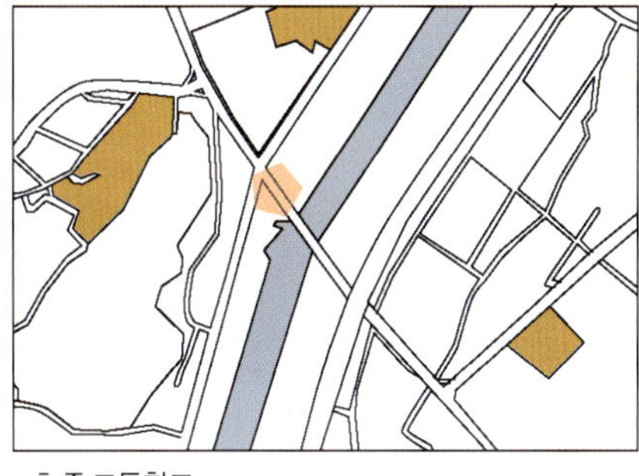

초,중,고등학교
-은정초등학교
-갈산초등학교
-신미림초등학교
-신목고등학교

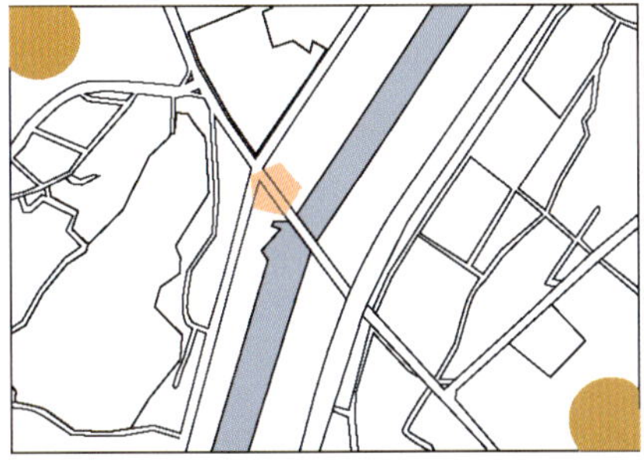

대중교통
-양천구청역(2호선), 구로역(1호선)
-마을버스 정류장

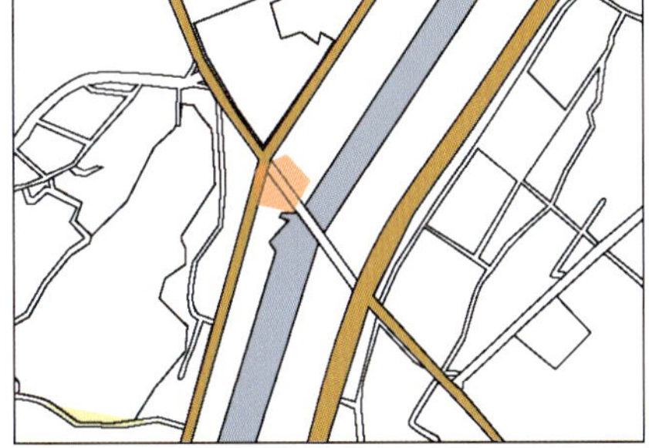

도로
-서부 간선도로
-안양천로
-목동로
-구로중앙로

● 오금교 관련 지역활동

-안양천 억새 단지
신정교와 오금교 사이에 위치하고 있으며, 안양천 수변의 억새들을 모아 단지를 조성하였다.

-flame grass complex, An-yang stream
Located between Sin-jung bridge and Oh-keum bridge, created with flame grass near by An-yang stream

-어린이 자연학습장
신트리 공원 중심에 자리 잡고 있으며, 자연, 원예, 덩굴, 농작물 등이 잘 관리되어 있는 체험 학습장이다.

-Natural exploration area for children
Sinteuri-park is located in the center, nature, gardening, vines, crops that are well managed and is experiencing learning center.

-양천구 어린이 교통교육관
유치원, 초등학교 어린이들에게 교통에 관한 교육을 실시하여 교통질서를 익히고 습관화할 수 있도록 하고 있다.

-Traffic education center for children, Yang-chun gu
By conducting traffic education to kindergartener and elementary school student, letting them to learn and habituate traffic order

-양천구 자전거 대행진
양천구민의 날을 맞이하여 기후 변화에 적극 대응하고 친환경 교통수단인 자전거를 생활화하도록 유도해 구민의 건강을 증진시키기 위해 개최되었다.

-Cycle march, Yang-chun gu
In celebration of people in Yang-chun gu, the Cycle march was held to react climatic change and to promote people's health using riding cycle which is environment-friendly means of transportation.

● 오금교 사용 목적

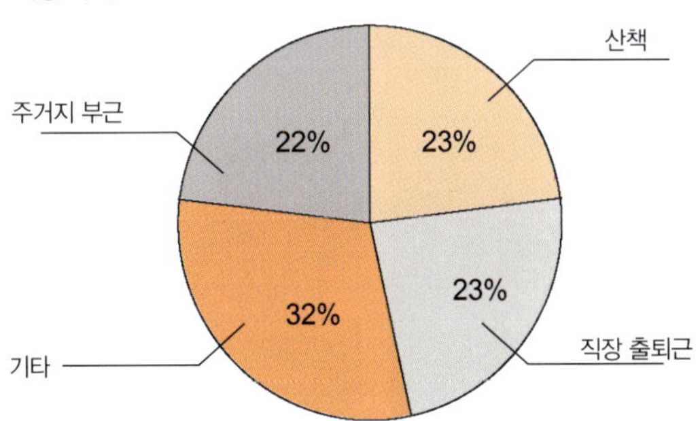

오금교를 지나는 대부분의 사람들은 특정 목적을 가지고 있지 않다.
People passing Ohgeum-bridge do not have a Specific purpose.

DESIGN CONCEPT

물체주머니 Material pouch

현재 40–50대 어른들이 초등학교에 다닐 적 가지고 다녔던 주머니.

Pouch that used to be brought by people in 70's and 80's during their elementary school life.

향수 자극
살아 온 삶의 투영

Stimulation of a sense of nostalgia
projection on past life

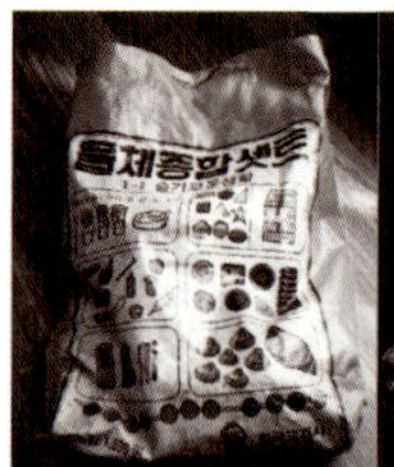

DESIGN

속이 보이지 않는 주머니와 다양한 촉감의 물건들

- 호기심을 유발하는 동선, 공간구성
- 한 가지에 국한되지 않은 다양한 성격의 공간 제시

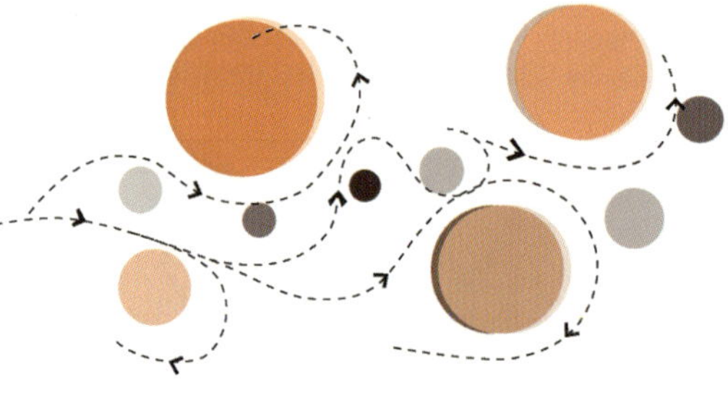

Cheat Invisible texture of a variety of pockets and stuff

-Spatial construction and movement that lead to curiosity

-Presenting several spatial characteristic that is not limited to only one concept.

DESIGN KEYWORDS

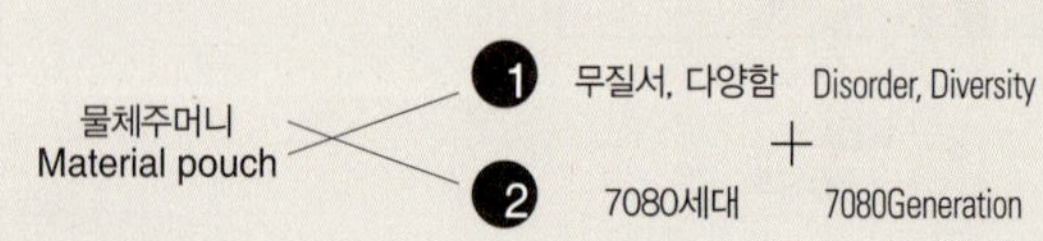

=

7080세대를 재현해 내는 다양한 공간 제공
다양한 높낮이의 길로 자연스럽게 재미를 이끄는 동선 유도

Various spaces that reproduced 70's and 80's culture.
Leading free movement with different level of height.

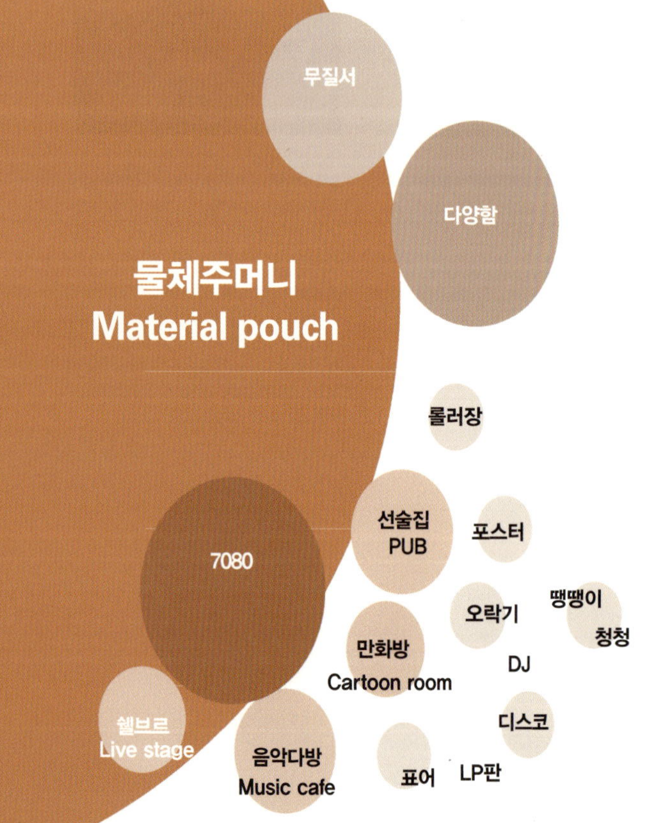

DESIGN PROCESS

MASS PROCESS

물체주머니의 형태적 특징을 전체 공간 디자인에 접목시켰다.

Characteristic of material pouch's shape is applied to entire space design.

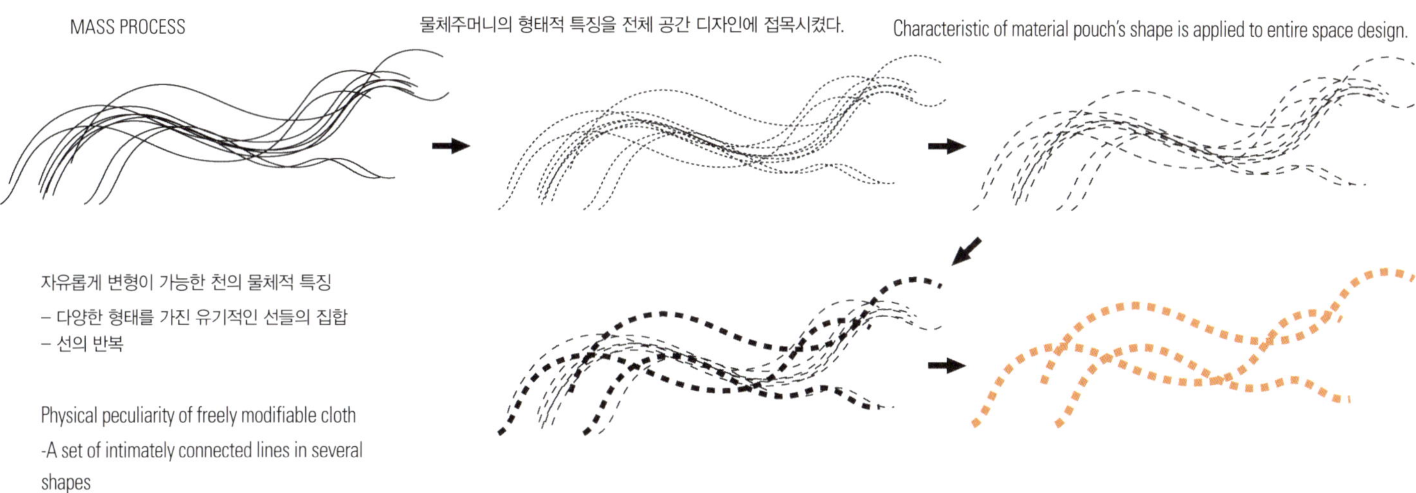

자유롭게 변형이 가능한 천의 물체적 특징
– 다양한 형태를 가진 유기적인 선들의 집합
– 선의 반복

Physical peculiarity of freely modifiable cloth
-A set of intimately connected lines in several shapes
-Repetition of lines

SPACE PROCESS

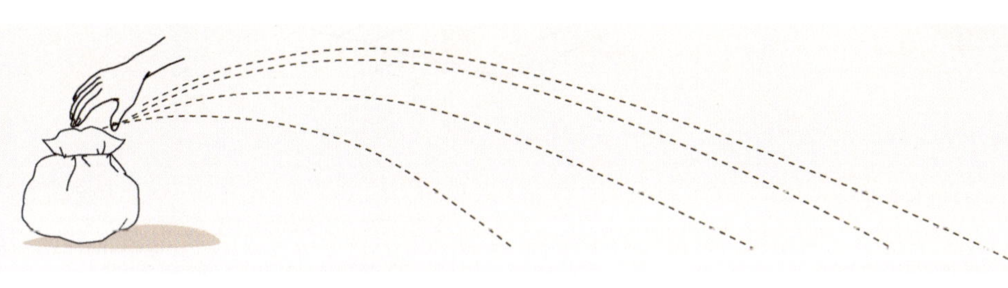

국민학교 시절 수업시간에 호기심 가득한 마음으로 물체주머니 안의 여러 가지 물건들을 꺼내 관찰했을 그 때를 상상하며 디자인 프로세스를 진행했다.

Start design process with reminiscent of science class in elementary school, searching the 'Material pouch' with curiosity.

〈물체주머니 놀이 실험 과정〉

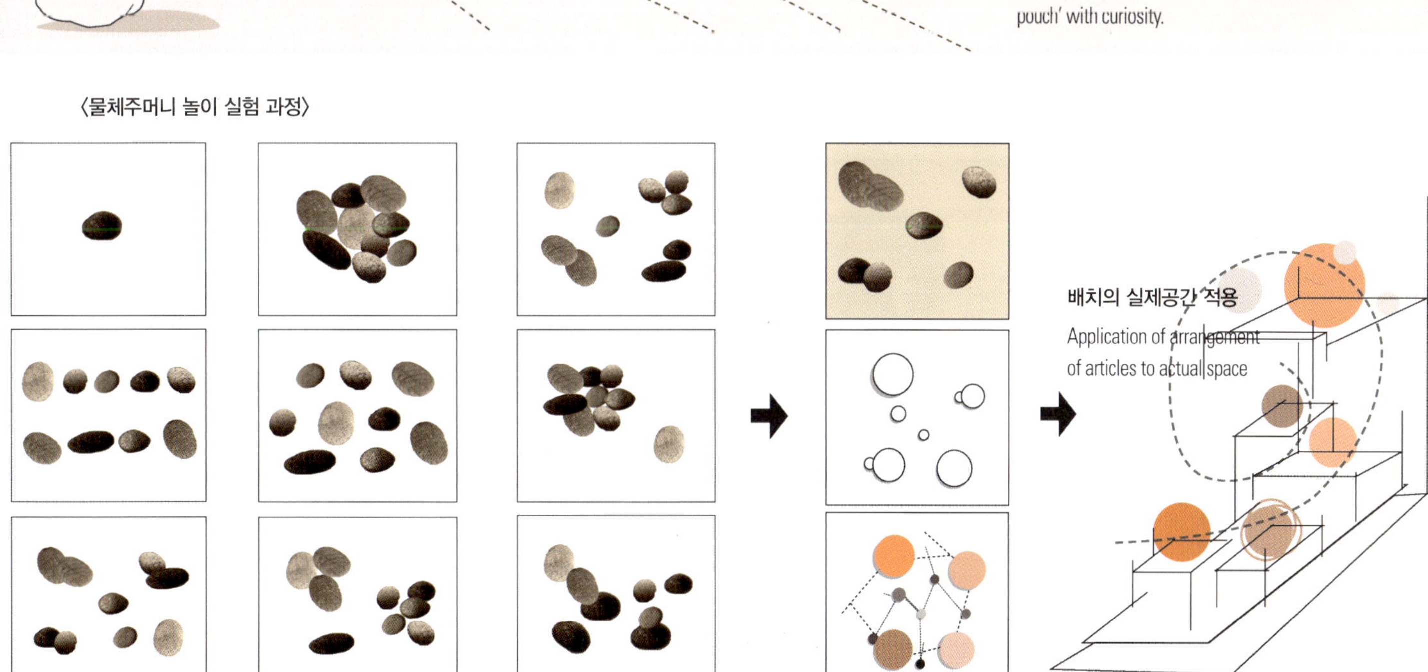

A. 주머니에는 여러 가지 잡동사니가 들어있다. 물체들을 하나씩 꺼내 이리저리 배열하였다.
B. 열을 맞춰 배열/ 물체들을 한때 모아서 배열/ 비슷한 성격끼리 배열/ 그 외의 배열
A.
B. Make arrangement in line/ gather the articles and arrange them/ arrange them with similar characteristic / the rest of free arrangement

물체가 흩어져 있는 것을 나타내기 위해 2층으로 공간구성
– 공간을 이동할 때 자유로운 동선이 나올 수 있도록 유도

For convenient use of expressing dispersed articles, two different stairways are constructed.
- This structure leads free movement while moving between spaces.

FLOOR PLAN

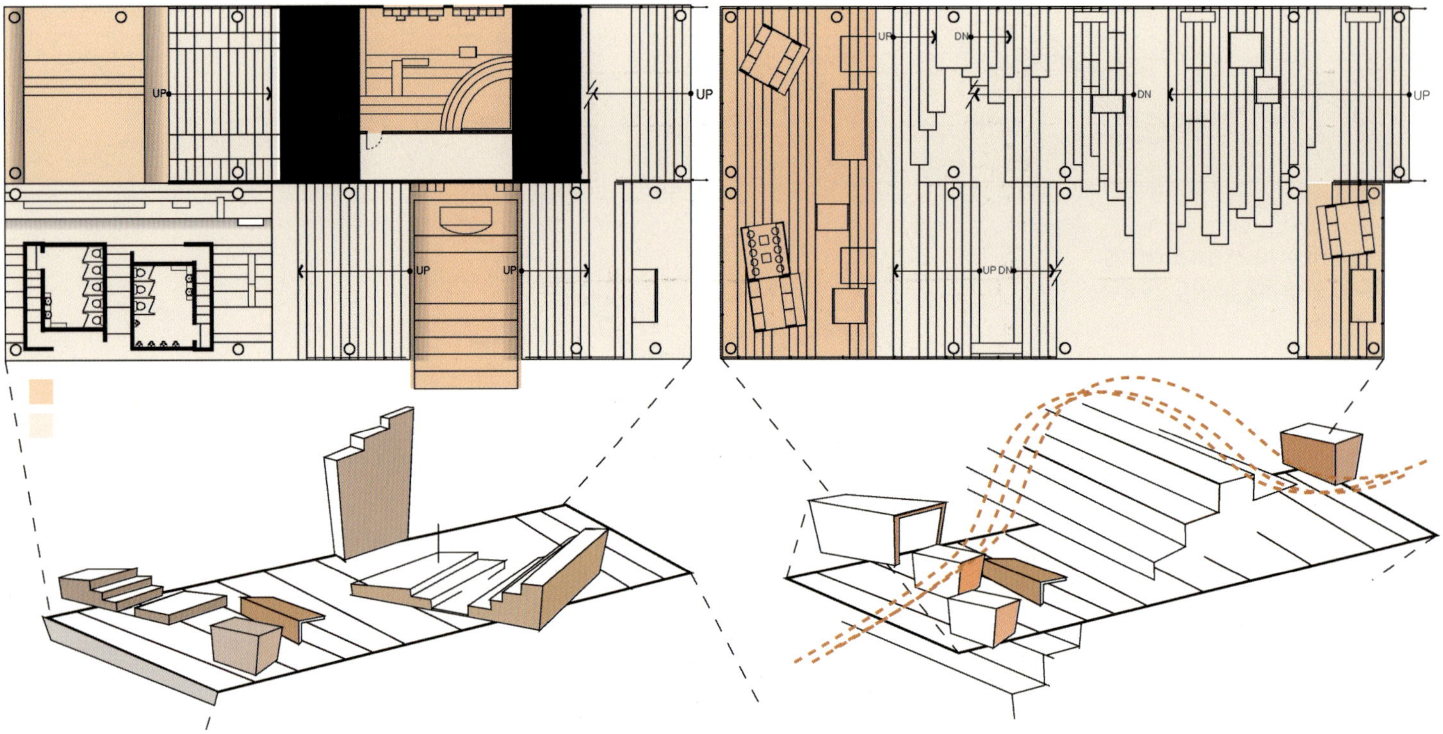

음악다방과 라이브 무대, 만화방과 같은 공간은 주변의 소리의 영향을 받지 않게 하기 위해서 독립적인 곳에 위치하였다.

Teahouse with musical performance, live stage and comic book shop are located in separated place to prevent from noises near by

그 외의 공간들은 자연스럽게 유도된 동선에 위치하였다.

Leading free movement with different level of height.

ELEVATION

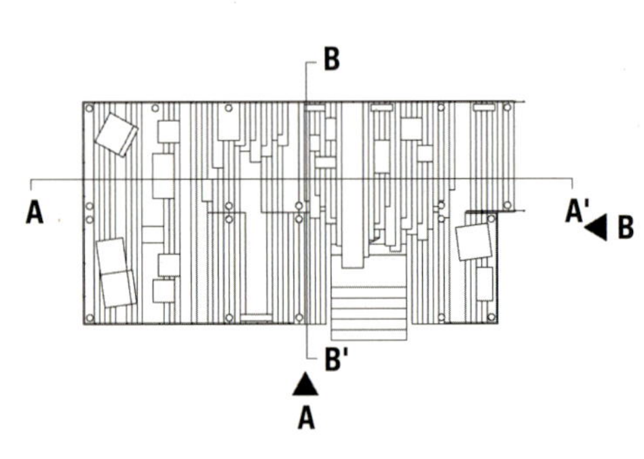

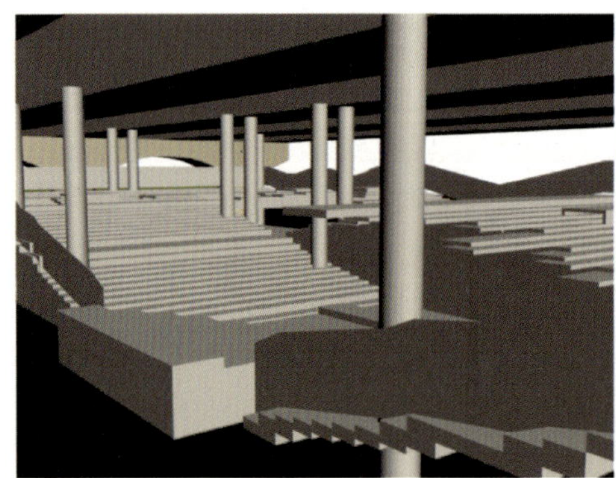

A view

B view

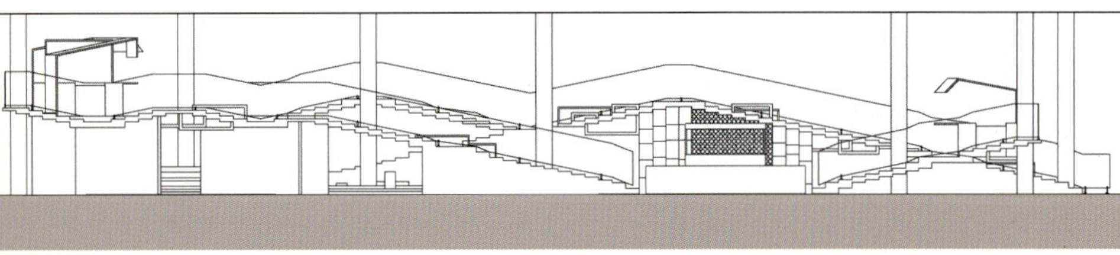

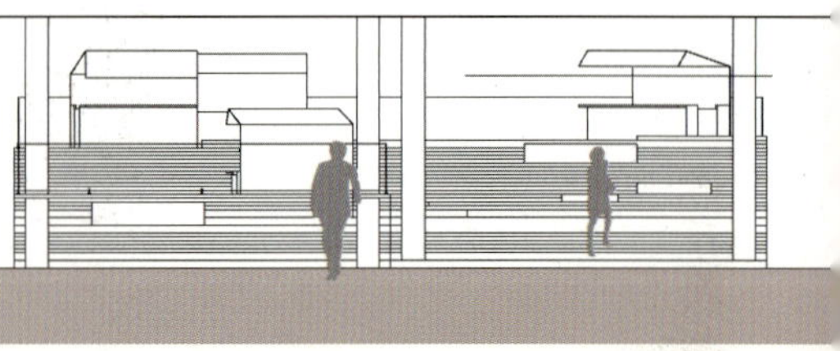

SPACE PROGRAM

음악다방 Music cafe · 라이브무대 Live stage · 선술집 Pub · 만화방 Cartoon room · 포스터, 표어 Poster & Watchword · 오락기 Amusement machine

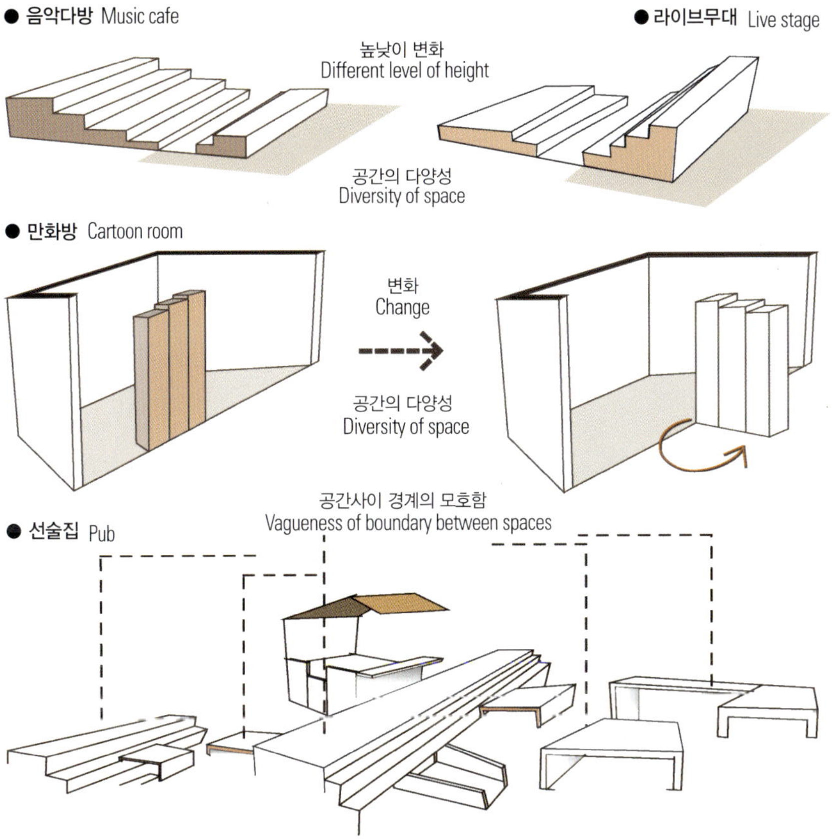

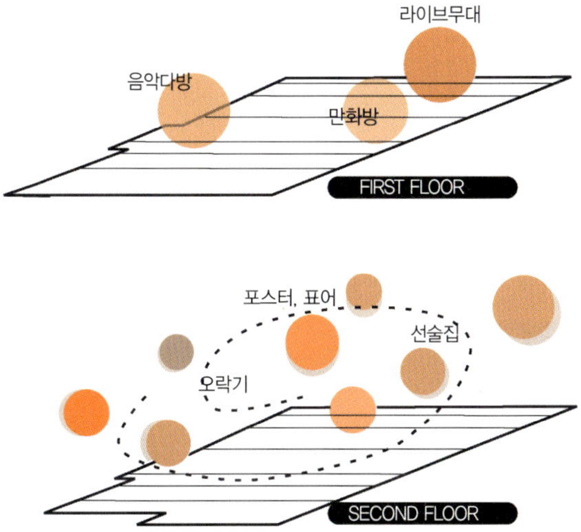

● 음악다방 Music cafe

높낮이 변화
Different level of height

● 라이브무대 Live stage

공간의 다양성
Diversity of space

● 만화방 Cartoon room

변화
Change

공간의 다양성
Diversity of space

● 선술집 Pub

공간사이 경계의 모호함
Vagueness of boundary between spaces

라이브무대
음악다방
만화방
FIRST FLOOR

포스터, 표어
선술집
오락기
SECOND FLOOR

곳곳에 배치된 각기 다른 형태의 좌석과 오락기, 7080시대를 재현해낸 물품, 포스터는 공간에 활동성을 부여하였다.

Different shape of Seats and game machines that placed in several spots, goods and posters that reproduce 70's and 80's give energetic feeling

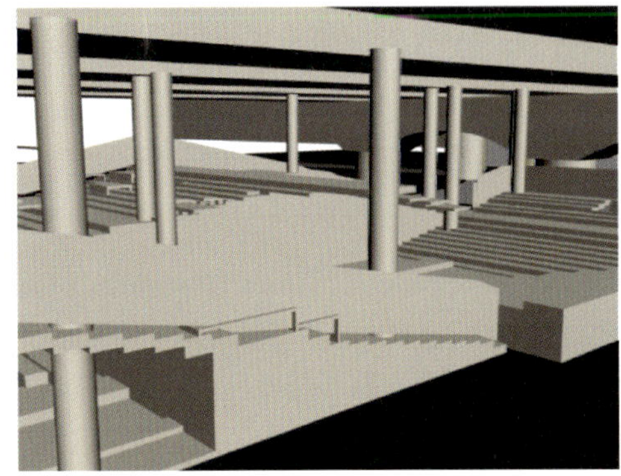

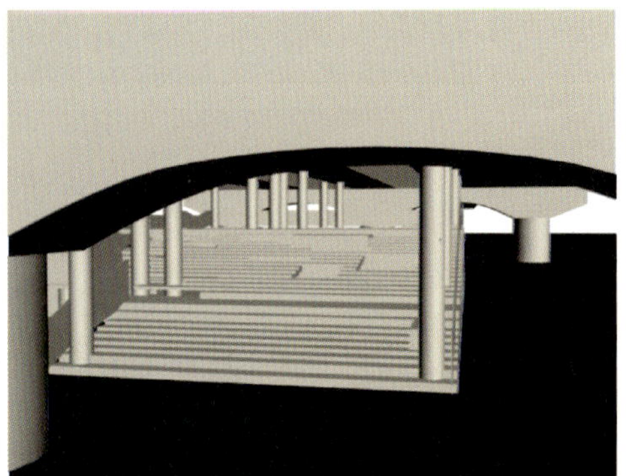

SECTION B-B'

SECTION A-A'

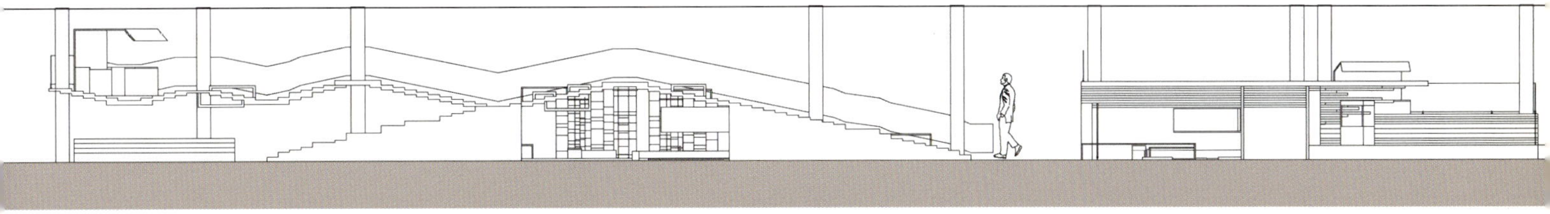

뚜렷하게 나누어 지지 않은 공간은 찾는 이들 또한 각자가 가지고 있는 벽을 허물고 서로를 즐기며 소통할 수 있기 위함이다.
이로 인해 세대 간의 문화교류가 이루어지도록 하였다.

Unclearly divided spaces have intention to let visitors break down walls in their hearts, enjoy and communicate each other.
Also, the spaces emphasize cultural communication between different generations

1층에 위치한 음악다방과 라이브 무대는 7080세대를 대표하는 문화이다. 대표하는 문화인 만큼 공간을 차지하는 비중이 크다. 라이브 무대는 주로 시민들의 참여로 이루어 지는 공간이다. 그 시대의 음악은 생활에 지친 그들에게 그 시대를 추억하며 휴식을 취할 수 있는 공간을 제공한다.

Teahouse with musical performance and live stage represent the generation of 70's and 80's culture. Therefore, they possess lots of spaces. Live stage is mostly made up of people's participation. 70's and 80's music offers place to rest and make them reminiscent of the past from tough life.

추억의 LP판으로 음악을 감상할 수 있는 뮤직박스.
Music box that we can listen golen oldies with LPs

과거와 현재에 유행하는 만화책을 모두 볼 수 있는 만화방
Comicbook shop that we can read popluar comicbooks in the past and now

2층에는 각기 다른 공간에 음식점들이 위치해 있다.
Different restaurants are located in 2ndfloor

음악다방 Music cafe

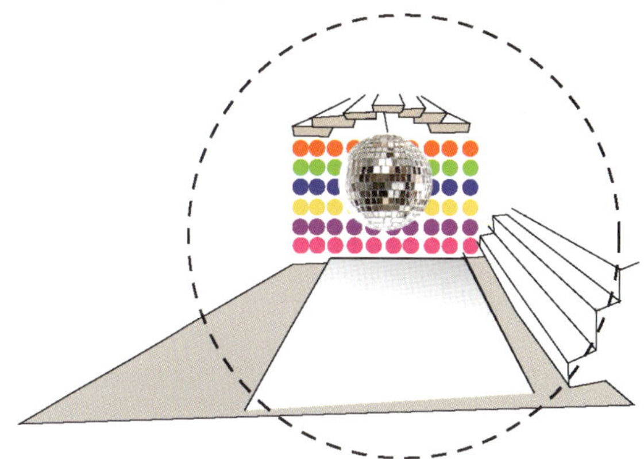

어릴 적 방과 후 들렀던 분식집

Snack bar that we used to drop by after school in the childhood

뮤직박스와 연결되는 공간에 위치한 다방

Teahouse is located in the space linked to the music box

현재 급격히 사라져가고 있는 포장마차 특유의 분위기를 재현해낸 선술집

Pub that is reproduced in particular mood of cart bar which faded away from modern scenes

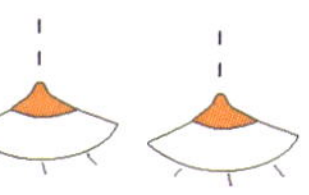

모든 공간에 단 차이를 둠으로써 공간 그 자체가 좌석이 될 수 있도록 하였다.
By adjusting level of height between each space, the spaces function as seats.

벽을 두지 않은 공간들은 자연스럽게 소통하게 된다.
Non-blocked spaces and other spaces are naturally connected.

돌아올 (회), 돌아올 (환)

(return) 품으로 다시 되돌아오다

Separated from the arms of those who were always with us come back again

현재 그들은 우리에게 무관심과 비난의 소리를 듣는다. 회환하다는 다시 되돌아 오다라는 뜻으로 다시 한번 우리의 곁으로 되돌아 오라는 뜻이다. 과연 이 공간 속에서 그들이 어떻게 재 탄생할 것인가? 이들에게 우리는 어떻게 다가가고, 그들은 또한 어떻게 달라질 수 있을까? 어떠한 방법으로 우리의 품으로 다시 돌아 올 수 있을까?

Current They are listen condemnation and apathy to us. return is means come back, They come back to be with us again means .In this space, how they will be re-birthing? How do we go to them approached and Also, how do they be different? we could come back what ways into the fold?

BACKGROUND

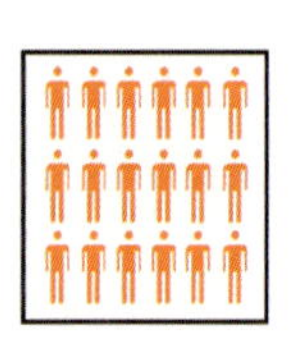

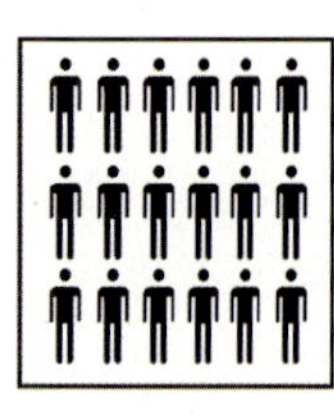

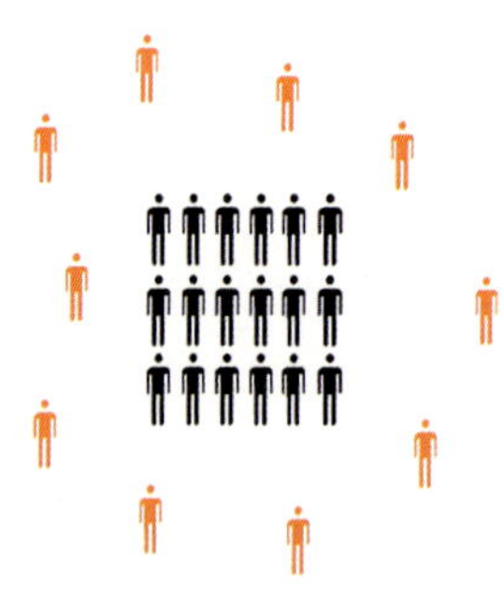

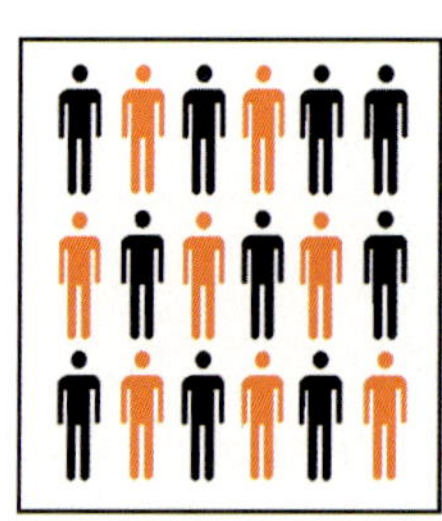

우리에게 일반적으로 사회적 약자는 노인,아동,여성,장애인을 생각한다. 하지만 다른 방향으로 생각해 보면 앞에서 예시한 사회적 약자보다 더 우리 현대사회에서 심각한 약자가 존재하고있다. 그들은 출소자다.

그들과 어울려 이 사회를 살아가는 모습을 보여주기 위해 숨어있지만 개방적인 공간을 통해 사람들과 조화를 이루고자한다. 이 개방적인 공간은 직업의 기회를 갖지 못하는 출소자들에게 직업의 기회를 주며, 좋지않은 인식속에 받은 심리적인 감정을 안정화 시켜 일상적인 생활을 할 수 있도록 기본적인 삶의권리를 주어지게 해줄 것이다.

We typically underprivileged children mostly elderly women with disabilities are thinking. but more socially underprivileged in our society there is a serious stands. We believe that the acronym of the prisoner.

They mingle with the society to live in hiding to show the appearance, but through an open space so that people will mingle with. This open space does not have an opportunity of a job opportunity gives him a job. So who in the bad perception of the psychological and emotional stability to do with our lives the basic rights of life, something that should be given

WHY?

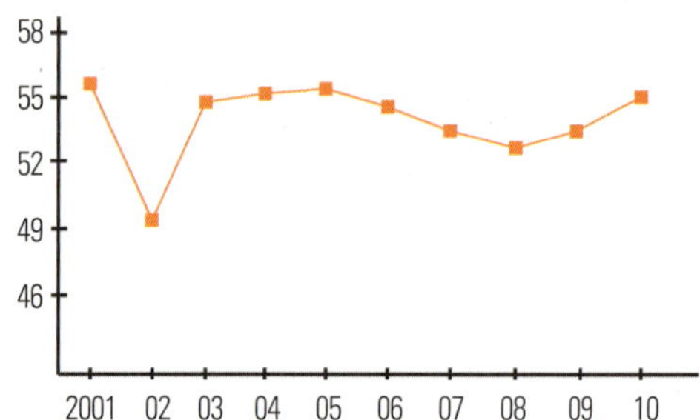

현재 출소자들은 출소 후 다시 범죄를 일으킬 확률이 50% 이상을 보이고 있다. 하지만 그들에게 우리가 조금만 더 관심을 기울이고 복지시설에서 삶의 권리를 제공한다면 그 재범률이 0.5%에도 미치지 못하는 결과가 나오고 있다.

Current inmates released from prison after a high chance of re-offense is showing more than 50%. But they are a little more paying attention to our welfare rights of life, if you provide facilities that represent 0.5% recidivism rate.

TARGET 출소자

CHARACTER	NEED	RESOLUTION
대인 기피 현상 Avoid interpersonal phenomenon	사람들과의 어울림 Comport with people	직접 커피숍 운영 Operating a coffee shop
개인주의 현상 Symptoms of individualism	공동체의 소중함 인식 It is a cherished community recognizedcommunity	소통의 공간 Space of communication
사회의 불필요한 존재 The presence of unnecessary social	직업의 기회 제공 Provide career opportunitiessocial	바리스타 자격증 취득의 기회 Barista is qualification

REQIREMENT

대화의 공간

바리스타 자격증

강연장

커피숍

SITE

경기도 수원시 장안구 우만동 효성사거리

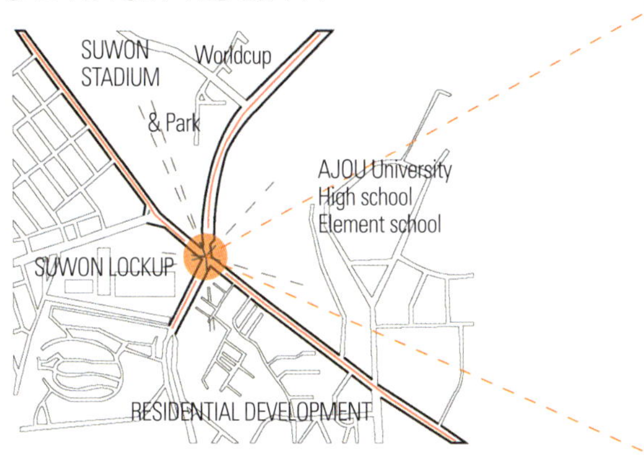

구치소

구치소가 존재한다. 시민들에게 직접적으로 위험하지는 않다. 하지만 그들은 불안하게 살고 있다.

There is a detention center. It is not dangerous to the citizens directly, but they are anxious to live.

월드컵 경기장 및 공원

월드컵 경기장이 위치하여 가족단체, 학생, 연인 등 수많은 유동인구가 이 곳을 유동하고 있다.

World Cup Stadium is located. Thus family organizations, students, couples, and many people are floating.

학교건물

이 곳은 초,고,대학교가 존재한다. 하루에 학생들이 약 800명이 이동한다.

This place elementary schools, high schools and universities exist. Approximately 800 people a day for students to move.

주거단지

이 곳은 아파트 단지들이 위치한다. 그래서 니 많은 유동인구를 존재하게 한다

This place is the location of an apartment complex, so that there is more to the floating population.

DESIGN CONCEPT

새의 둥지 Bird's Nest

우리에게 무관심과 상처를 받은 그들의 은신처가 되어준다. 그들을 새로운 사람으로 만들어주고 세상에 적응 할 수 있도록 해주는 목적에 맞게 새의 둥지를 컨셉으로 잡았다. 새의 둥지는 자신이 어떤 수단으로 만들어 낸 공간이다. 이 공간에서 자신의 노력에 따라 유용하게 사용 될 수 있는 기능들로 채워져있다.

We received the indifference and hurt will be their refuge. They made a new man and the world That allow you to adapt to suit your purpose as a bird's nest I got the concept. Bird's nest in any way which it is a made-up space. Depending on their efforts in this space that could be useful is filled with features.

DESIGN KEYWORD

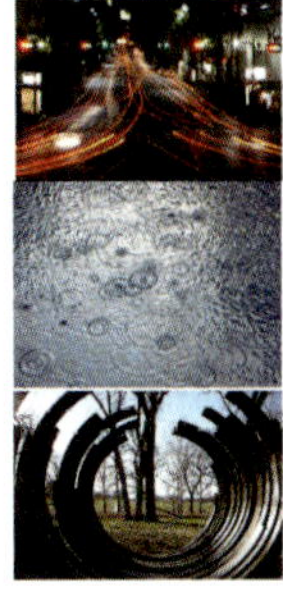

분기 (BIFURCATION)

두 갈래로 나눠지는 것 – 동선의 표현
의미적 분기 – 공간 용도의 다양성

Be divided into two streams - the show route
Semantic bifucation- the use of various spatial.

겹치다 (OVERLAP)

원의 형태들이 겹쳐서 공간의 매스를 구성
형태들의 겹침에 의한 디자인 구성

Overlapping circles in a configuration space
By overlapping the design configuration

둘러싸다 (ENCLOSE)

둘러싼 원의 형태를 가지고 FACADE 디자인
둘러싸인 반복된 느낌으로 인한 디자인 구성

Circled in the design configuration has the form is enclosed.

MASS PROCESS

1. 의미적 분기 – 사용자들에게 따른 공간의 용도
Semantic bifucation-users according to the use of space.

흐름 속에 계단	+	강연장	바리스타 자격증 실기장	+	커피숍	통로구간	+	음악감상
Stairs in the flow	+	Auditorium	Barista Certification Practice	+	Coffee Shop	Section Chair	+	Listening to music

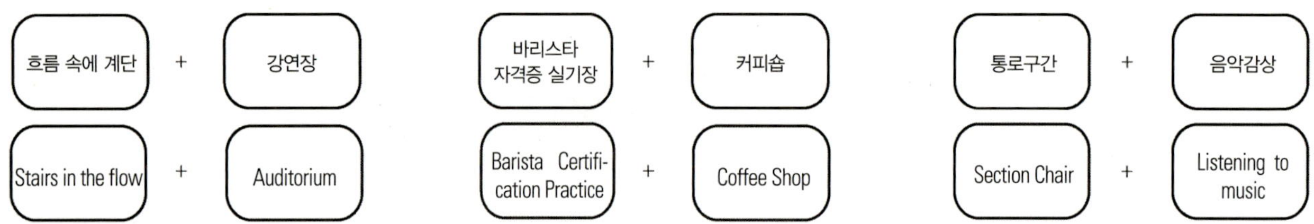

2. 겹침에 의한 공간 구성들
By the spatial configuration of the overlap

(1) 곡선에 의한 겹침

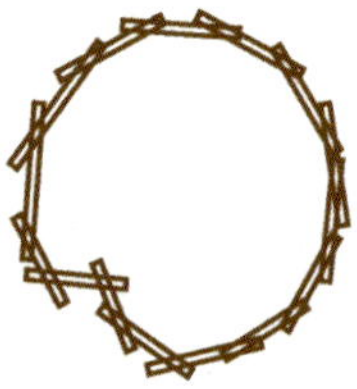

둥지를 구성하는 선(가지)들의 결합
Lines that make up the nest (twings) a combination of.

1. 가지들의 겹침 과정
Process of branch the overlap.

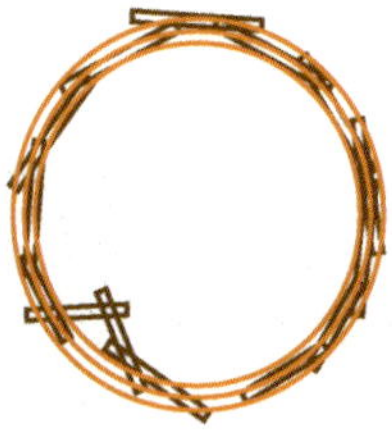

선에 의한 나타나는 원형의 형태
By the overlap of lines in the form of a circular.

2. 원들의 겹침 과정
Course circles that appear to overlap.

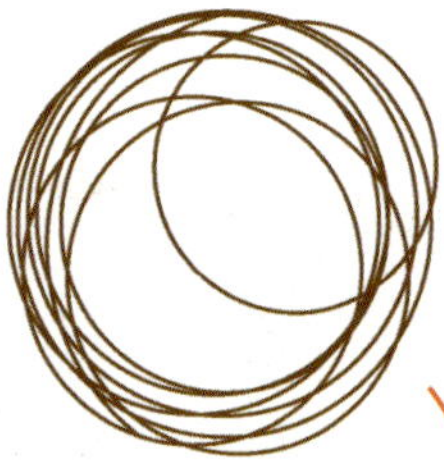

원들의 다양한 크기가 겹침
The overlap circles of different sizes.

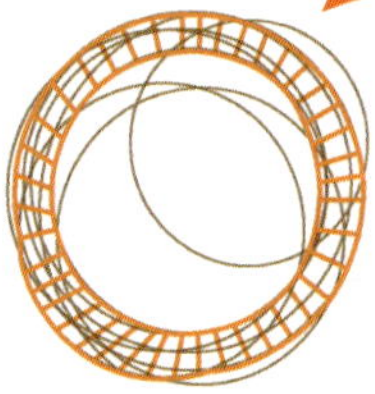

원들을 선택하여 하나의 원으로 전체적인 공간을 구성하였다.
Select the circles of one circle was the whole configuration space.

3-2. 겹침에 의한 공간속의 공간 구성
Spatial configurations by an overlap in space.

3-3. 원형형태의 조명과 단의 생성
Stage lighting and creation of a circular form.

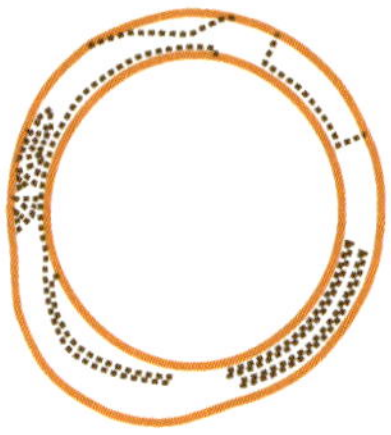

실선은 겹침에 의한 전체적인 공간을 구성하였고 점선은 공간속의 공간 및 조명, 단을 구성하였다
The solid line the overlap with the whole configuration space. The dashed lines represent the space was generated with the lighting and stage

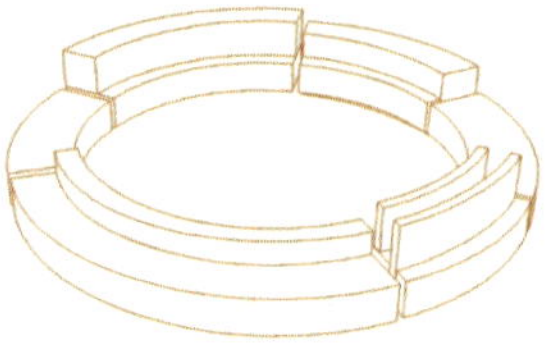

원형 형태의 공간에 또 다른 열린 공간과 닫힌 공간을 구성하였다
Another round in the space of forms of open space and has formed a closed space.

(2) 직선에 의한 겹침

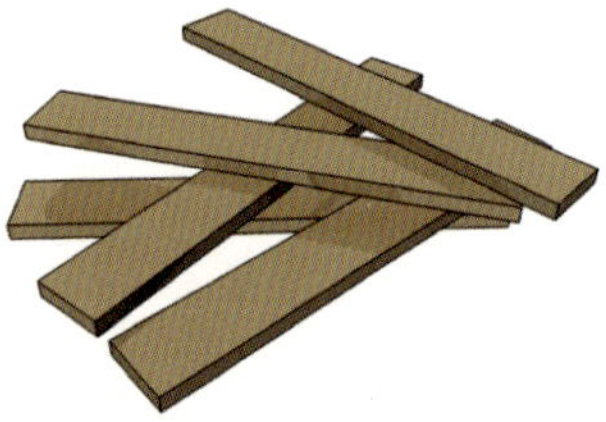

둥지를 구성하는 나뭇가지를 도형으로 표현했다.
That make up a nest of twigs were expressed in the form.

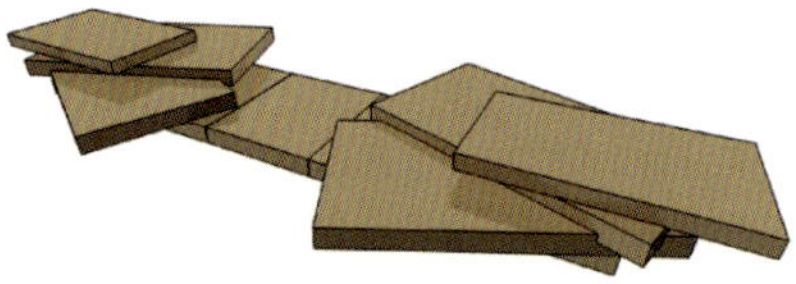

수직으로 나뭇가지들을 겹쳐 단을 형성하였다.
Overlapping vertical columns formed of branches.

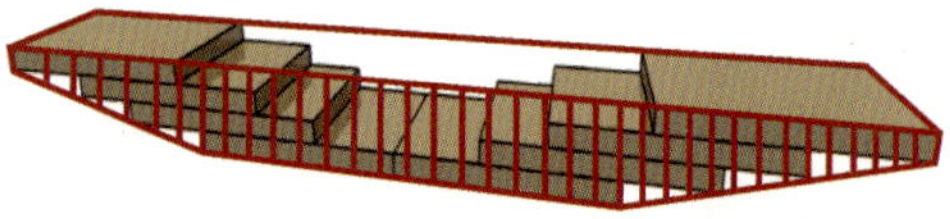

공간 내의 단의 겹침으로 오르막과 내리막을 나타내었다.
Uphill and downhill to the overlap in space stage showed.

DESIGN PROCESS

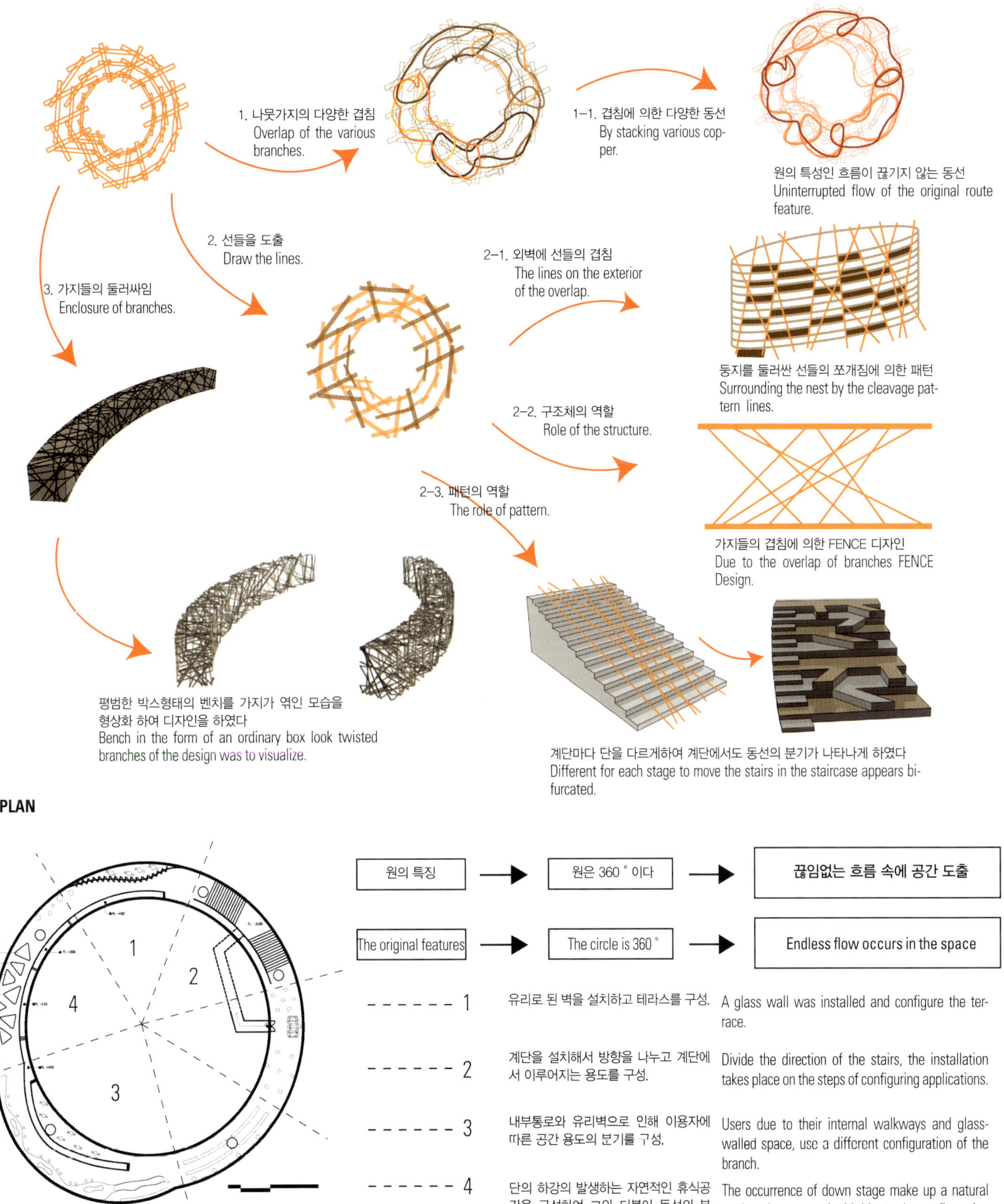

SPACE PROGRAM

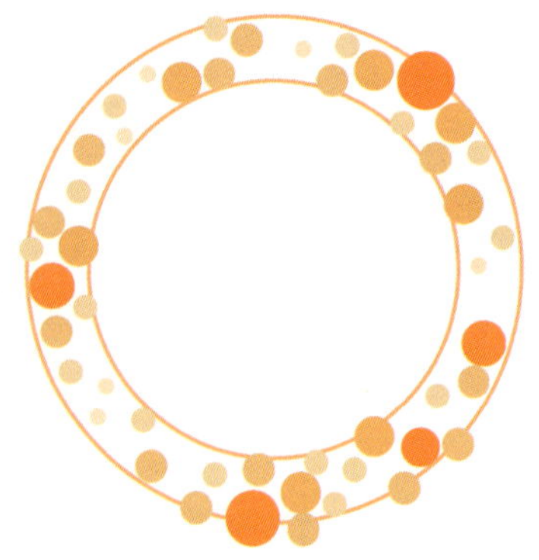

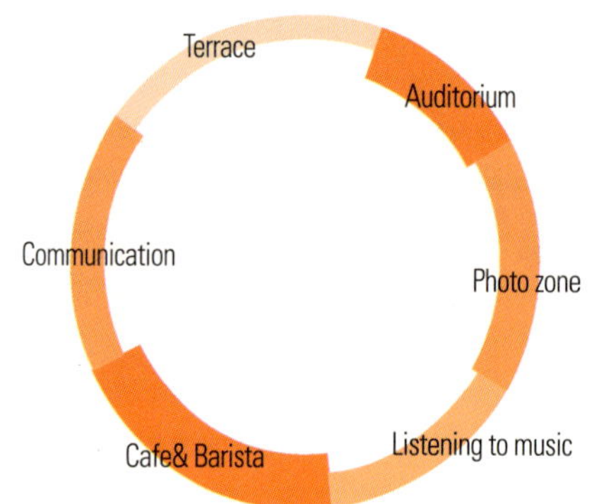

공간 내의 유동 밀집 분포

공간의 형태에서 나타나는 사람의 밀집분포이다. 주로 위쪽과 아래쪽에서 많은 유동이 발생하여 일상적인 생활을 하면 지낸다.

Floating density distribution in space

Spatial distribution in the form of a bunch of people are appearing. Many flow from the top and bottom usually occurs when you come on a day-to-day life.

공간 속의 다양한 용도

공간에서 그들이 행할 수 있는 용도와 위치이다. 구분은 되어있지만 흐름속에서 하나의 공간으로 나타나고 있다.

Floating density distribution in space

They can do in space is the purpose and location. Are separated by a space, but in the flow are shown.

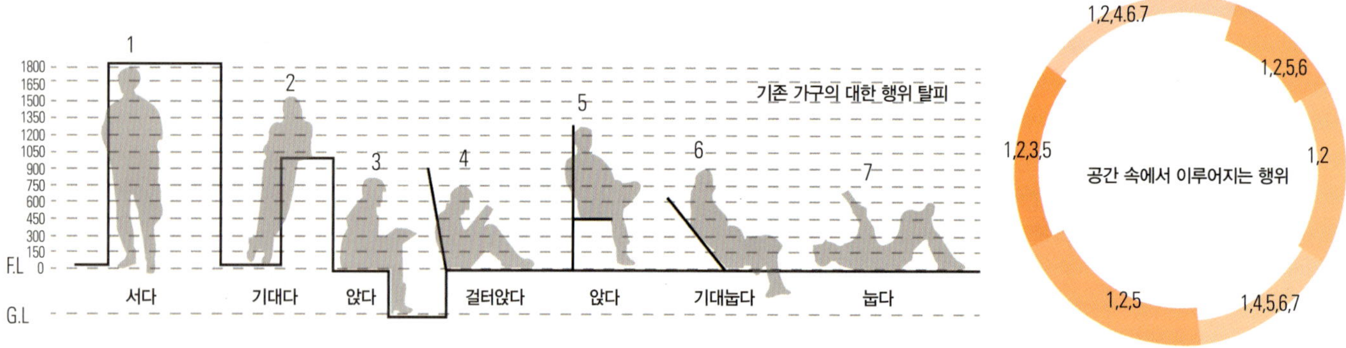

기존 가구에 의존하여 공간내에서 행위가 이루어 졌으나 이번 우리의 공간에서는 자유롭게 어디서나 서거나 기댈 수 있다. 기본 바닥보다 낮은 장소에선 걸터 앉을 수도 있으며, 쿠션이라는 물체 하나에 의해 누울 수도 엎드려 행동 할 수 있다.

By relying on existing Furniture made in the space but its activity in this area freely anywhere in our stand, I could count. In places the floor is lower than the default may be sitting straddles and is cushioned by a single object can lie down you can act.

ELEVATION

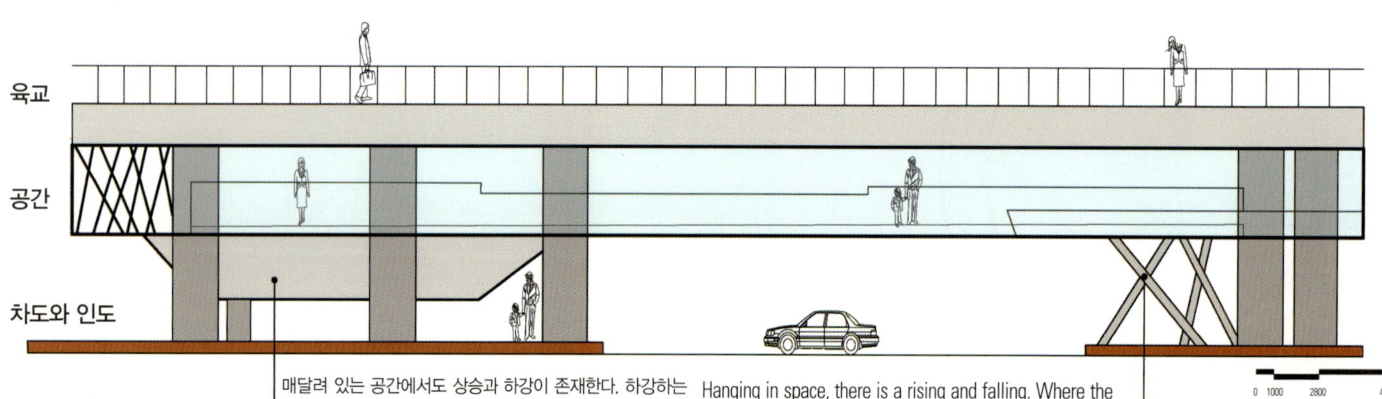

매달려 있는 공간에서도 상승과 하강이 존재한다. 하강하는 곳은 땅과 연결이 되어 보행상의 안전 문제가 없다

Hanging in space, there is a rising and falling. Where the land is connected to the falling pedestrian safety, there is no problem

구조체로 지지대를 설치하였다

The support structure was installed.

Bird's eye view
Barista practical view
Soo CAFE
직업을 가지기 위해 미리 바리스타가 되어 사람들과 만나 거리감을 없애는 공간
Barista's job is to get a pre-meeting people and eliminate the space distance
Cafe view
Soo&Soo
바리스타 자격증을 취득하기 위해 실기교육을 받으며 커피를팔아 사람들과 좀 더 가까워지는 공간
Barista qualification to receive practical training more we get closer to some people, coffee is the place to sell

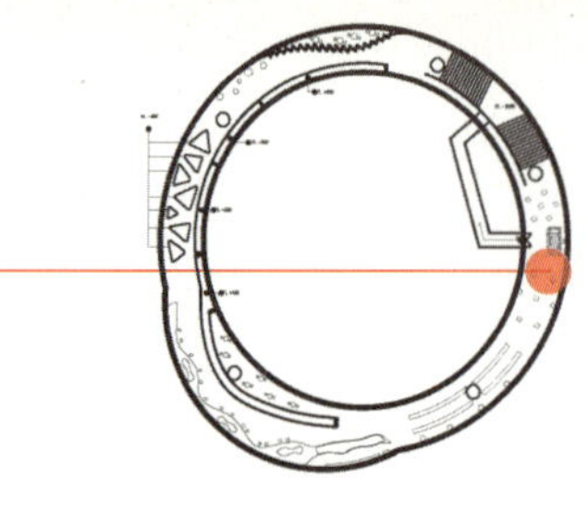

입구부분으로 처음에 들어와 사진전시대로 인해 이 공간에 대한 궁금증과 그들의 대한 거리감을 해소시켜주는 공간.

As part of the entrance to the first exhibition period due to come into this space for questions and their spatial distance that help relieve.

유리벽으로 하나의 벽을 설치해 테라스와 공간의 구분을 지어서 떠 있는 공간의 특징을 강조.

The glass walls of a segment of the space to install the terrace and a could build a floating emphasized the characteristics of the space.

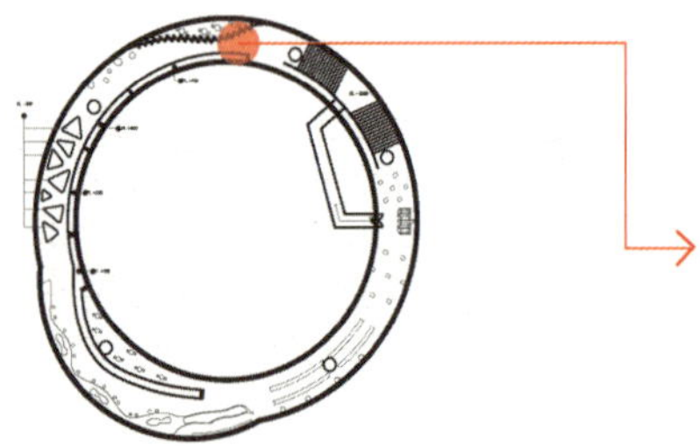

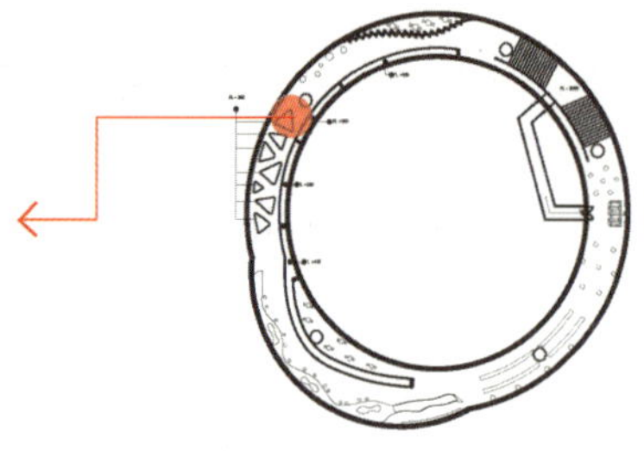

흐름 속에서는 단순히 계단이지만 이 곳에 프로젝트 빔을 쏨으로써 강연장이 될 수 있는 두가지 성격을 가지는 공간이다.

You simply place the stairs in the flow, but this project speaks to the beam has been installed with two basic characteristics is the space.

일정한 장소에 테이블 대신에 흐름 속에서 설치된 테이블에 앉아서 자연스럽게 이야기하고, 또한 커피를 마시고 즐길수 있는 공간.

On the table instead of a regular place to sit at your table is installed in a natural flow to the story, and also a place to enjoy a cup of coffee.

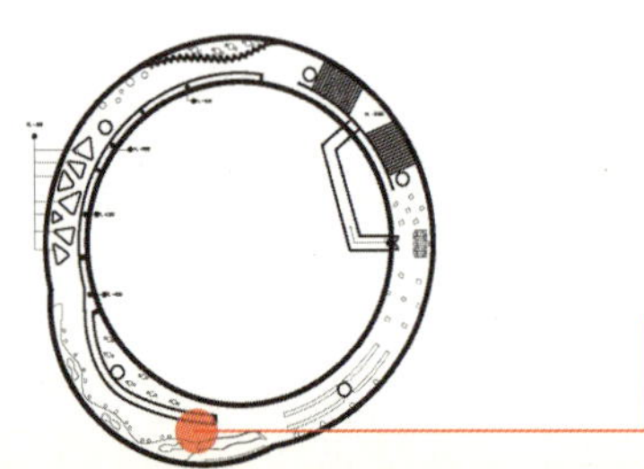

Stair hall view

흐름 속에서는 단순히 계단이지만 이 곳에 프로젝트 빔을 쏨으로써 강연장이 될 수 있는 두가지 성격을 가지는 공간
You simply place the stairs in the flow, but this project speaks to the beam has been installed with two basic characteristics is the space

Relaxation view

그들에게 책과 음악을 들으며 심리적인 안정감을 가지게 되는 공간
Their books and listening to music to have the peace of mind that the space

TALE OF TALES

이야기 속의 또 다른 이야기

대림역은 서울 구로구에 위치해 있으며 직장인들과 학생, 외국인들의 유동인구가 많다. 서울 메트로 2호선과 7호선이 교차하고 있으며 역 밑의 공간은 도림천이 흐르고 기둥이 굉장이 많다. 또 자전거 길과 조깅을 즐길 수 있는 길뿐이다. 하지만 이러한 시설조차 직장인들이 퇴근한 저녁시간이나 주말에 밖에 사용하지 않는 실정이다.

Located in Seoul, Guro daerimyeok and students, office workers, a lot of floating population of foreigners.
Line 2 and Line 7 has many dorimcheon flows below the pillar.
And the road that you can ride a bike long enough to enjoy jogging, but even these facilities do not use except for the evening or on weekends.

BACKGROUND

옛날의 도림천은 빨래터로서 이용되어왔다. 빨래터는 사전적으로 시내나 샘터에서 빨래할 수 있게 만든 일정한 장소를 일컫는데, 예로부터 빨래터는 혼자가 아닌 여럿이 이용하는 공간으로 마을 여인네들의 사랑방 역할을 하였다. 사람이 모여 사는 곳이면, 으레 빨래터가 생기기 마련이었고, 빨래터는 서민의 생활과 밀접한 그들의 정신적, 생활적 중심공간이자 공공적인 대중의 장소였었다.

Long time ago, Dorimcheon was used as a public wash place. The dictionary definition of wash place is a stream or a spring for certain place of doing laundry. Those places had been treated as women's site for chattering. Once a town was formed, a public wash place would be created so far. A wash place is a spiritual, life site of ordinary people and public place.

그러나 오늘날 현대사회는 오프라인만남보다는 온라인상에서의 만남이 중점이 되고 있다. 사람과 사람의 만남이었다면 오늘날에는 익명과 익명의 만남이 되고 있다. 온라인상에서의 만남을 옛날의 빨래터의 기능처럼 오프라인으로 만날 수 있는 공간, 즉 오프라인 공간을 만들어 보고자 한다.

However, meeting in online rather than off-line is being emphasized in nowadays. Meeting in past is more like man-to-man which is totally opposite to today's anonymous meeting. We want to make the online meeting like the public wash place as offline site.

구로디지털단지는 2000년대에 들어 산업구조가 바뀌면서 첨단산업,즉 중소기업들이 많이 입주하게 되었다. 옛날의 공장들은 단층이었다면 오늘날은 아파트 형 공장으로 층을 많이 쓰게 되었는데 그로인해 인력수가 많이 필요하게 되었다. 공장이 많은 데에 비해 인력이 부족하여 인력난이 일어나고 있는 것이 현실이다.

Smaller businesses are moving in Guro Digital Complex thus industrial structure has been changed since 2000. Most of factories in old days were one-storied compared to apartment style factory in nowadays. Therefore plenty of workers are needed because of huge number of factory.

TARGET

공간을 필요로 하는 모든 사람들
All the people who need the space

온라인 상에서 만남을 오프라인 만남으로 이어주는 공간을 제시한다. 온라인 상에서는 남녀노소를 구분하지 않고 만남이 이루어지고 있으며 이 만남이 오프라인 만남까지 이어질수 있는공간은 앞으로 더욱 요구될 것이다.
또한, 중소기업 구인구직 공간 및 커뮤니케이션 공간 을 제시하여 구인정보를 알아보거나 커뮤니케이션 할 수 있는 모든 사람들의 요구를 들어줄수 있는 공간을 제시한다.

We suggest a site which links between online meeting and offline meeting. In online site, meeting is done without distinction of sex or age. Places for those meetings are link to offline will be expected more and more. Also we propose job opportunities for smaller business and communication place to satisfy people who want those information and communication.

SITE ANALYSIS

서울특별시 구로구 구로6동 125-66
6-dong, Guro-gu, Seoul 125-66
대림역
Daerim station

대림역 위치
Site Location

직장인 분포현황
Tracking the distribution of workers

교통을 이용한 접근
Access to transportation

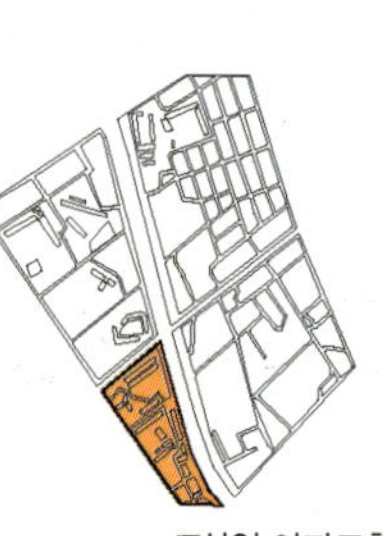

IT산업 아파트형 공장
IT industry, apartments, factories

종합분석
Comprehensive Analysis

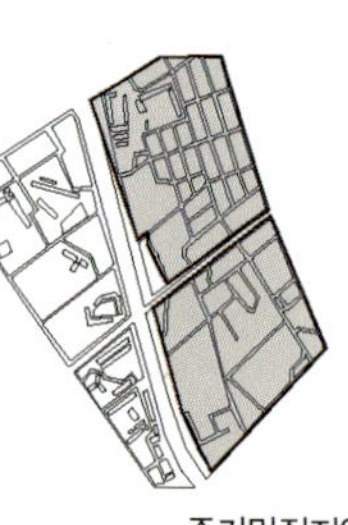

주거밀집지역
Residential area

WEEKLY PLAN

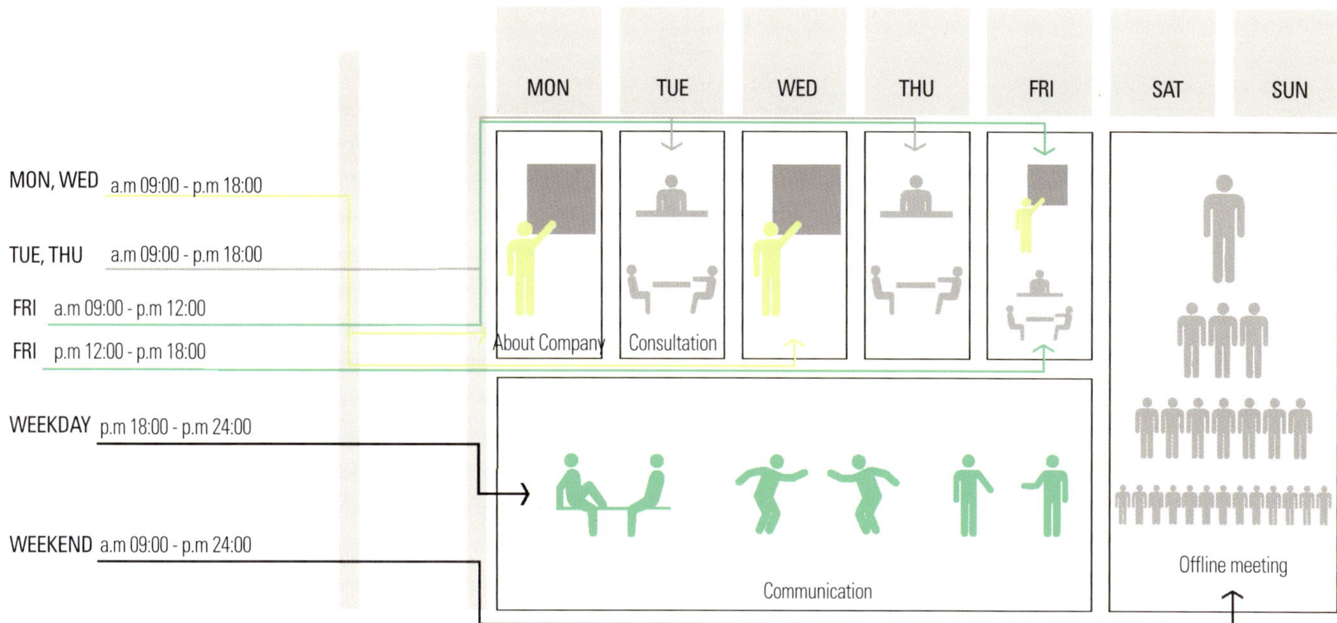

DESIGN CONCEPT

MONTAGE

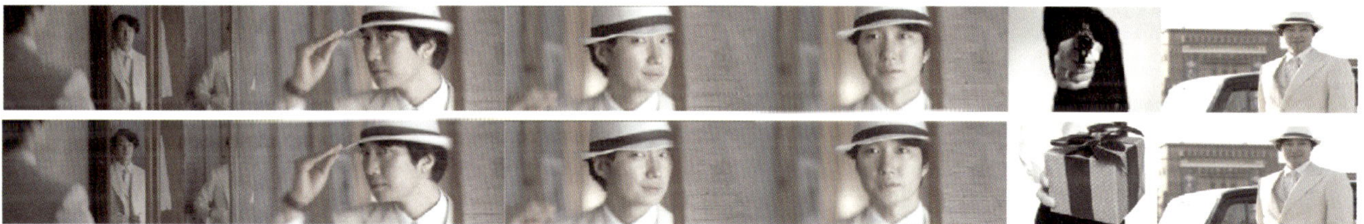

프랑스어로 모으다, 조합하다라는 뜻에서 사용되어 온 건축용어였으며 영화나 사진 편집 구성의 한 방법 따로따로 촬영한 화면을 적절하게 떼어 붙여서 하나의 긴밀하고도 새로운 장면이나 내용으로 만드는 일 또는 그렇게 만든 화면

Converge in French, is the combination was an architectural term that has been used as means.
Edit movies and photos as a way to split the configuration by editing the footage shot scenes. Or other information or making a scene that is created.

DESIGN

1. 장면과 장면을 붙이거나 적절하게 장면과 장면사이에 다른 장면을 넣음으로써 생기는 새로운 의미 추출
2. 영화를 만들기 전까지의 편집을 통한 과정 추출

1. Scene and the scene is inserted. Or in between scenes to add another scene occurs is a new scene.
2. Making films until the extraction process by editing.

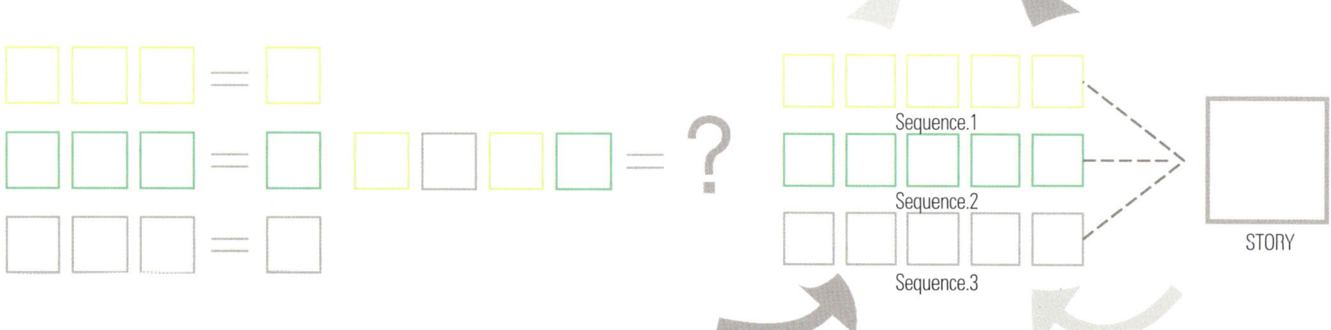

DESIGN KEYWORD

Shot

촬영의 기본 단위로서 한 번에 촬영한 장면
Basic unit of photography shoot once

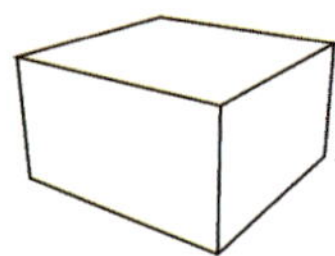

쇼트를 하나의 매스로 봄
shot snap as a mass

Scene

몇 개의 쇼트가 모여서 이룬 장면
a scene gathered some shot snaps

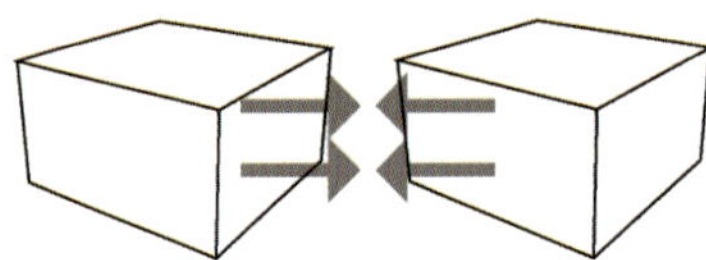

매스와 매스를 충돌시킴
crashing mass.

Sequence

몇 개의 신이 한 시퀀스를 이룸
A sequence gathered some scenes

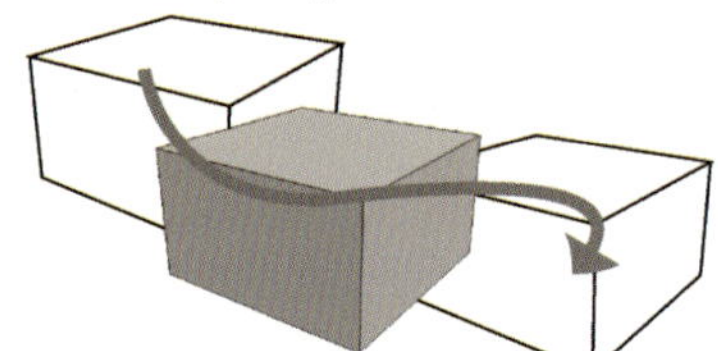

매스와 매스를 충돌시켜 매스간의 충돌이 새로운 공간을 만듬
Collision between mass make new space by crashing mass.

EDITING

선조적 편집
Metrical Montage

그 자신의 분류에 의하여 이 상충의 몽타쥬는 장면간의 내용에는 관계없이 커팅의 템포에 따라 병치되는 편집방법.
By his own classification of the sketch of a conflict between the contents of the scene, regardless of the cut, depending on the tempo of the editing methods are juxtaposed.

긴매스를 하나의 시퀀스로 봄
long mass as a sequence

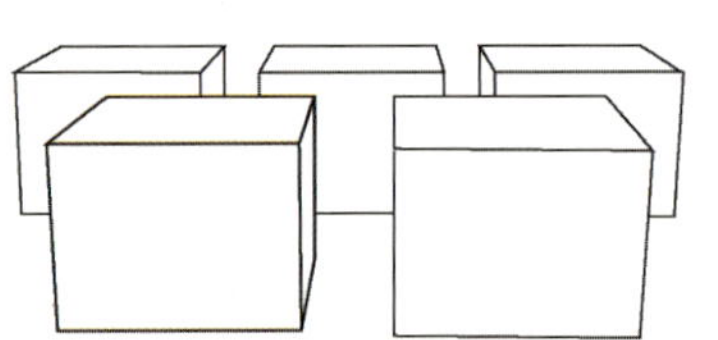

시퀀스안에 들어갈 각 개별의 매스들
juxtapose and arrange each mass'
by tempo of cutting which will be in sequence.

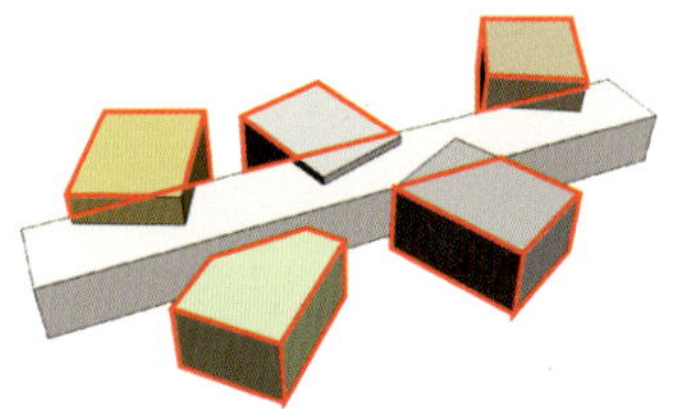

시퀀스안에 들어갈 각 개별의 매스들을 커팅의 템포에 따라 배치및 병치시킴
juxtapose and arrange each mass' by tempo of cutting which will be in sequence

격조 편집
Tonal Montage

멜로디는 물론 색채나 조명에 따라 정서적 반응을 유도하는 편집방법.
an editing methods by not only color or lighting but melody.

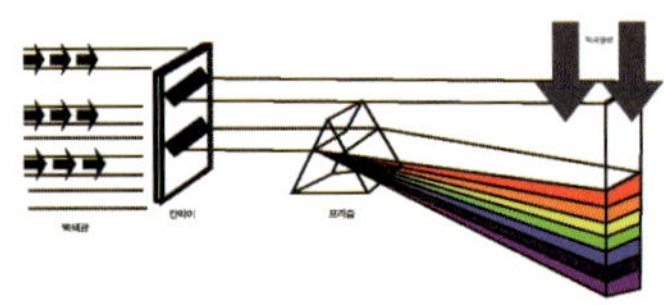

빛을 셀로판지에 비추었을때 생기는 색감.
Cause when illuminated with light caused color selropanji.

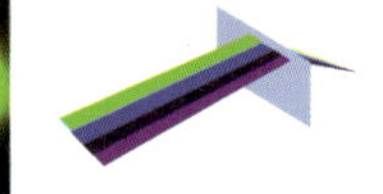

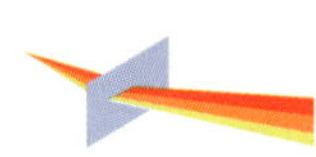

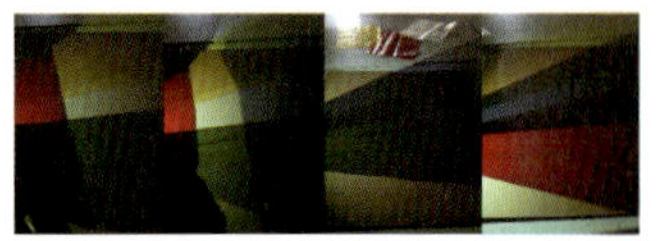

빛을 비추었을때 색과 색이 충돌함으로써 생기는 전혀다른 색감.
Cause when the light illuminated completely different color and color saturation caused by the conflict.

지적 편집
Intellectual Montage

장면간의 충돌이 새로운 지적 개념을 낳고 이것이 다시 다음 장면과 충돌하여 또 다른 지적 개념을 갖게하는 편집방법.
an editing methods which continuous collision between scenes makes new intellectual concept.

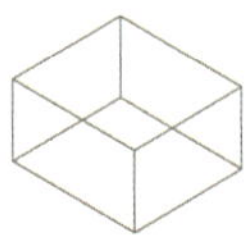

매스

Mass

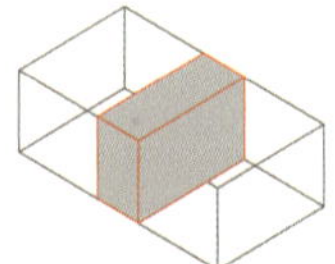

매스와 매스충돌.

Mass and mass collision.

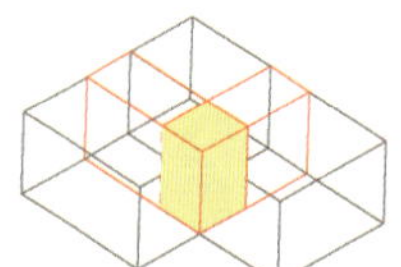

매스 충돌 후 다른 매스와의 충돌.

Collisions with other mass after mass crash.

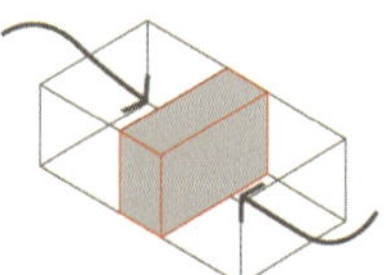

매스간의 충돌 후 중첩되는 새로운 공간(관점) 형성.

The new space after a collision between the mass nesting (view) form.

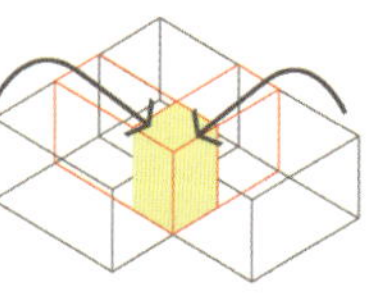

충돌 후 다른 매스와의 충돌하여 또 다른 새로운 공간 형성.

Collisions with other mass after a collision with another new space (perspective), forming.

SPACE PROCESS

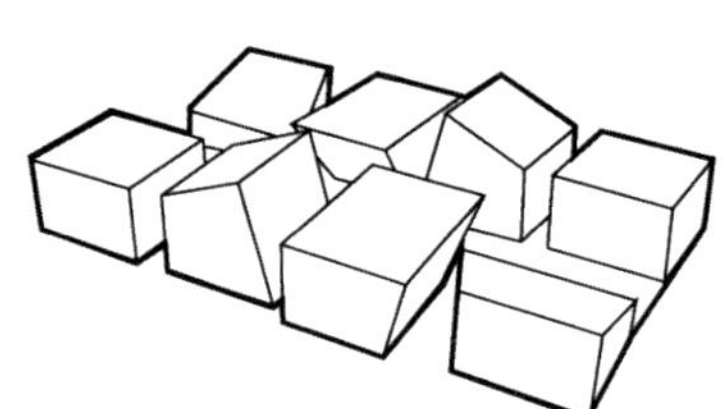

공간안에 들어갈 각 개별의 매스들
Each mass will be in space

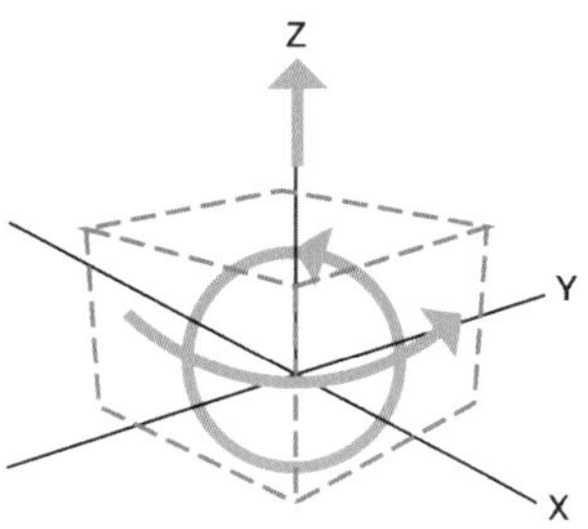

공간안에 들어갈 각 개별의 매스를 X, Y, Z 축으로 회전 및 움직임
Movement and rotation of each mass will be in space, by x, y, z axis

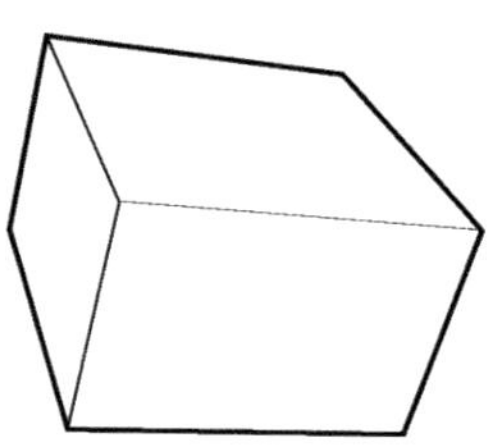

매스들을 회전 및 움직임으로서 생기는 공간의 변화.
Change of space caused by rotating and moving mass .

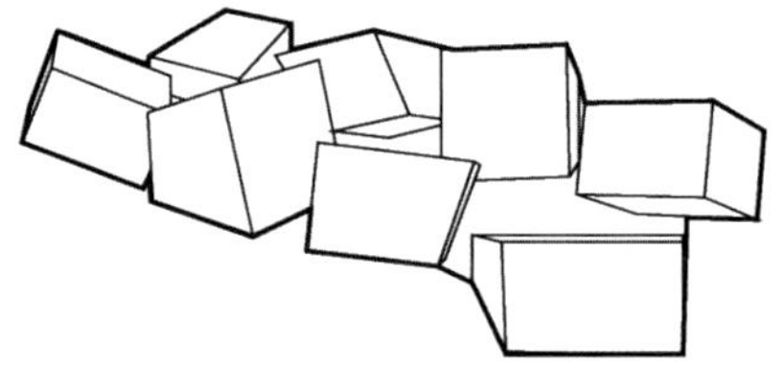

공간안에 들어갈 각 개별의 회전 및 움직임된 매스들.
Each rotated and moved mass ' will be in space.

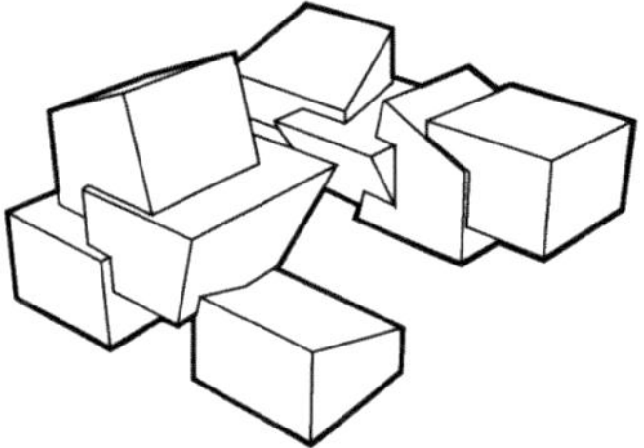

회전 및 움직임된 매스들을 충돌함으로서 공간의 위계가 없어짐.
Grade of space is disappear by crashing rotated and moved mass.

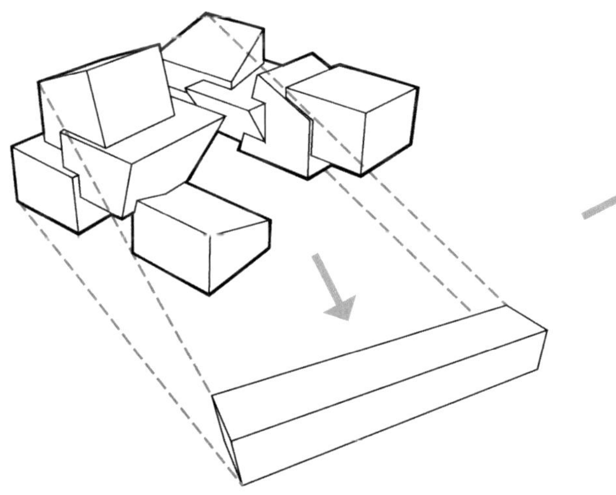

각 개별의 회전 및 움직임된 매스들을 공간안에 결합시킴.
Combine each rotated and moved mass' into space.

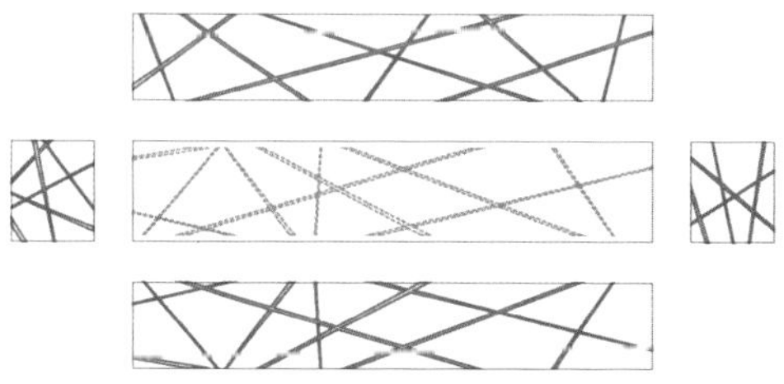

결합된 매스들의 외곽선을 이용하여 외부프레임 및 내부조명 사용.
Use outer frame and inner illumination by using combined each rotated and moved mass' contour line.

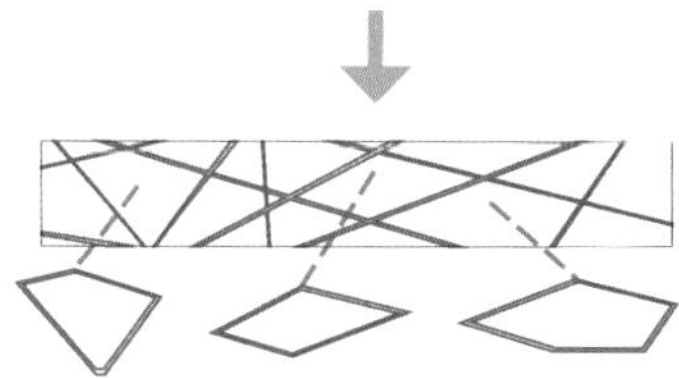

프레임 사이에 매스조각에 프리즘을 이용하여 색깔과 색깔을 충돌시킴.
Crash color and color using prism of mass piece between frame.

DESIGN KEYWORD

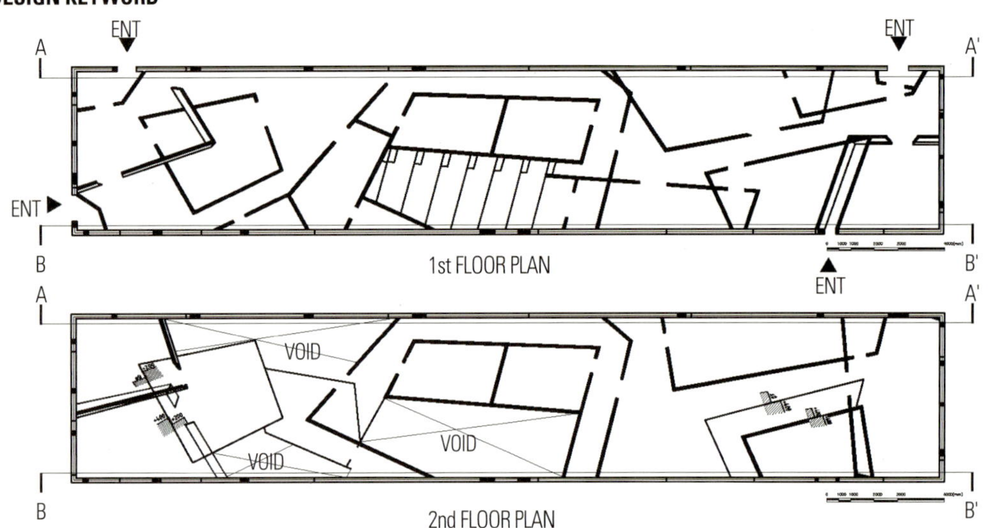

1. 강의실 및 상담실
2. 창고
3. 화장실
4. 계단강의실

1. Classroom and Counseling
2. Warehouse
3. Toilets
4. Stairs classroom

1. 오프라인 만남실 (주말)
1. Offline Meeting Room (Weekend)

SECTION

SECTION A - A'

SECTION B- B'

EXPLODED VIEW

TRAFFIC LINE

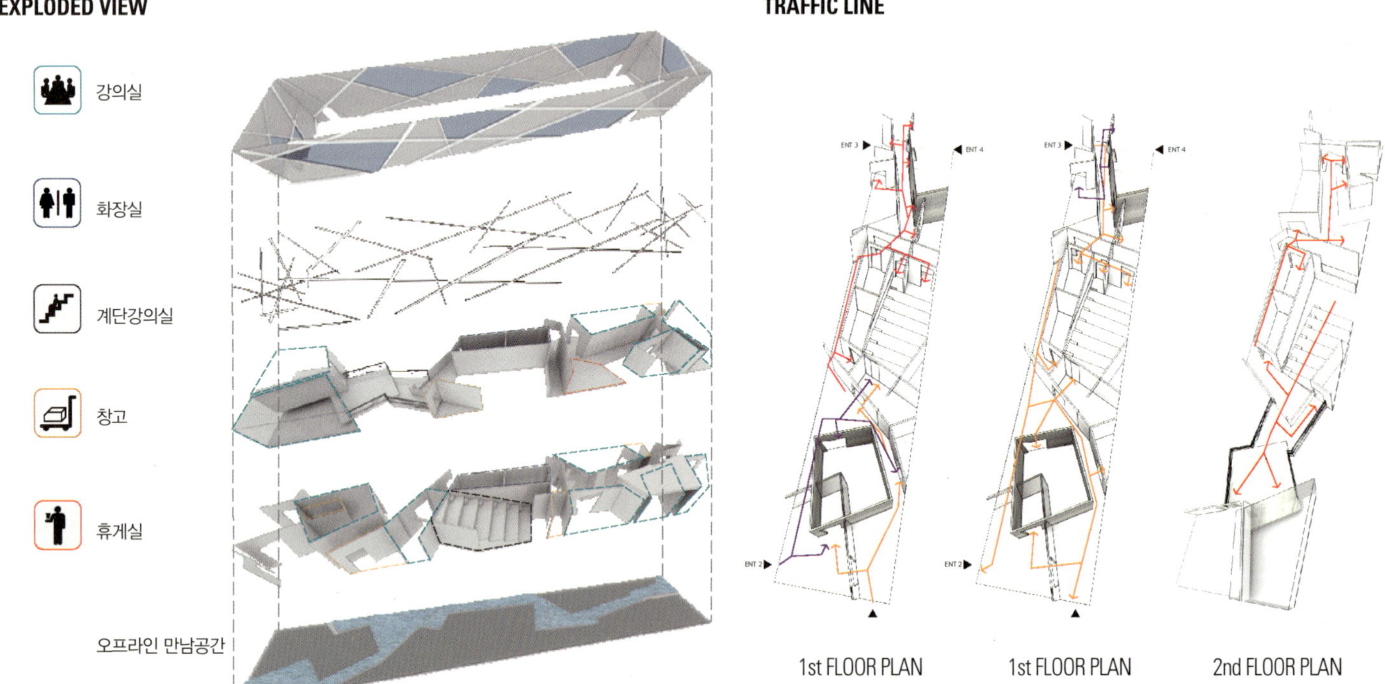

void
void
void
2ST FLOOR

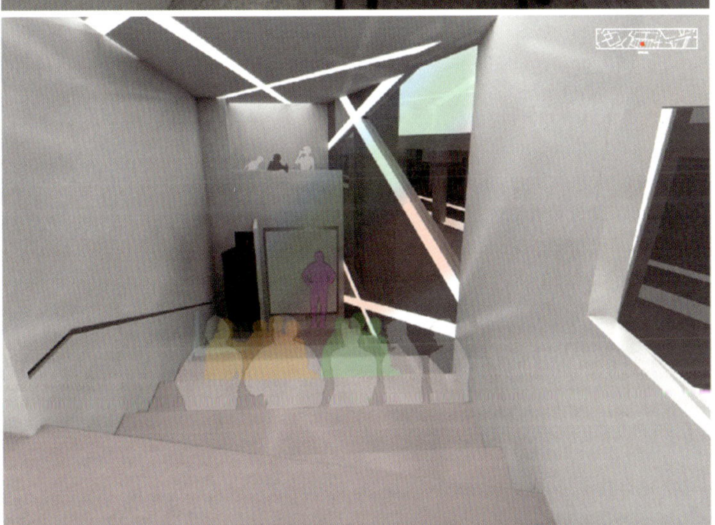

Life's too
short for the
wrong job.
jobsintown.de

02

間加樓

사이 [간] 더할 [가] 누락 [누]

다리 사잇길에 새로운 공간을 더해 만든 새로운 개념의 누.

현재 우리나라에서는 어린이 놀이터 등 기존의 어린이를 위한 놀이공간이 더 이상 어린이들만의 공간이 아니게 되었습니다. 야간에는 어린이놀이터가 비행청소년들의 아지트로 변모되어 아이들의 안전까지 위협받고 있는 것이 우리나라의 현실입니다. 이로인해 왕성한 활동을 해야할 아이들을 위한 운동 공간은 더욱 찾기 힘듭니다. 저희는 한창 활동해야할 아이들에게 안전하고 자유롭게 운동하거나 뛰어 놀 수 있는 공간을 만들어 주기 위해 다리 밑 하부공간을 사용하고자 합니다.

Located in Seoul, Guro daerimyeok and students, office workers, a lot of floating population of foreigners.
Line 2 and Line 7 has many dorimcheon flows below the pillar.
And the road that you can ride a bike long enough to enjoy jogging, but even these facilities do not use except for the evening or on weekends.

SITE

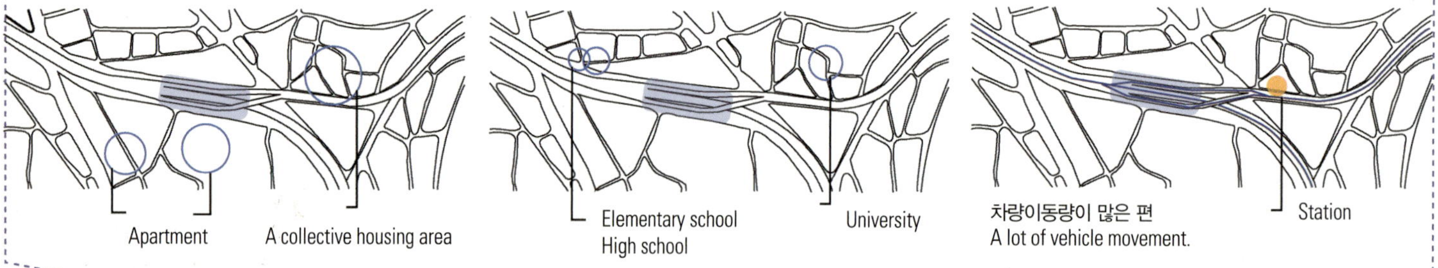

Apartment
A collective housing area
Elementary school
High school
University
차량이동량이 많은 편
A lot of vehicle movement.
Station

SITE CHARACTER

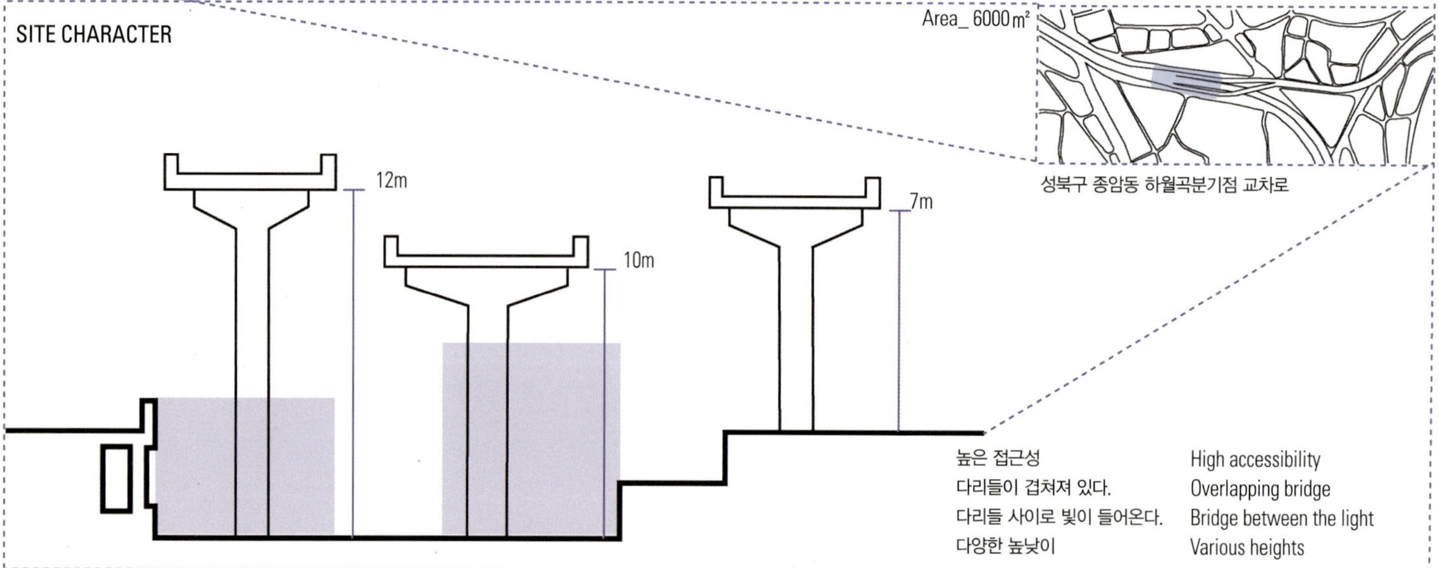

Area_ 6000㎡
성북구 종암동 하월곡분기점 교차로
12m
10m
7m
높은 접근성
다리들이 겹쳐져 있다.
다리들 사이로 빛이 들어온다.
다양한 높낮이
High accessibility
Overlapping bridge
Bridge between the light
Various heights

아이들과 어른들 Children and adult

1-7세
8-13세
14-19세
20-35세
36-50세
51-70세
71세이상
DAY-TIME
NIGHT-TIME

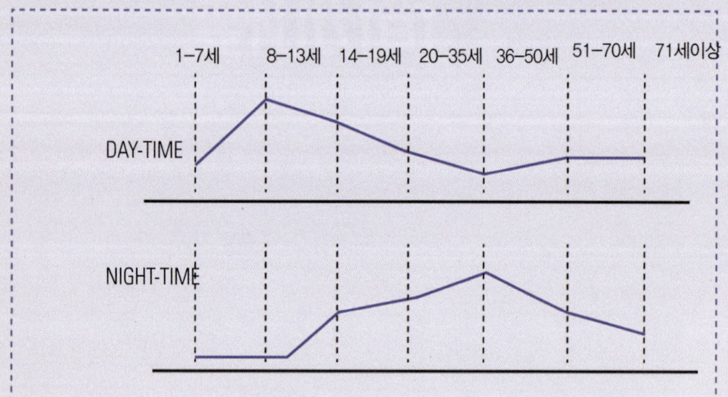

7세부터 13살 초등학생을 대상으로 아직 어른들의 보호가 필요한 아이들.

어린이들은 자신이 속해 있는 환경에 민감하며, 감성적 인지와 공간과의 상호작용을 통해 지적 성숙을 이해하며 달성해 나갑니다. 어린이 공간은 많은 경험적 지식을 얻을 수 있는 장소입니다. 그러므로 어린이들의 인지 발달에 긍정적인 영향을 줄 수 있도록, 우수한 놀이 환경뿐만 아니라 다양한 환경으로 경험을 확장시켜 민감하고 적극적인 상호작용이 이루어지는 공간을 제공해야 합니다. 이로 인해 아이들의 지적 · 육체적 발달과 함께 사회적 경험의 폭도 넓혀질 것입니다. 그러나 미성숙한 어린이들이기 때문에 보호자의 관리와 보살핌 안에서 행동을 해야 합니다. 저희는 이러한 특성들을 통하여 부모와 유기적으로 연결될 수 있게 공간과 공간 사이사이에 이러한 요소를 넣어 디자인 할 것 입니다.

7 to 13 years old elementary school students still need adult protection.

Children are used to be very sensitive to their own environment. There will be an interaction between them and the environment and through this interaction, they will grow up and become a matured person. This is the place where the children gains experience and learn many new things. Therefore, we need to create a positive environment that can make the children gains a lot of experience in a healthy way but at the same time, preventing the existent of negative environment. Creating a positive environment will helps in the development of children's mental, physical as the children gaining more experiences from healthy environment. However, the young children which are still immature need to be take care in the parents or the irresponsible person. As the children are still immature, the parents will able to develop the children's character in the correct way more easily.

SPACE USE

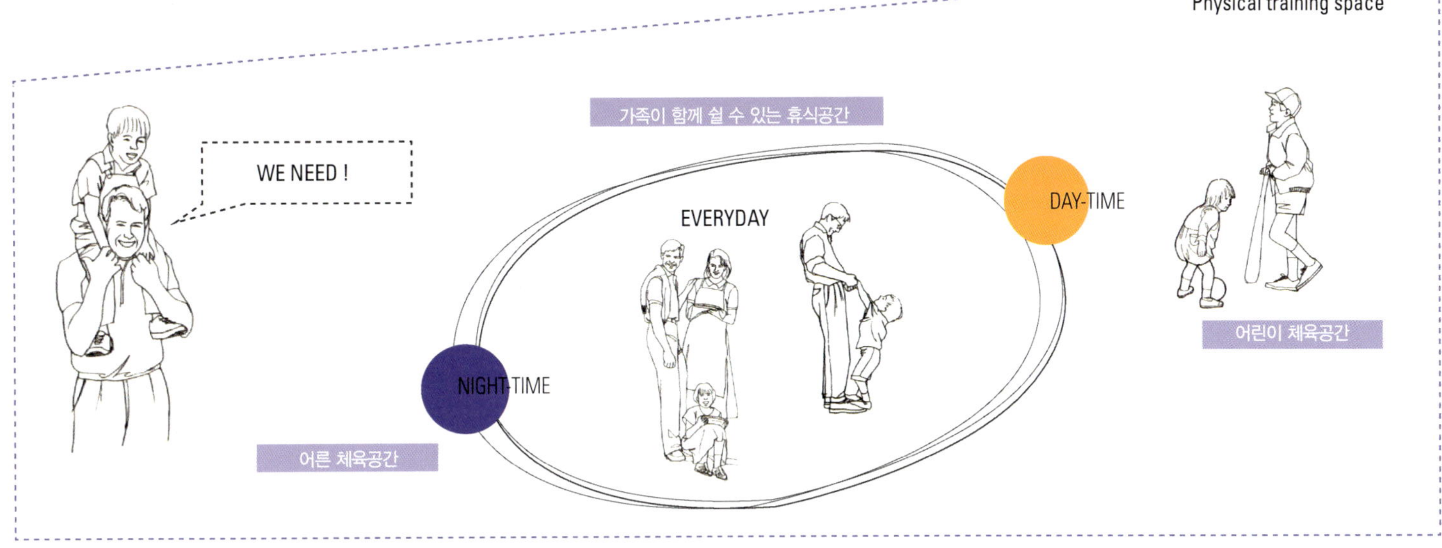

낮에는 주로 아이들이 사용하고,
새벽과 밤에는 어른들이 사용한다는 걸 알 수 있다.

Mostly used by kids during the day and
The morning and at night, you know that adults can use.

낮에는 아이들의 사이즈에 맞는 크기로 가벽이 세워진다.
그리고 밤에는 가벽이 내려가면서 성인들의 사이즈에 맞는 크기로 바뀐다.

Children during the day, the walls are built in sizes that match the size.
And at night, the walls all the way down to the size of adults is changed to the correct size.

신체 분석.
Body analysis

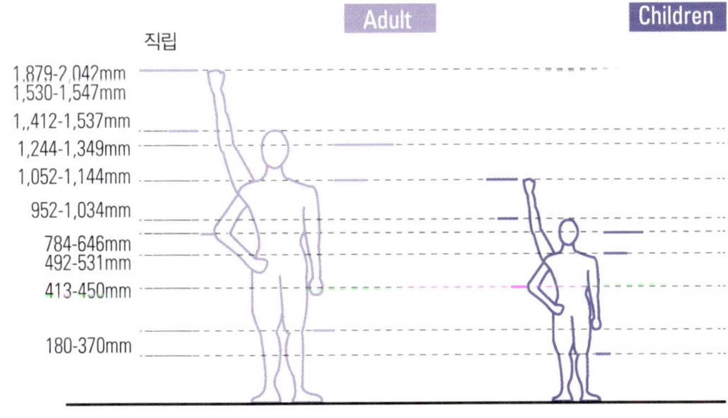

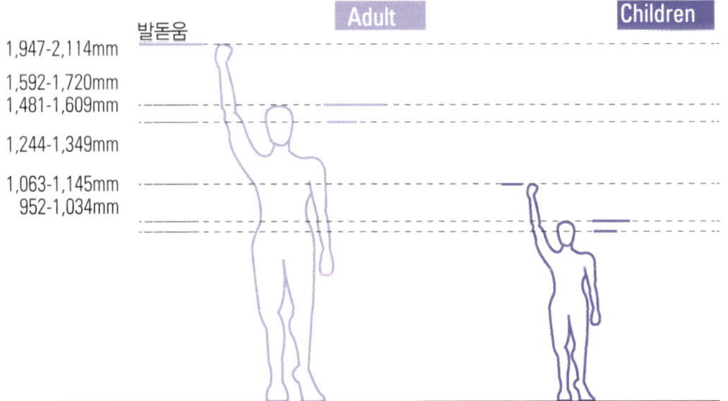

▶ 이러한 신체 분석을 통해 공간을 구성해 본다면,
Through the analysis of this body composed bondamyeon space,

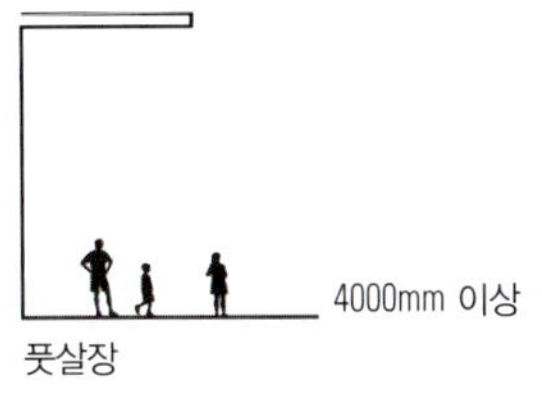

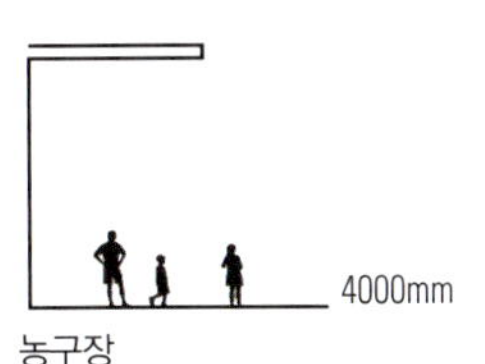

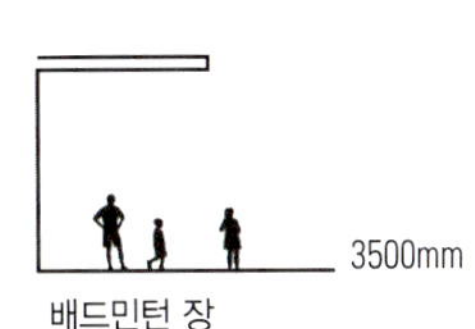

각 공간별로 규격높이가 있으나,
아이들이 사용하기엔, 풋살장은 4m이상을 주어야하고,
농구장은 4m, 배드민턴 장은3.5m로도 충분하다.

Standards for each area by the height,
but is merely the children, most of the 4m at least
give futsal and basketball courts of the 4m,
3.5m Badminton section is sufficient.

체육 공간 규격.
Physical space standards

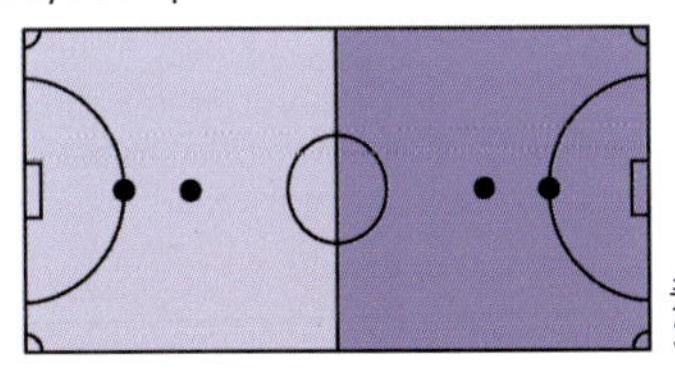

풋살장
38000*18000

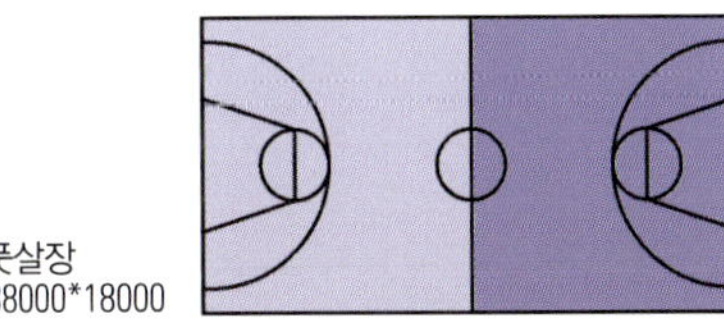

농구장
28000*15000

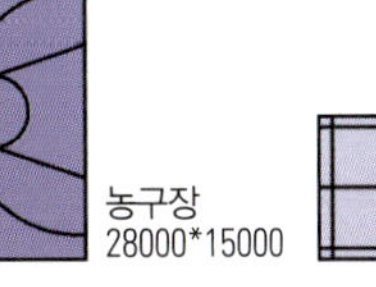

배드민턴 장
11340*6100

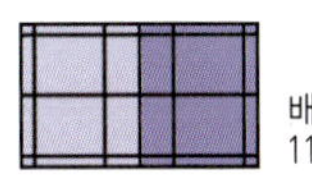

Adult | Day-time
Children | Night-time

CONCEPT

KANGAROO

부모와 아이가 함께 여가를 보낼 공간.

아이들이 부모의 보살핌아래 노는 모습은 캥거루를 연상시킵니다.
저희는 캥거루를 컨셉으로하여, 캥거루의 성장 과정과 특징들을 통해 공간을 구성할 것 입니다.

Children under the care of their parents at play is reminiscent of a kangaroo.
Our concept of the kangaroo and the kangaroo and the characteristics of the growth process that is configured through the space.

KEYWORDS

캥거루의 유기적인 선들.
Lines coming from the organic body

JUMP

캥거루 점프시 다양한 높낮이.
A variety of jump height

HUG

캥거루가 아이를 돌보는 모습.
Under the care of a parent

MASS PROCESS 01 사이트 특징 Features of the site

Base floor

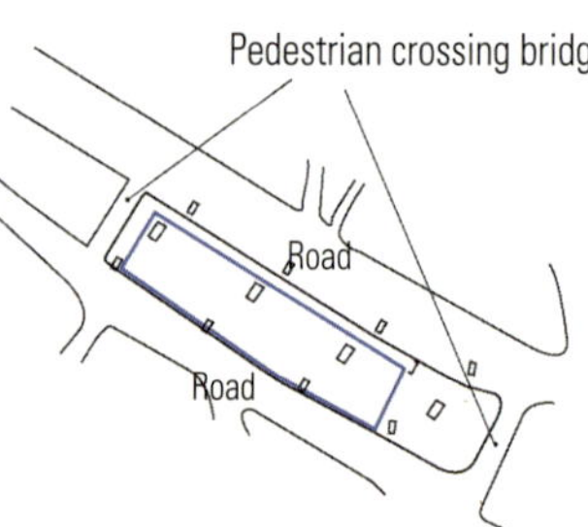

Divided by light

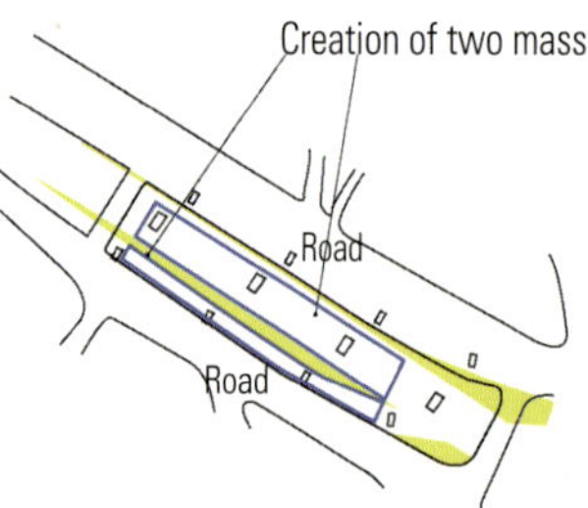

Mass-1
Mass-2
Road
Road

Flyover의 다른 높낮이들을 mass에 적용해본다.
Applied to the mass of the different pitch flyover

Have a limited view of the mass 2.
Adjust the height of mass1.
Thus, the view of mass2 diverse.

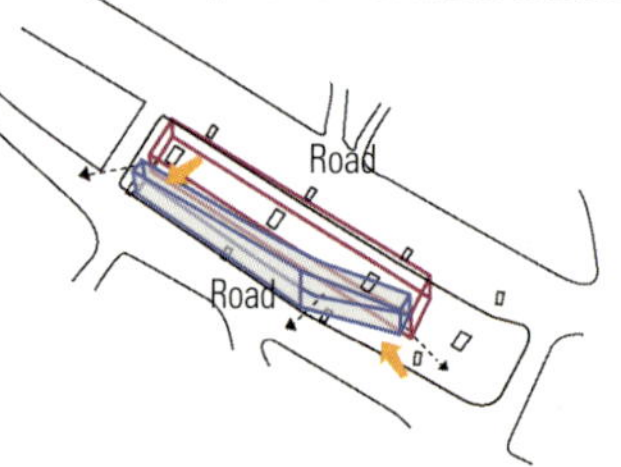

mass-2의 뷰는 한정적이다.
mass-1의 높이를 조절하여,
mass-2의 뷰를 다양화 시칸다.

MASS PROCESS 02 공간간의 관계 Between space and space relation

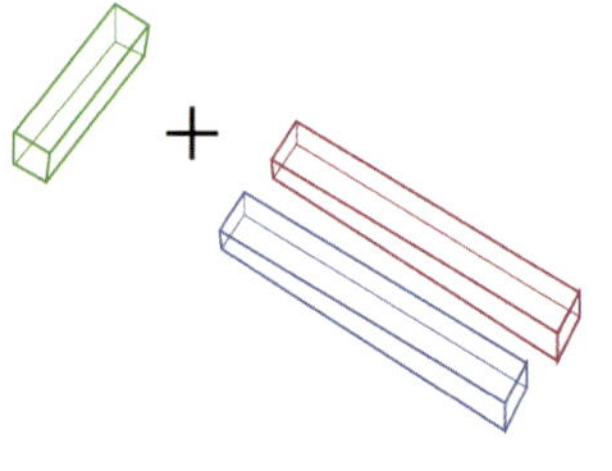

하나의 매스를 추가하여,
다른 두 매스를 연결시킨다.

각각의 공간들은
하나의 공간으로 연결이 된다.

By adding the mass of one, and two other associates mass.

Each room is connected to a single space.

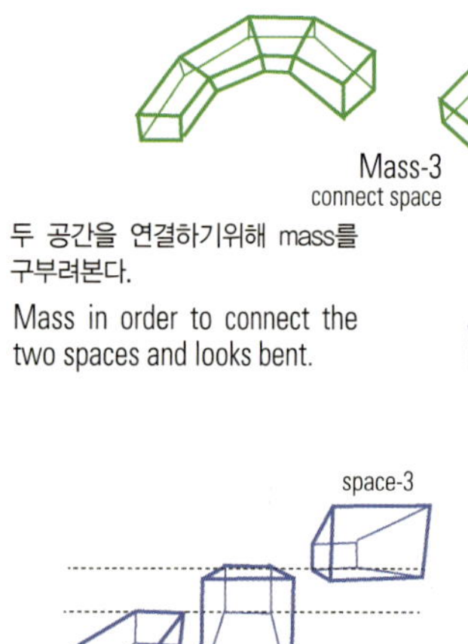

Mass-3
connect space

두 공간을 연결하기위해 mass를 구부려본다.

Mass in order to connect the two spaces and looks bent.

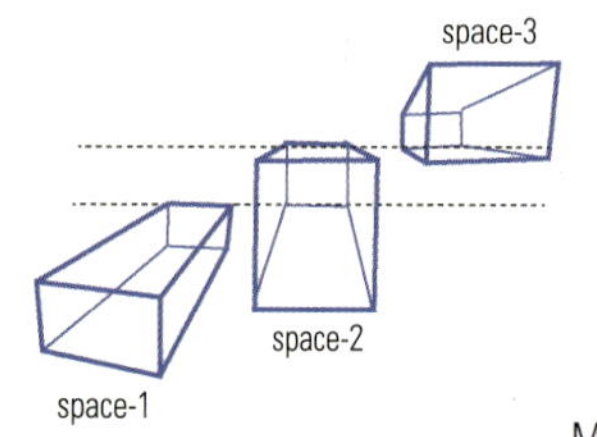

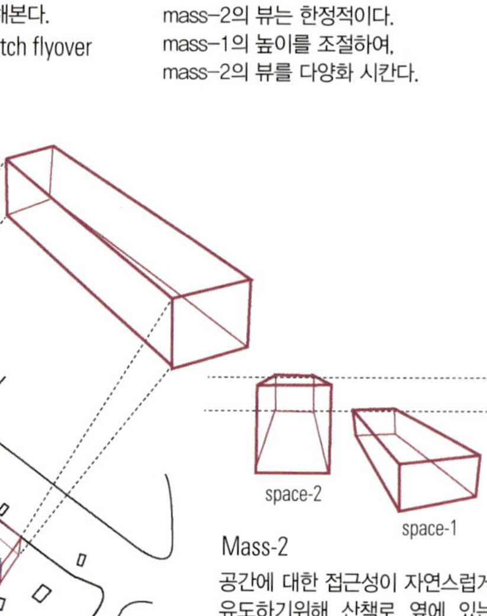

Mass-1

각 공간에 높낮이를 준다.
이를통해 다양한 뷰가 형성된다.

Gives the height of each space.
It is formed through the various views.

Mass-2

공간에 대한 접근성이 자연스럽게 유도하기위해 산책로 옆에 있는 mass에 두 개의 mass만 넣는다.

Spatial proximity to the promenade next to the mass in order to induce naturally two put ten thousand mass.

ZONNING

배드민턴장
교육공간
사무공간
휴식공간
농구장
탈의실
화장실

LEVEL-3

야외	실내
	배드민턴장

LEVEL-2

야외	실내
휴식공간	사무공간
	교육공간
	농구장

LEVEL-1

야외	실내
풋살장	사무공간
	부모공간
	탈의실
	화장실

DESIGN PROCESS

01

아기 캥거루의 시점에서 바라본 육아낭
Seen from the viewpoint of a baby kangaroo pouch

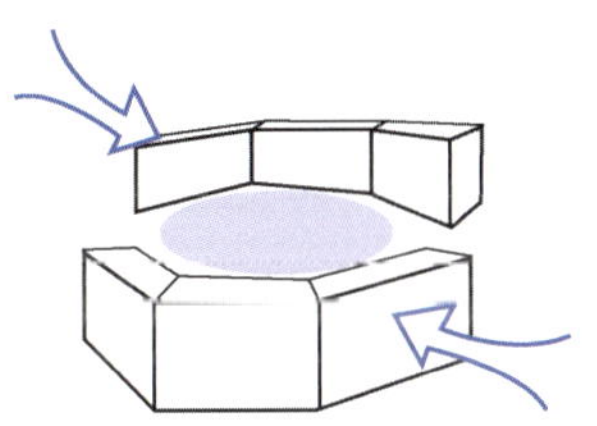

소음과 매연이 있는 FLYOVER 공간에서 벗어나기 위해 건물외관을 덮어서 이를 차단시킵니다.

To avoid the soot and the disturbing noise from the flyover, The building between the flyover can be cover and this should stop this problems.

차단된 외관과는 반대로, 내부를 열어놓으므로, 각 공간들을 유기적으로 연결시켰습니다.

There should a space at the opposite direction of the building So it can be connected in more naturally looking-way

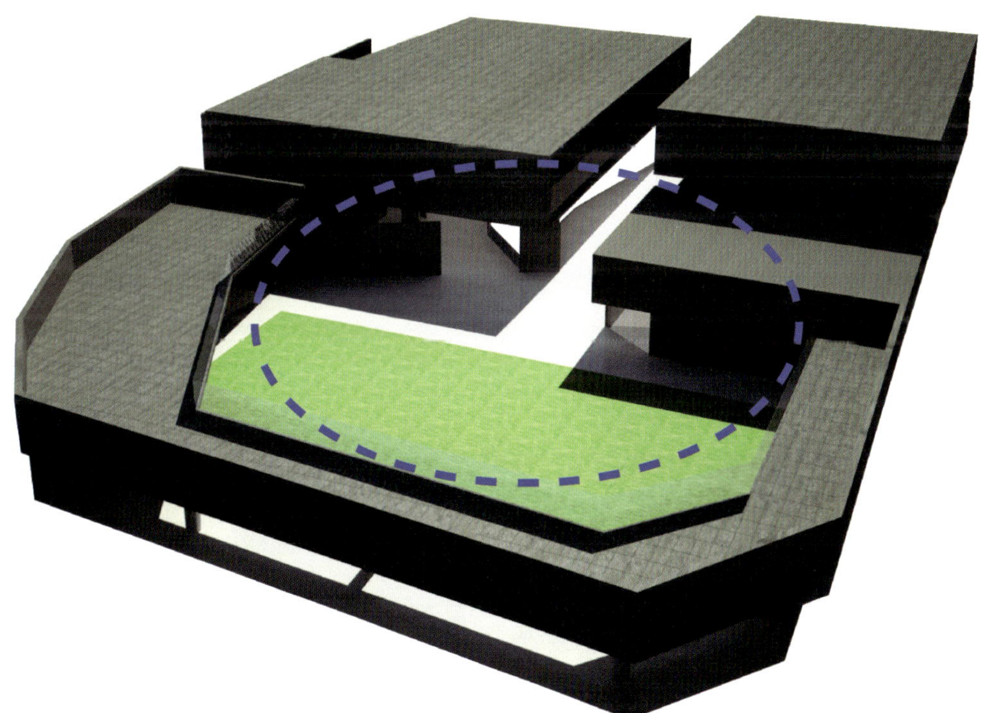

아기는 태어나자마자 아기 주머니 속으로 들어갑니다.

Right after the baby is born, the baby will be place in the mother's pouch.

아기는 따뜻하고 안전한 주머니 속에서 6개월을 지냅니다.

For 6 month, the baby will live in the warm and safe protected pocket.

6개월이 지나면 주머니를 들락날락하는데 몸이 커져도 위험이 느낄 땐 주머니 속으로 들어갑니다.

6 month after that, the baby will come out from the mother's pocket as its body grow up and become bigger but if the baby sense danger, it will get inside the mother's pocket for protection.

02

부모와 아기 캥거루의 관계

The relationship between parent and baby kangaroos

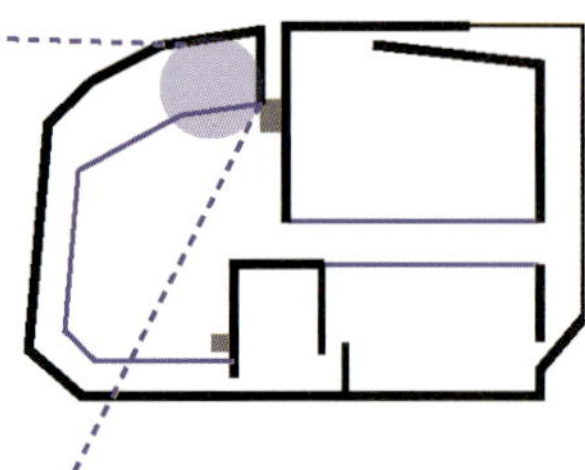

떨어져 있는 각 공간들을 하나로 연결하여,
그 연결고리에서는 아이가 어디에 있든지 알 수 있다.

Off by connecting each space together,
the link no matter where the child can see.

DETAIL
캥거루의 활동패턴을 공간에 넣어보기.

캥거루는 주로 밤에 활동하고 낮에는 쉬거나 잠을 잡니다.
Kangaroo is nocturnal and sleeps during the day, rest or sleep.

낮에는 아이들의 체육 공간이 되고, 밤엔 동네 어른들의 담소 공간.
Physical space and children during the day, night, neighborhood chat room for adults.

03

열린 개념의 캥거루 육아낭

The concept of open kangaroo pouch

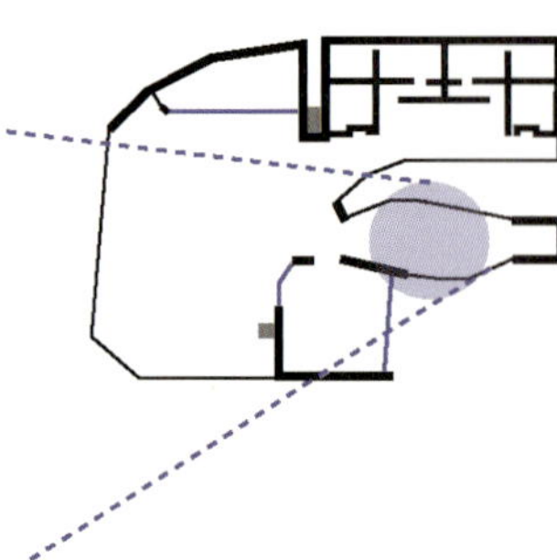

다리사이를 통해 빛이 들어오는 공간을 열어본다.
Try putting marsupial in the external wall.

열려 있는 공간에 육아낭을 넣어본다.
Natural light is the light come in pouch.

DETAIL

아기 캥거루가 육아낭 속으로 들어가기.
Go into a baby kangaroo pouch.

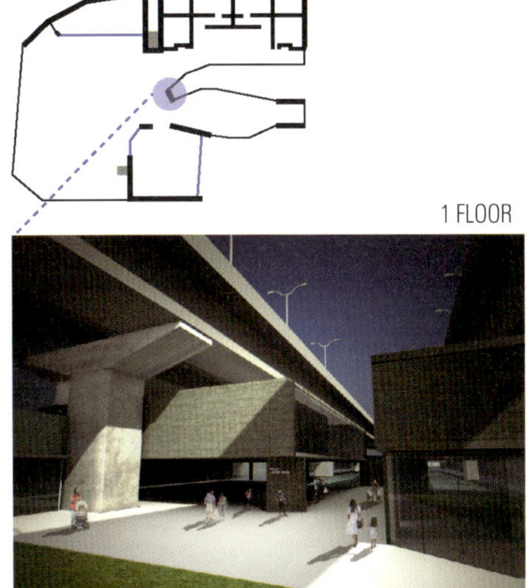

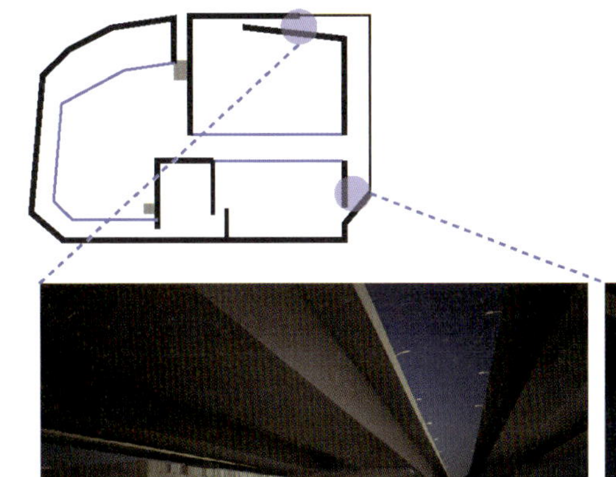

어미 캥거루는 새끼가 태어나기 전에 주머니 속을 깨끗이 닦아 놓습니다
새끼에게 주머니가 있는 곳을 알려 주기 위해 자기 배 주위에 침을 묻힙니다.
그러면 새끼는 침이 묻어 있는 길을 따라 주머니 속으로 들어갈 수 있습니다.

열린 벽을 활용한 구조체 형식의 방향 표지판.

뛰어난 시인성을 연출합니다.

Mother before the birth of a baby kangaroo pocket, drop it, clean it inside.
To young to tell you where pockets around the spit on his ship buried.
The pups in the saliva may enter into a pocket along the way.

Directional signs on the walls of a structure type utilizing open.

To produce excellent visibility.

공간 속의 또 다른 공간 연출하기.
Space in the space is directing

아이들의 보육공간에 캥거루 주머니를 연상시키는 큰 오브제를 넣어봅니다.
Kangaroo pouch of children in child care space, reminiscent of the view put large objects.

다양한 높이의 열려 있는 구멍을 통해서 보여지는 공간은 흥미롭습니다.
Various heights of the open space is visible through a hole is interesting.

04

FLOOR PLAN

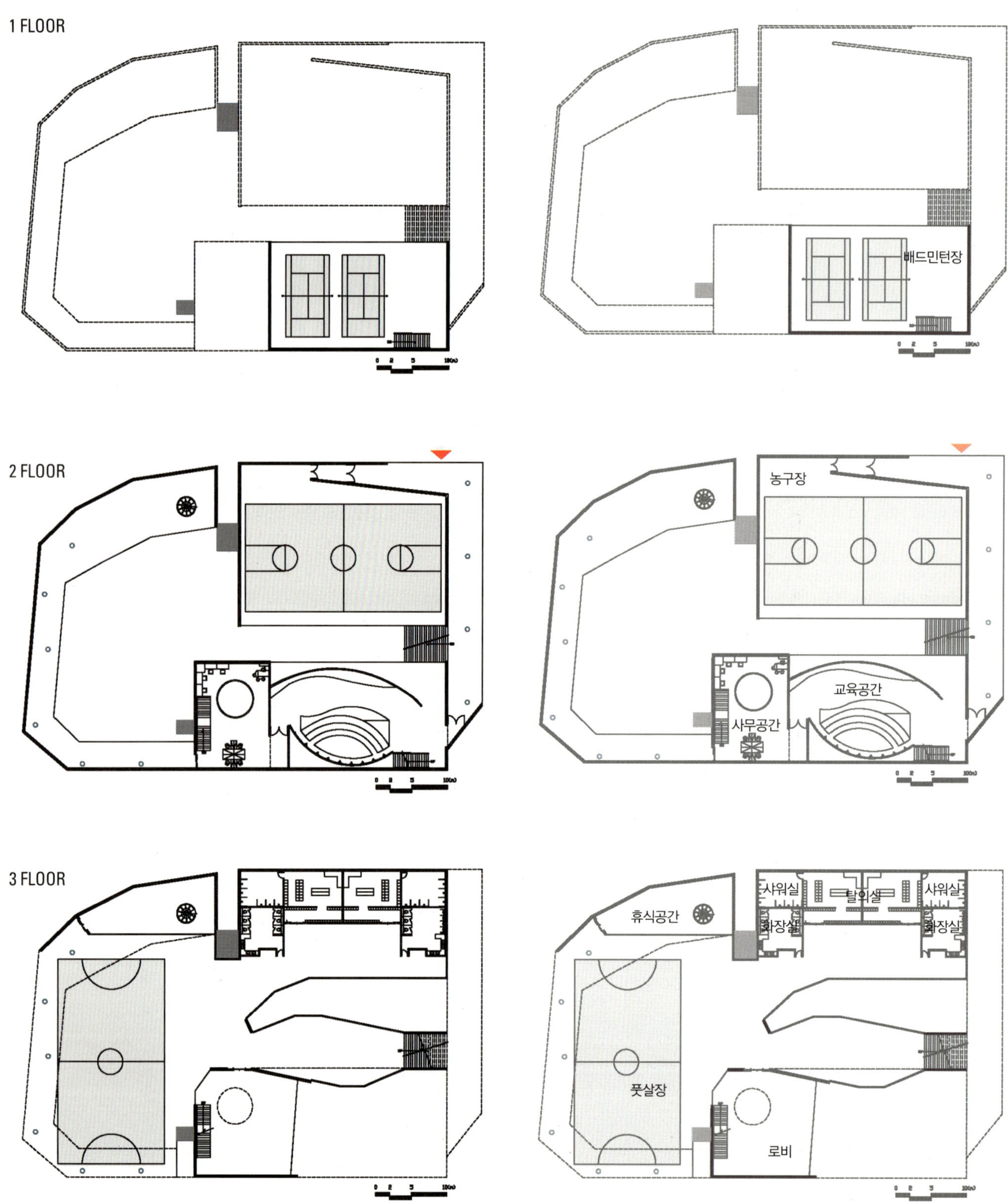

05

캥거루들의 점프
Kangaroo's jump

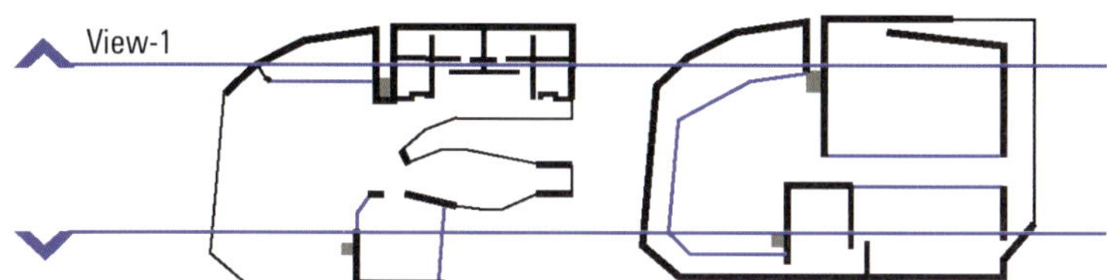

캥거루의 점프를 공간에 적용하여, 각 공간들의 높낮이를 달리해본다.
Applied in the space of kangaroo jumps, sees dalrihae the height of each room.

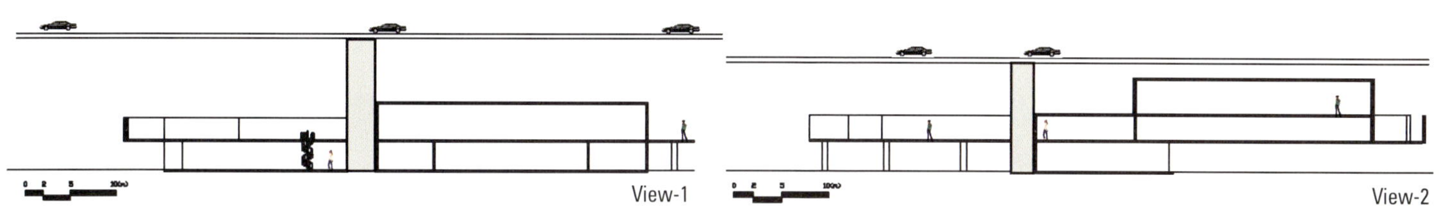

캥거루의 점프를 공간에 적용하여, 각 공간들의 높낮이를 달리해본다.

Stumble on microdot

미세 점을 우연히 발견하다

작은 틈 사이에서 발견된 자연의 메세지를 들려주는 공간

Space with the nature's message from little crack.

작은 것은 잊혀지거나 소외되기 쉽지만 작은 소재가 하나, 둘 모여들어 새로운 형태가 만들어지고, 그 과정 또는 결과에서 처음과는 다른 무언가를 발견할 수 있는 공간을 제공하여, 작은 소재에 대한 우리들이 가지고 있던 인식의 방향을 달리 생각해 보기를 바란다.

Small thing can be forgotten easily or alienationed but the little things can be gathered and make new shape and those process or consequence can make the different space that can be found new thing compare to beginning.

PROPOSAL

언뜻 생각하면 곤충은 사람에게 해로울 뿐 아니라 불필요한 존재라고 여기기 쉽다. 그러나 우리가 징그럽다고 생각하는 지렁이는 식물이 잘 자라지 못하는 땅에 식물이 자랄 수 있는 양분을 배설하여 영양이 풍부한 흙이 되게 한다. 또한 땅속에서 부지런히 움직이면서 땅을 부드럽게 만들어 주기도 한다. 이처럼 지구상에서 서식하는 모든 생명체는 서로 도움을 주고 받으며 조화를 이룸으로써 생태계를 유지시킨다.

따라서 우리 인간 역시 자연의 순리를 따라 모든 생명체를 존중하고 보호함으로써 생태계를 유지하고 보존하려는 태도를 가져야 한다.

현대의 아이들은 도심속 일정한 틀 속에서 반복되는 생활을 함으로써, 그들이 가지고 있는 무한한 가능성이나 창의력을 발전시키기 어렵다. 이들이 자라가는 성장 과정에서 곤충을 몸소 체험하면서 무한한 호기심과 창의력을 보존하고 발전시킬 수 있는 곤충 체험 학습관을 제안하고자 한다.

People easily think the insects are harmful and unnecessary for humans. But the earthworm that we think disgusting makes the nutrition for the earth and it helps the plants growth. Also their moves make flexible earth. All of creatures are help each other and make harmony so they maintain the nature system.

So, Human also need to follow the nature's law and respect and protect all the creatures so protect nature system.

In these days, because children's lives are repeated in frame of urban so it's hard to develop their creativity or possibilities. So we purpose the experience space with insect that children can experience insects so they can keep and develop their curiosity and creativity.

'도시' 라는 틀
The frame named Urban

아이들
Children

곤충
Insect

대자연
Nature

SITE ANALYSIS

서울시 마포구 서강대교

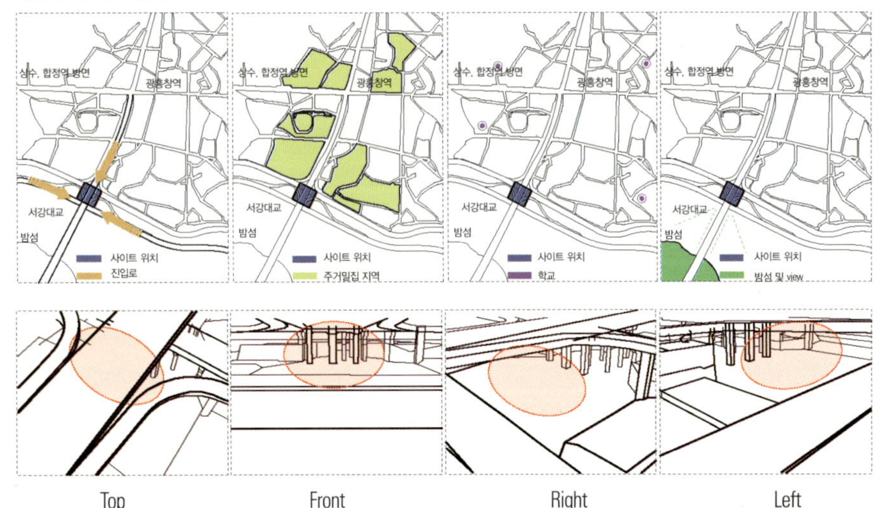

여의도 샛강 상류에서 서강대교 강변 남단까지 자리잡고 있으며, 길이는 3.8km이고, 밤섬, 샛강 등이 자연상태 그대로 잘 보존되어 있어 주변 경관이 아름답다. 전면에 한강이 위치하고 있어 open된 view가 형성되고, 주변에는 다수의 초-중-고등학교가 위치하고 있다.
인근에는 광흥창역이 위치하고 있으며, 주거밀집 지역과 근접해 인구의 유동성이 좋고, 상수나들목과 현석나들목 등이 주변에 있어서 접근이 용이하다.

It is located upper 여의도 샛강 from Southern 서강대교 강변
Length : 3.8m the nature's well protected so the landscape is wonderful. it is located in front of Han river so it can make opened view and their's some elementary, middle, high school at near. there's 광흥창 station at near and the residential area is not far from it so the flow of people is good. and this site easy to access from 상수나들목 and 현석나들목deploy Asia's cultural similarity in the Immigrant Facility, we are striving to compose the place by using the law of similarity and continuity in Gestalt theories and by incorporating the concept that has the characteristics of Korean background and traditional architecture which is Bunhapmun, Madang, Maru.

TARGET

Family & Ked

자연과 격리되어 살아가는 현대인들의 문제점은 무엇인가?
톱니바퀴처럼 돌아가는 흐름 속에서 하나의 부품처럼 생활하다보니 일관된 생각들만을 하고 지내게 된다. 어른과 아이들의 소중한 것들이 잊혀지고 있는 실정이다. 어른들에게는 순수한 동심의 기억을 아이들에게는 상상의 공간을 재현할 수 있는 호기심 가득한 공간을 제안하고자 한다.

what's the problem of modern people who is saperated from nature?
People trappted in the flow like serrated wheel they can think different thing.
The important things of adult and children are forgotten. For adult, it will give the childhood memory. For the children, it will make the space that children can explode their imagination.

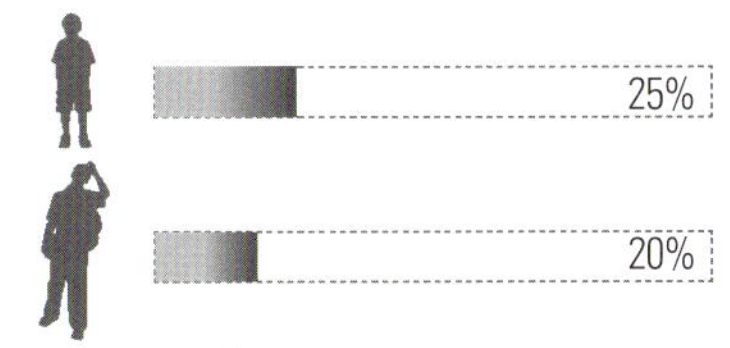

어린이집 9 약 1,035명	Nursery 9 approximately 1,035 people
초등학교 8 약 9,251명	Approximately 9,251 people eight elementary schools
중학교 6 약 7,734명	,Approximately 7,734 people six junior high schools
고등학교 5 약 6,245명	Approximately 6,245 people high five

CONCEPT

허물

허물은 파충류, 곤충류 따위가 자라면서 벗는 껍질을 말한다.
파충류 또는 곤충류들은 허물을 벗음으로써 그 전 보다 낳은 상태로의 모습으로 발전이 된다.
아이들이 허물을 벗고 나옴으로서 잊고 지내던 그들만의 상상력을 발휘하기를 바란다. 공간 자체만을 바라보지 않고, 공간과 사람이 하나 되었을때의 모습을 바라보고자 한다.

Fault

The insect's slough is the taken off skin from 파충류 and 곤충류
They are developed them selves with taken off slough.
We hope children take off their slough so they can explode their imagination.
We tried to imagine the space with human.

KEYWORDS

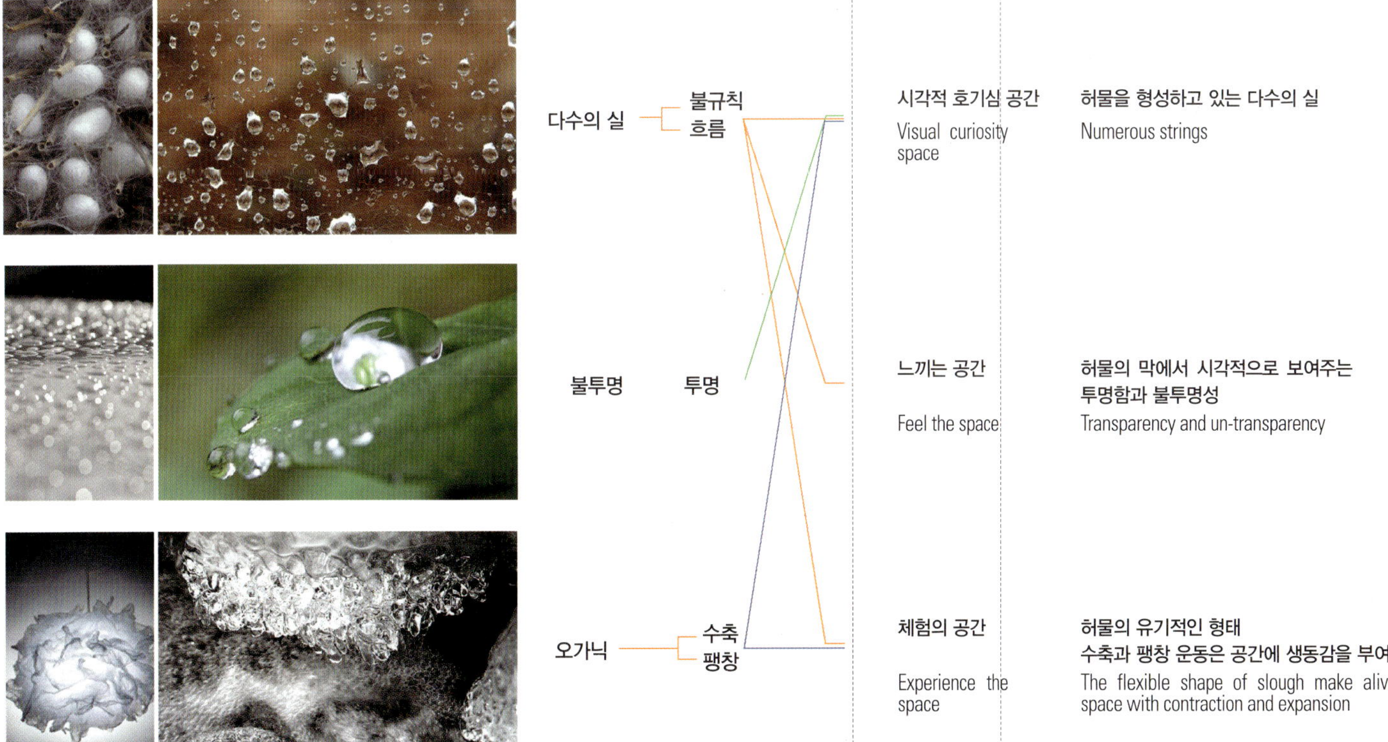

DESIGN PROCESS 1

누에의 실을 이용한 공간전개과정

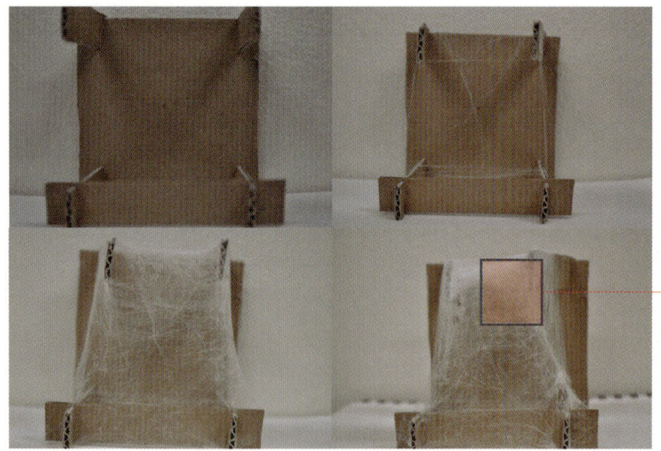

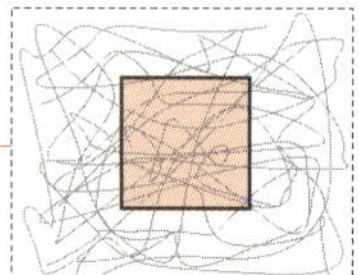

다수의 실 배열
An array of multiple rooms

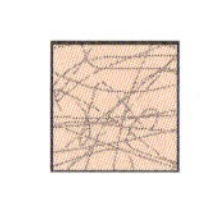

내부 속 공간
In the space inside

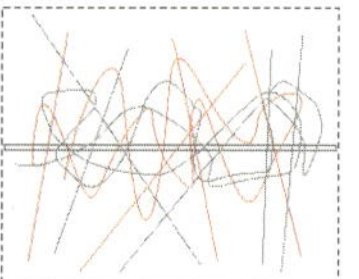

공간의 형태
Space in the form of

누에가 실을 이용해 집을 짓는 과정을 보면, 처음에는 가느다란 실들이지만 수많은 실과 그 실들이 엇갈리고 겹쳐 배열되면서 한 지점에서 어떠한 형태가 완성됩니다. 그 완성된 형태의 내부는 다양한 모습의 공간을 형성하게 됩니다.

If you see the process of silkworm's house making, at first it is just thin strings but when it gathered and arrayed, at some point, it makes certain shape.

허물의 막을 이용한 공간전개과정

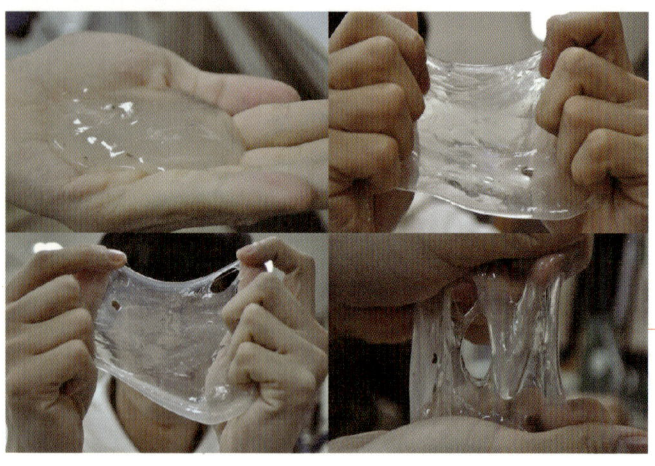

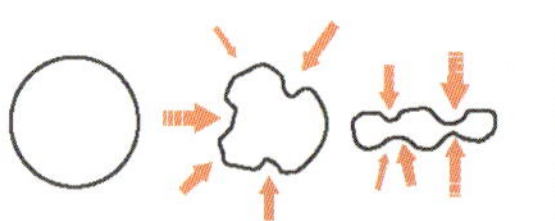

막(공간)의 수축
Membrane (space), the contraction of

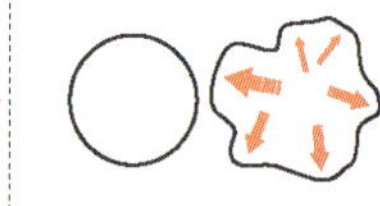

막(공간)의 팽창
Membrane (space), the expansion of

허물 속 유충의 움직임으로 인해 허물 막에서 수축과 팽창현상이 일어나게 됩니다. 수축과 팽창 현상으로인한 내외부의 다양한 형태의 변화와 그 변화속에서 유기적인 새로운 공간이 형성되고, 공간과 공간간에 호기심을 자극시킬 법한 현상이 일어나게 됩니다.

Because of the movement of worms in slough, in the slough, there's contraction and expansion.

곤충의 성장 과정을 통한 공간전개과정

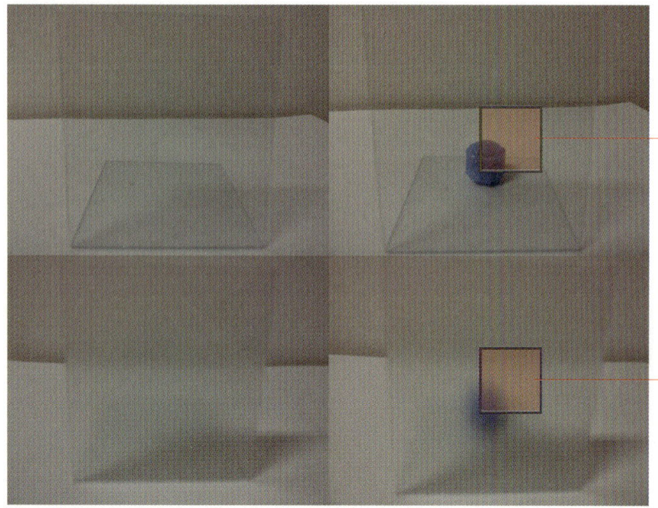

시간과 공간의 흐름
The flow of time and space

처음 허물이 형성되었을 때는 그 내부가 불투명하거나 보이지 않지만, 시간이 흐르고 곤충의 탈피 시기가 가까워 질수록 허물막이 약해져 내부가 어느정도 비춰지게 됩니다. 시간의 흐름속에서 공간도 같이 흐르게하여 사람들로 하여금 시각적 호기심을 유도시키게 됩니다.

At first, when slough just made it's not transparent but after few hours, when the time of molting is close, the slough is getting transparancy and weaker then before.
In the flow of time, we made the flow of space so gathered people's curiosity.

DESIGN PROCESS 2

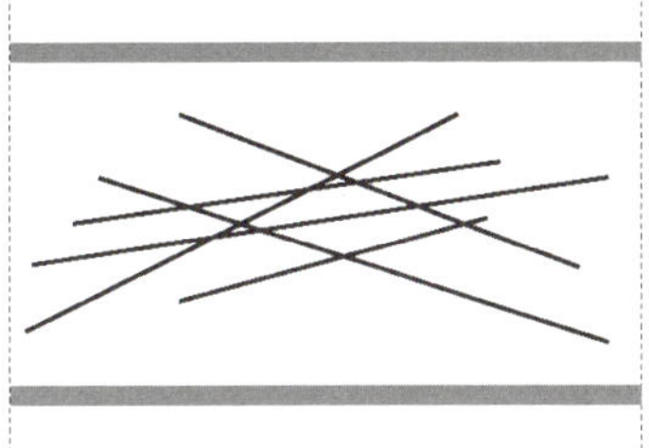

불규칙한 선의 배열
irregular array of line

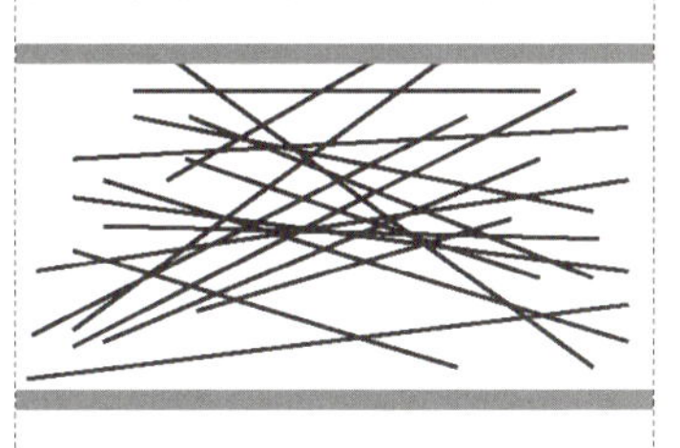

여러번에 걸친 선들의 엇갈림
numerous crisscross with line

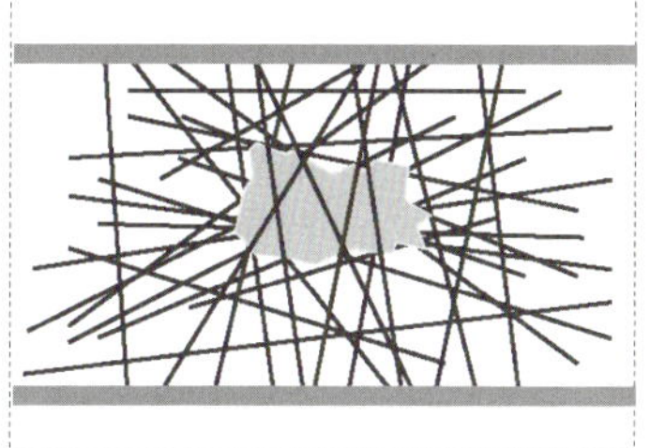

중첩된 한 지점에 덩어리 형태 형성
the point of node make the mass shape

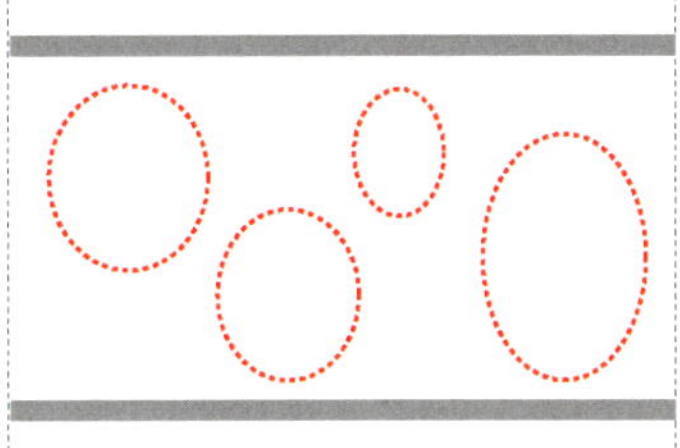

일반적인 공간 형태
the shape of normal space

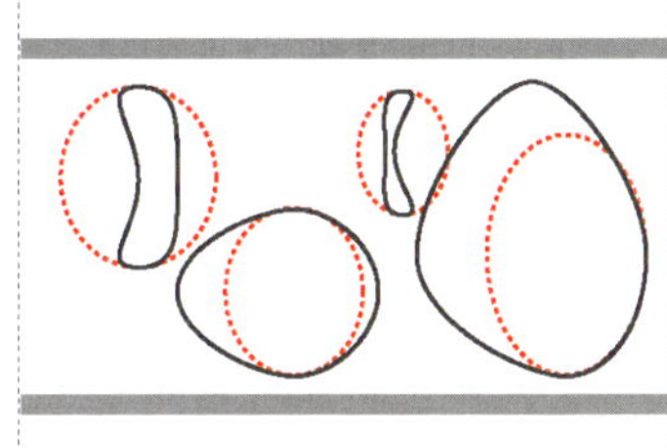

공간 속 수축과 팽창
the contraction and expansion in space

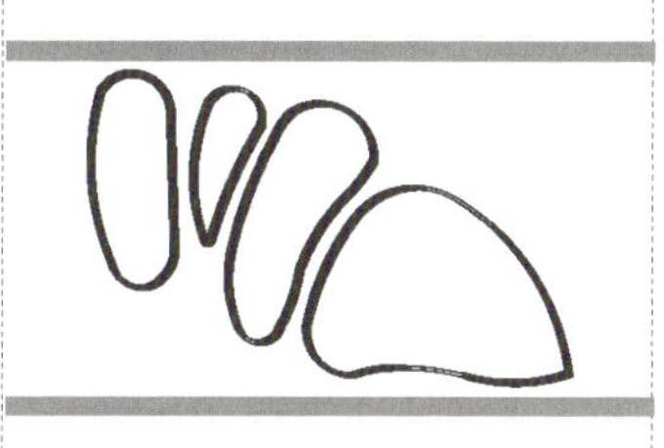

solid와 void가 반복되는 동선 흐름 형성
Formation of circulation flow with repetition of solid and void

허물의 형성
formation of slough

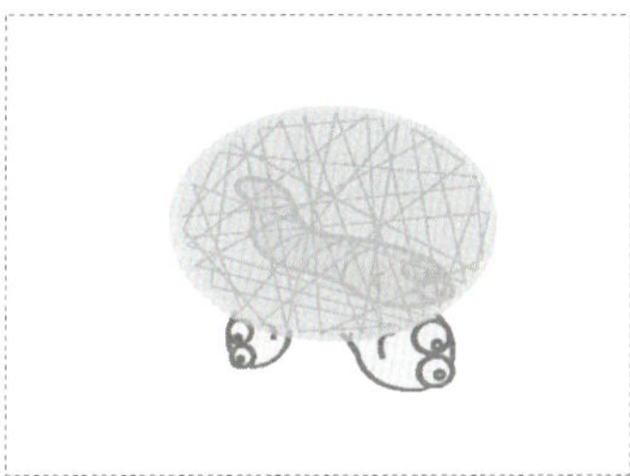

곤충의 탈피
the molting of insects

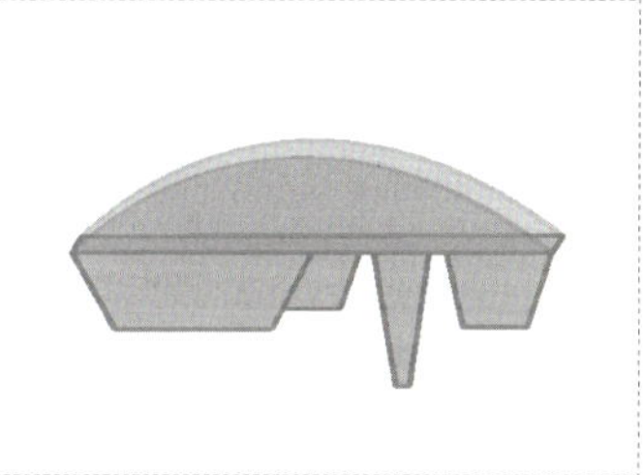

매스의 레벨차로 형태적 호기심 유도
Gather curiosity with level difference of mass

허물 – 공간 속에서의 외부에 대한 그리움
Slough - the longing for outside from inside

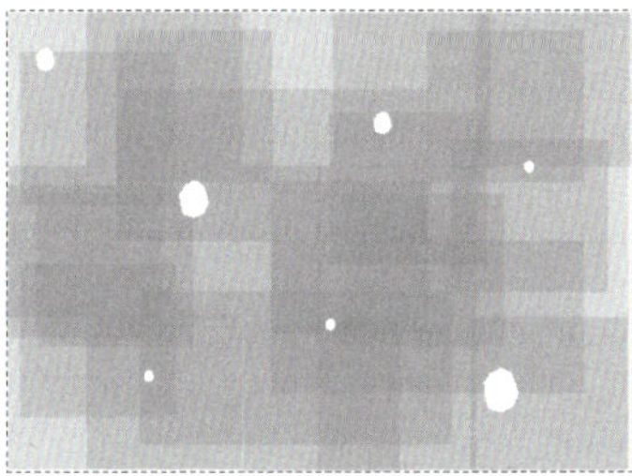

허물 – 시간적 흐름에 다라 일부 공간 open
Slough - at flow of time, some part can be opened

공간 – 허물 속에서 외부에 대한 갈증 해소
Space - quench thirst with outside

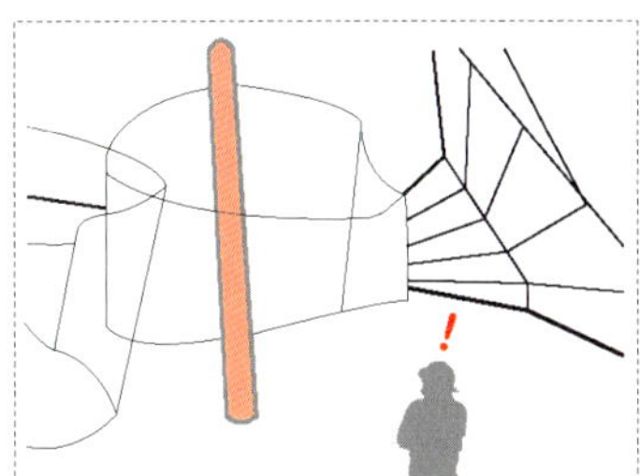

공간 속 선의 출현
appearance of line in space

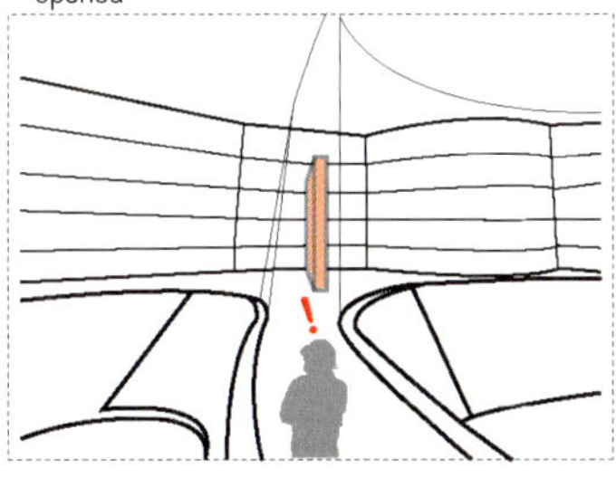

공간 속 선이 융합 면의 출현
appearance of compound of face

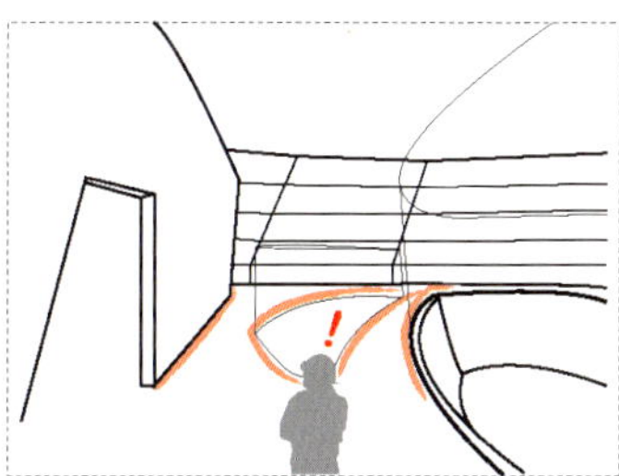

공간에 대한 호기심 유도 및 흡입력 발생
gather curiosity and concentration about space

FLOOR PLAN & LINE & PURPOSE

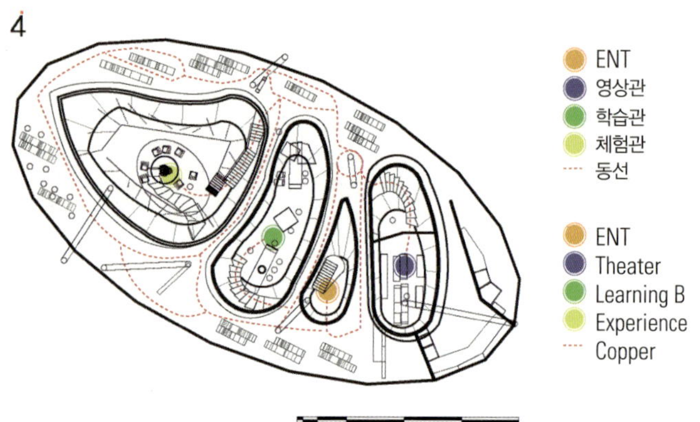

ZONING

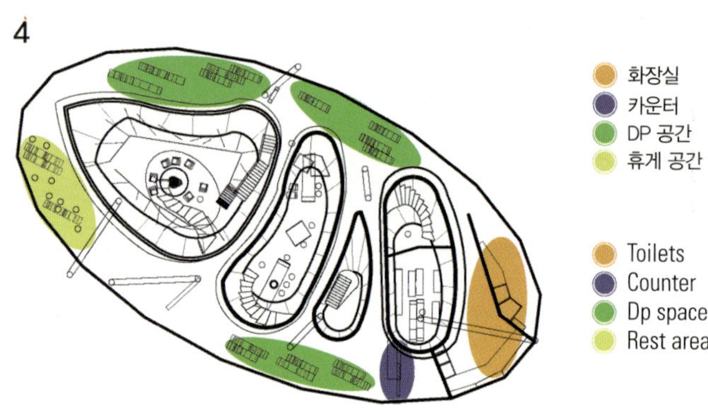

FURNITURE PLACEMENT

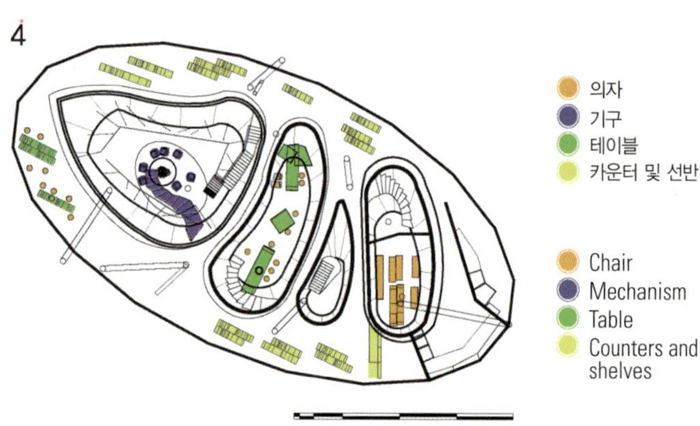

FLOOR PLAN & TRAFFIC LINE

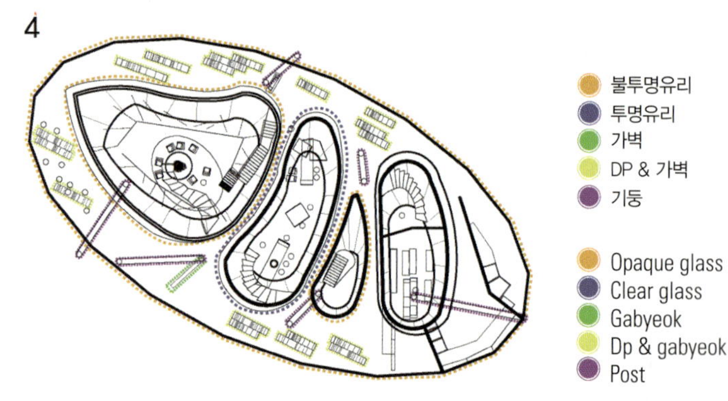

ELEVATION

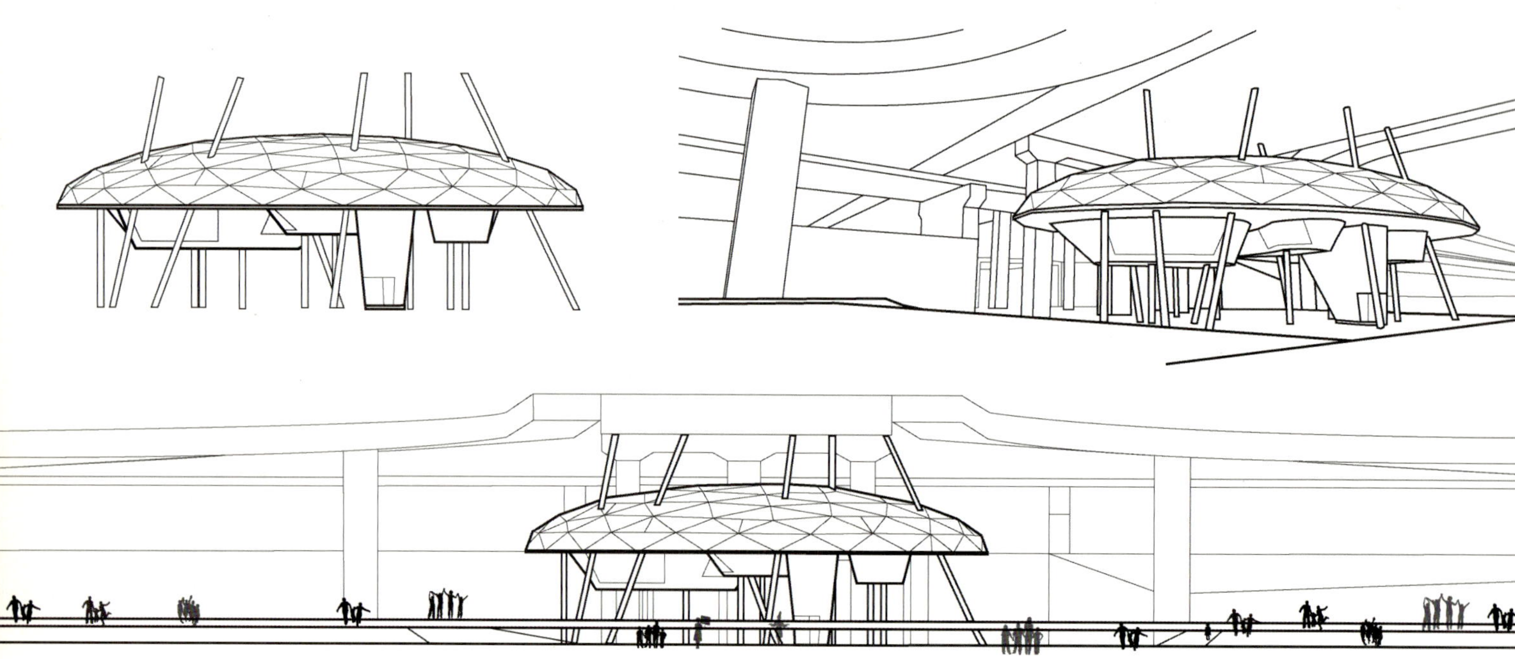

entrance and counter view (the start of circulation)

curvy circulation line started from counter

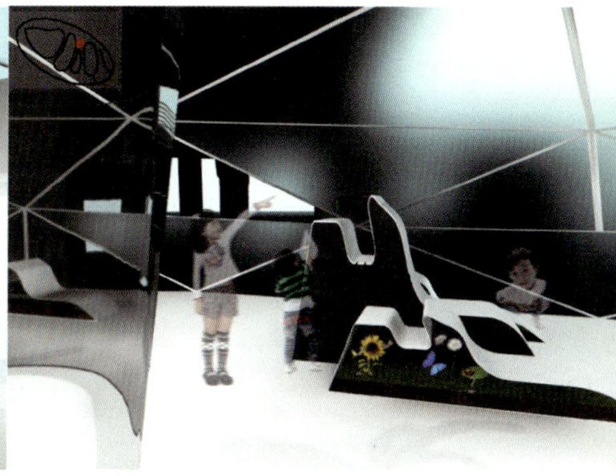

DP space at the start and end of circulation

gather curiosity with 3 visual feelings

Open space and border of double space

amplificate curiosity about two spaces

DP space - find flow in irregular

DP space - create the formation with contraction and expansion in space of slough

Rest space - opened view in double space

Education spcae - open space - shelter in slough

Theater - entrance to the slough spcae

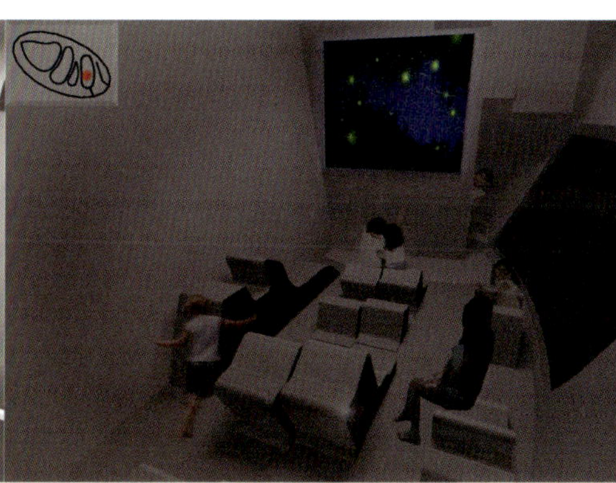

Theater - make the flexible furniture's shapes with contraction and expansion

각종 기구를 통하여 체험자 스스로가
허물 속에 있는 듯한 상상을 유도
Induce people's imagination with many equipments.

열린 공간(투명유리)
Open space (clear glass)

0%

이중적 공간(불투명 유리)
Dual space (opaque glass)

50%

닫힌 공간(벽)
Closed space (wall)

100%

호기심의 연속성

상상 그 이상의 공간
The space beyond the imagination.

세가지 COLOR의 공간 구성을 통하여 공간에 대한 지루함을 해소시켜주고, 체험자에게 다음 공간에 대한 호기심을 계속해서 부여함으로서 다음 공간으로의 자연스러운 동선을 유도한다. 각 공간이 가지는 고유의 COLOR는 체험자의 창의력과 상상력을 발휘할 수 있는 동기를 부여한다.

Quench the boring by the space composition of 3 color and gather people's curiosity so induce circulations.
Each color of each space give the motivation for creation and imagination to people.

진입 동선(닫힌 공간) – DP 공간을 살짝 노출시켜 호기심 형성 – 상부 도착시 닫힌 공간에서 오픈 공간으로 진화

Circulation of entrance(closed space) - expose DP space and gather curiosity.
when people arrive at the upperside, the closed space changed to open space.

-13
B-14
A-11
B-08
B-09

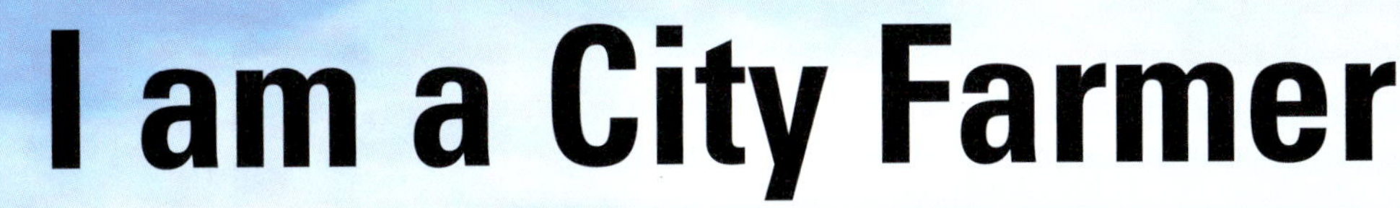

I am a City Farmer

나는 도시농부다.

도시화가 가속되면서 현재 우리나라 사람들의 90%이상이 도시에 살고 있다.
갈수록 각종 공해가 심해지고 먹거리에 대한 불신이 늘어나고 있으며 스트레스와 성인병 등 각종 현대 질병이 우리를 괴롭히고 있다.
그러나 이렇게 편리한 도시의 삶을 접고 평안한 농촌으로 돌아간다는 것은 어려운 일이다.
믿을 수 있는 건 바로 나 자신 뿐이기에 텃밭형 주말농장에서 적지않은 비용과 시간을 들여 직접 재배를 해보지만 도시화에 따른 문제를 완전히 해결할 수는 없다.

I am a City Farmer.

As urbanization gets accelerated, 90% of the current population in Korea is living in the cities.
Pollution is getting worse and distrust over food is increasing. Stress and various diseases are threatening the humanity.
However it is a difficult task to go back to the rural community by leaving the convenient city life.
Problems caused from the urbanization are not easy to be resolved completely by harvesting from individual garden farm during the weekends.

BACKGROUND

도시농업의 역사 – 본디 농업과 도시는 하나

인류의 역사 속 도시는 농업을 중심으로 형성되기 시작했다. 이러한 사례는 페루의 공중도시 마추픽추에 존재했던 테라스형 농지나 프랑스 베르사유궁전에 있었던 왕비의 텃밭과 오두막에서 찾을 수 있다.
그러나 현재 다시 농업이 도시로 들어오는 경향이 있다.
과거의 자연 환경을 인간이 그대로 이용하였듯 새로운 형태적 어휘를 통해 자연과 인간이 건축물 내에서 공존할 수 있다.

The history of urban farming – Originally farming and city happened collectively

Cities in human's history were formed and developed around farming.
The examples can be found from a terraced farm in Machu Picchu, the city above sea level in Peru and
vegetable garden and cottage that the queen of France owned in Palace of Versailles.
However the current trend shows that the agriculture is expanding in the city.
As the humanity utilized the natural environment in the past, human and nature can co-exist in architecture through newly created vocabulary.

PROPOSAL

일상에서 지친 현대인들에게는 도시속 삶이 유쾌하게는 다가오지 않는다.

The urban life is not delightful idea to modern society.

편리한 도시의 삶을 접고 편안한 농촌으로 돌아가 새 삶을 시작한다는 것은 익숙한 많은 것들을 포기해야 한다.

BStarting a new life in peaceful farming rural area means giving up different things that are embedded in our lives.

PROBLEM

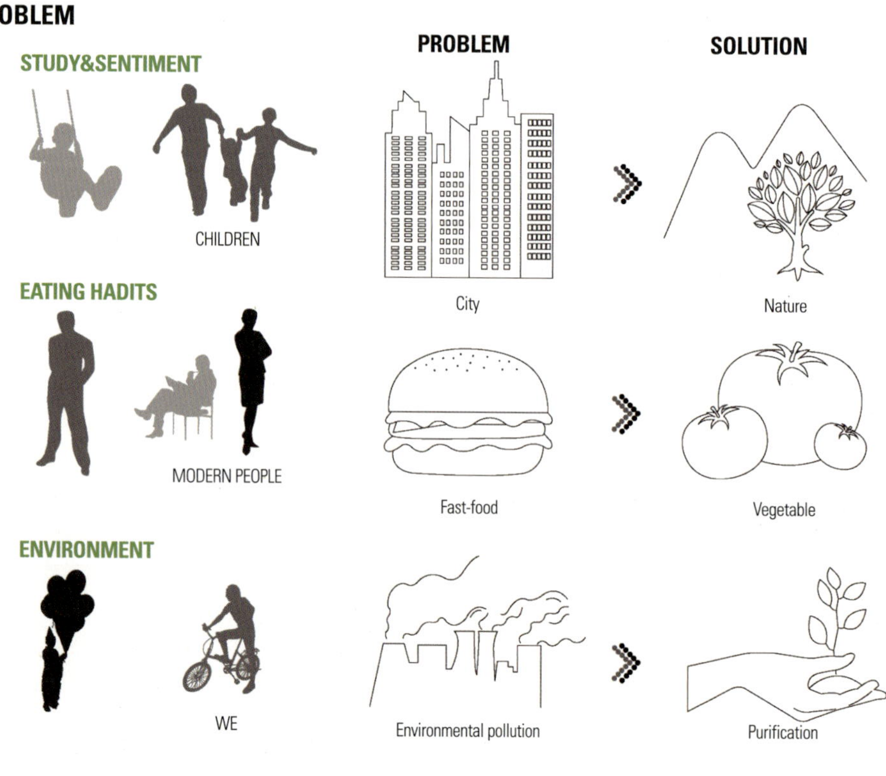

도심 속 아이들에게는 현재 주어진 답답한 공간에서 공부하는 것보다는 자연속에서 자연과 어울리면서 공부를 하는 것이 더욱 효과적이다.

It is more effective to study in nature with active interaction rather than studying in current suffocating space in cities.

현대인들은 패스트푸드의 다량섭취로 인해 비만인구가 증가하고 있다. 이로 인해, 다이어트와 친환경적 농산물에 많은 관심이 커지고 채소섭취량이 늘어나고 있다.

The population with obesity is increasing due to high intake of fast-food. Thus the interest towards organic agriculture products and healthy diet of eating vegetables are increasing.

오염되어 있던 안양천을 더 자연적인 환경으로 조성함으로써 사람들의 인식을 바꾸어 나가고자 한다.

Re-creating the polluted Anyang River into more eco-friendly environment is intended to change people's perception.

SITE ANALYSIS

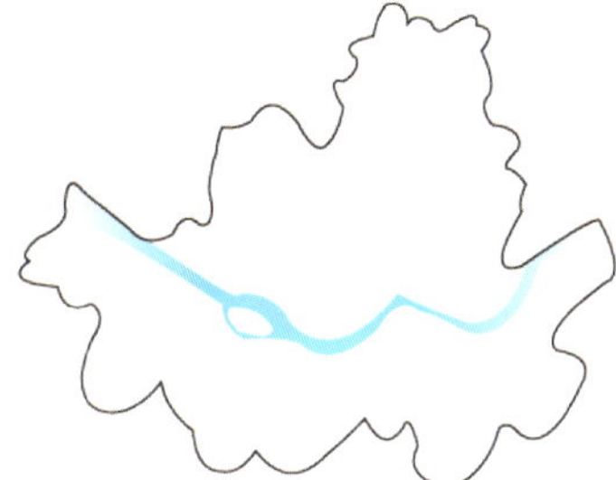

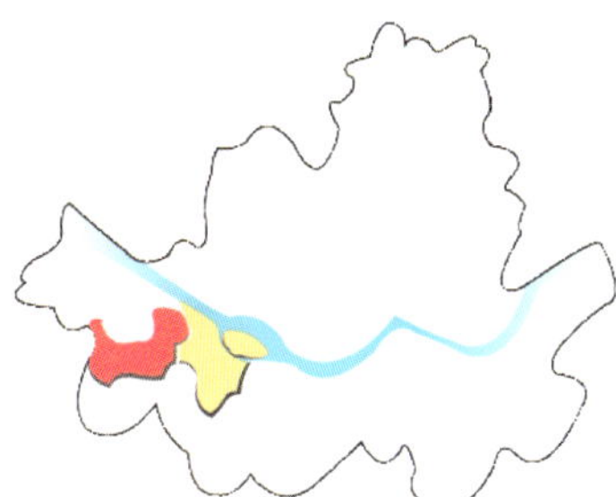

도림역
Dorimcheon Station
오목교역
Ohmokgyo Station
양천구청역
Yangcheon-gu Office Station

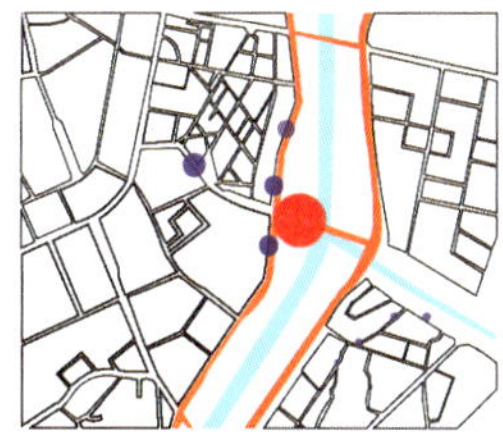

신정교 입구
Entrance sinjeonggyo

삼성아파트 정문, 후문
Samsung front door of the apartment, the back gate.

현대아파트 정문, 후문
Hyundae front door of the apartment, the back gate.

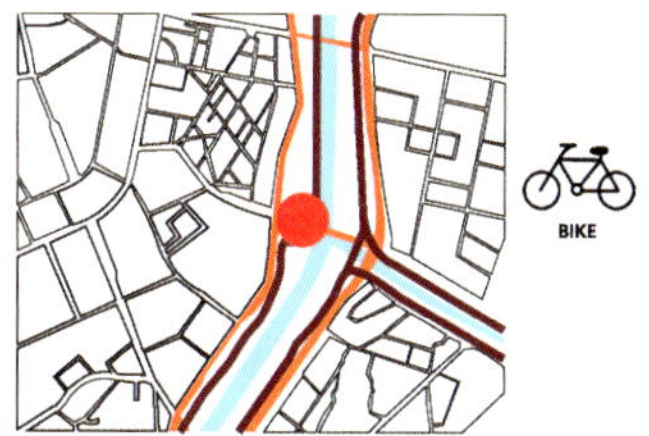

3m 정도 떨어진 곳에 위치하고 있다.
3m much less binary is located.

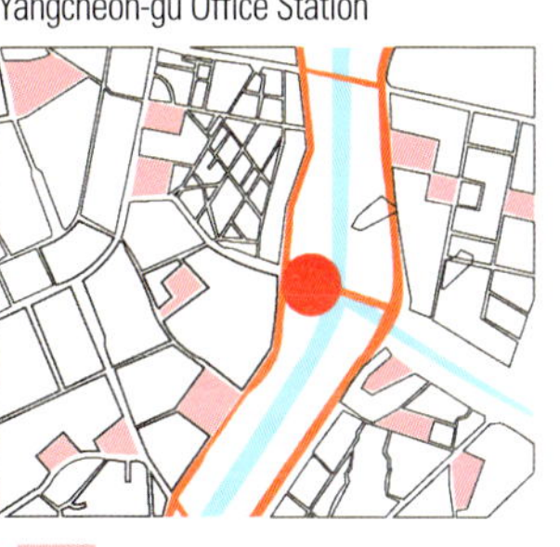

School

Rsidential area

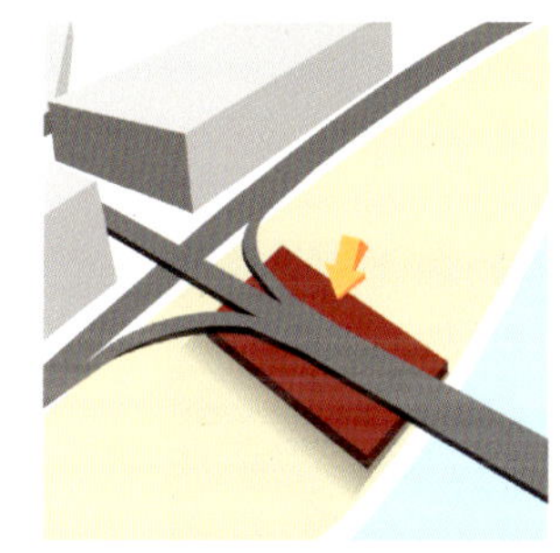

위치_ 서울특별시 양천구 신정동과 영등포구 문래동을 연결하는 다리 하부

Location: Seoul, Youngdeungpo Mullae-dong, Yangcheon sinjeongdong and a bridge connecting the sub-space.

TARGET

FAMILY

INDIVIDUAL

CHILD

MAIN USER

가족 단위의 이용객 단위에서 각각의 개인 단위로 세분화한 사용자들이 주된 사용자들로 잡았다.

The main target users are the individual users who are divided into specific unit rather than users in family unit.

SUB USER

신정교 근처에 인접해 있는 학교의 학생들과 선생님들을 또 다른 이용자로 선정하였다

The students and teachers from the schools near the Shinjung Bridge are the another targeted users.

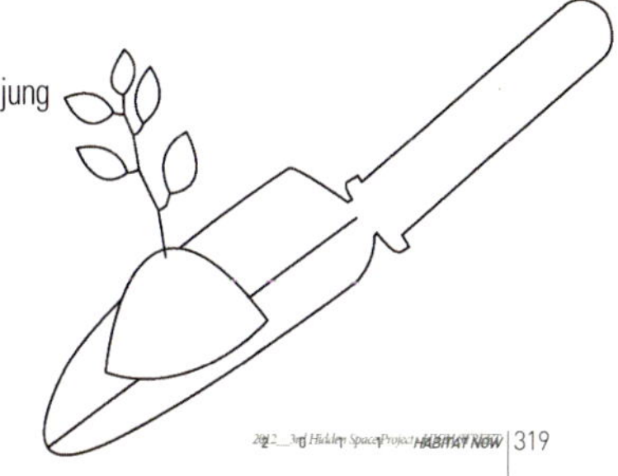

CONCEPT

잎(Leaf)

식물의 줄기에 붙어서 광호흡과 탄소동화 작용을 하는 녹색의 기관이다.
생장에 가장 중요한 에너지를 얻는 작용은 광합성 과정이다.
이 광합성 작용을 위해 잎은 햇빛을 잘 받도록 커다랗고 위를 바라보게 되어있다.
이러한 잎의 특성으로 비가 오는날에는 나무 밑에들어가 비를 피할수도 있다
잎은 자신의 성장을 위한 기관이기도 하면서 동시에 다른 생물들에게도 많은 도움을 준다.

It is a green organ that acts as functions of **광호흡** and **탄소동화** by the stem of a plant.
The function that creates the most energy for growth is photosynthesis.
The leaf is designed big and head above in order to receive the sunlight and to enable the function of photosynthesis.
One can take advantage of this characteristic of the leaves in order to avoid the rain under the trees.
The leaf is an organ for its growth that provides benefits to other creatures at the same time.

MOTIVE

KEYWORD

Vein of the leaf

식물의 잎에 있는 관다발을 말하며, 잎 속의 물과 양분의 이동통로

One of the vascular bundles of a leaf and moving path for water and nutrient in the leaf.

Tinted Autumnal leaf

기후의 변화로 식물의 녹색 잎이 빨간색, 노랑색, 갈색 등으로 변하는 현상

Phenomenon of leaves changing its green color to red, yellow or brown, etc due to change of climate.

Phototropism

빛쪽을 향하거나 빛에서 멀어지기 위해 식물의 줄기와 잎이 굽어 자라는 현상

The phenomenon of plants' stem and leaves to the stimulus of light, producing a bending towards or away from the light

Spiral phyllotaxis

줄기마다 잎을 한 장씩 피우되 일정한 개도를 유지하면서 서로 어긋나게 피우는 방식

Criss-crossing spiral arrangement of leaves with constant number with alternative leaves in rotation.

04/20 A.M 8:00 07/20 A.M 8:00 10/20 A.M 8:00

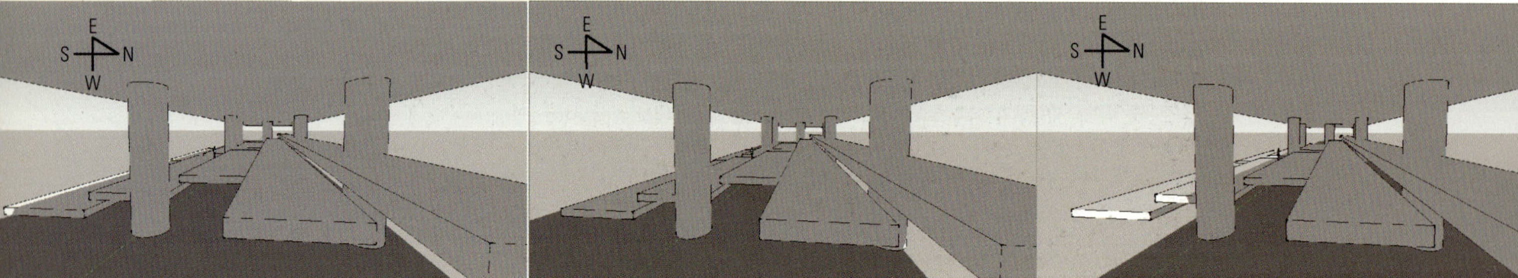

SPACE PROCESS

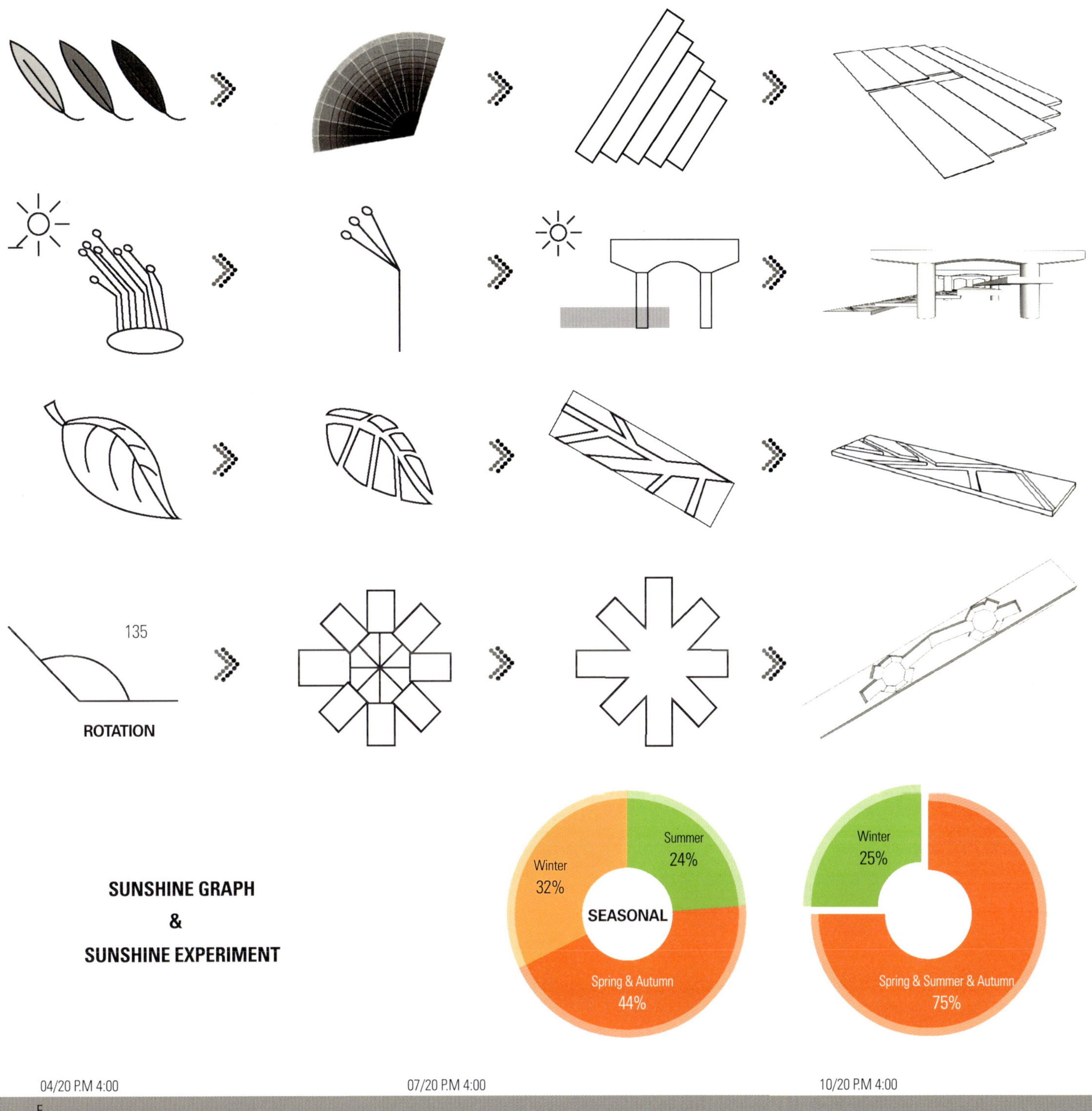

EXPLODED VIEW

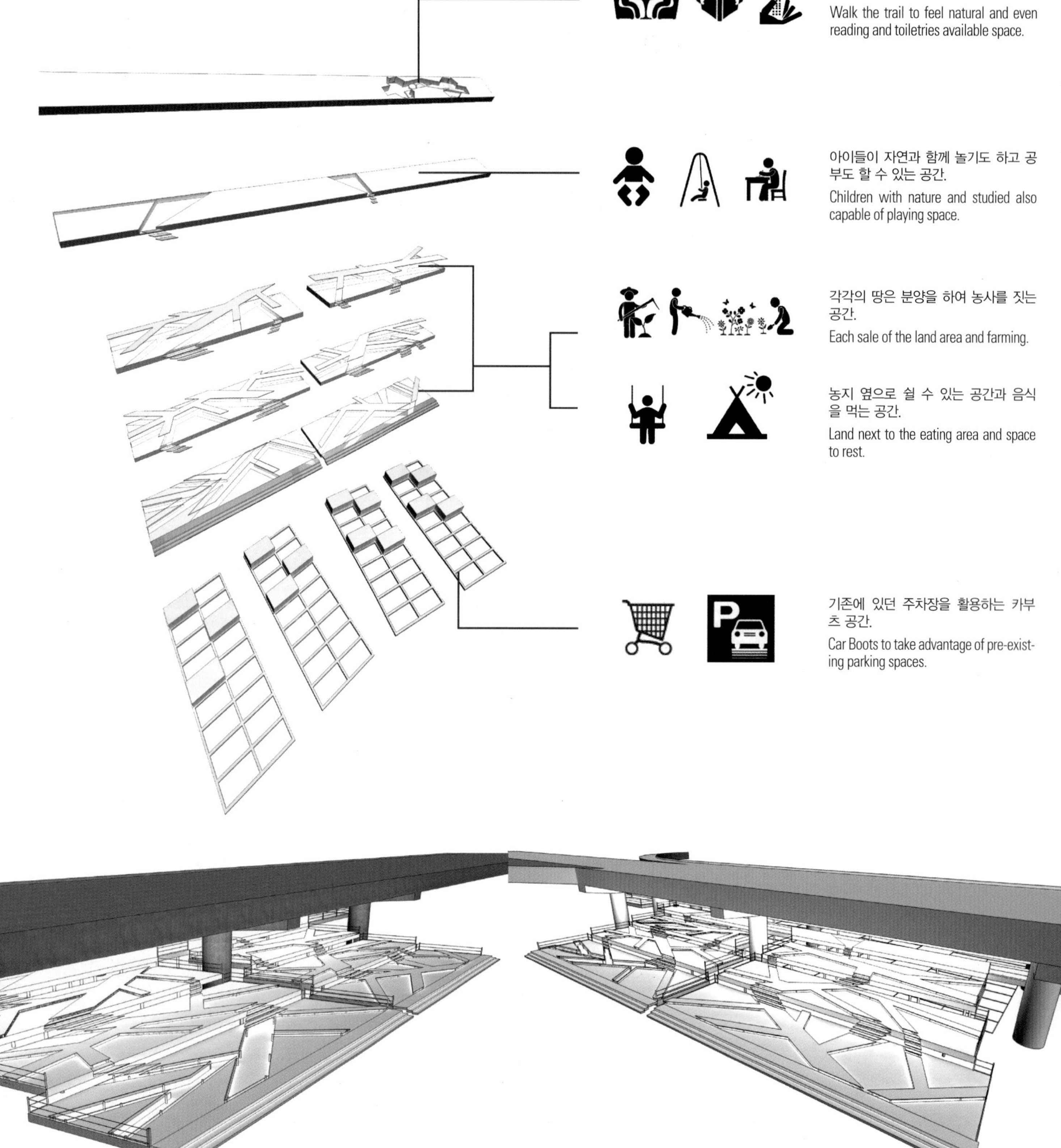

4F

아이들과 어른들 모두를 위한 조그만한 독서 공간.

For both children and adults with little reading room.

많은 식물 속에서 편안한 마음으로 독서를 즐길 수 있는 공간.
In many plants with peace of mind is a great place to enjoy reading.

개방된 공간.
Open space.

아이들의 작은 모래 놀이터, 그 옆엔 이 공간만의 특징인 징검돌들이 있는데, 이를 통하여 좀 더 자연적인 요소에 다가가고자 하였다.

Small children playground sand, and beside the space unique characteristic which jinggeomdoldeulyi, thus creating a more natural element was designed to reach.

4F

이 공간은 모든 사람들이 들러 쉴 수 있는 공간으로 오솔길 느낌을 살렸으며 잠시 들러서 벤치에 앉아 휴식을 취할 수 있으며, 농장에서 직접 재배하는 허브와 꽃으로 만든 티도 즐길 수 있는 공간.

Trail & tea room walk the trail to go to relax while sitting on a bench or a raising on the farm made up of herbs and flowers where you can enjoy a tea.

저녁에 자신의 농장을 찾은 사람들을 위해 각 구역의 팻말에서는 빛이 나오도록 하여 쉽게 자신의 공간을 찾을 수 있게 하였으며, 펜스 맨 윗줄도 안전을 위해 불이 나오도록 설계하였다.

In the evening for those who found his farm in each district
In the light green signposts naohdo easily be able to find their own space. the top line of the fence was safe to come out for the fire seol

THE GROWING SEASON

Spring & Summer (3-7month)

Fall (8-11month)

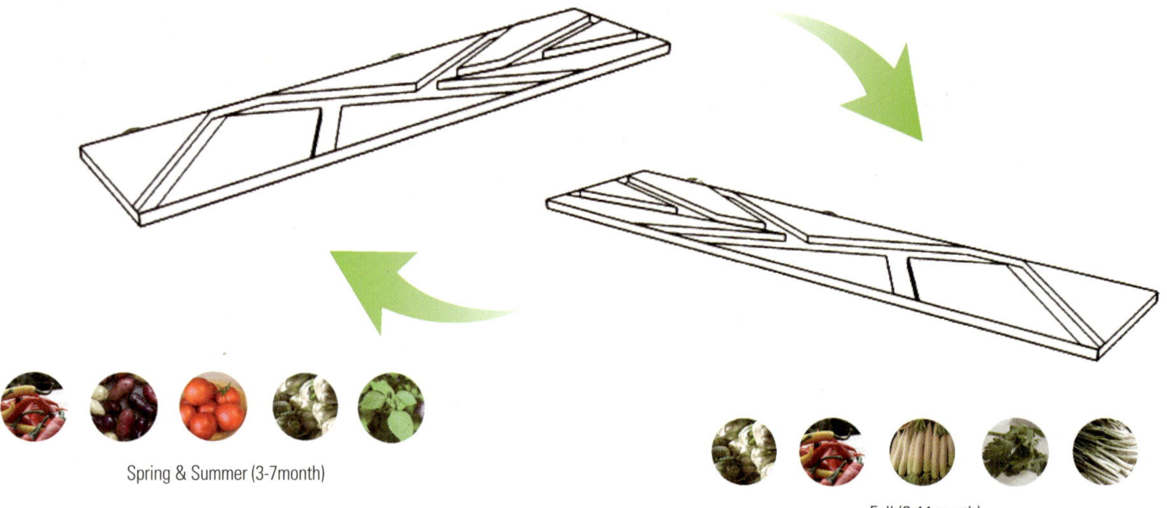

CAR BOOTS

계절별 채소를 활용하여 한 해에 이모작이 가능하며 주차를 하지 않은 남은 공간에 카부츠를 열어 자신의 채소를 판매 할 수 있도록 하였다.

자신이 생각한 것보다 많은 양의 채소를 수확한 경우 주말에 열리는 카부츠를 통하여 물물교환 또는 채소를 필요로하는 사람에게 판매가 가능하며 자신이 판매를 원하는 주의 주말에 카부츠를 이용하면 된다.

Double cropping in a year by taking advantage of seasonal vegetables, and parking is not available kabucheureul remaining open space to allow their vegetables were sold.

Than they thought, if large amounts of vegetables harvested kabucheureul held on weekends through barter or vegetables can be sold to people who need and want to sell their note for the weekend is kabucheureul you.

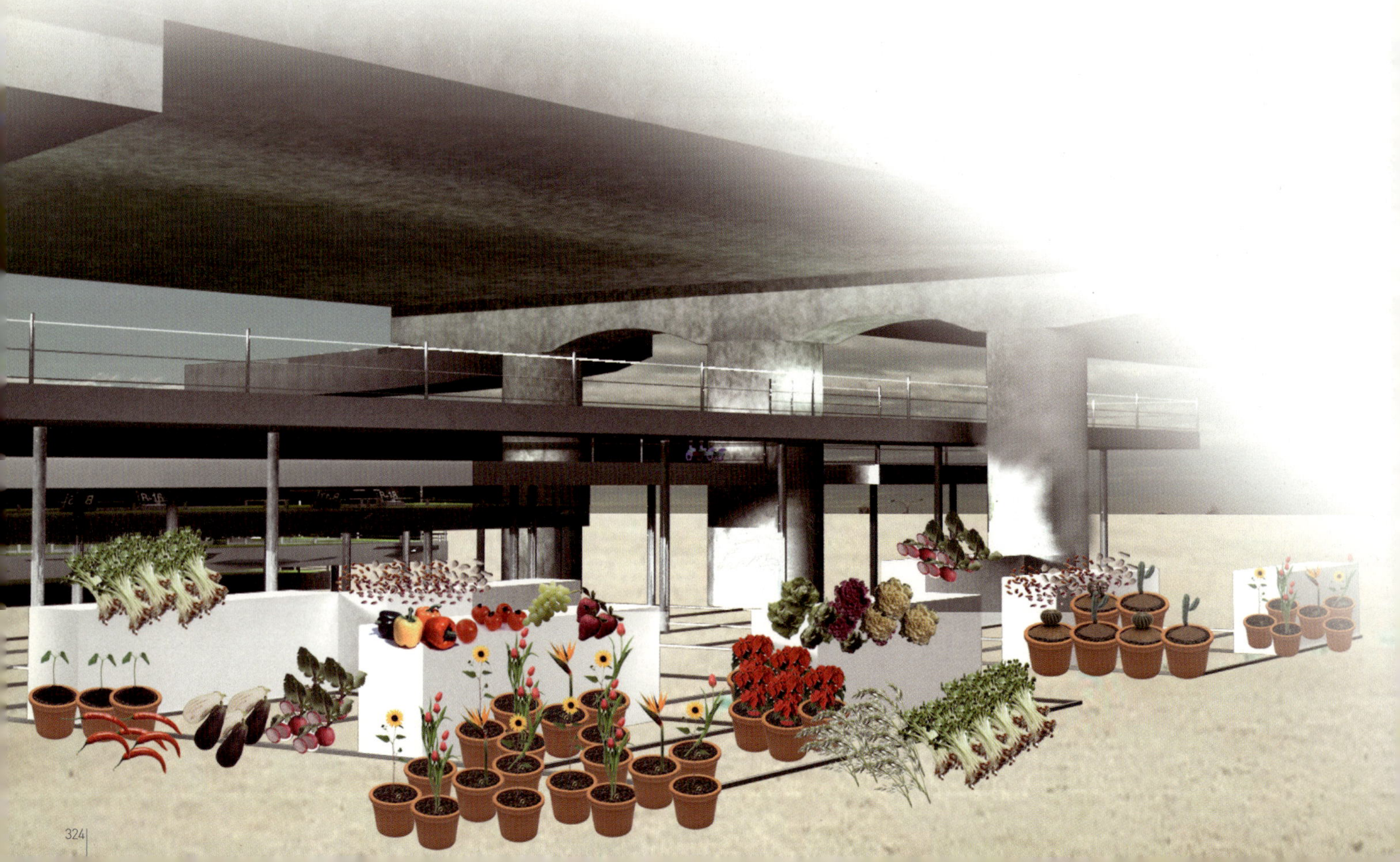

MAHATMA
LIVE IN DAEJEON
2008.12.6.PM6:00
DAEJEON RS HALL
GUEST : BURNING HEPBURN, HELLION

Enjoy in the cat's cradle.

다양한 사람들이 공연의 정보를 습득하는 공간

우리의 공간은 한강진역의 블루스퀘어 앞에 위치하는 고가밑의 사람들의 이동통로를 좀 더 활동적인 공간으로 이용하기 위해 공연관람객 및 차도 보도를 이용하는 사람들에게 이중적으로 문화활동을 권장하기 위해주변 공연 및 전시 정보를 시각적 촉각적으로 받아들일 수 있도록 공간을 표상하였다.

Han gang-Jin of our space station high in the Blue Square is located in front of the base path of movement of people .To use space more active performances for visitors and those who use a car reported double the To encourage cultural activities and performances around the visual information, tactile exhibits unacceptable Representation so that the space was.

BACKGROUND

현대인의 생활수준이 높아지면서 문화생활에 대한 중요성도 증대되었지만 공연에 대한 정보는 취약하기만 하다. 공연장의 홈페이지가 없는 경우도 있으며 있어도 이뤄지고 있는 공연에 대한 정보가 제대로 기재되어 있는 경우가 드물다. 우리가 모르게 하루에도 무수히 쏟아지는 공연, 문화 정보를 보다 쉽고 재미있게 접할 수 있는 공간을 제시한다.

관람객은 주변에 위치하고 있는 공연장이나 문화공간에서 개최되는 공연 및 전시에 대한 정보를 얻으며 비교, 분석하여 공연을 관람할 수 있다. 공연일정과 정보는 물론, 바로 티켓팅을 할 수 있는 티켓부스, 공연에 대한 간단한 퍼포먼스를 볼 수 있는 공연장, 공연을 기다리거나 끝난 후에 쉴 수 있는 쉼터도 마련 될 것이다.

사이트는 불특정 다수가 많이 지나 칠 수밖에 없는 공간이기 때문에 문화생활에 관심이 없는 사람들도 문화생활을 권장할 수 있는 공간이 될 수 있다. 또한 공연을 하는 사람들에게 관람객을 끌 수 있는 홍보의 공간이 되기도 하며, 모든 상호할 수 있는 불특정 다수들이 서로 교류하며 소통하는 공간의 장이 될 것이다.

Modern people are improving our standard of living. So we are awaking to the need for cultural life. But Imformation of performing arts are inadequate. In large part There are no Performance Hall. Although there is a web site, there is no imformation of performing arts. We provide places making these accessible easily for performing arts and imformation of performance. Audience can compare and analyze performance or display in cultural complex. Then they choose one of them. Moreover There are entertainment events, schedule, ticket booth for ticketing fastly, concert hall for simple performance and places for relaxation after the show. Because unspecified individuals pass the web site, this place can encourage people who be indifferent about culture to enjoy the cultural life. besides, It takes a role to promote a performance and becomes a form of communication to everyone.

SITE

사이트는 한남동 한남1고가차도 하부공간이다. 이곳은 공연장 바로 앞에 위치하고 있으며 주변에 교육기관, 지하철, 버스정류장, 문화 공간, 주거공간이 고루 분포하고 있다. 유동인구가 많은 편이며 주로 지나치는 통로로 이용되고 있다. 주변에 여러 공연 및 전시장이 분포해 있어 비교적 문화생활을 하기 편리한 지역이다.

It's placed at the front of the auditorium and there are bus stations, entertainments, residential areas evenly distributed in the surroundings. This area has a large floating population and it's often used as just passing by place. Because there are many shows and exhibitions nearby it's convenient to enjoy the culture life.

서울특별시 용산구 한남동 727-56
Hannam-dong 727-56, Yongsan-gu, Seoul

DETAILED VIEW

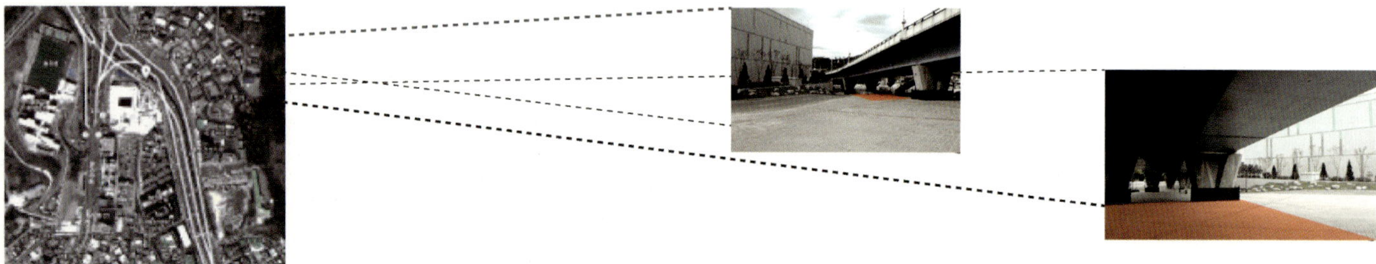

TARGET

가까운 곳에 지하철 6호선 한강진역이 위치하며, 반대편에는 버스정류장이 있어 유동인구가 많은 편이다. 인종, 성별과 상관없이 전 연령층이 고루 분포하고 있다. 사이트 앞에 위치한 공연장 관람객들과 이곳을 지나가는 보행자들이 주요 타겟이 될 것이다.

Hangangjin Station(Subway Line 6)is located on neck of the wood and On the opposite side, there is a bus stop so This area has a large floating population. The population is evenly spread regardless of age , race, and sex. pedestrian on the street and audience could be the main target.

A FLOATING POPULATION

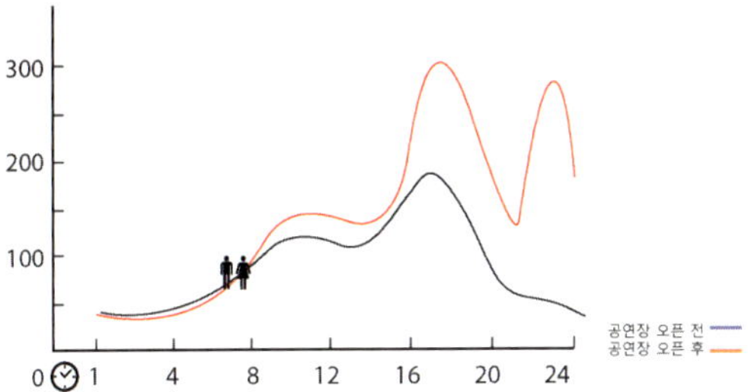

CONCEPT

실뜨기
Cat's cradle

실뜨기는 한 가닥의 실을 끝을 묶어서 하나의 고리형태로 만들고 두 사람이 서로 번갈아 가면서 실 사이에 손가락을 넣어 안으로 혹은 밖으로 뒤집어 실이 맺히지 않도록 하면서 모양을 짜는 놀이이다. 이는 상호간의 소통과 교류가 있어야 하며 움직일 때 마다 예각삼각형, 마름모, 둔각삼각형 등 다양한 형태가 나타난다. 이 과정에서 추출되는 여러 가지 다각형의 모양을 공간에 적용하며, 움직일 때마다 모양이 변하는 점을 이용하여 공간이 용도에 따라 형태가 변하도록 가변성을 더하였다.

Cat's cradel is a game by making a loop tying the end of a thread and by putting fingers inside or flipping outside making patterns, being careful not to entagle the thread with two people. This can be played with communication and exchanges between them and while the game is continuing variety of patterns will appear like an acute, obtuse angled triangle,rhombus. By applying these patterns and using the changes of the patterns while continuing it added the variability of the changes as it move.

DESIGN KEY WORDS

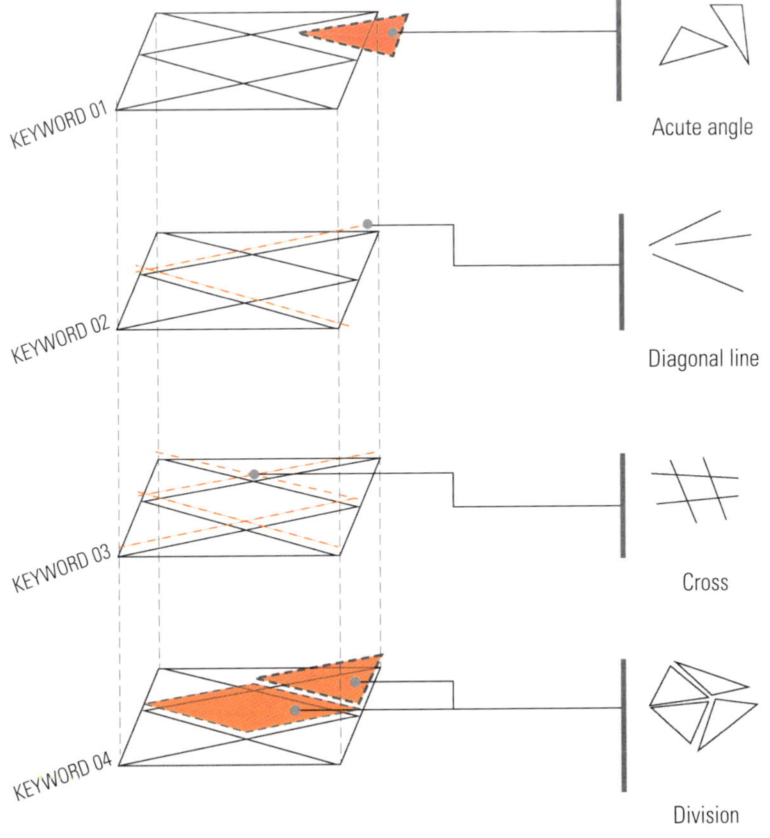

선에 의해 생기는 예각은 날카로운 긴장감을 준다.
Acute angles by lines give us a sharp sense of tension.

사선을 공간에 적용하여 역동적인 느낌을 준다.
Diagonal lines give us dynamic aspects.

실뜨기에서 선은 일차적이지 않고 위아래로 입체적으로 교차한다.
In the cat's cradel, Lines aren't monotonous. in other words they are multi-level crossing.

여러가지 선에 의해 분할되는 다양한 형태의 면을 볼 수 있다.
We can look on the various side that partitioned by the lots of lines.

DESIGN PROCESS

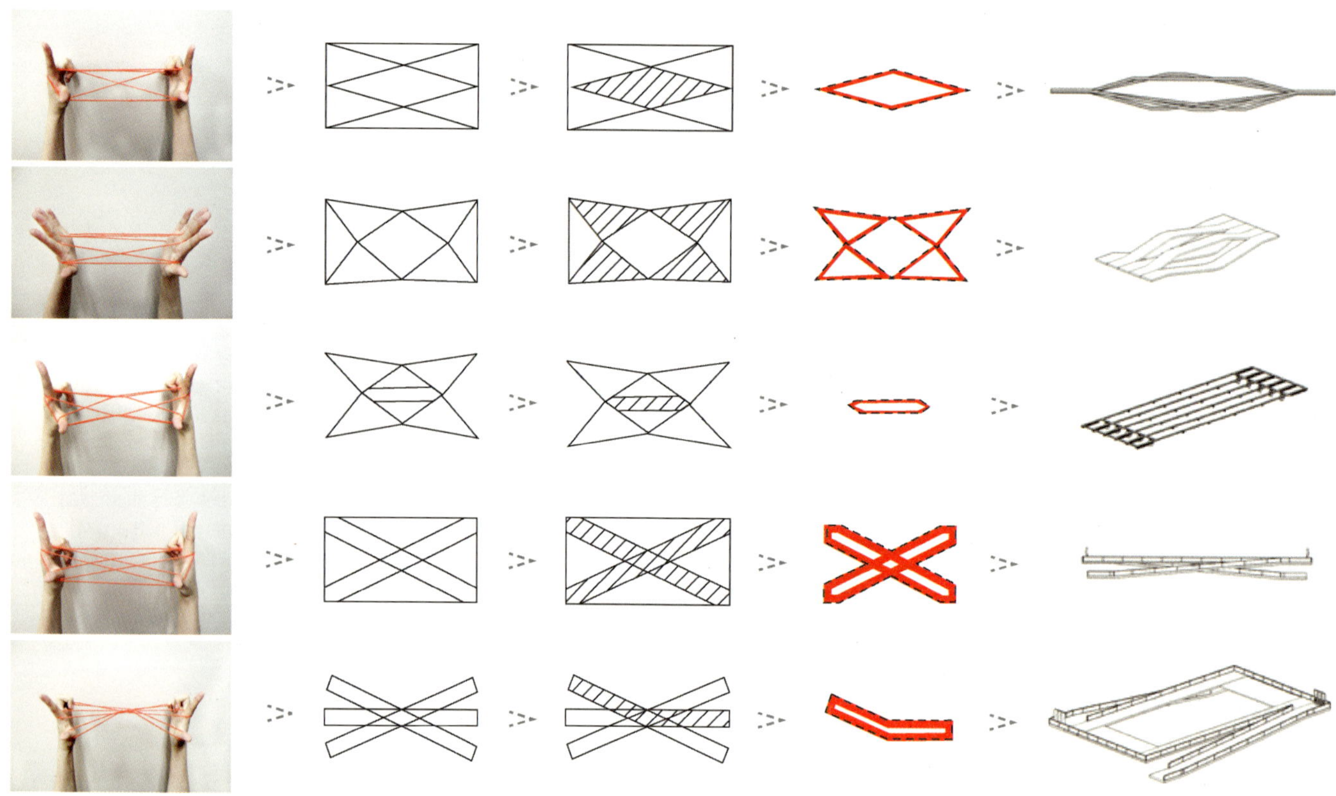

SPACE PROGRAM

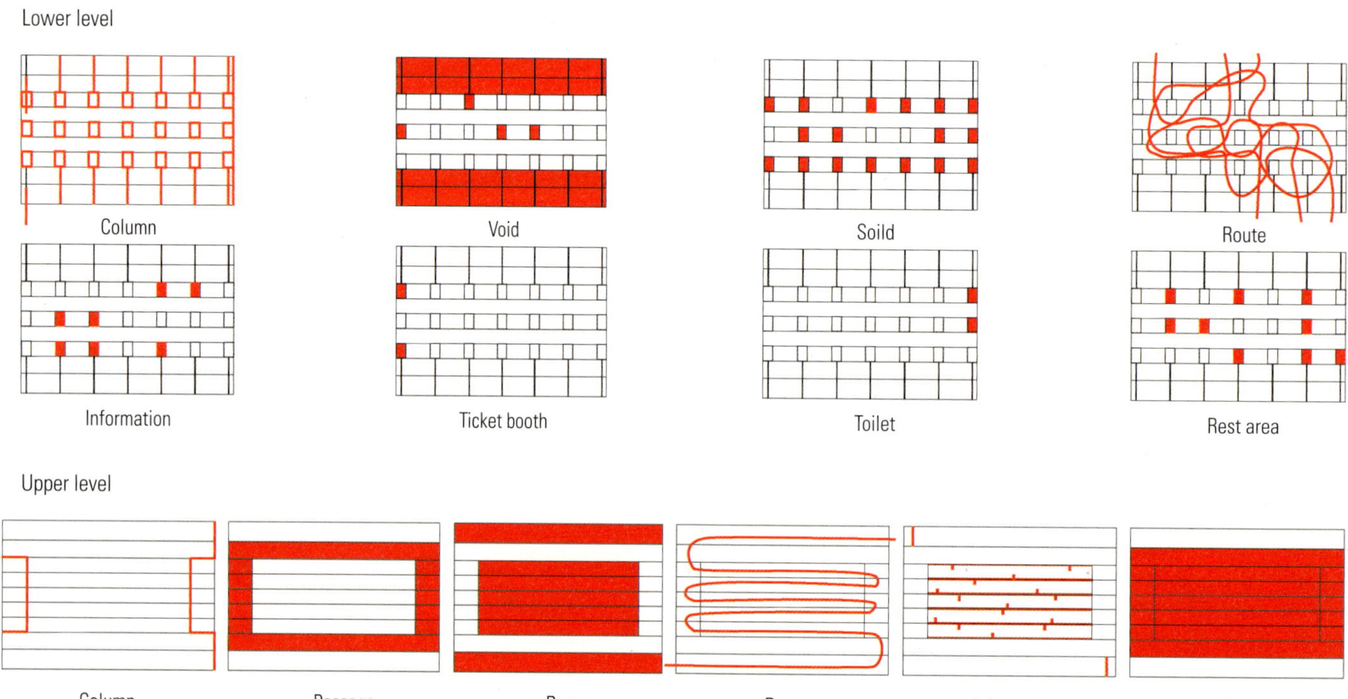

TRAFFIC LINE

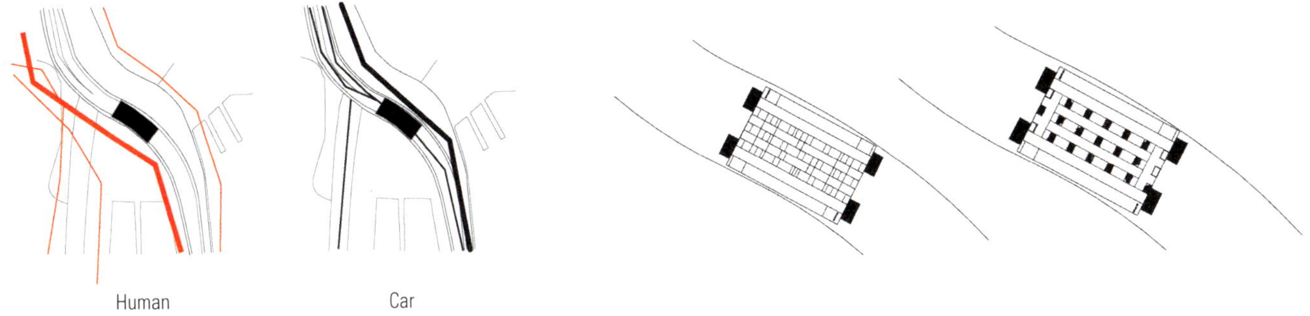

FLOOR PLAN

Lower level

Upper level

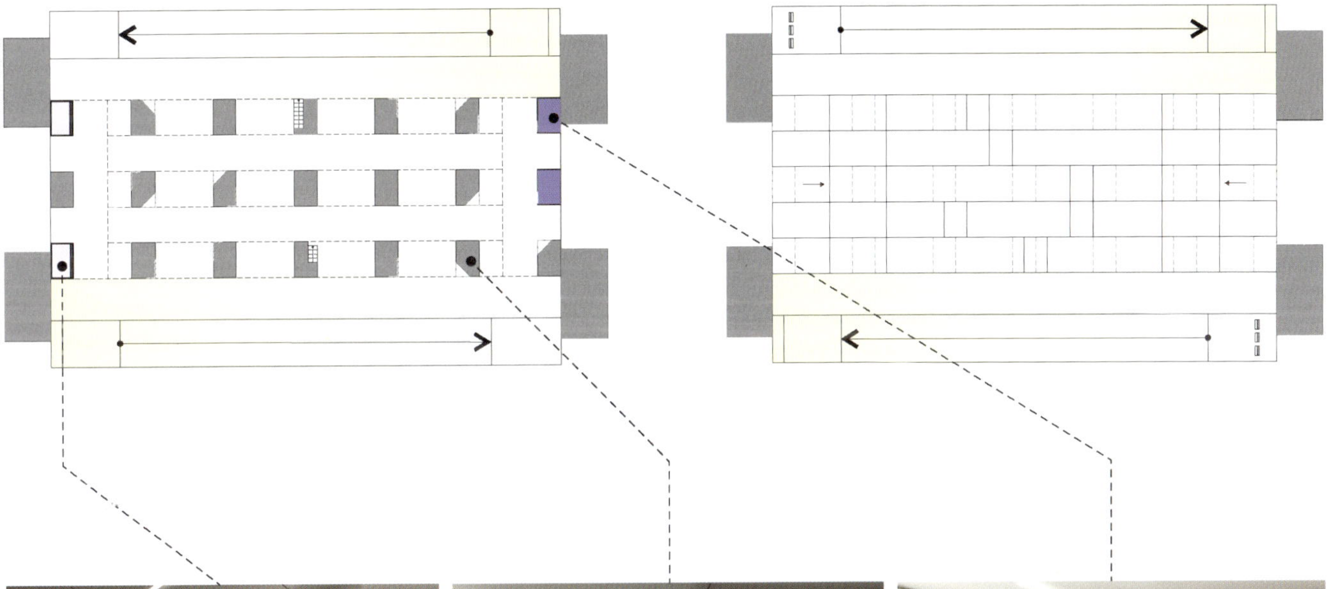

Ticket booth

SITE가 위치하는 근처의 공연 티켓을 바로 예매 및 구매를 할 수 있으며 간단한 공연 소개 및 티저 동영상 등의 정보를 볼 수 있다.

In this place, People order tickets for a performance near here and they are provided with imformation for example a listings magazine, teaser video and so on.

Rest area

공연 관람 전후에 관객들이나 지나가는 사람들이 편히 쉴 수 있는 공간이다.

This place provides places to relax to provide places to audience and pedestrian.

Public toilets

공연 관람객 및 지나가는 보행자들이 이용할 수 있는 남녀 공중화장실을 마련하였다.

This is toilet that audience and pedestrian are available.

TRANSFORMATION

FASE 01

길이 평평할 때에는 유리파티션이 다리하부 공간에 고정되어 있어 공연 무대로 쓰인다.

When the floor is flat, partition of glass is fixed under the feet. so It plays a role in performing.

FASE 02

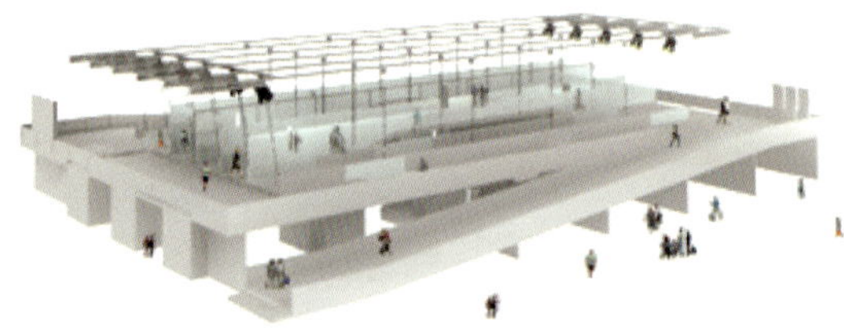

5갈래의 평평한 길에 구간마다 관절이 있다. 각각의 길은 관절에 의해 움직여 경사로가 되며 각도와 높이가 다르다. 다리 하부 공간에 고정되어 있던 유리파티션이 내려와 각각의 길의 난간이 된다. 유리파티션에는 공연에 대한 영상이 나와 길을 지나가면서 볼 수 있다.

five - forked flat road have joints in each section. each road moves by joints. then it becomes a slope that have different height and angle. Partition of glass fixed under the feet moves down and becomes railing of each road. Partition of glass plays videos. so people watch these when they pass by.

ELEVATION

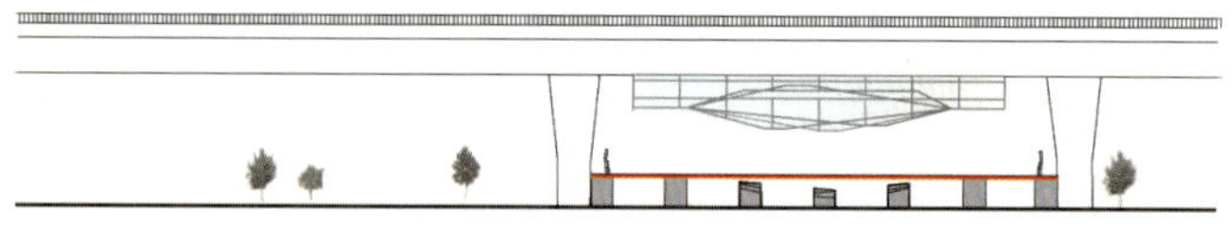

FASE 01

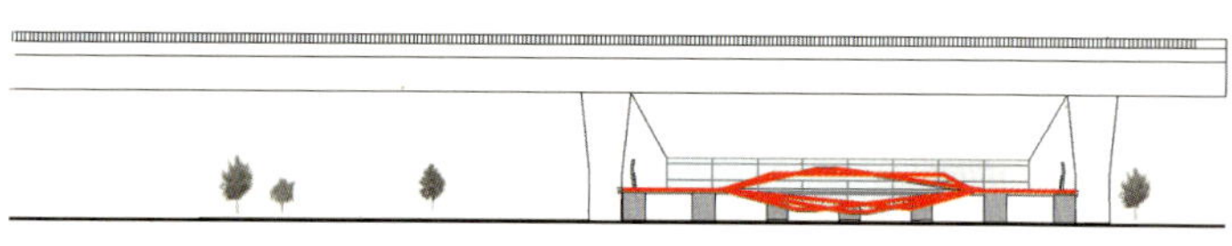

FASE 02

TRANSFORMATION

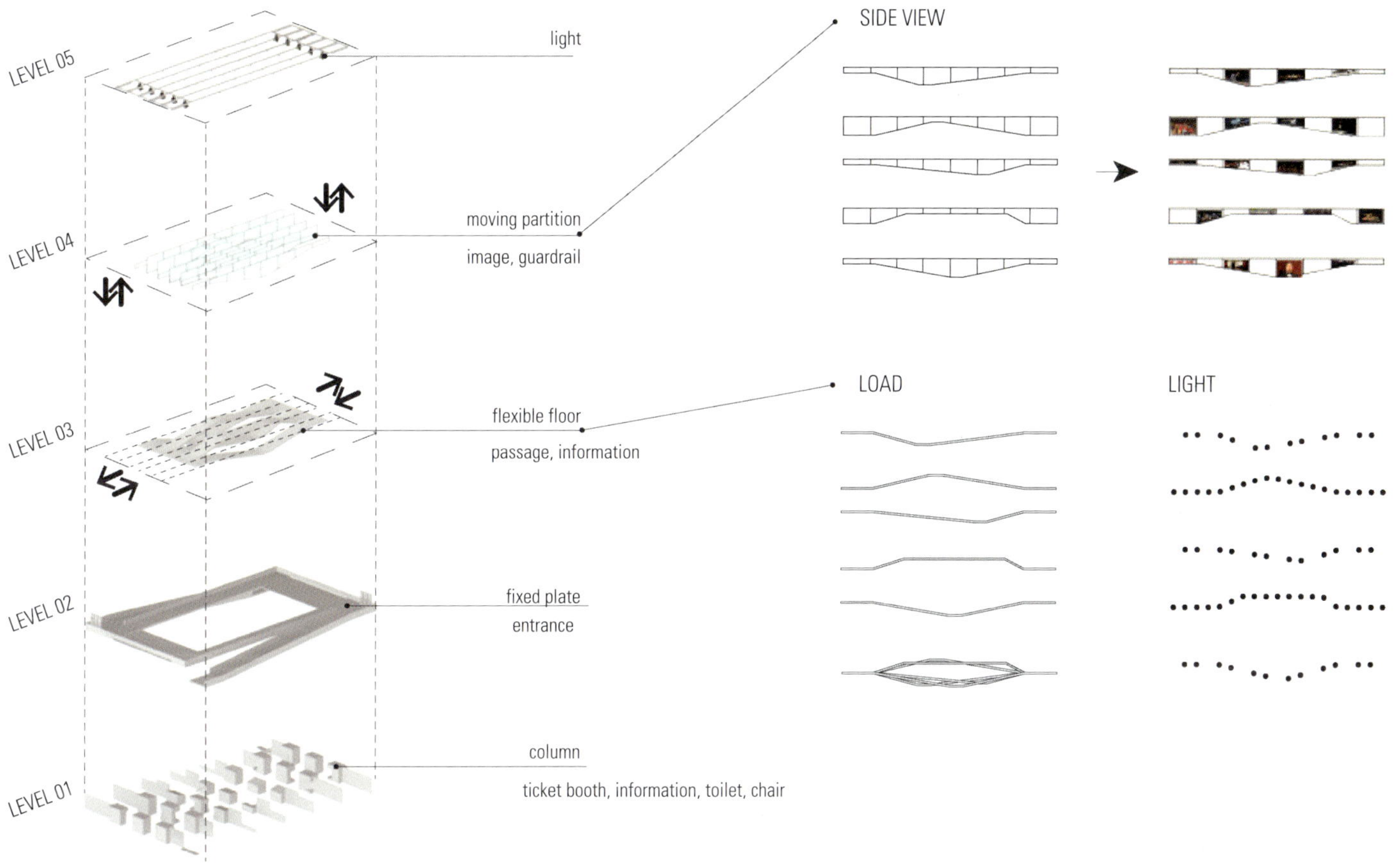

LEVEL 05
light

LEVEL 04
moving partition
image, guardrail

LEVEL 03
flexible floor
passage, information

LEVEL 02
fixed plate
entrance

LEVEL 01
column
ticket booth, information, toilet, chair

SIDE VIEW

LOAD

LIGHT

MOVING LOAD (DAY TIME/ NIGHT TIME)

공연정보 전달 장소 Show where information delivery

공간을 지나가는 차량 및 보행자들이 공연에 대한 열정이나 내용에 관한 정보를 쉽게 알 수 있는 곳이다.

Space for passing vehivles and pedestrians on the show schedule and information about the content of the space is easy to see.

휴식공간 Resting place

공연 관람 전후에 관객들이나 지나가는 사람들이 편히 쉴 수 있는 공간이다.

This place provides places to relax to provide places to audience and pedestrian.

퍼포먼스장
Resting place

공연에 관한 간단한 퍼포먼스를 보여줌으로써 관람객들의 흥미를 유발할 수 있다. 시청각을 이용한 퍼포먼스를 통하여 보다 실감나게 공연 내용을 이해할 수 있다.

Because of a poor environment, Street musicians that don't sign a contract with a record company are in trouble. so this place will provides a nice space to perform art to street musicians.

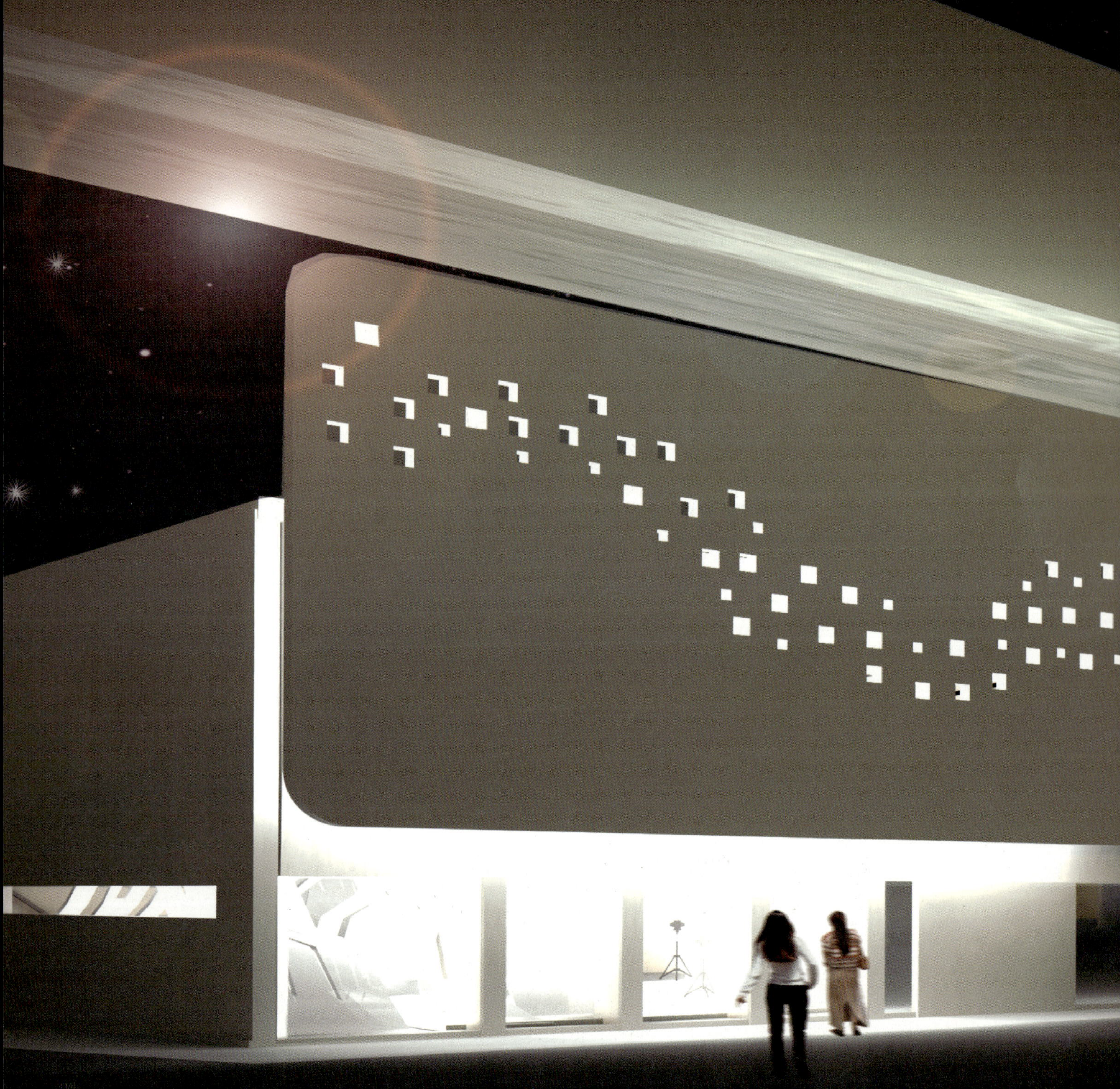

별이 빛나는 밤에

감성을 전하다

The Starry Night

PROPOSAL

모든 것들이 빠르게 변해가는 요즘, 우리들은 그에 맞춰 바쁘게 살아간다. 그러다 보면 많은 것들을 쉽게 접하고 쉽게 잊어 버리는 일상에 익숙해지게 된다. 그래서일까 정말 중요한 것을 잃고 있는 것은 아닌가 의구심이 들 수도 있다. 그래서 우리는 변해가는 사회의 모습에 초점을 맞춘 것이 아니라 변하기 전의 예전의 것들을 보다 많은 사람들에게 보여주고자 한다. SLOW LIFE의 추구가 늘어난 현재 너무 바쁘게 사는 와중에 예전 것들의 향수를 느낄 수 있는 공간이 될 수 있도록 도와 줄 것이다.

All of the changing quickly these days, we get busy living accordingly. Faces a lot of things easier when you're easy to get used to the routine is Forgetful. accordingly, really beginning to wonder whether what is important is that of losing can be.The changing appearance of the four meetings so we focused on the variable before the old ones but more people want to show. The increase in the pursuit of slow life live in the midst of the current is too gappeuge can feel the nostalgia of the old ones can be a space will be able to help.

MULTI INFORMATION SPACE

예전의 모습을 보존하고 있는 만리동의 구두거리.

In the past to preserve the appearance of Manridong verbal distance.

시각에 의존하는 일차원적의 전시에서 탈피.

Never rely on one-dimensional perspective of the Move away from display.

정보, 문화적 추억, 홍보가 동시에 접목된 공간.

Information and cultural memories, promotion Grafted area at the same time.

TARGET

MEMORIES REMINDER

TRAVELER

7080GENERAL

FAMILY

BACKGROUND

아날로그 시대가 지나가고 디지털 시대가 온 지도 어언 10년이 넘었다. 그 동안 새로운 매체들은 계속 쏟아져 나왔고 더욱 더 빠르게 더욱 더 다양하게 증폭되어 갔다. 빠른 것에 익숙해지다보니 사람들이 아날로그 시대의 감성을 그리워 하게 되었다. 이메일 대신 손편지를 쓰고 MP3 대신 카세트 테잎을 혹은 그 전의 LP판을 그리워 한다. 그런 그들에게 감성을 되살릴 수있는 공간을 전해준다.

On the analog era is passing and the digital era was over 10 years. In the mean time, continue in the new media was a blinding even faster, became more and more variety of amplification. I get used to being fast all the emotions of people missed out on was the analog era. Instead, write an email sonpyeonjilreu instead of MP3 or cassette tape will miss its previous edition LP. You can tell them the sensitivity to the space provided.

SITE ANALYSIS

서울역 Seoul Station

서울역이 위치하여 여행객들과 많은 유동인구가 밀집. 다국적,다계층,다문화의 사람들이 모인다.

Many travelers and Seoul with this position the floating population density. Multinational, multi-tiered, multi-cultural people congregate.

서울역 광장 Seoul Square

서울역이 위치하여 여행객들과 많은 유동인구가 밀집. 다국적,다계층,다문화의 사람들이 모인다.

Many travelers and Seoul with this position the floating population density. Multinational, multi-tiered, multi-cultural people congregate.

만리동 Manridong

왕초 거지왕 김춘삼, 염천교 밑에서 등 여러가지 문학 작품과 대중매체에 등장했던 만리동이 위치해 있어 주거민들이 있다.

kimchunsam homeless boss, and various literary works under yeomcheongyo Manridong that appeared in the mass media are located, there are civil housing.

구서울역사 Old Seoul Station

구 서울역사는 복원 작업을 거쳐 박물관과 문화공간의 참여가 가능한 복합 문화공간으로 재탄생 하였다

Old Seoul Station through the restore operation of museums and cultural spaces Participation in multi-cultural space that was reborn.

SITE

여행사, 법무사, 어학원이 가장 높은 수치를 보였다. 분석 결과 서울역 부근에 분포해 있는 사람들의 연령대는 30, 40대가 가장 많은 것으로 나타났으며 그 다음이 20대로 집계되고 있다. 즉 7080 세대가 주를 이룬다는 것이다.

Agency, beopmusa, schools had the highest levels. Analysis of people Seoul near the dotted ages thirty, forty, appeared to be highest consideration and then the aggregate is twenty. 7080 generation will be the week that are affected.

USER ALALYSIS

서울역은 여행객뿐만 아니라 주변에 회사들도 밀접하고 있어 직장인들이 많다. 인근에는 여행사, 유학원, 증권사 등의 비율이 높다.

Seoul, as well as the travelers around the company are also close and there are a lot of workers.Near the agency, agents, brokers, and a high proportion.generation will be the week that are affected.

TRAFFIC LINE

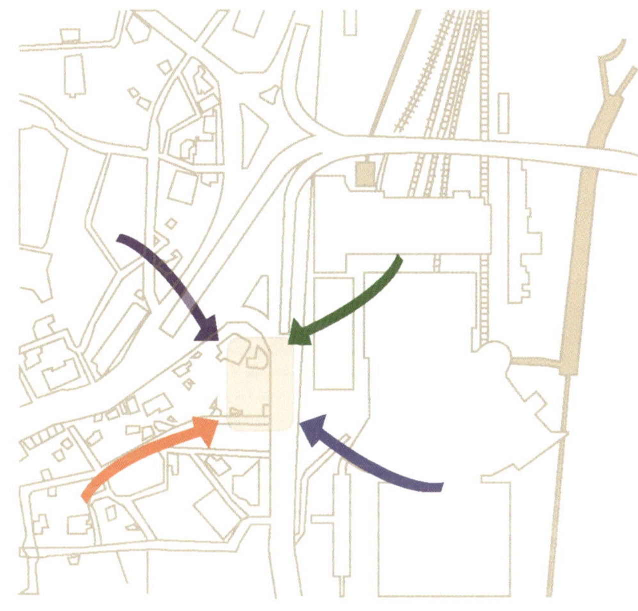

NEEDS

TRAVELER

세계화가 가속되면서 한국에 대해 관심을 갖는 외국인들 늘어 났다. 따라서 한국을 여행하는 여행객들도 늘고 있는 추세이다. 관심은 있으나 한국을 잘 모르는 외국인 여행객들에게 한국의 현재뿐만 아니라 과거를 접할 기회를 제공함으로써 한국에 대한 관심도를 높인다.

As globalization accelerates, foreigners with an interest in Korea increases. Therefore, tourists are traveling to Korea in the trend is also increasing. Interest to tourists, but foreigners are not familiar with Korea. Korea's current as well as providing the opportunity to encounter the past in Korea and increase interest.

FAMILY

부모는 자식에게 자신들의 과거를 보여주고 싶어하고 자식은 부모의 과거를 알고 싶어하기 마련이다. 하지만 말로만 설명하기엔 어쩐지 부족하다. 그러한 부족한 점을 채워줄 수 있는 공간을 제공함으로써 서로를 더욱 더 이해하고 가족관계의 친밀감을 높인다.

Parent to the child wants to show their past history of the child and the parent wants to know is established, but somehow is not enough to explain in words. A space to fill those gaps by providing a better understanding of each other more and increase the intimacy of family relationships.

7080GENERAL

현대 사회에서 바쁜 일상을 살아가는 7080세대에게는 늘 옛것에 대한 향수가 있고 그런 그들을 위해 타임머신을 탄 듯 과거를 보여줄 수 있는 공간이 필요하다. 따라서 7080세대의 향수를 충족시켜줄 공간을 제시한다. 같은 과거를 공유한 세대끼리 함께 즐기며 공감할 수 있다.

Busy life to live in modern society, three for 7080, and then I always have a nostalgia for old ones to their own past was like riding a time machine that can show you need space, so the nostalgia of 7080 to meet three presents astute space.

ARCHIVE

1. 기록 보관소	2. 기록 보관소에 보관하다	3. 파일을 보관하다
Archives	It is kept in the archives	Files are kept

7080세대의 추억이 담겨 있는 물건이나 자료들을 보관하고 볼 수 있는 공간을 제시.
Memories of 7080 that contained three items or materials stored and provide a place to see.

ARCHIVE + 세대간의소통 = **ARCHIVE** 공간의 생성

USER NEEDS

TEENAGER

엄마아빠 세대를 이해하고 한발짝 다가갈 수 있는 기회를 주는 공간 제공.

Mom and Dad's generation to understand and giving the space a step and provides an opportunity to reach.

TRAVELER

한국의 현재는 물론 과거까지 동시에 느낄 수 있는 공간 제공.

South Korea's current and past at the same time to provide a space to feel.

7080GENERAL

사라진 옛자취를 보고 느낄 수 있는 공간 제공.

Old trace can see and feel lost in the space provided.

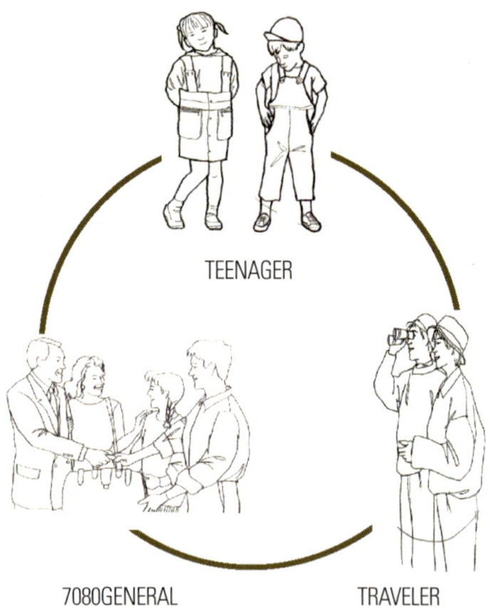

함께 공감하고 소통할 수 있는 공간

Together sympathize and Communication that can be space

CONCEPT PROCESS

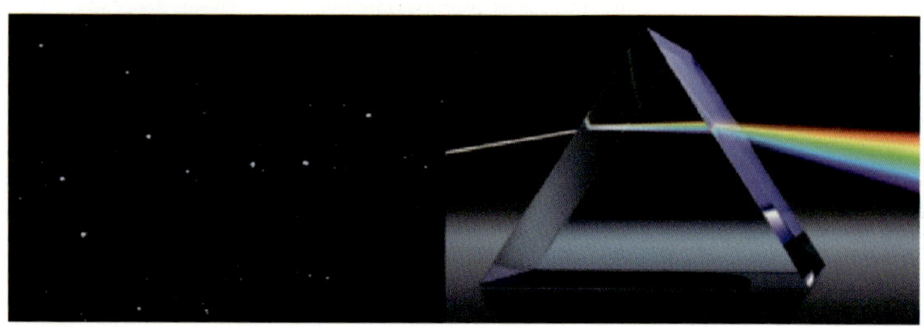

별자리 (CONSTELLATION) + 스펙트럼 (SPECTRUM)

별자리는 모양이 변하지 않고 항상 그 자리에 있어 길을 잃은 사람들에게 지표가 되어주고, 여행자들이 주류를 이루는 서울역의 특성과 적합하다. 스펙트럼은 빛의 파장에 따라 분해하여 배열한 것으로 별의 색이 각각 다른 이유는 스펙트럼과 연관이 있다.

Does not change the appearance of the constellation are always on the property has been lost, and those indicators, Dominated by travelers and is suitable for the characteristics of Seoul.Depending on the wavelength of the light spectrum to be arranged by decomposing the color of a star is associated with the spectrum for different reasons.

ABOUT DESIGN

DESIGN

마치 과거로 이동하는 듯한 연속성 있는 공간 제시. 지정한 SITE로 얻은 다양한 사람들의 행태와 SITE의 특성을 근거로 나타나는 공통점을 이용한 공간 제시.

It seems to go into the past in the space provided continuity.Obtained as specified SITE chara cteristics of a diverse group of people based on behavior and appear to show similarities with the space.

DESIGN CONCEPT

다양한 방문자들이 옛자취를 느끼고 우리나라의 여행지를 소개할 수 있는 공간.

Visitors feel a variety of the old trace and introduce Korean travel space.

DESIGN PROCESS

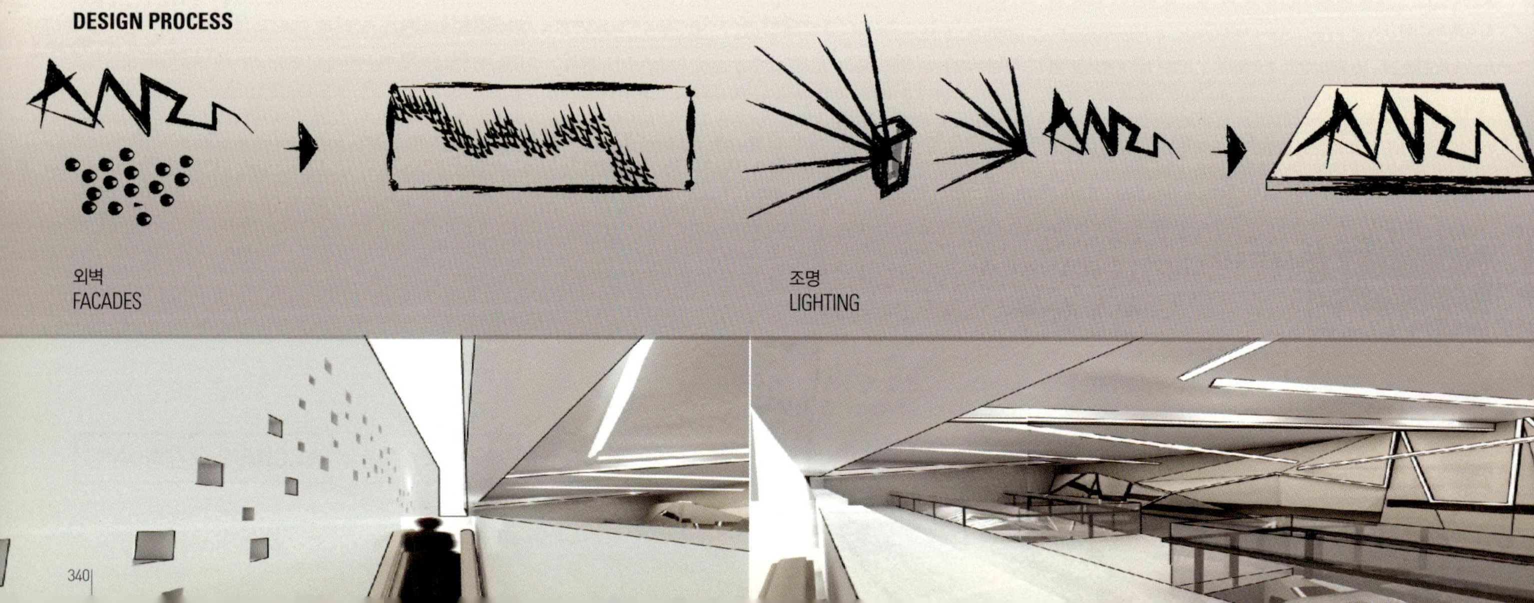

SPACE INFORMATION

ARCHIVE SPACE 사람들의 감성을 이끌어 내는 공간 Space to elicit people's emotions

PRIVATE & PUBLIC 공적인 공간 안의 개인적인 공간 Personal space in public spaces

SEPARATE AREA FROM CURRENT 현재에서 분리되어 과거의 시간으로 이동한 듯한 느낌을 받을 수 있는 타임캡슐 같은 공간
Now past time to move to separate from the feeling of space like a time capsule

TRAVEL INFORMATION 떠나거나 복귀한 여행자들에게 더욱 더 다양한 여행지에 대한 정보를 알려 주는 공간
To leave or returning traveler has more information on a variety of spaces to tell

DESIGN KEYWORD

선
LINE

빛
LIGHT

스펙트럼
SPECTRUM

별자리
CONSTELLATION

각각의 별들을 별자리로 엮을 때 생기는 별자리의 선을 키워드로 정하였다.

Constellation of stars in each constellation lines occur when rolling was defined as keywords.

우리가 별자리를 잘 볼 수 있는 이유는 각각의 별들이 빛나고 있기 때문이다. 별자리에서 중요한 요소 중 하나인 빛을 키워드로 정하였다.

We can see well why the constellation of stars, each shining because of Constellation important One of the elements of light was defined as keywords.

스펙트럼으로 인해 별들의 색깔이 각각 다르며 각각의 공간을 특색있게 만들어 주는 역할을 한다.

Each color is due to the stellar spectrum is different for each trait that makes the space plays a role.

이 공간의 컨셉이기도 한 별자리는 여행자들의 지표 역할과 7080 세내의 추억의 지표 역할을 한다.

Constellation is also a concept of space as an indicator of the traveler and the generation of 7080 will serve as an indicator of memories.

DESIGN PROCESS EXPLANATION

외벽
별이 하나하나 모여 별자리의 형상을 나타낸다.

FACADES
One by one stars together represent the shape of the constellation.

조명
바닥과 같은 패턴의 별자리를 바탕으로 천장에 각도를 바꿔 입체적으로 배치함으로써 생동감을 준다.

LIGHTING
Floor and ceiling based on a constellation of the same pattern, the angle will change to life by placing three dimensional.

내벽
입체적으로 별자리의 선을 표현한다.

THE INNER WALL
To three dimensional representation of the constellation lines.

장식
별자리의 선의 모양과 흡사한 장식을 만든다.

DECORATION
Line like shape of the constellation makes decorating.

바닥
별자리 패턴을 본따 부분부분 높이를 다르게 하여 밋밋한 바닥에 변화를 준다.

FLOOR
Chosen as a reminder of the constellation pattern by varying the height of the pieces flat on the floor will change.

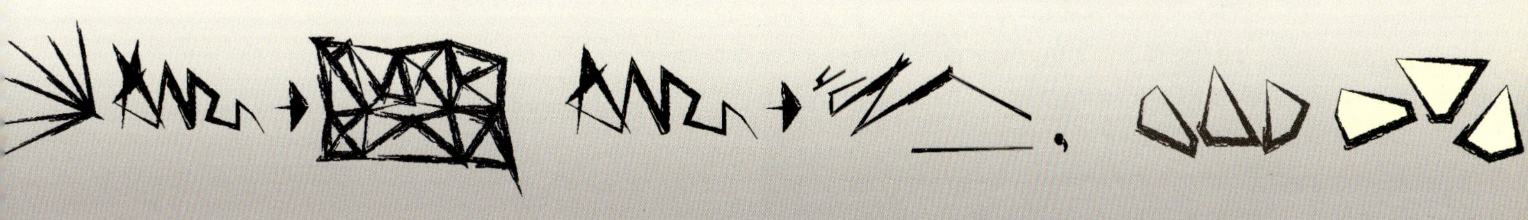

내벽
THE INNER WALL

장식
DECORATION

바닥
FLOOR

한공간, 두가지 정보 = 일석이조 (一石二蔦)

One place, killing two birds with one stone, two pieces of information

예전의 것에서 잃어버린 감성을 찾는 시간
From old time to find the lost sensitivity

여행자들을 위한 정보를 알려주는 공간
Offer information for travelers in a space

편리해진 현대사회에서 우리는 가끔 옛 문화를 그리워하기도 한다.또한 외국인 여행객들에게도 우리나라의 현재뿐만 아니라 지나간 과거를 동시에 보여줌으로써 흥미를 유발시키는 공간이 필요하다. 이러한 요구를 채워주며 서울역의 성격이 드러나는 공간을 제시한다.

With everyone in the modern world is we often have old cultures missed. In addition,To foreigners korean at the same time, as well as current and interesting, showing that the space is needed. Seoul to fill these needs, the nature of the exposed area jumyeo presents.

PLAN

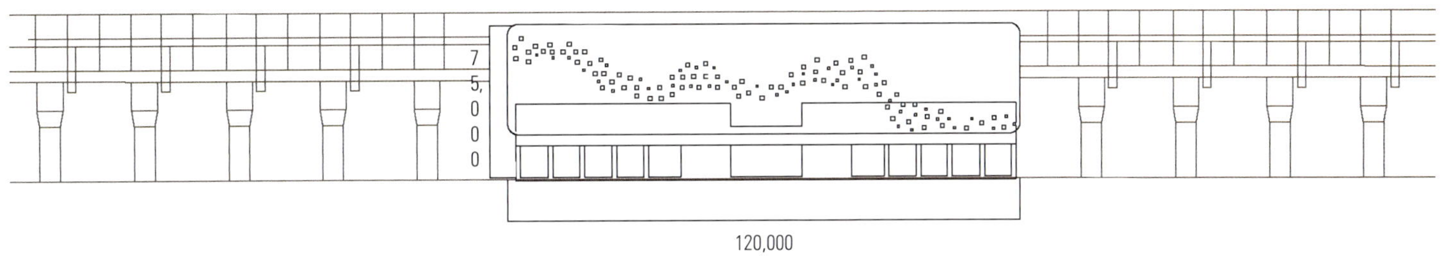

TOP

FRONT

LEFT

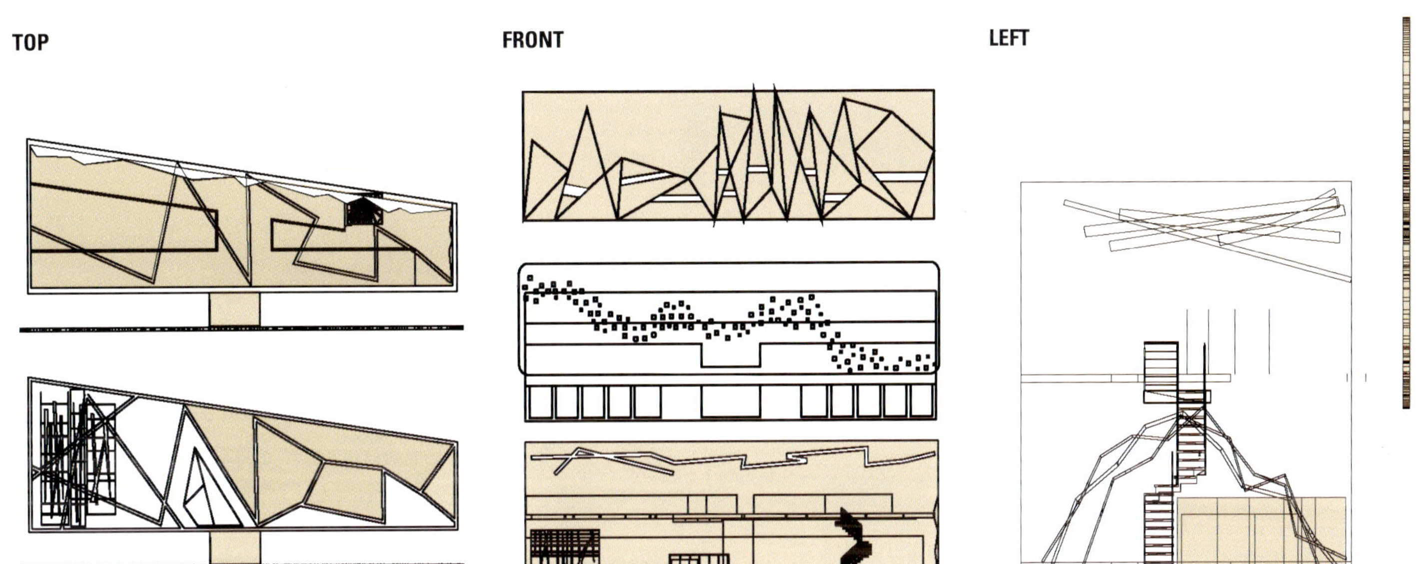

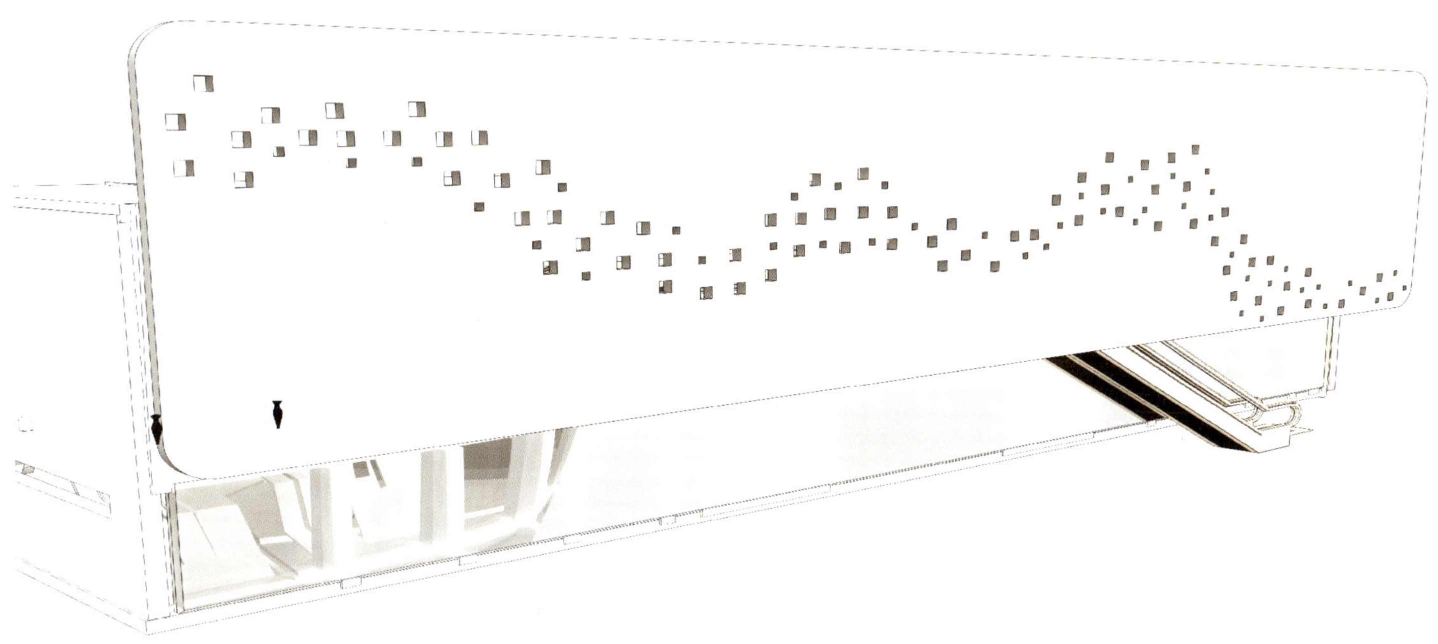

각 지역별로 펼쳐지는 이벤트
The events unfolding in the local

만리동 & 서울역의 분기별 행사
Quarterly events manridong & Seoul

소소한 체험과 이벤트 속에서 느끼는 경험
Little experience in the event experience

7080세대의 감성을 찾아주는 행사
7080 three events of the sensibility that finds

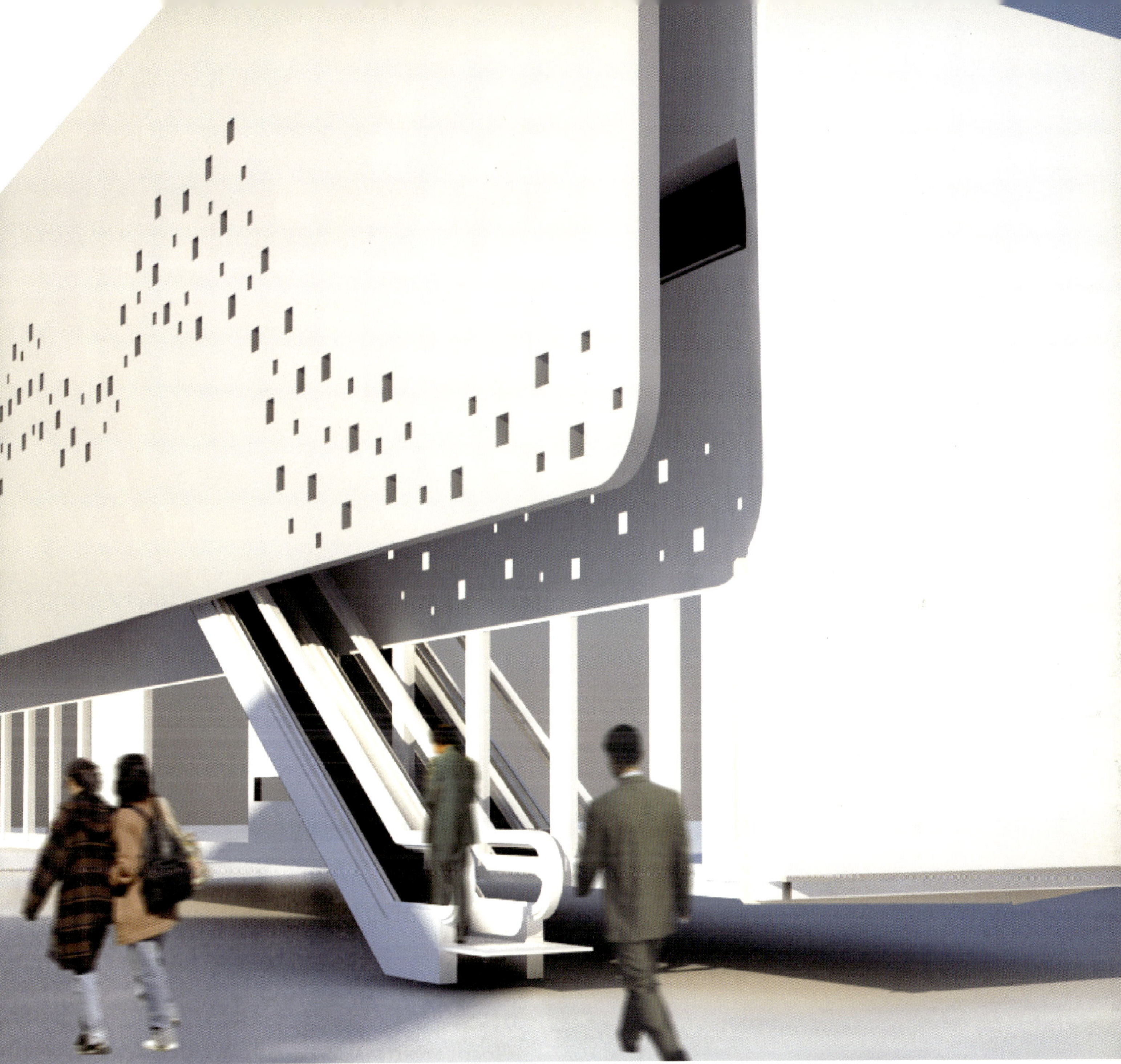

이 곳에서는 정보를 주는 것뿐만 아니라 이 공간 속에서 펼쳐지는 이벤트들도 느낄 수 있다. 대표적으로 세가지 정도의 성격을 띠고 있는 행사가 마련될 것이다. 첫 째로는 7080세대의 감성을 찾을 수 있는 통기타 가수들의 공연이나 그 시절의 놀이, 영화 등을 체험할 수 있는 행사가 마련이 될 것이다. 두 번 째로는 만리동과 서울역에서 분기별로 행사가 개최된다. 이전에는 행사를 할 수 있는 공간들이 성루역 광장에 한정되어 있었지만 이제는 이 곳에서 행사를 더욱 더 편하고 쾌적하게 관람 할 수 있을 것이다. 세번 째로는 여행객들에게 더욱 더 생생한 정보를 주기 위한 행사로 전국 각 지역의 특색을 좀 더 가까이서 볼 수 있는 이벤트이다. 실제로 그 지역 사람들이 와서 지역의 특징을 알려주고 농산물이나 그 곳에서 하는 행사, 그 지역의 성격에 대하여 체험할 수 있는 행사이다.

Here, in this space, as well as giving information also can feel the events unfolding. Ttigo typically in three groups: the nature of the event will Be prepared. Firstly to find the emotions of the 7080 three artists' performances and the acoustic guitar playing of those days, where you can enjoy movies and events will be arranged. second time and the event will be held on a quarterly basis in Seoul and Manridong. Prior to the event that space is limited to the seongruyeok Square, but now Here, even more comfortable and pleasant events will be able to w atch. The third, a more realistic information for travelers to events nationwide intended to Characteristics of each region more closely to see is an event.In fact, the local people come to know the characteristics of the agricultural area and that there, events, and the local nature of the event is about to experience.

Off-line Network

개인 또는 이웃 간의 추억 공유를 위한 오프라인 소셜 네트워크 공간

Personal memories of offline social networking space for sharing

BACKGROUND

사람과 사람을 연결해 주는 '네트워크(Net work)'의 개념이 정보기술(IT)의 발달과 함께 변화하고 있다. 미니 홈피가 주된 시대를 넘어 '소셜 네트워크 서비스(SNS)'의 시대가 자리 잡고 있는 것이다. 하지만 이러한 발달로 현대 사회에서 오프라인상에서의 만남은 줄어들어 새로운 문제로 대두되고 있다. 특히 지역 주민 이웃 간의 소통과 교류의 단절이 그 예이다. 이러한 소통의 문제점을 해결하기 위하여 지역 주민들 간 오프라인상의 교류 공간과 더불어 유대감 형성 공간을 제안한다.

`Network that connects with people (Net work) 'The concept of information technology (IT) is changing with the development of. Mini hompi beyond the primary period 'social networking services (SNS)' is located in a part of the era will. However, these advances in modern society, reduced off-line meeting new problems are emerging. In addition, local residents, for the busy modern society communication and exchanges between neighbors has been severed. This communication addresses the problem of offline dialogue guidance for the local people need space.

TARGET

- 친구, 연인, 가족 등 추억, 기념, 자신의 감정을 오프라인 상에서 공유 하고자 하는 사람들.
- 이웃 간의 교류가 부족한 구로구 시민들.
- Friends, lovers, family, memories, memorial, his feelings of people who are willing to share off-line.
- of interaction between neighbors Guro-citizens.

SITE

서울특별시 구로구 구로동 안양교

Guro-gu, Seoul anyanggyo

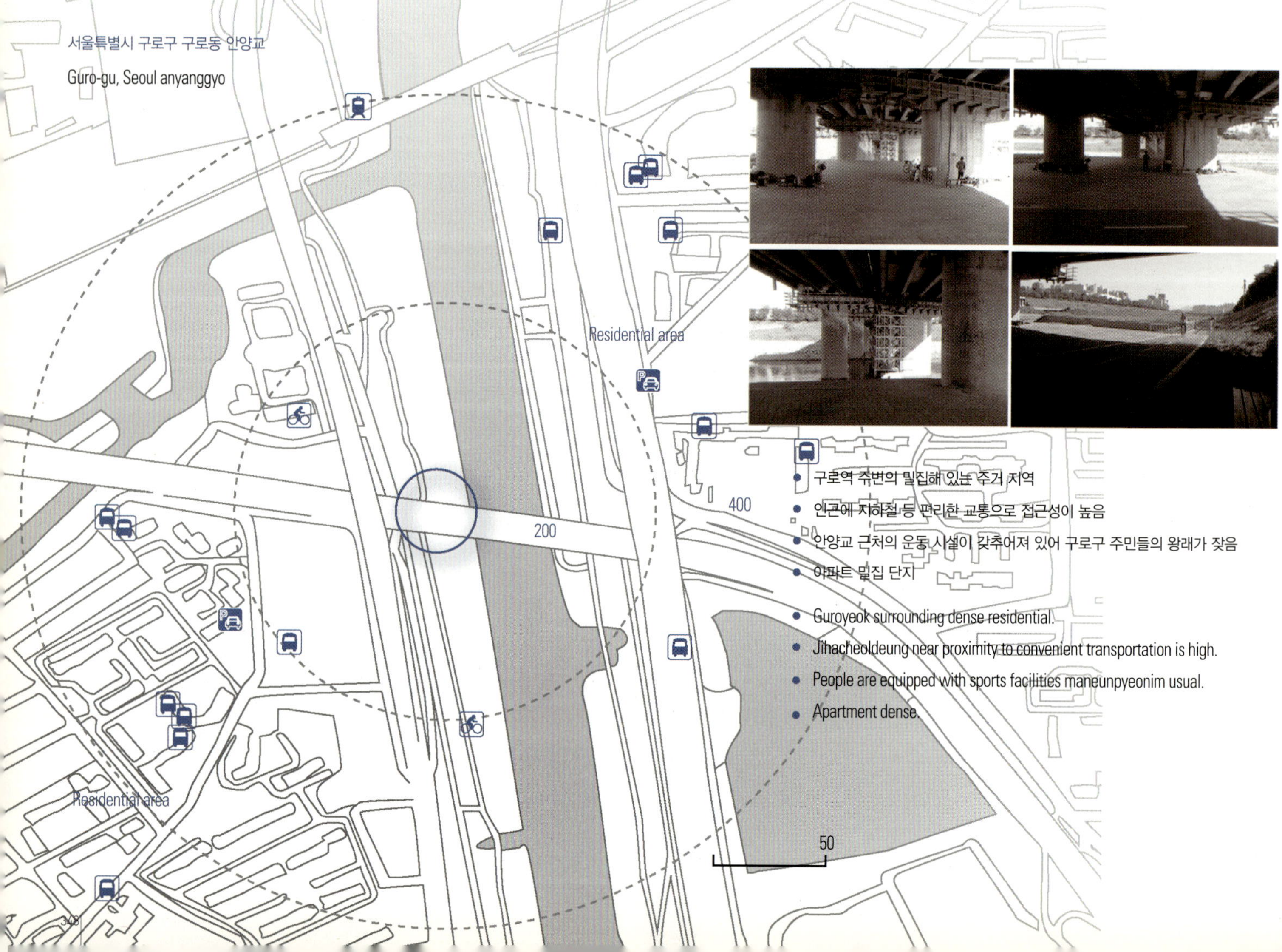

Residential area

200

400

50

Residential area

- 구로역 주변의 밀집해 있는 주거 지역
- 인근에 지하철 등 편리한 교통으로 접근성이 높음
- 안양교 근처의 운동 시설이 갖추어져 있어 구로구 주민들의 왕래가 잦음
- 아파트 밀집 단지

- Guroyeok surrounding dense residential.
- Jihacheoldeung near proximity to convenient transportation is high.
- People are equipped with sports facilities maneunpyeonim usual.
- Apartment dense.

SITE ANALYSIS

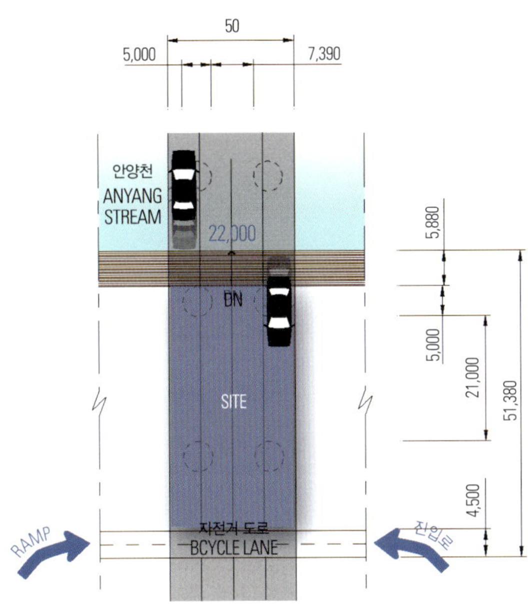

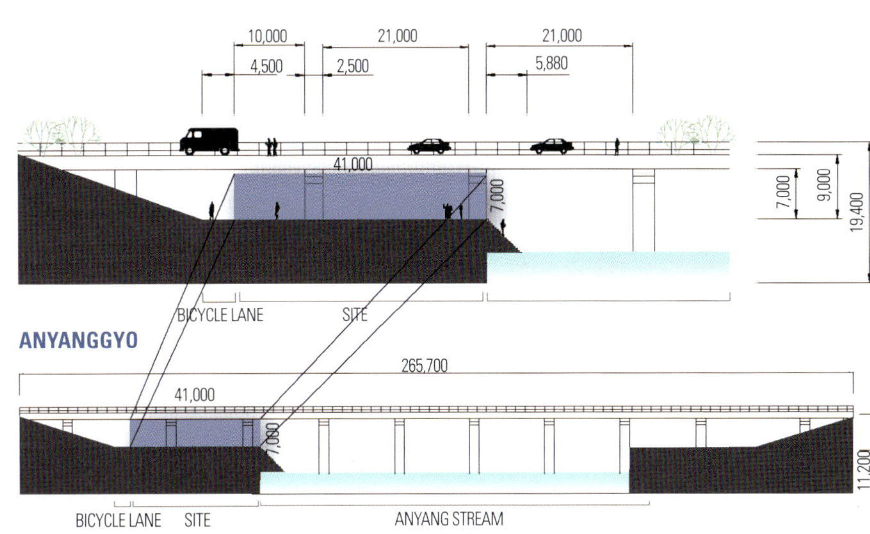

PROPOSAL

어떠한 장소에 자신의 흔적을 남기고 싶은 욕구 No place in their The desire to leave a trail	+	자신의 과거의 추억을 회상하는 버릇 Back on their past The habit of reflection	+	현재 자신의 상태, 기분을 공유하고자 하는 욕구 Their present status, and mood The desire to share	+	지역 주민, 이웃 간의교류, 대화의 부족 Between neighbors Ex-change, and the lack of dialogue	+	오프라인상의 사회적 네트워크 결핍 Lack of an offline social network

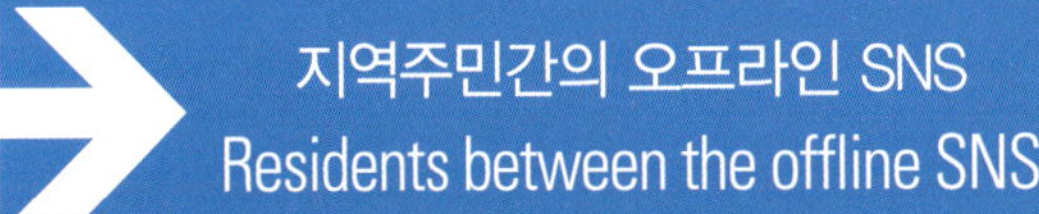

오프라인상의 개인 공간에 자신의 남기고 싶은 추억, 현재 자신의 상태 등을 남기고 공유하며, 이 후에 추억들을 회상할 수 있는 공간. 지역 주민, 이웃이 모여 서로 소통과 교류를 하는 만남의 광장이다.

Would you like your own personal space, off-line memories, like the current status, and share their memories to recall after the space. Local residents, gathered together neighbor communication and dialogue, to exchange a meeting space.

CONCEPT

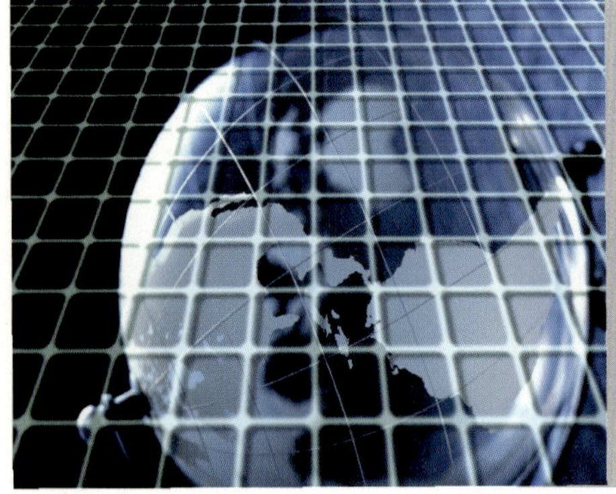

NET (網)

1. 그물처럼 만들어 가려 두거나 치거나 하는 물건을 통틀어 이르는 말.
2. 그물처럼 얽혀 있는 조직이나 짜임새의 뜻을 나타내는 말.
3. 같은 말 : 네트워크(랜(LAN)이나 모뎀 따위의 통신 설비를 갖춘 컴퓨터를 이용하여 서로 연결시켜 주는 조직이나 체계)
   [유의어] 네트워크, 그물, 통신망

망의 네트워크라는 뜻은 지역주민간의 오프라인 SNS을 의미하는 것으로 지역 주민들의 기억하고 싶은 추억, 현재 자신의 상태, 사진 등을 기재하고 보관하며 지역 주민들간의 소통과 대화의 기능을 뜻하는 것이다. 그리고 망의 또 다른 뜻인 그물망은 서로 다른 끈이 여러 형태로 서로 엮이면서 커다란 망을 형성하는 것이다. 이것은 기능적으로 공간의 용도를 설명 할 수 있는데 서로 다른 끈이 개개인을 나타내며 그 끈이 이 공간에서 엮이어 이 지역의 구성원임을 인식시켜주고 서로 엮이고 함께 있어야만 완성되는 그물망의 특징인 지역 주민 구성원간의 소통과 단결의 중요성을 의미한다.

That means the network of networks between local residents, which means that the offline sns want to remember the memories of local residents, the current state of their own, photographs, archive and disclose the communication and dialogue between local residents, which means the feature will And mesh networks implying a different strap to another in many forms, is to form networks with each other yeokyimyeonseo large. This feature can explain the purpose of the space, which represents a different individual ties between its twenty yeokgieo strap is a member of the region to recognize one another and must be completed yeokyigo mesh with the characteristics of communication between members of local residents as refers to the importance of unity.

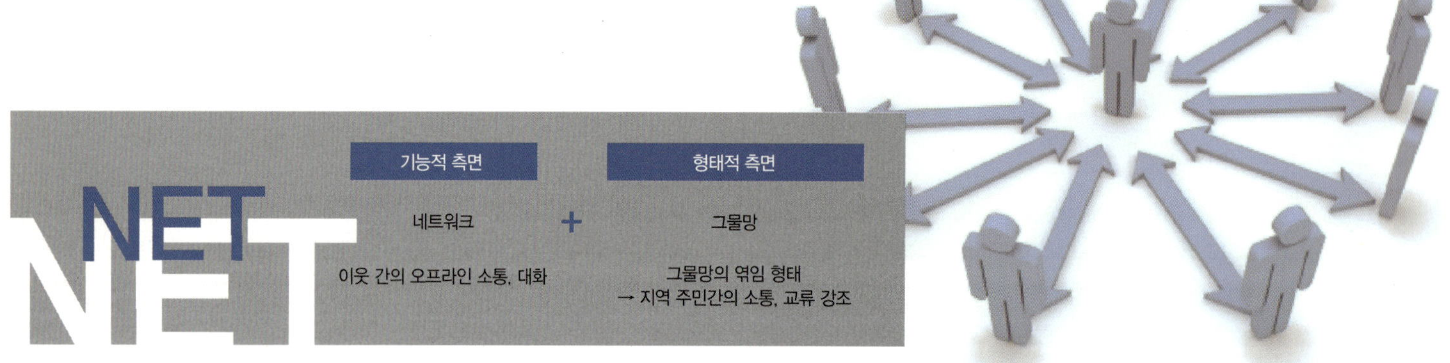

DESIGN KEYWORD

선	그물망의 직선의 형태를 추출하여 전체적으로 직선의 곧은 느낌이 두드러지게 구성	**LINE**	LINE Net in the form of a straight line by extracting the whole feel of the straight configuration significantly.
면	선과 선의 만남으로 만들어지는 면의 특성을 활용 → 선들이 모여 면을 구성하고, 면들이 모여 공간을 구성	**SIDE**	SIDE Meet the face of the line created by lines and take advantage of the characteristics → Side lines, and gathered together the configuration space plane Forming.
교차	망의 선과 면의 교차, 선과 선의 교차, 면과 면의 교차를 활용하여 공간을 구성	**CROSSING**	CROSSING Lines intersecting networks face, good line, intersection, faces to the utilization of the space configuration, cross sides.
연결	그물의 서로 다른 선들이 하나로 연결되어 그물이 형성되는 특징 → 천장에서부터 벽까지 연결되어 내려오는 가벽	**CONNECTION**	CONNECTION Nets of different lines are connected to one of the features formed in the net. → Gabyeok down from the ceiling connected

기둥의 용도

다리 구조적으로 꼭필요한 기둥을 전시 D.P 공간으로 구성하여 기둥을 실용적으로 활용 하도록 한다. 컨셉인 망을 적용하여 기둥에 상자 모양의 공간을 뚫어서 활용한다.

USES PILLAR

Pillars bridge is structurally kkokpilyohan exhibition space consists of DP to take advantage of the pillars are practical. By applying the network concept in the box-shaped pillars bore the space is utilized.

개개인의 상자
Individual's Box

사진이나 글 등 자신의 추억을 회상할 수 있는 물건을넣어 개인의 상자에보관하며, 언제든지 상자를 꺼내어 추억을 회상할 수 있다.

Photos, text, and recall their memories of the things that you can put a person's Store it in a box and the box taken out at any time you can recall memories.

주변에 주거지가 밀집되어 있는 안양교에 위치하여 이웃 간의 소통의 부족 문제를 해결하고, 개인의 추억을 공유하여 지역 주민 간의 교류를 활성화 시킨다.

Housing is concentrated around the school is located in Anyang lack of communication between the neighbors solve, and share personal memories of the interaction between the local population activated

전시공간_1

그물망 컨셉으로부터 선을 추출하되 선에 두께를 주어 여러개의 선을 나란히 함으로써 선이 모여 면을 만들어 공간을 구성한다. 이러한 면으로 공간을 감싸고 서로 겹침으로서 공간에 재미를 주었다. 또한 한쪽에 가벼운 선 소재와 대조되는 매스감이 느껴지는 벽을 두어 공간에 균형을 이로도록 한다.

EXHIBITION SPACE_1

Extract lines from a net concept, but given the thickness of the lines the lines of several things to line side by side to create a space configuration. These planes overlap each other, as the space surrounding the area and gave fun. In addition, on one side, as opposed to re-light seonso sense I feel the mass balance on the walls of this room to be placed

전시공간_2

옆쪽 전시공간과 대조적으로 이곳에서는 추출한 선에 세로로 두께를 주어 정면에서 보았을 때는 얇은 선들이 규칙적으로 배열된 느낌을 받을 수 있는 공간이다. 얇은 두께의 선들을 일직선으로 배열하여 면을 만들고 면이 맞닿는 곳은 겹치게 됨으로써 공간에 재미를 유도했다. 컨셉인 망의 겹침과 서로를 이어준다는 주안점을 착안하여 공간에서도 느낄수 있게 서로 다른 선들이 겹치고 결국 맞닿아 이어져 컨셉인 망에 충족시키는 공간을 고안해 냈다.

EXHIBITION SPACE_2

Here, in contrast to the side of the exhibit space to extract the vertical lines give jukkereul looked at the front of a regular arrangement of the filaments can be felt is the place. The thin straight lines to create an array of side facing the side where the fun in space induced by becoming overlap. Networks overlap with each other after concept jundaneun conceived emphasis makes the space feel different from the lines overlap and eventually leads abuts keonsepin Mans devised to meet the space.

SPACE PROCESS

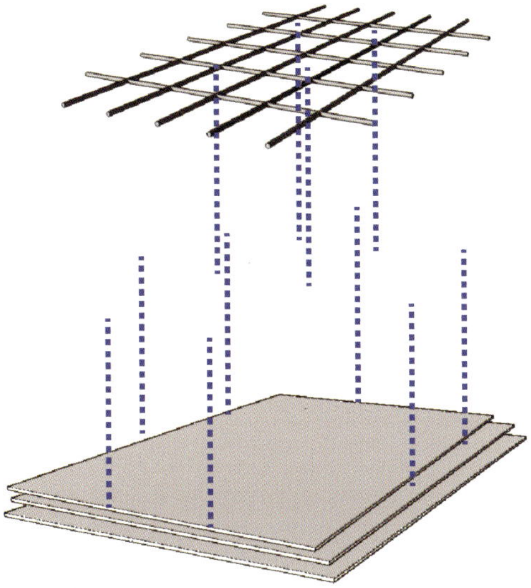

그물의 형태에서 그리드 형성

그리드의 세분화 → 사람의 동선 구성

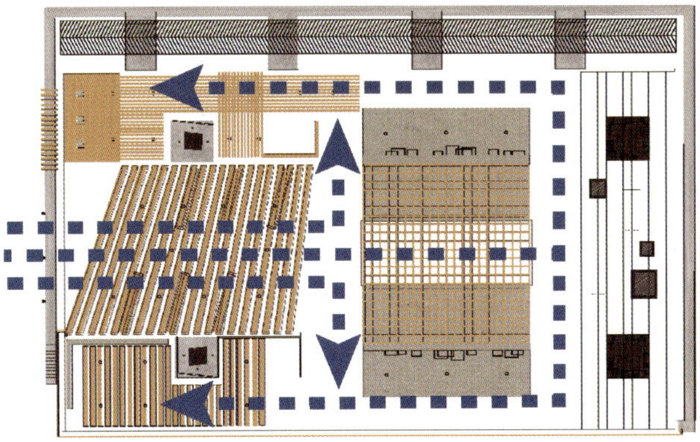

TRAFFIC LINE

망의 그리드 형성 → 공간의 구성 형성

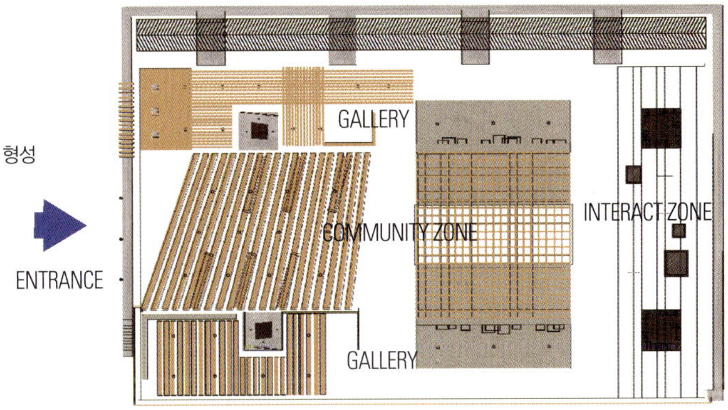

ENTRANCE

GALLERY

COMMUNITY ZONE

INTERACT ZONE

GALLERY

ZONING

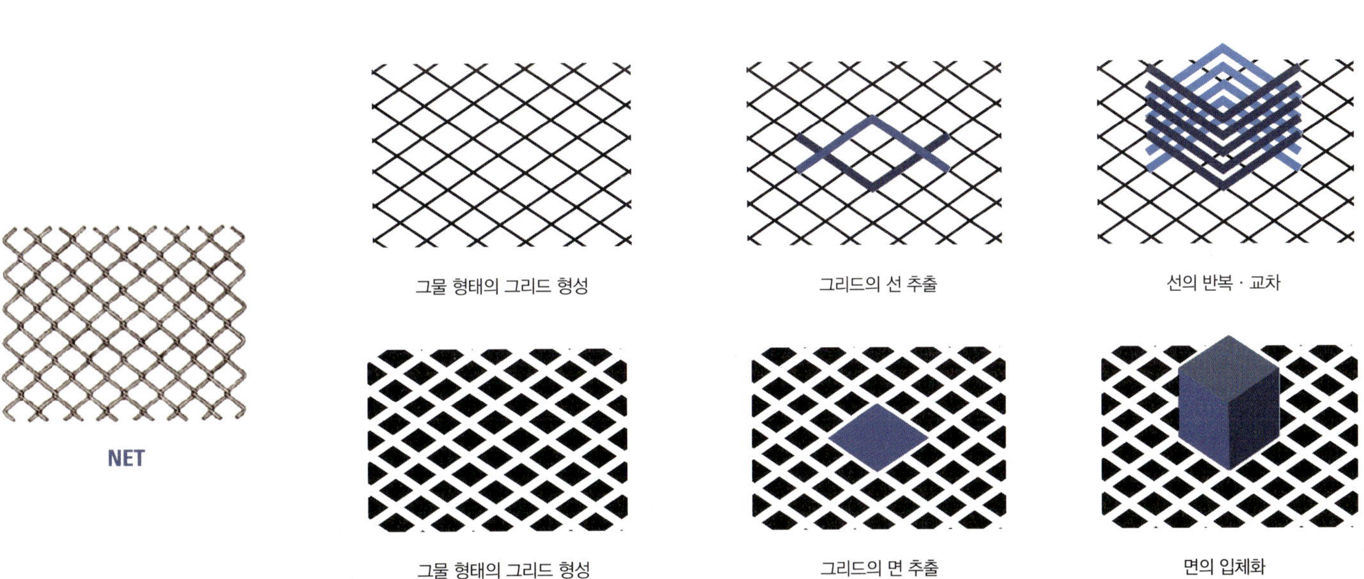

NET

그물 형태의 그리드 형성

그리드의 선 추출

선의 반복 · 교차

그물 형태의 그리드 형성

그리드의 면 추출

면의 입체화

입구 부분으로 컨셉인 망을 활용하여 선들이 여러형태로 입구를 감싼다.

As part keonsepin entrance lines into various forms by leveraging network covering the entrance.

망의 그리드를 활용하여 벽에 네모난 상자를 돌출시켜 물건들을 전시하고 보관한다.

Networks by utilizing a grid on the wall extruded using a square box to display and store things.

선으로 감싼공간에 굵은 기둥을세워 그안에 물건을 전시하여 전체적으로 조화롭게 연출하였다.

Bold lines wrapped pillars erected in the space stuff in it was rendering the exhibition as a whole johwaromge.

한 쪽 벽면에는 직선들이 돌출되어 벽과 천상을 감싸고 그 선들이 반대편 벽을 파내는 형식의 형태로 공간에 재미를 주었다. 바닥에는 벽으로부터 선들이 이어져나와 의자가 되는 형식을 취했다.

In a wall surrounding the straight lines protruding walls and ceilings geuseondeulyi opposite wall space in the form of digging up the format was fun. Lines on the bottom of the wall leads to a type that I took and chairs.

倉洞-下里

서민들의 문화로서 자리 잡힌 포장마차가 거리의 환경 문제란 이유만으로 사라지고 있는 것이 지금 우리의 현실이다. 지역 개발과 문화 보존의 두 갈랫길에서 포장마차의 개선이 필요하다.

Another Cart Bar Community

PROBLEM SOLUTION

01 사라지고 있는 포장마차.

02 '환경오염, 교통불편, 도시개발' 의 이유로 철폐

03 '생존권, 한국문화, 서민애환' 대립 의견

04 개발과 보존의 사이에서 '개선' 을 찾다.

Pollution
A trffic problem
urban communities

WHY?

BUT.

Livelihood
One of Korea vultural
Commonalty's joys and sorrows

SITE ANALYSIS

서울시 도봉구 창제5동에 위치한 창동 강북지역의 대단위 아파트 주거지역으로 유통, 편의시설, 상설의류매장 등이 많으며 1,4호선 의정부, 당고개, 노원역이 인근에 치하고 있다. 1차 상권지역에 위치하고 있으며, 역 주위에 대학교가 자리하고 있어 대학생의 이용이 잦고, 직장인 또한 많이 드나드는 곳이다.

This is a coastal residence complex of the area north of the Han river which placed in Chang-je 5Dong, Dobong-Gu Seoul. This is placed around line 1,4 Uijeongbu, Dang Gogae and Nowon station and it has distribution, convenient facilities and several outlets. This place is located the first business district and near the university. Because of that reason, a lot of university student and those are frequently take advantage from here.

Residential development

TARGET

창동 FLYOVER를 찾는 주된 사용자로는 직장인, 대학생들로서 대단위 주거지역으로 인한 창동역에서의 사람들의 이동이 원활하다. 창동역을 드나들면서 FLYOVER에 위치한 포장 마차촌을 접근하기에 용이하며 아침에 출근길, 또는 저녁시간대의 퇴근길에 역은 가장 붐비게 된다. 역 주위에서 일어나는 사람들의 활발한 동선을 정리하여, 역을 이용하는 사람들을 창동FLYOVER에 위치한 포장마차 촌으로 끌어들인다. 그럼으로써 단지 거리의 먹거리 장터로 느껴질 수 있는 역할을 사람들 간의 소통과 정을 나누고, 새로운 형태의 스트리트 퍼니쳐를 제공함으로써 사람들에게 호기심을 불어 일으키며 FLYOVER 속 또다른 세계를 느낄 수 있다.

Those two groups of the people are the main customer of the ChangDong FLYOVER, and also smoothly people used this place from the residential area. The ChanDong station is easily to approach the wrap carriage area especially, the time of round trip to work is the most crowded time. It draw the customers and the station users into the wrap carriage area of Chan Dong FLYOVER, by the make clear the brisk moving line around the station, thereby this street becomes as the food market. So this place can provide the new facts of the street culture, comes to a mutual understanding between people and excites curiosity in the same time.

Educational facilities

Public institution

Green zone

05 교통 동선의 정리. 06 시간에 따른 한정된 활성화. 07 상부 공간의 활용 필요성. 08 수납 공간의 부족으로 생기는 환경문제.

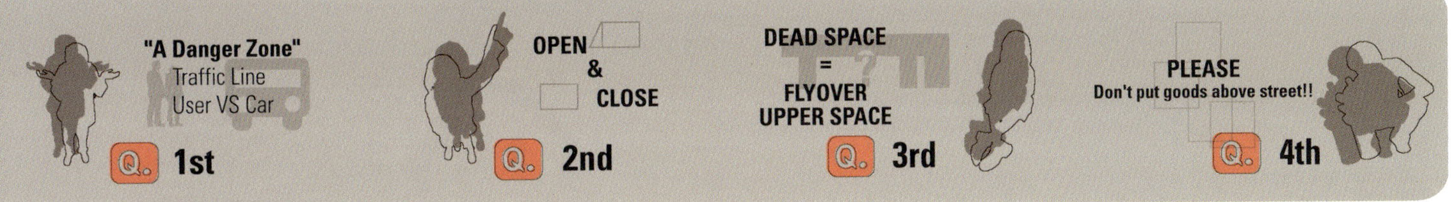

SOLUTION

동 선 : 포장마차를 찾는 사람들의 이동 경로와 차도의 동선 충돌로 생기는 문제점.
→ 이용자들의 동선을 안으로 이끌어 들임.

Traffic line : The trouble of the crush the moving line between the wrap carriage users with street for car.
→ To lead the moving line of the customers.

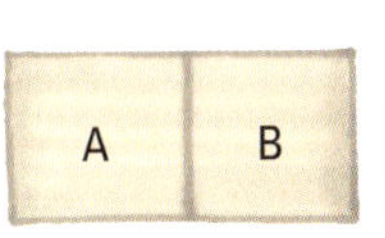

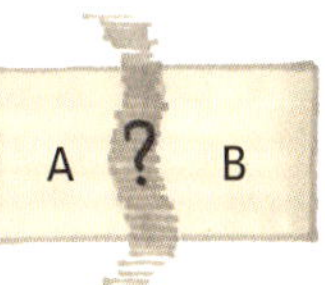

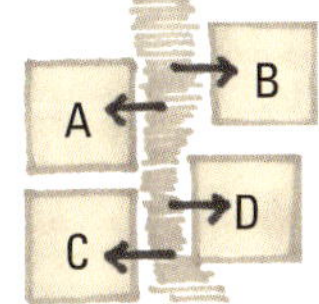

Flyover : 이미 포장마차로 활성화가 된 flyover가 밤에는 많이 이용되지만 밤에만 활성화 되는 한정적인 문제로 Hidden space가 부각되는 낮.
→ 낮에도 활용 될 수 있는 street furniture를 설치.

Flyover : The wrap carriage place used more frequently at night than day time.
→ Make the possibility of day use by set up a street furniture.

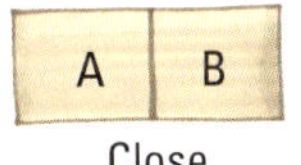

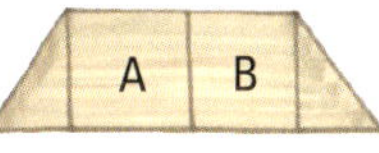

상부공간 : 7m의 높은 Flyover의 dead space.
→ street furniture 또는 flyover와 조화를 이룰 수 있는 object design.

Overhead Space : 7m dead space of the FLYOVER.
→ It need something object design for that empty area of the street furniture or FLYOVER.

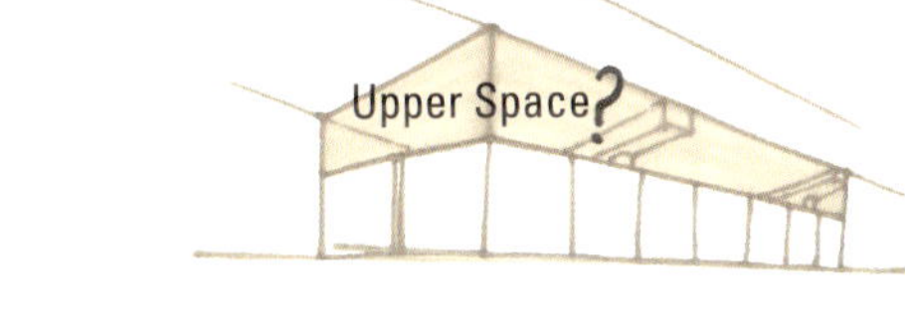

수납공간 : 포장마차 내부에 둘 수 없는 물건들을 바깥 공간에 내 놓으면서 생기는 문제점.
→ 별도의 수납 공간의 설치.

Storage space: The problem of the stuff on the street cause of no enough space inside of the wrap carriage.
→ Need extra storage.

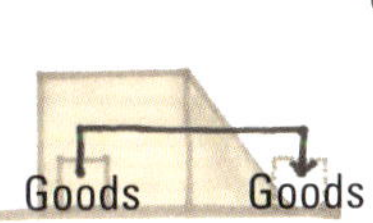

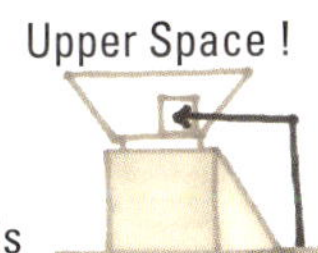

USER NEEDS

회색 빛의 건물이 가득한 현대 속에서 바쁘게 살아가는 사람들은 시간에 쫓기다보면 때때로 사람들과의 끈끈한 정을 잊을 때가 있다. 그럴 때일수록 더더욱 사람들과의 따뜻한 마음들을 더 원하게 되고 찾게 된다. 길거리 포장마차들은 겨울철 우리의 발목을 붙잡고, 이야기를 나누는 하나의 공간을 만들어준다. 우리는 이러한 사람들의 감성과 그들이 느끼는 정에 대해 포장마차촌이라는 공간 안에서 디자인을 하였으며, 보존과 개선을 통해 그들에게 새로운 디자인을 선보인다.

Sometimes, people are forgotten the strongly attached mind if they living in hurry reclusion modern society. That moment peoples are more want to belongings and 'Jung'. The wrap carriage on the street will offer the place for conversation with people. We designed based on the people's these sensitivity and 'Jung' into a warp carriage spaces, and this space will present through the preservation with improvement.

SEQUENCE IN THE STORY

#1. 낚시하러 나온 어부는 물고기가 더 많이 잡히는 산 속으로 들어간다.
#1. Fisherman moved into the mountain for catching huge amount of fish.

#2. 어디선가 나는 복숭아 나무 향기에 이끌려 더 깊게 들어가게 된다.
#2. To move into the deepen places be led by peach tree scent from somewhere.

#3. 어부는 동굴을 발견하는데, 그 동굴은 매우 비좁고 어두웠는데 빛이 나는 곳을 따라 계속 걸어들어갔다.
#3. Fishermen discovered the cave, that place was really dark and narrow. He kept up the light.

#4. 시야가 넓어지면서 동굴에서 빠져나온다.
#4. When his visual field became wide then he escaped from the cave.

#5. 복숭아 나무와 어우러진 작은 마을을 발견한다.
#5. He arrived to a small village mingled with peach trees.

#6. 벗들과 노래하며, 술을 마실 수 있는 행복한 마을에서 어부는 즐거워한다.
#6. He was happy with drinking and singing with friends in the village.

#7. 마을에서 나온 어부는 그 곳을 무릉도원이라고 불렀다.
#7. He called that place to 'Mulengdowon' seems like fairyland.

During the walk down the stair which placed in front of the station which is blocked with steep slope can see street vendors.
The wrap carriage places have seen when people passed the street vendors.

급경사로 앞이 막혀 있는 역 입구.

계단을 내려가는 동안에도 시선의 막힘.

역을 빠져 나가면 노점상이 보인다.

노점상을 지나면 포장마차 촌이 나온다.

PLANS

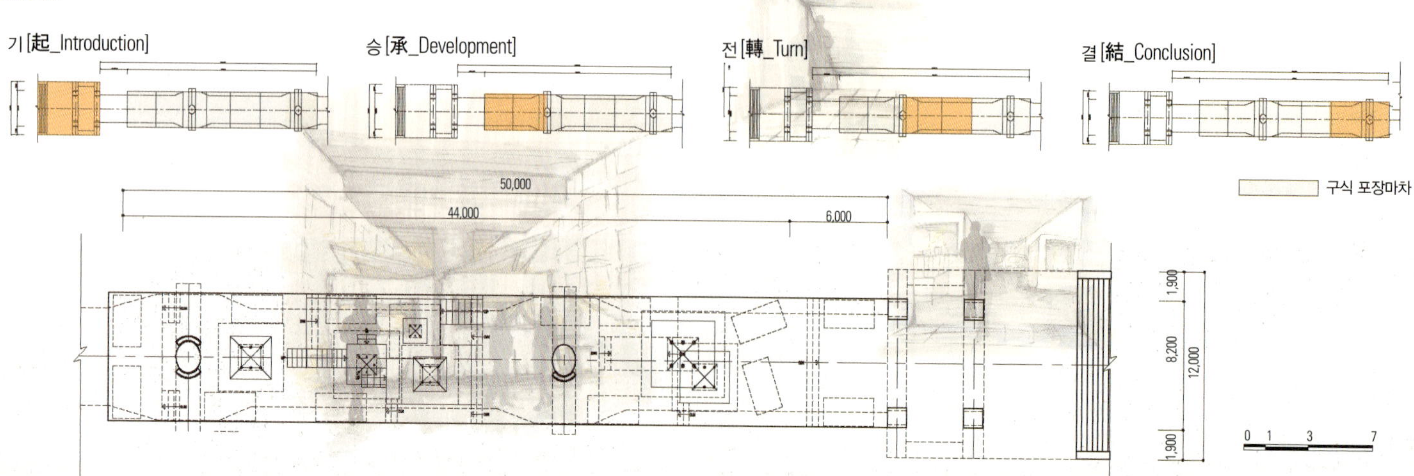

CONCEPT

창동 flyover에서는 서민들의 무릉도원을 빗대어 그들이 바라는 행복에 대해 이야기를 나누고 정을 느낄 수 있는 감성적인 public design을 볼 수 있다. 무릉도원을 다녀왔던 도원명의 시에서는 노래하며, 벗들과 어울려 술을 마시는 세상을 나타냈다. 현세를 벗어나 무릉도원을 찾아 가고 싶은 사람들의 마음과 포장마차 촌을 찾는 사람들의 근심과 생각, 꿈에 빗대어 공간을 표현한다.

In Chang Dong FLYOVER can represent sensitivity public designs a allude 'Mulengdowon' of citizen hope as happiness. In Dowonmyung's poetry represented the world as singing and drinking with friends. The same as this, we are representing the space for citizens thought, worry and dream.

작은 마을 속 행복 Happiness in a small town		포장마차 촌 Cart Bar Village
사람들과 어울리는 장소 A gathering place for people	➔	스트리트 퍼니쳐 Street furniture
개인이 품은 소박한 꿈 Individuals with a simple dream		현세에 살아가고 있는 사람들 People who live in this world
현세를 벗어나고픈 사람들의 마음 Beyond the mundane minds of hungry people		포장마차 촌 Flyover로 찾아오는 사람들 Directions to the Village People Cart Bar Flyover

'기승전결'을 공간에 넣다. 'Four Steps In Composition' Put In The Space

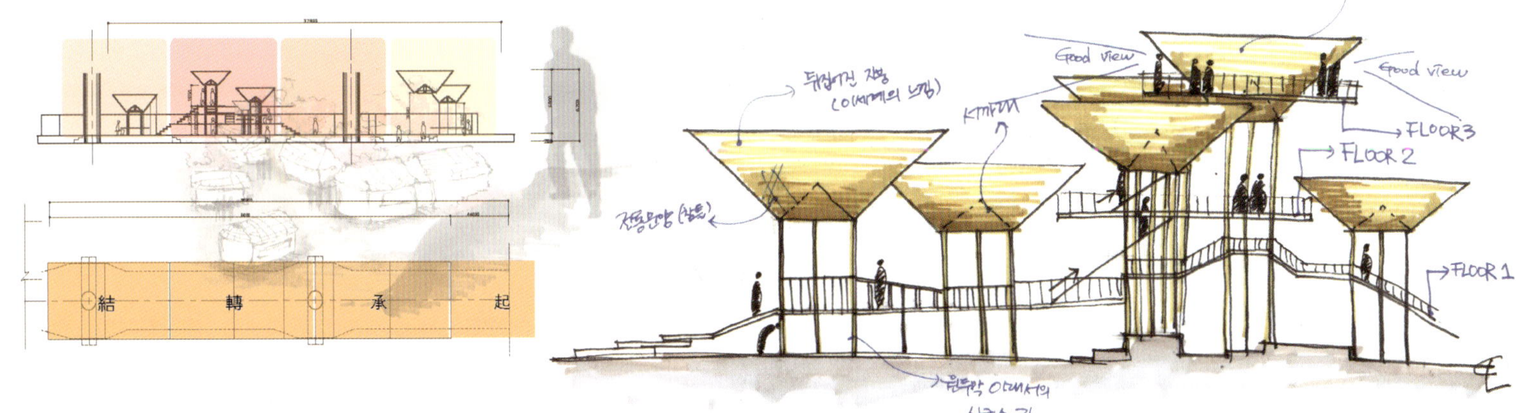

3D STORY
기 [起_Introduction]
단차이로 공간을 구분하다.
승 [承_Development]
옛 선조들의 유희, 포석정을 옮기다.
전 [轉_Turn]
지붕을 조명에 반짝이는 술잔에 비유하다.
결 [結_Conclusion]
어디서든 먹고 쉴 수 있는 공간을 제시하다.
起 Floor space shall be divided by the difference in level.
承 The old ancestral play, "Poseokjeong" has been moved.
轉 The roof Was likened to a light sparkling Glass.
結 Anywhere that you can eat and rest space is presented.

술잔 형태의 지붕은 붓 글씨나 그림 등으로 동양화를 보는 것과 같은 느낌을 연출하였다. 각각 다른 높이의 퍼니쳐는 위치에 따라 공간을 다르게 느낄 수 있다. 또한 가벽에 사람형태의 구멍을 뚫어 공간에 재미를 더 한다.

Brush with the glass-shaped roof in the form of text, images, etc., were directed oriental feel like. Furniture of different heights, where you can feel the other space depending on the location. And free-standing Wall holes in the shape of people in space adds to the fun.

PROPOSAL

보일 듯 말 듯 블라인드 "Seems to me seemed to show Blind"

무릉도원은 꿈꾸는 사람에게 보여지는 이세계로 무방비로 노출 되어있는 곳이 아니라 가려져있는 곳이다.
The Mulengdowon is a hidden place then wide open to this world for dreamer.

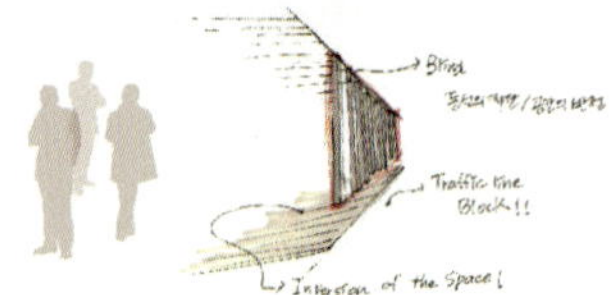

낮에는 다소 흉물스러워 보이는 포장마차를 정리해주고, 위험한 동선의 차단과 이용자들에게 공간의 반전 효과를 준다.
The public designs can organize the hideous things during the day time, give reversal effect to space for citizen and it shuts out the dangerous moving line.

원두막 + 주막 STREET FURNITURE "Lookout Hut On Stilts + Old Inn (street furniture)"

각각 다른 높이의 퍼니쳐는 위치에 따라 공간을 다르게 느낄 수 있을 뿐만 아니라 늘 분주한 창동역 flyover에서 잠시나마 여유를 찾아 볼 수 있는 공간을 제공 한다.
The different height furniture can give the leisurely place to Chan Dong station FLYOVER which is always busy.

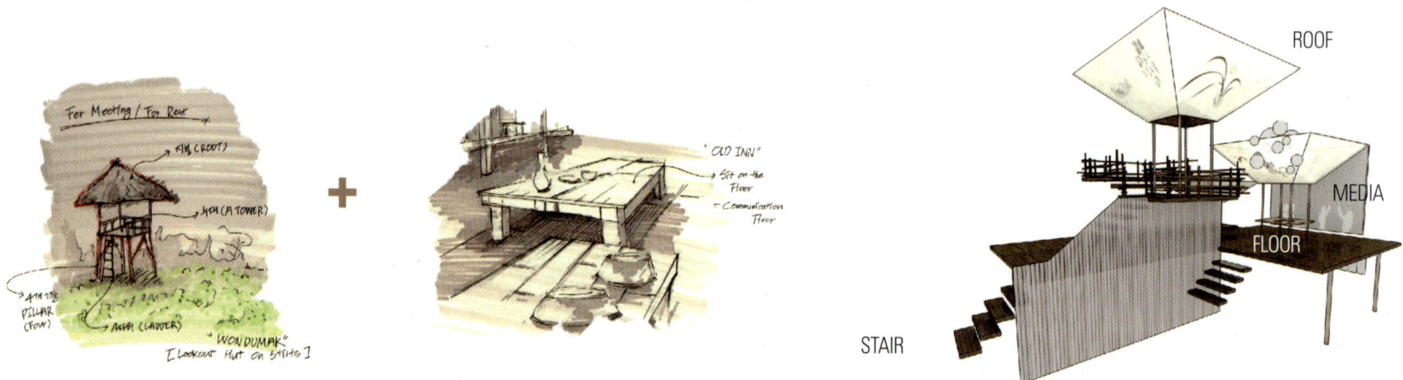

옛 주막의 마루 기능에 원두막의 휴식 공간을 더한 공간이다. 내가 원하는 장소에 앉아 음식을 먹을 수 있고 휴식을 취할 수 있다.
This is the place can take a rest and eat something with sitting wherever people wants.

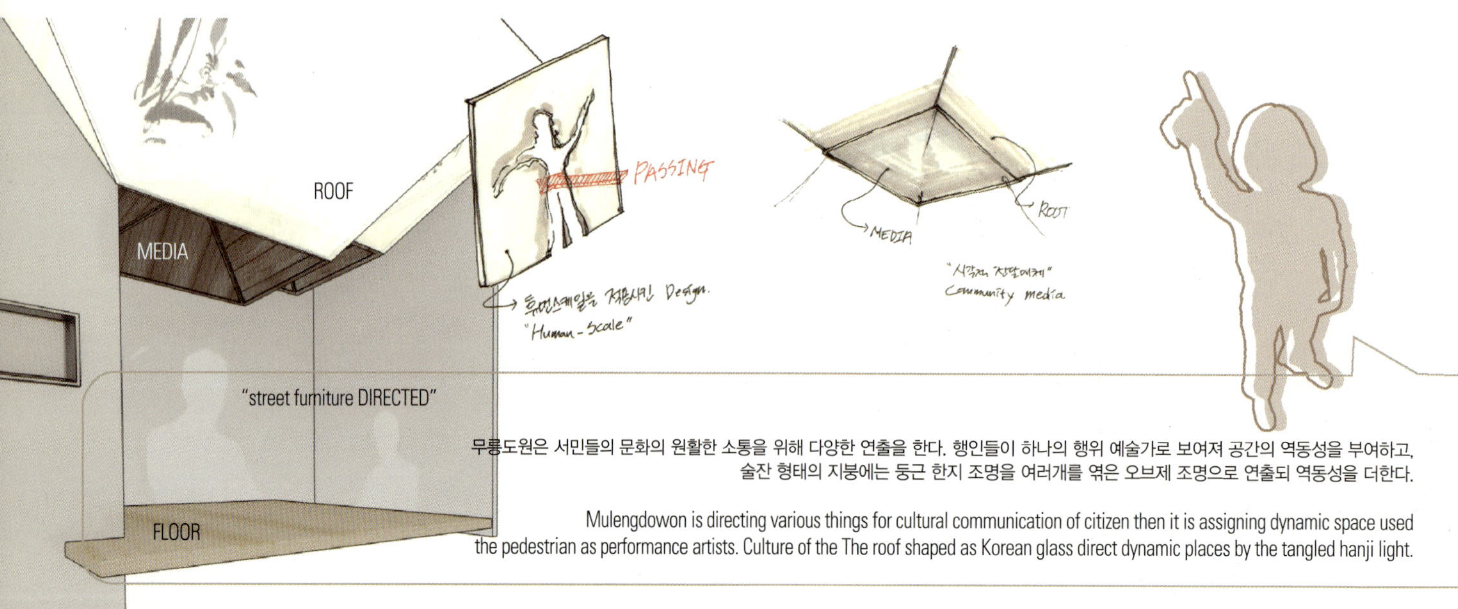

무릉도원은 서민들의 문화의 원활한 소통을 위해 다양한 연출을 한다. 행인들이 하나의 행위 예술가로 보여져 공간의 역동성을 부여하고, 술잔 형태의 지붕에는 둥근 한지 조명을 여러개를 엮은 오브제 조명으로 연출되 역동성을 더한다.

Mulengdowon is directing various things for cultural communication of citizen then it is assigning dynamic space used the pedestrian as performance artists. Culture of the The roof shaped as Korean glass direct dynamic places by the tangled hanji light.

레벨 차이를 준 기단　"Podium level, giving a difference"

전통건축의 기단을 만들어 빗물과 지하수로부터 건물을 보호하고 동선의 차단을 돕는다.
The traditional architecture will help the blocking of moving line and also protect the building from rain and subsurface water.

기단 레벨 차이는 무릉도원의 이야기 속으로 빠져들게 하는 역할도 작용한다.
Then the different level of the stereobate works a position for fall into the Mulengdowon story.

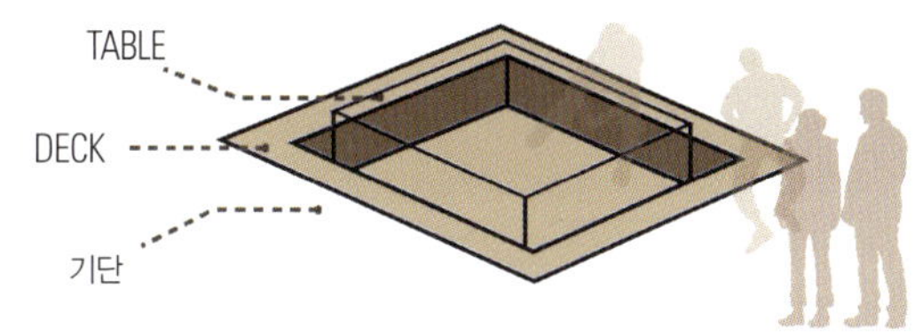

포장마차　"Cart bar Design"

주방과 홀이 모두 있는 비효율성 한국식 포장마차를 탈바꿈한다.
포장마차가 최소 규모로 줄어들어 Flyover 내부의 공간을 확보하고, 내부는 주방과 테이블을 이어 간단하게 먹고 갈 수 있는 선술집 형태로 구성된다.

We transformed the inefficient Korean wrap carriage both included the kitchen with hall.
It will make reduce the space as smallest so, inside of wrap carriage will become consist as stand-up bar.

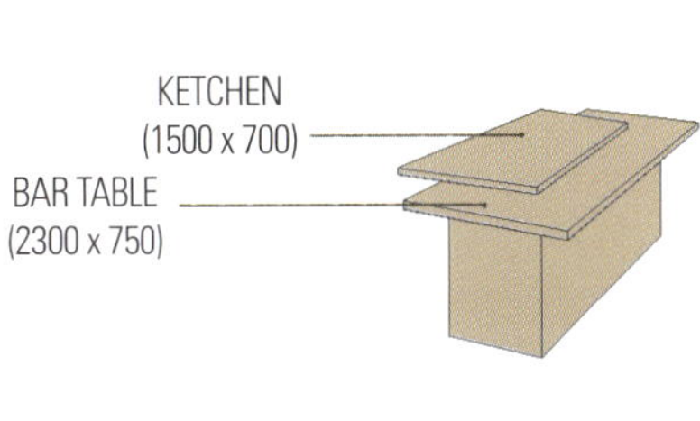

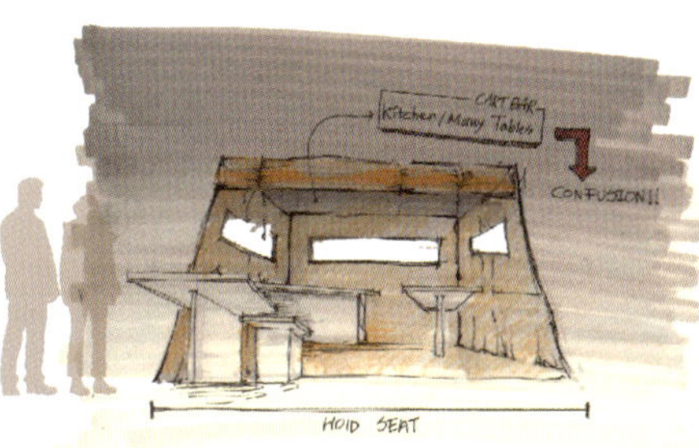

Inefficiency Korea cart bar
비효율성 한국식 포장마차.

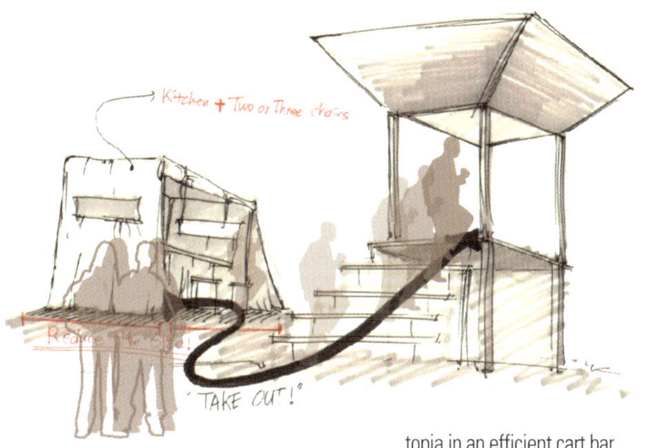

topia in an efficient cart bar
무릉도원 속 효율적인 포장마차

휴식처　"Roost"

무릉도원 속 사람들이 모이는 광장의 공간으로 테이블 공간이 조금 더 넓고 많다.
휴식처 공간은 역사 속 주(酒) 문화 중 하나였던 포석정을 넣어 또 하나의 재미있는 요소를 더한다.
The space for tables becomes wider to gather the people into the square in Mulengdowon.
Rest place added one more interesting fact by included 'Posuckjung' which was one of the Korean traditional drinking culture.

ta-da(h) [tədɑ́ː]
짜잔~! 나오신다, 기대하시라 (사람 · 물건을 처음 소개할 때 앞세우는 소리)

타다 [동사]
– 탈것이나 짐승의 등 따위에 몸을 얹다.
– 도로, 줄, 산, 나무, 바위 따위를 밟고 오르거나 그것을 따라 지나가다.
– 그네나 시소 따위의 놀이 기구에 몸을 싣고 앞뒤로, 위아래로 또는 원을 그리며 움직이다.
– 악기의 줄을 퉁기거나 건반을 눌러 소리를 내다.

To ride
- to ride on a vehicle or an animal.
- to climb up or pass by a road, a rope, a mountain, a tree or a rock.
- to ride a seesaw or an amusement facility and move by making circles up and down or back and forth.
- to pluck at the musical instrument strings or strike the keys to make a sound.

TA-DA

BACKGROUND

21세기는 세계화, 정보화시대인 동시에 문화의 시대이다. 이 시대를 살아가는 아동들에게 다양성과 창의력이 요구됨에 따라 미술교육의 중요성은 점점 더 커지고 있다. 글로벌 경쟁력의 척도가 되고 있는 상상력과 창의력은 생각의 자유로움에서 꽃을 피울 수 있다. 특히 두뇌와 감성발달이 가장 왕성한 영, 유아기에 즐거운 활동을 통하여 자유로운 생각이 자라 날 수 있다. 집에서 접하기 힘든 여러 가지 재료를 직접 만지고 체험해보는 이른 바 '퍼포먼스 미술교육'은 자유롭고 개방적인 환경 속에서 다양한 표현 활동을 통해 그리기 중심의 정적이고 획일적인 기존의 미술교육과는 다른 방법을 제시한다. 스스로 그리고 만들어가는 표현과정 속에서 문제를 해결하고 더 새로운 활동을 하며 함께 모여 체험하고 학습함으로써 사회성과 협응력을 길러가며 창작과 창의성을 발휘할 수 있는 공간을 제공하고자 한다.

The 21st century is the ear of globalization, informatization and culture. As children are required to have diversity and creativity in this era, the importance of art education becomes greater. Imagination and creativity, the measure of global competitiveness, can blossom in freedom of thoughts. Especially, freedom of thoughts can be cultivated through exciting activities in children and infants when the intellectual and emotional development is most active. 'Performance art education' where you can directly touch and experience various materials that can be hardly accessed at home presents a way that is different from the existing static and uniform art education through various expressional activities in a free and open environment. The aim is to provide space where you can develop social and collaborative abilities and exhibit creativity and creation by solving problems, taking part in new activities, experiencing and learning together in the process of self-drawing and making expression.

SITE 고척교

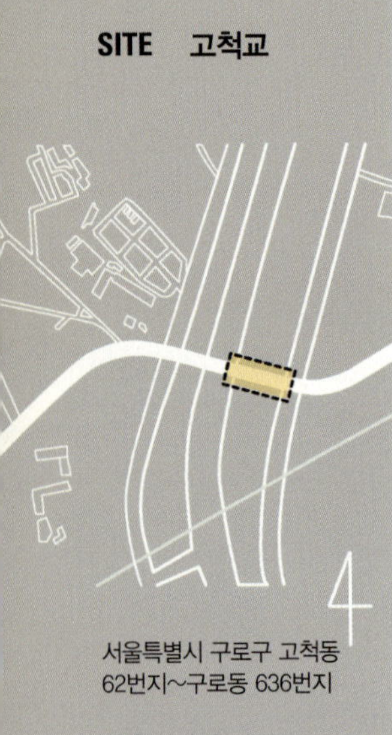

서울특별시 구로구 고척동
62번지~구로동 636번지

62 Gochuk-dong ~
636 Guro-dong Guro-gu, Seoul

SITE ANALYSIS

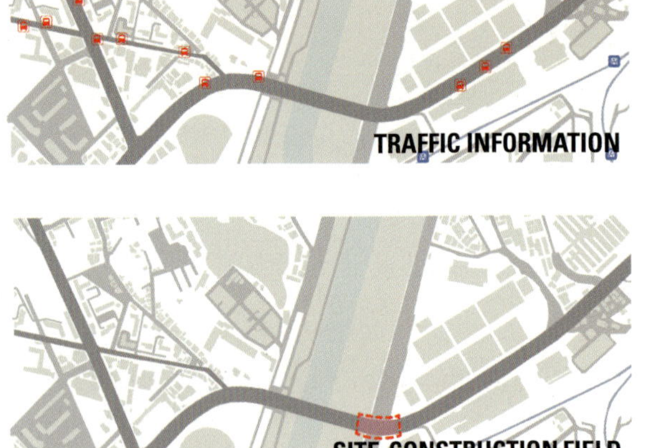

구로동과 고척동을 연결하는 다리인 고척교 주변에는 영향력 있는 부도심으로 성장한 영등포와 신도림이 위치하고 있다. 업무용 고층 빌딩과 아파트의 건설, 서울과 경기를 잇는 교통의 발달로 인한 유동 인구의 증가로 상업 시설도 많이 생겨나고 있는 추세이다. 그러나 아동들을 위한 공간은 어린이 공원이라는 놀이터 몇 개뿐이고 그 외의 시설은 부족한 게 현실이다. 고척동 주민들이 쉽게 접할 수 있는 고척교 하부를 대상지로 하여 고척교 주위의 밀집한 주거지에 주거중인 아동들에게 예술참여공간을 만들어주어 아동들의 상상력과 창의력을 키워줄 수 있도록 한다.

Youngdeungpo and Shindorim, which have developed into influential subcenters of the city, are located around the Gocheok bridge that connects Guro-dong with Gocheon-dong. There are many more commercial facilities due to the construction of skyscrapers for business and apartment houses and the development of traffic connecting Seoul and Gyeonggi areas which has led to the increase of a flowing population. However, as for children's space, there are only a few playground including children's parks while other facilities tend to remain insufficient. The aim includes the generation of space for residential children around and under the Gocheok bridge to participate in art activities, which will develop their imagination and creativity.

ELEVATION

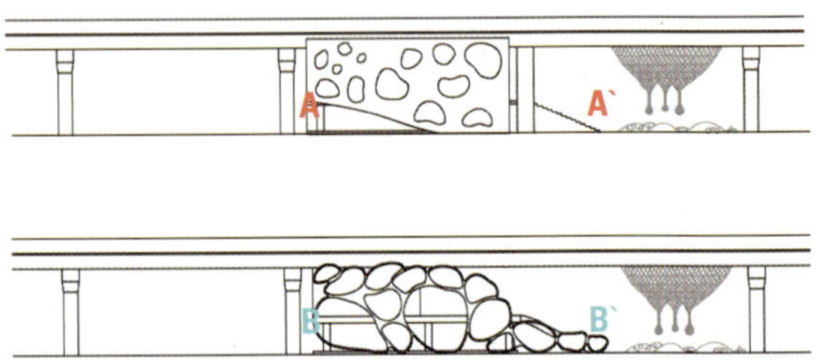

A A`

B B`

PLAN

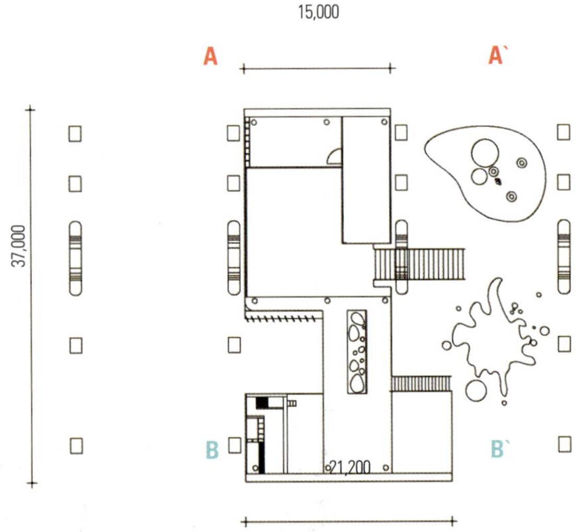

15,000

A A`

37,000

B B`

21,200

SITE

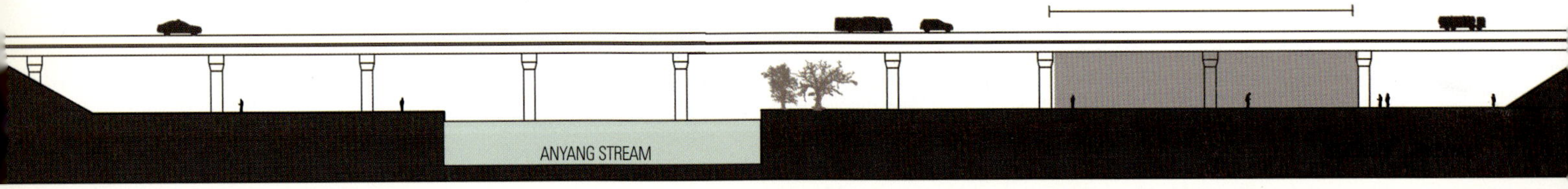

ANYANG STREAM

TARGET

보고 듣고 쓰는것 보다는 만지고 움직이며 느끼는 활동이 필요한 4~6세 영/유아.

Touching, moving, and feeling rather than seeing, listening, and writing will not restrict creativity of children and infants aged 4-6.

DALIY LIFE

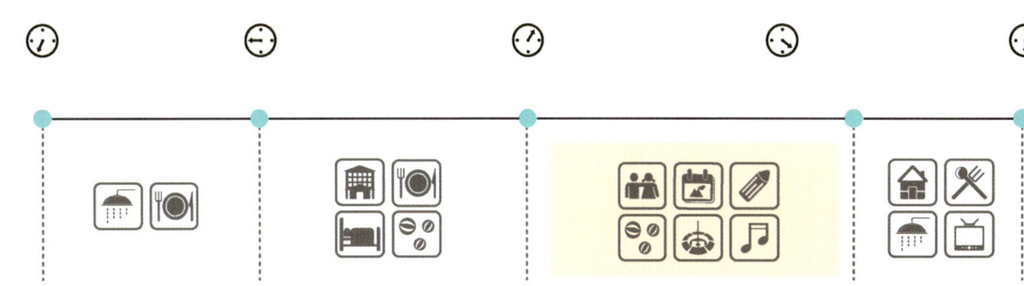

EFFECT OF ART EDUCATION

PURPOSE OF ART EDUCATION

Rignt-Brain Functions
Holistic thought
Intuition
Creativity
Art and music

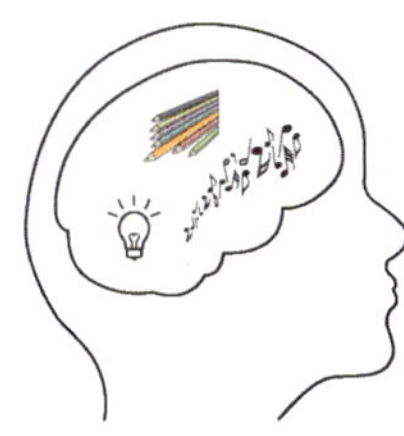

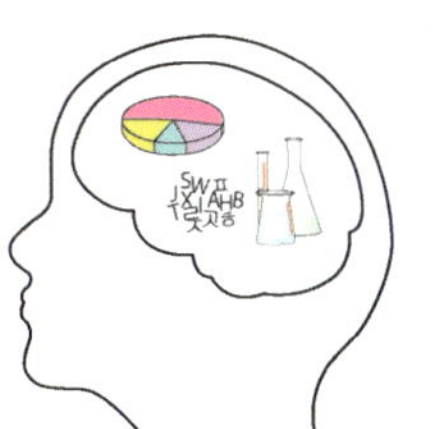

Left-Brain Functions
Analytic thought
Logic
Language
Science and math

사람들이 청소년 이후에 미술표현 능력이 거의 발달하지 않고 어린 시절보다 오히려 조잡한 그림을 그리는 것은 그들이 좌뇌의 주도로 그림을 그리기 때문이다. 학교교육이 좌뇌 계발에 편중됨으로써 사물을 총체적으로 파악하는 능력과 예술적 능력, 창의력 등이 우세한 우뇌가 제대로 계발되지 못하고 있는 것이다. 우뇌 발달에 도움이 되는 미술 교육을 통하여 어릴 때 부터 평소에 대상을 잘 관찰하고 보이는 대로 그리는 습관과 생각하면서 창의적으로 그리는 습관을 길러야 한다.

People tend to draw coarser pictures than they used to as children while their art expression decreases after adolescence because they draw by using left brain. School education is centered around the development of left brain and the development of right brain which provides excellent abilities of overall identification of objects, art expression and creativity takes place improperly. Based on art education helping with the development of right brain, children need to develop their habits to draw creatively by observing daily objects and drawing them as they are.

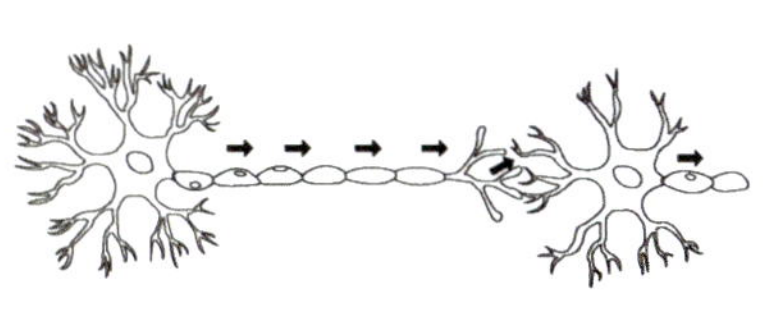

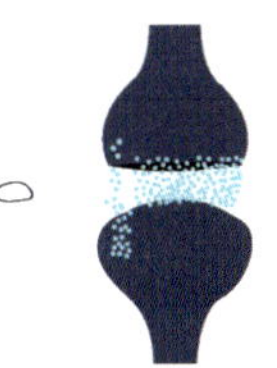

synapse

신경세포인 시냅스 고리가 복잡하고 정교할수록 지능의 수준이 높아진다. 이러한 시냅스의 증가와 뇌세포의 활발한 분열은 유아기 발달과정에서 성인과 다르게 월등히 높다고 한다. 때문에 유아기에 미술교육을 통해 두뇌 발달을 도울 수 있다.

More complex and elaborate synapses as nervous cells lead to higher intellectual capacity. The increase of synapses and active division of cerebral cells are known to be much higher in infants' development stages than that of adults. Hence, art education for infants can help with brain development.

CONCEPT

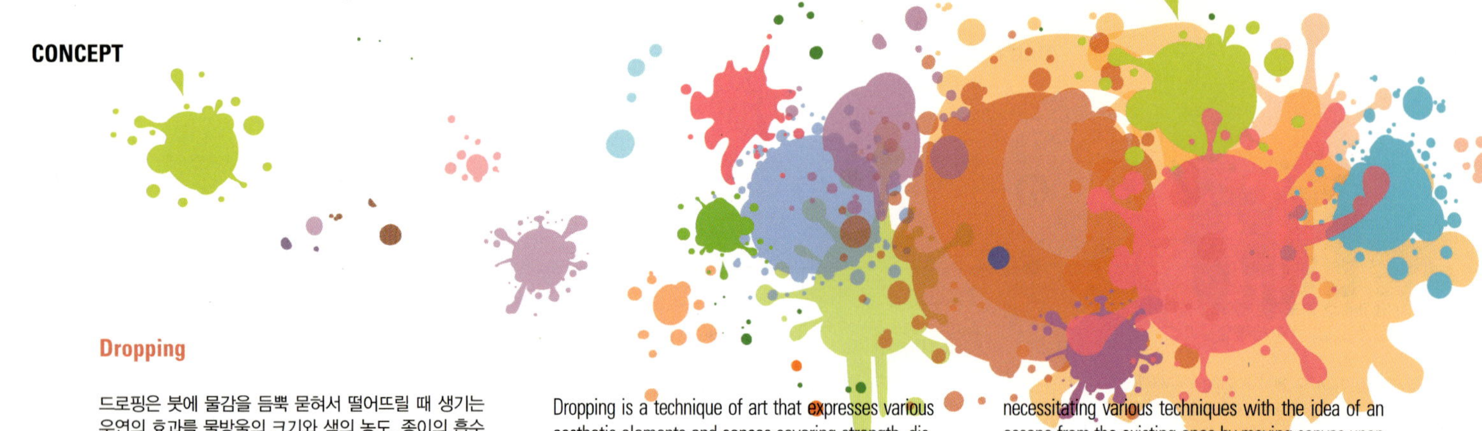

Dropping

드로핑은 붓에 물감을 듬뿍 묻혀서 떨어뜨릴 때 생기는 우연의 효과를 물방울의 크기와 색의 농도, 종이의 흡수력, 점의 밀도 등에 따라서 강약, 원근, 방향, 리듬, 변화 등의 여러 가지 미적 요소와 감각을 표현 할 수 있는 미술기법이다. 붓에 묻은 물감을 뿌리는 방법과 캔버스를 움직였을 때 물감이 경사를 따라 움직이며 남겨지는 자국을 통해 그리는 방법이 있다. 일반적으로 붓과 같은 도구를 사용하여 그리는 방법이 아닌, 뿌리거나 그림이 그려지는 캔버스를 움직이는 행위를 통하여 기존의 틀에서 벗어나 다양한 방법이 필요한 퍼포먼스 미술교육을 위한 공간을 표현하고자 한다.

Dropping is a technique of art that expresses various aesthetic elements and senses covering strength, distance, direction, rhythm and changes according to the sizes of water drops, density of colors, absorption of paper and density of points based on accidental effects coming from dropping paints fully applied to brushes. You may sprinkle paints applied to brushes and draw through traces on a slant of canvas as it moves. The aim is to produce space for performance art education necessitating various techniques with the idea of an escape from the existing ones by moving canvas upon which a picture is drawn rather than drawing by using brushes and other tools.

KEYWORDS

Water Drop Gravity

Bounce

Concentration

Momentary Accidental

Stretch

Print

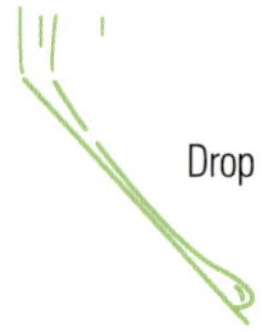

Drop

Flow Slope

Quantity

Shape Rhythm

KEYWORDS ANALYSIS

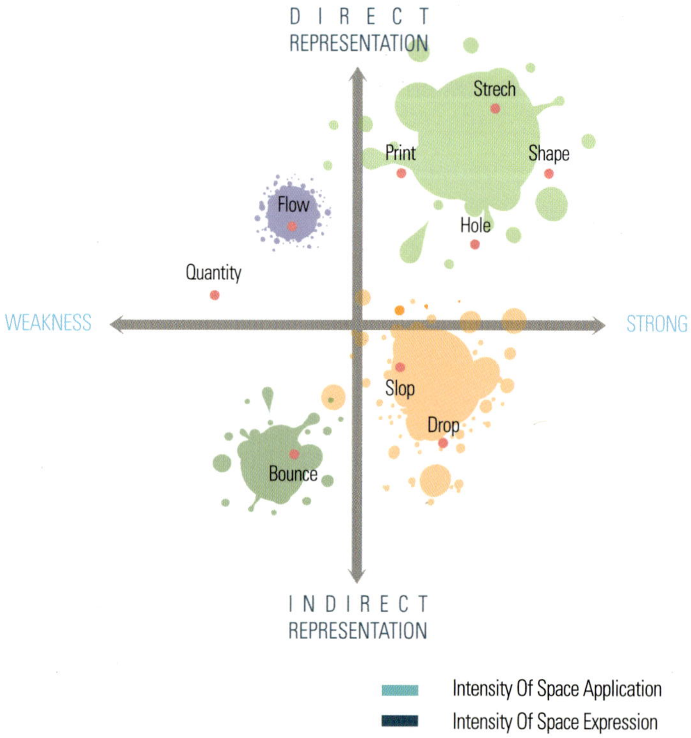

DIRECT REPRESENTATION

Strech

Print

Shape

Flow

Hole

Quantity

WEAKNESS

STRONG

Slop

Drop

Bounce

INDIRECT REPRESENTATION

Intensity Of Space Application

Intensity Of Space Expression

DESIGN PROCESS

PART 1. Momentary Accidental

드로핑 행위를 통해 얻어지는 우연적 효과
Incidental effect is obtained by dropping the behavior

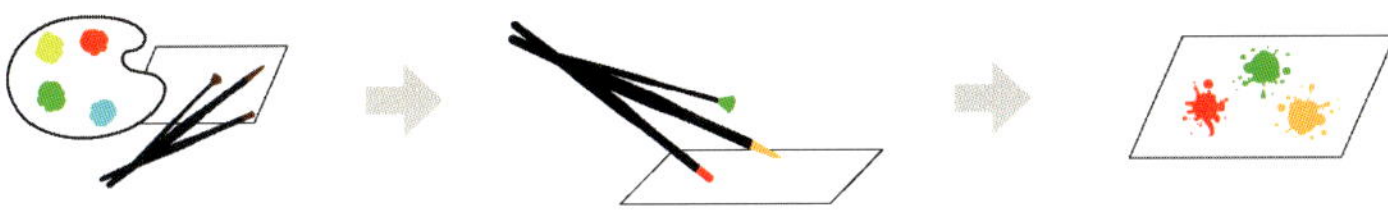

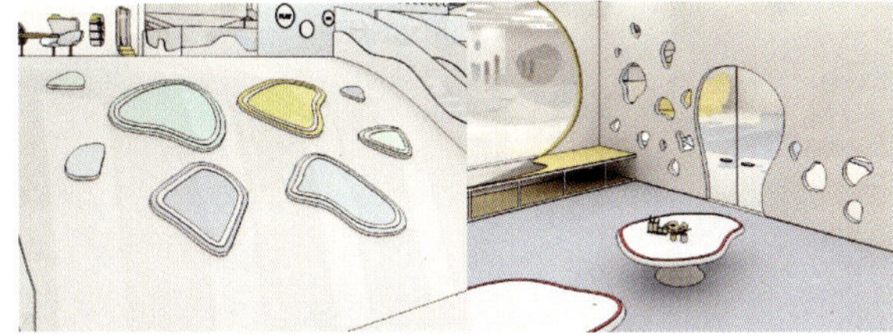

PART 2. Stretch

늘어지는 물감의 특징을 적용한 외부 놀이공간 생성
Loose paint, create a play area outside of the characteristics of the applied

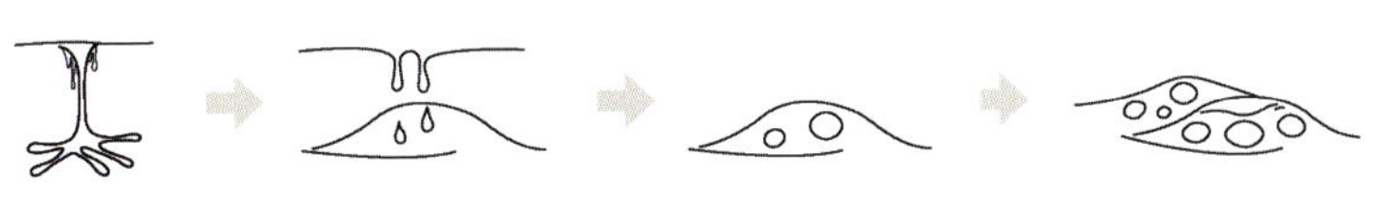

PART 3. Flow slope

공간에 경사를 주어, 경사를 따라 움직이며 남겨지는 자국을 라인으로 추출
Given the slope in the space, moving along the slope of the line, its extraction is left

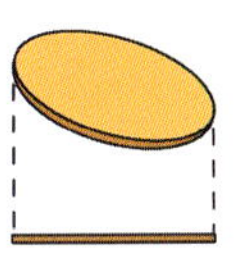

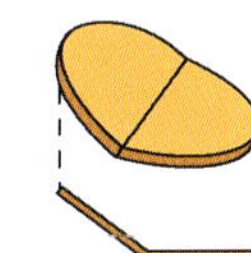

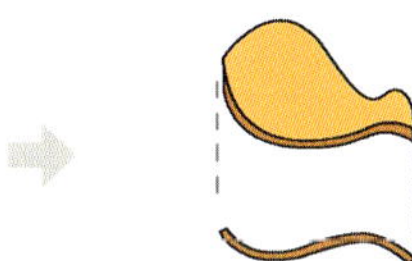

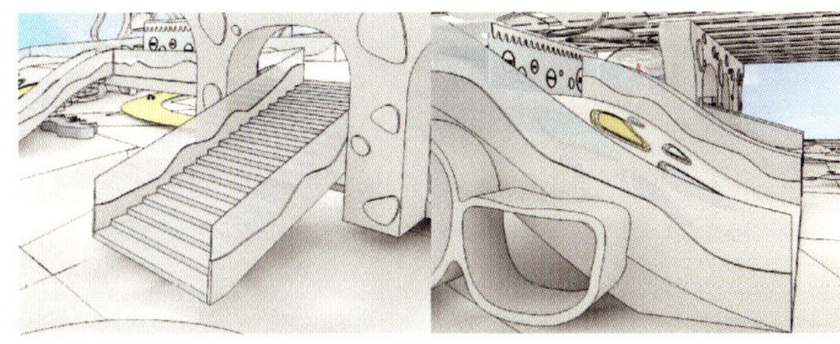

PART 4. Print

물감이 흘러내리는 모양을 통한 패턴 형성
Pattern formation through the appearance of two paint flowing

PASSIVE ANALYSIS

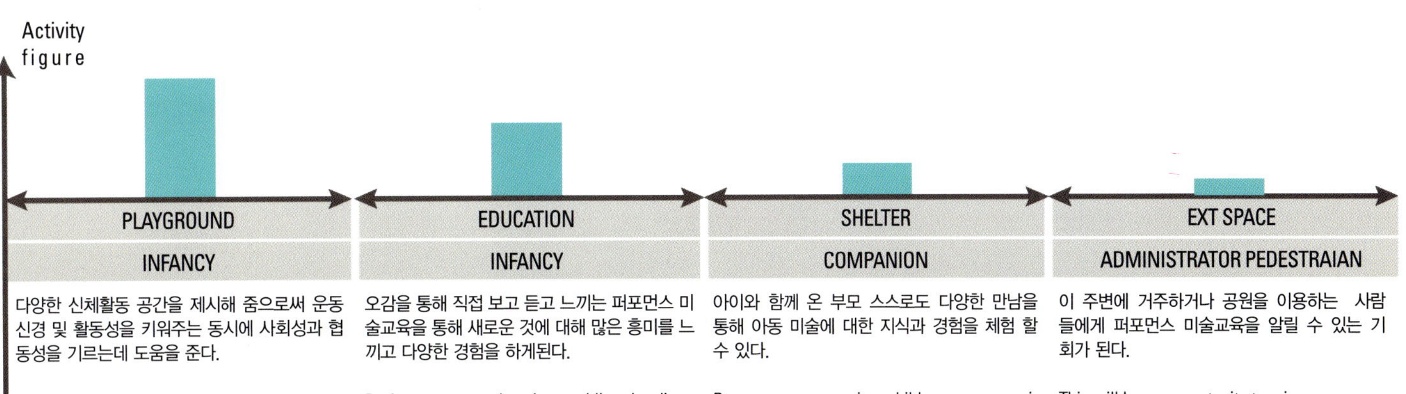

다양한 신체활동 공간을 제시해 줌으로써 운동 신경 및 활동성을 키워주는 동시에 사회성과 협동성을 기르는데 도움을 준다.

The presentation of space for various physical activities will grow motion nerves and activities in addition to development of social and collaborative abilities.

오감을 통해 직접 보고 듣고 느끼는 퍼포먼스 미술교육을 통해 새로운 것에 대해 많은 흥미를 느끼고 다양한 경험을 하게된다.

Performance art education enabling the direct listening and feeling though the five senses will lead to more interest in and various experiences with new things.

아이와 함께 온 부모 스스로도 다양한 만남을 통해 아동 미술에 대한 지식과 경험을 체험 할 수 있다.

Parents accompanying children can experience knowledge of art education for children through various encounters.

이 주변에 거주하거나 공원을 이용하는 사람들에게 퍼포먼스 미술교육을 알릴 수 있는 기회가 된다.

This will be an opportunity to raise awareness of performance art education among people living or passing under the Gocheok bridge.

SPACE PROGRAM

Convex Mirror

아이들의 호기심을 자극하는 동시에 인지능력을 발달에 도움이 되는 공간을 제시해준다.

This is to suggest space that stimulates children's curiosity and helps with the development of their cognitive abilities.

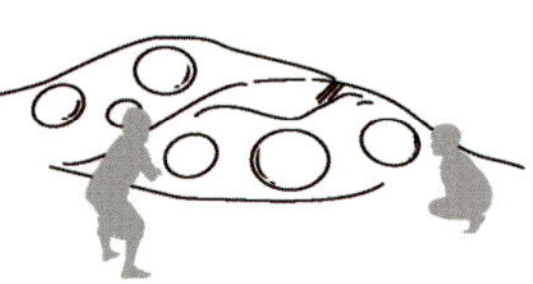

Net

올라가거나 매달리는 행동을 통해서 운동신경을 기를 수 있는 신체활동 공간을 제시해준다.

As they climb up or cling to something, the space will provide a chance for physical activities which can develop motor nerves.

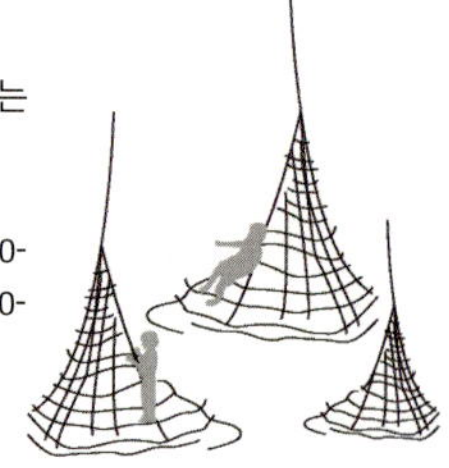

Hole

활동적인 아이들에게 구멍과 그 구멍 속 숨은 공간을 보여줌으로써 호기심을 자극하고 공간감을 익히게 해준다.

By showing holes and hidden space in the holes to active children, you can induce children's curiosity and help them to secure spatial senses.

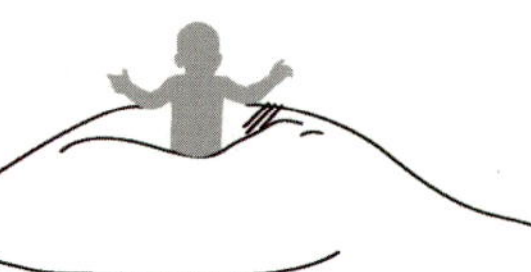

Rotational play

직접 밀어 회전시키고 지나 다니는 놀이를 통하여 흥미를 유발시키고 지능발달을 촉진시킨다.

Plays where children can directly push, rotate, and pass by, their interest and intellectual development will be promoted.

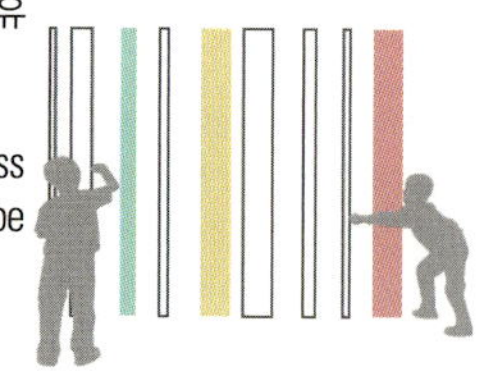

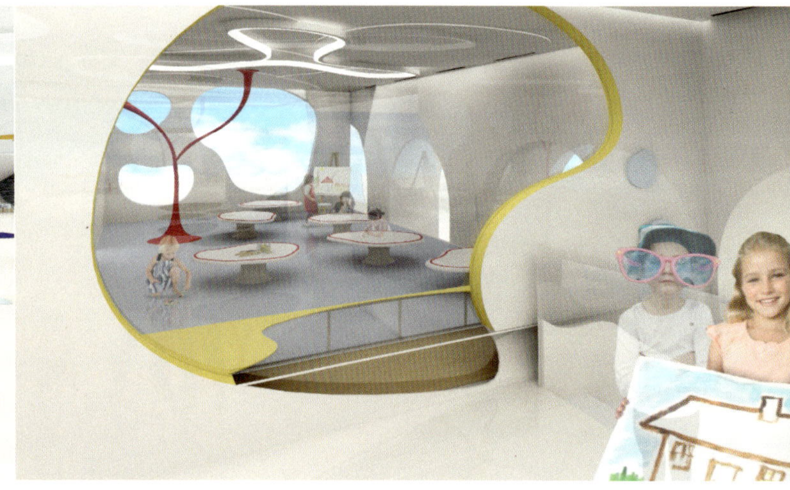

Touch

직접 터치하여 반응을 바로 체험할 수 있어 아이들의 감성과 호기심을 자극하여 교육 효과를 증진시킨다.

Children can directly touch and experiences reactions, leading to the stimulation of emotions and curiosity as well as improvement of educational effects.

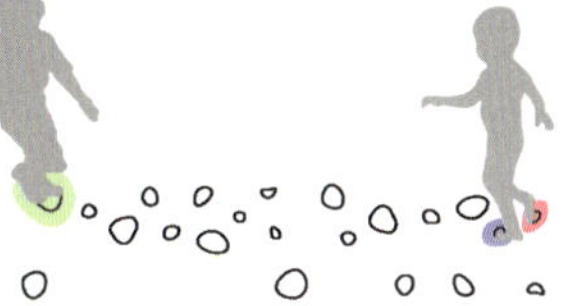

Drawing Activity

무언가 손에 잡고 그리는 행동을 통해 손의 힘을 길러줌으로써 손 글씨를 빠르게 접할 수 있다.

Their activities gripping something and drawing, children's grasping power can increase and they can develop handwriting much faster.

Stacking Blocks

장난감을 좋아하는 아이들에게 블럭 놀이 공간을 제시하여 블럭을 쌓고, 형태를 만드는 공간을 제시함으로써 사고력과 집중력, 동시에 사회성과 협동성을 기를 수 있다.

If a child loves to play with toys, the space for block playing and building can be presented so that he/she can manipulate blocks to make forms, which will develop the capacity to think, concentrate, socialize and cooperate.

Hand Printing

다양한 색을 인식할 수 있고, 손을 찍는 행동을 통해서 성취감을 느낄 수 있다.

As they can recognize various colors and apply palm printing, their feeling of achievement will be generated.

외부 공간에 아이들의 호기심을 자극하는 놀이공간을 제시해주어 다양한 신체활동을 통해 운동신경을 기를 수 있는 공간을 제시해준다.

In outer space, stimulate the curiosity of childr en presented a play area given Physical activity and movement through a variety of neurological and will suggest a place to cultivate.

아이와 함께 동반한 부모에게 휴식공간을 제공해주는 동시에 다양한 만남을 통해 아동미술을 접할 수 있는 기회를 제공해준다.

Accompanied by a parent with a child that provides a relaxing atmosphere at the same time through a variety of children's art to keep you in meeting provides an opportunity.

자기들이 직접 그린 그림을 직접 전시하고 관람하는 것은 '전시회'의 딱딱한 공간이 아닌, 체험공간으로서 다가가기 때문에 아이들이 쉽게 즐길 수 있는 기회를 제공한다.

Exhibit to his own paintings their own and to watch 'Fair' is not a rigid space, because of experience as a space approachable Children provides an opportunity to easily enjoy.

PLAY
ART

Recycle

새롭게 태어나다

WEST FLYOVER

· SITE 선정 이유

율전동 중심상가에 위치하고 있어 유동인구가 많고, 접근성이 매우 적합하다. 또한, 고가도로 및 하부공간이 비어있으며 사용하고 있지 않는 공간이다. 주로 불건전한 청소년들이 공간으로 사용되고 있다.

· Site 해당공간 특징

주변에 율전중, 천천중, 초등학교 등 많은 학교들이 밀접해 있다. 또한, 율전 주민센터와 놀이문화를 즐길 수 있는 상점들과 주택, 아파트단지 등이 있어 특정 세대가 아닌 모든 사람들이 이용 가능하다.

· Reason for selection this site

This flyover is located nearbyh Yuljeon shopping centerm. So this flyover has many of floating population and propers to access. Also under the this flyover was emptied and rarely used. However this space was mainly using by unwholesome youth.

· Feature of site

It is concentrated in Yuljeon middle school, Cheoncheon middle school and many elementar schools. In addition there are the Yuljeon Pupulation Center, enjoyable cultural places and apartments. Therefore all non-specific generation of people is available.

· SITE 접근 방법

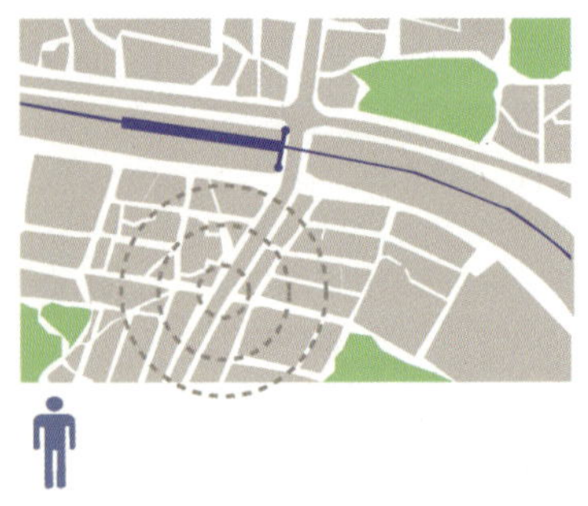

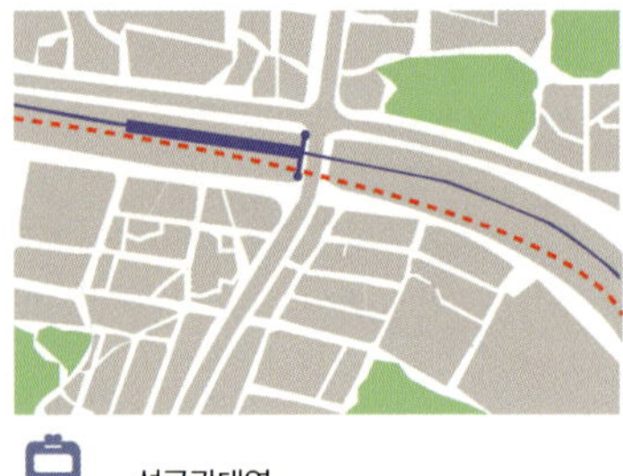

성균관대역

버스승강장(62-1)

위락시설

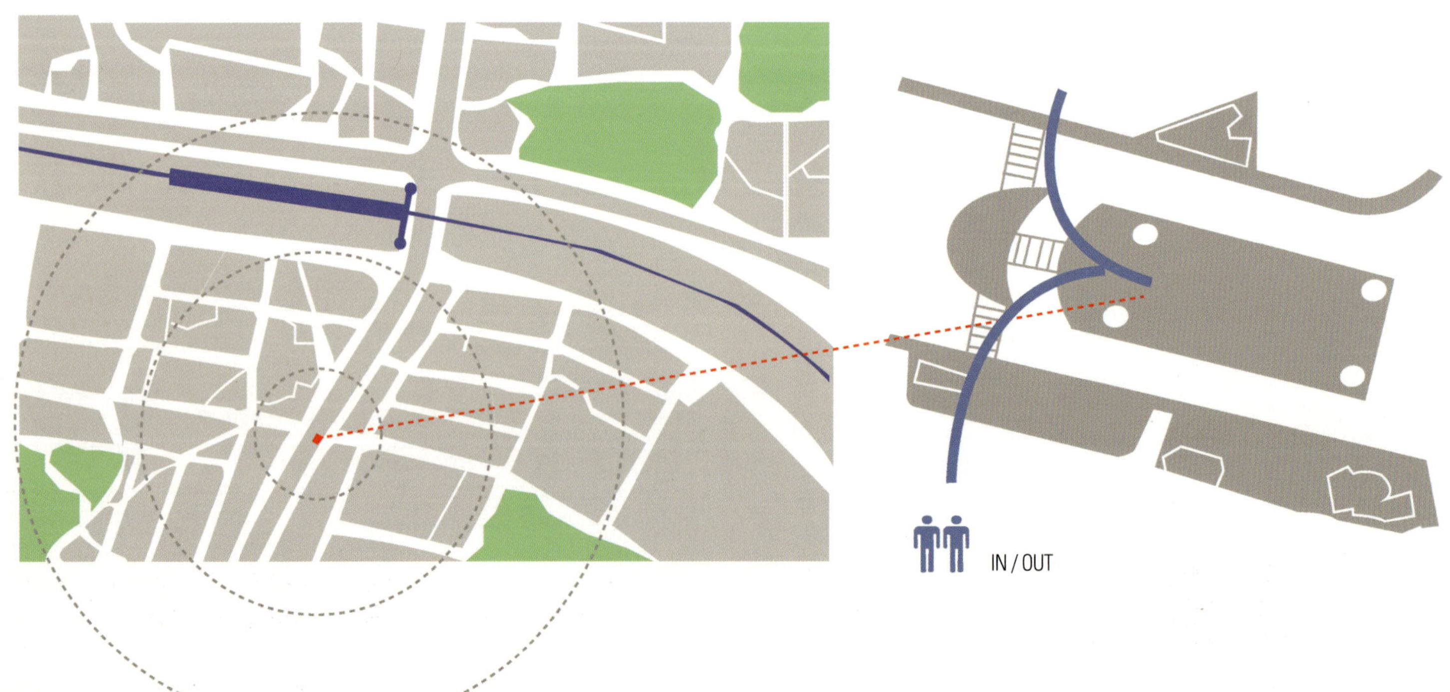

BACKGROUND

전 세계가 자연환경에 대한 걱정과 더불어 쓰레기 문제에 대하여 고민하고 있다. 하지만 현재 우리나라는 길거리에 쓰레기통도 있지 않아 아무곳에나 쓰레기를 버리고 있다. 그래서 아무 곳에서도 사용하지 않고 버려지는 쓰레기들을 여러 가지 방법으로 재활용하여 사용할 수 있는 물건을 만들거나 장식용으로 전시할 수 있도록 하는 체험을 통하여 환경의 소중함을 일깨워 지기 위한 공간으로 사용하려 한다.

The natural environment around the world worry about the trash along with the door Jee is worried about. But now the country No trashcan on the streets because there is not used anywhere. Without waste of garbage in several ways recyling that can be used to create or decorative items. Through experience in the environment that allows you to showcase is trying to use the space .

공간활용방법 (How to take advantage of the space)

· 광장이 있어 주민들에게 휴식공간을 마련해 준다.

· 재활용품을 가지고 오면 직접 공간을 꾸밀 수 있는 재미를 준다.

· 관리자인 시민단체는 주민들이 가져오는 쓰레기를 재활용품 재료로 이용하여 만들 수 있는 제품을 개발한다.

· 주민들은 재활용품으로 새로운 물건을 만들고 이용하여 재활용품에 대한 인식을 바꾸고 배울 수 있도록 한다.

· 관리자는 환경보호와 관련한 강사를 초청하여 주민들이 강의를 들을 수 있도록 하고 그와 관련한 여러 프로그램을 만들어 지속적으로 주민들이 관심을 갖을 수 있도록 유도하다.

· Square will arrange a realaxing atomsphere for residents.

· Decorate the space comes with recycling that can be directly gives fun

· Civil organixations to get the people who manage the recyling materials shall be developed that can be made of it.

· As residents recycled to create new stuff using recycled materials to learn to change the perception of recycling.

· Managers with respect to protecting the environment and the people invited lecturers associated with it. So you can hear lectures and several programs con tinously made to allow residents to have an interest in is induced.

TARGET

· 주 이용층

율전동 서부고가도로 근처에 거주하고 있는 유아부터 청소년, 유아를 데리고 오는 주부를 주 이용 층으로 잡는다.

· Mainly user

From children and teenager to housewife who live nearby Yuljeon west overpass road.

CONCEPT

· 분자

고체, 액체, 기체 상태로 존재 할 수 있으며 분자간의 거리가 변화하변서 상태가 변한다. 분자는 쪼개져 다시 원자로 될 수 있으며 원자 조성의 변화에 따라 수많은 물질을 만들어 낼 수 있는 분자의 수도 증가하고 있다. 분자는 온도와 압력에 따라 고체, 액체, 기체 상태로 존재할 수 있으며 상태가 변하더라도 분자 내의 원자 간의 결합의 길이는 변하지 않고 분자간의 거리가 변화하면서 상태가 변한다.

· Molecule

Molecule can be solid, liquid, or gas, distance between molecules as the state changes. Molecule can be Atom again as broken up.
According to changes in the atomic composition of substances that can produce many of the molecules may have increased. Molecule can be solid, liquid, or gas depending on the temperature and pressure. Although the state and saw that the length of the coupling between atoms within molecules change, while the distance between moecules without changing the status change.

SITE 운영방식

· 시민단체

대부분 공동선과 공공의 이익을 위해 봉사나 구호 활동을 펼치고 기본적으로 국가나 자본으로부터 독립해서 활동하는 비정부기구를 뜻한다. 시민사회단체로도 불린다.

· Ngos

Most of the common good, public relief for the benefit and volunteer work were basic move. Also the country engaged in actibites independent of capital. That means The non-governmental organixations and activites. And it was called The civil society organizations.

SITE PLAN

· Elevation

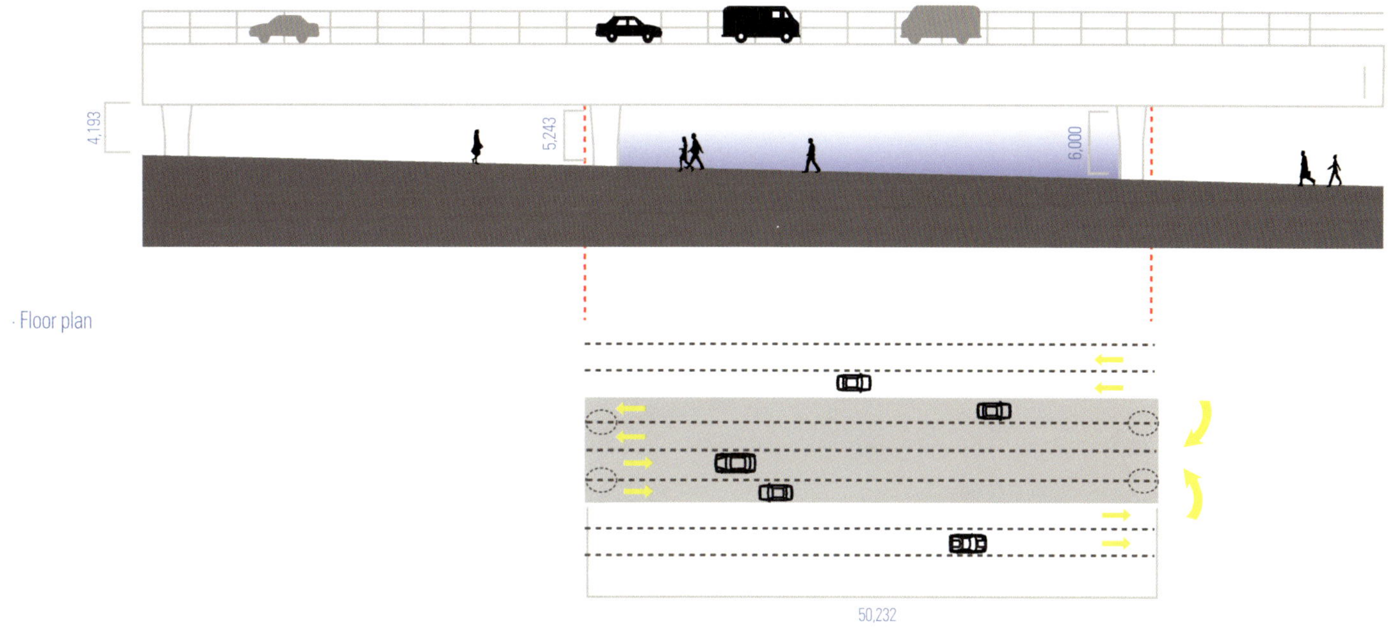

공간평면도, 조닝 (Flight of the space, the zoning)

· Floor plan

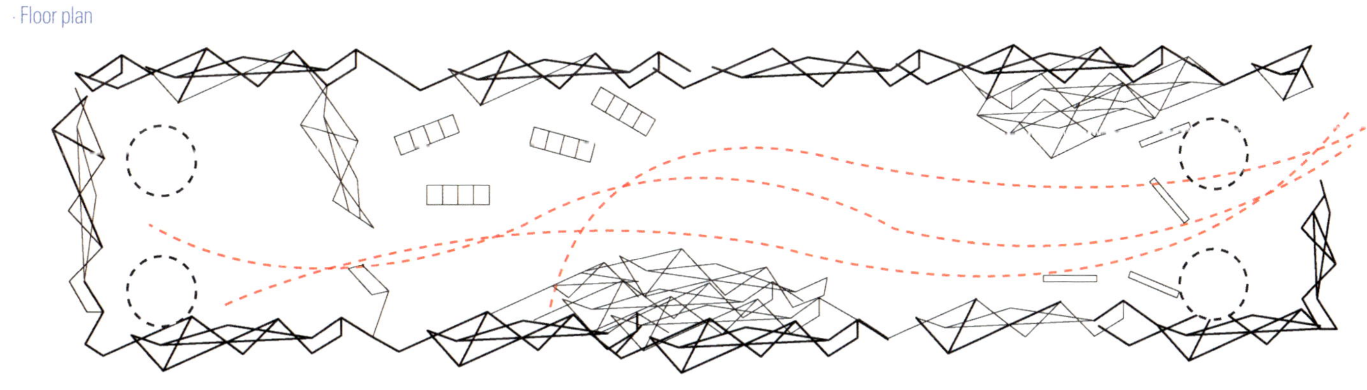

· Floor plan

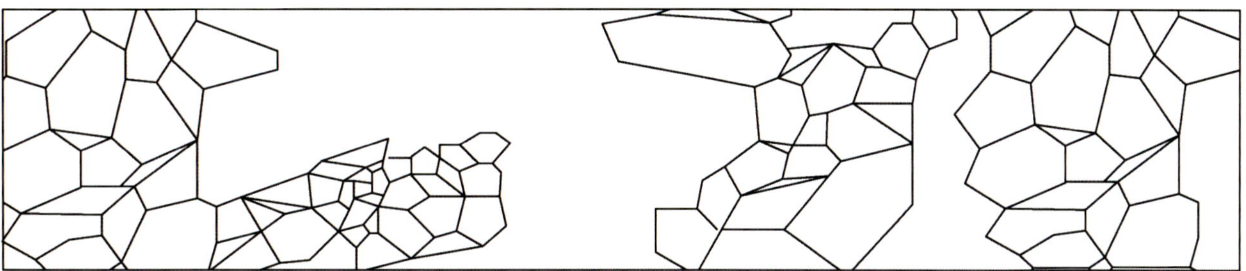

PROCESS

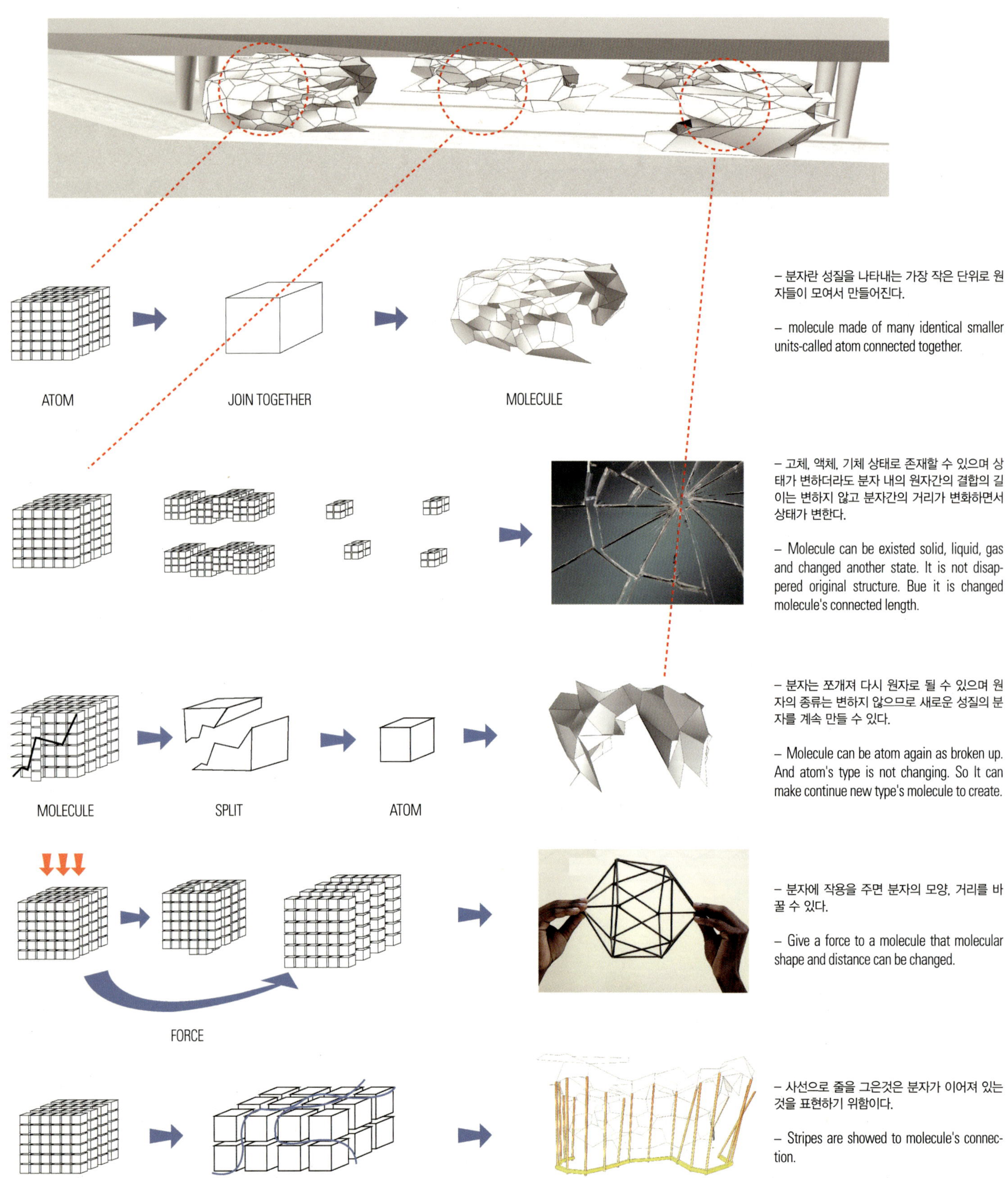

– 분자란 성질을 나타내는 가장 작은 단위로 원자들이 모여서 만들어진다.

– molecule made of many identical smaller units-called atom connected together.

– 고체, 액체, 기체 상태로 존재할 수 있으며 상태가 변하더라도 분자 내의 원자간의 결합의 길이는 변하지 않고 분자간의 거리가 변화하면서 상태가 변한다.

– Molecule can be existed solid, liquid, gas and changed another state. It is not disappered original structure. Bue it is changed molecule's connected length.

– 분자는 쪼개져 다시 원자로 될 수 있으며 원자의 종류는 변하지 않으므로 새로운 성질의 분자를 계속 만들 수 있다.

– Molecule can be atom again as broken up. And atom's type is not changing. So It can make continue new type's molecule to create.

– 분자에 작용을 주면 분자의 모양, 거리를 바꿀 수 있다.

– Give a force to a molecule that molecular shape and distance can be changed.

– 사선으로 줄을 그은것은 분자가 이어져 있는 것을 표현하기 위함이다.

– Stripes are showed to molecule's connection.

SKETCH

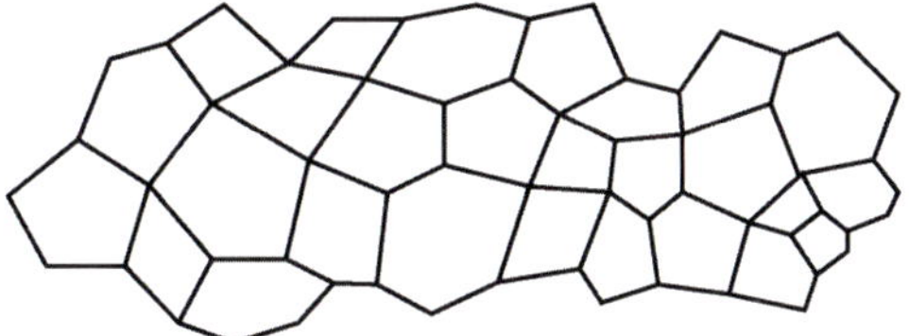

· 시설물 외피

건물 외피에는 비닐재질과 철을 사용하여 밖에서 안을 볼 수 있어 궁금증을 유발 할 수 있고 비닐재질을 사용하여 빛이 들어왔을 때 굴절이나 반사를 이용할 수 있다. 외피에 종이재질을 사용하여 앉을 수 있도록 하였다. 분자들이 서로 연결시켜주는 선들로 모형을 제작하였고, 외피에 의자가 될 수 있는 공간들이 있어서 늘 똑같은 외피모형에서 변형될 수 있는 모습을 보여줄 수 있다. 분자가 이동하여 변하는 모습을 외관에서 표현했다.

· Building Envelope

Building envelop by using the recyled color paper. It can ot see inside from outside and induce curiosity. Color paper will be interested by children mainly to use. Building Envelope can be used by chair inside. So building envelope isn't same appearance that the model can be transformed by people to use. It is funny to people who are not contacting our construction.

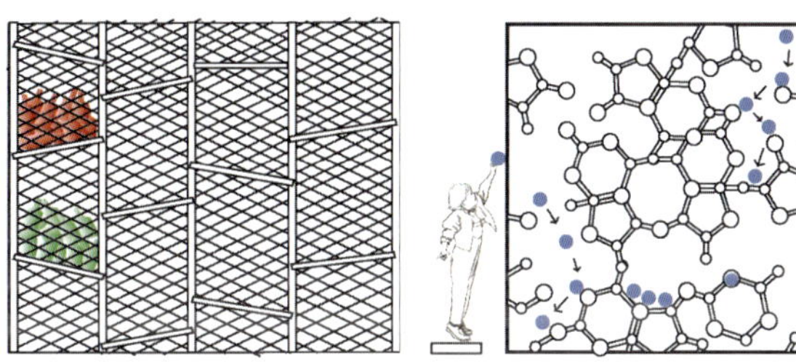

· 시설물 내부공간

프레임 자체가 움직여서 아무것도 재활용품을 넣지 않았을 때 일직선으로 직선 형태를 보여지고 있다가 재활용품을 안에 넣으면 프레임 자체가 움직이는 형태여서 아래로 내려가는 모형을 나타내고 있다. 분자가 상호작용을 통해서 변화하는 모습과 늘 새롭게 모형이 바뀌는 것을 나타내고 있다.

· Inside the building space

Put recyclable material in the net of the frame itself is moving shape, next to the model shows. Molecule through interaction with the ever changing appearance indicates that a change in the new model.

벽체를 이용해 내부에서는 시민들이 원하는 자리에 앉을 수 있는 의자로 사용할 수 있도록 변할 수 있다. 동시에 외부에서 보면 일률적이었던 모양이 시민들이 어떠한 힘을 작용하면서 그에 따라 계속해서 변화하는 모양을 볼 수 있을 것이다. 분자가 어디로 떨어지거나 붙어져서 새로운 상태로 변하는 것을 보여준다.

Citizens want to be seated in a chair that can be changed so that you can use. At the same time look from the out side appearance was uniform. While the citizen act accordingly no power to continue will be able to see the changing shape.

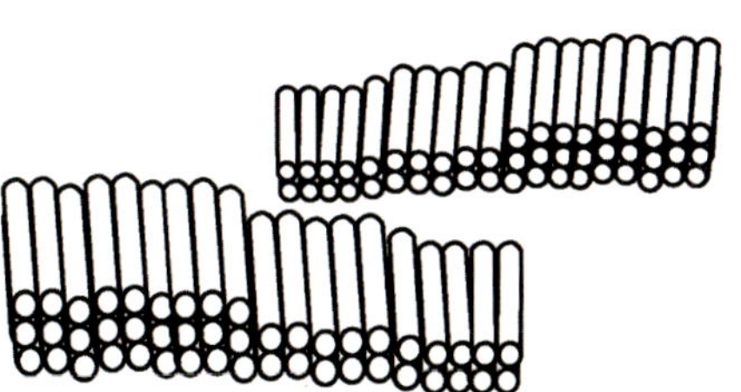

· 시설물 내부가구

두꺼운 종이로 앉을 수 있도록 만들고 그 안은 빈공간이어서 페트병을 넣을 수 있어 아이들이 꺼내서 난타를 치듯 장난치며 놀 수 있도록 한다. 이것은 분자가 서로 떨어지고 붙어서 새로운 상태로 변하는 것을 보여준다.

· Furniture inside the building

Thick paper, making it sit there, you can put plastic bottles in the empty space. Take the kids to create and hold bottles have to be to play the nanta. Where molecules are attached to each other new fall shows that the state changes.

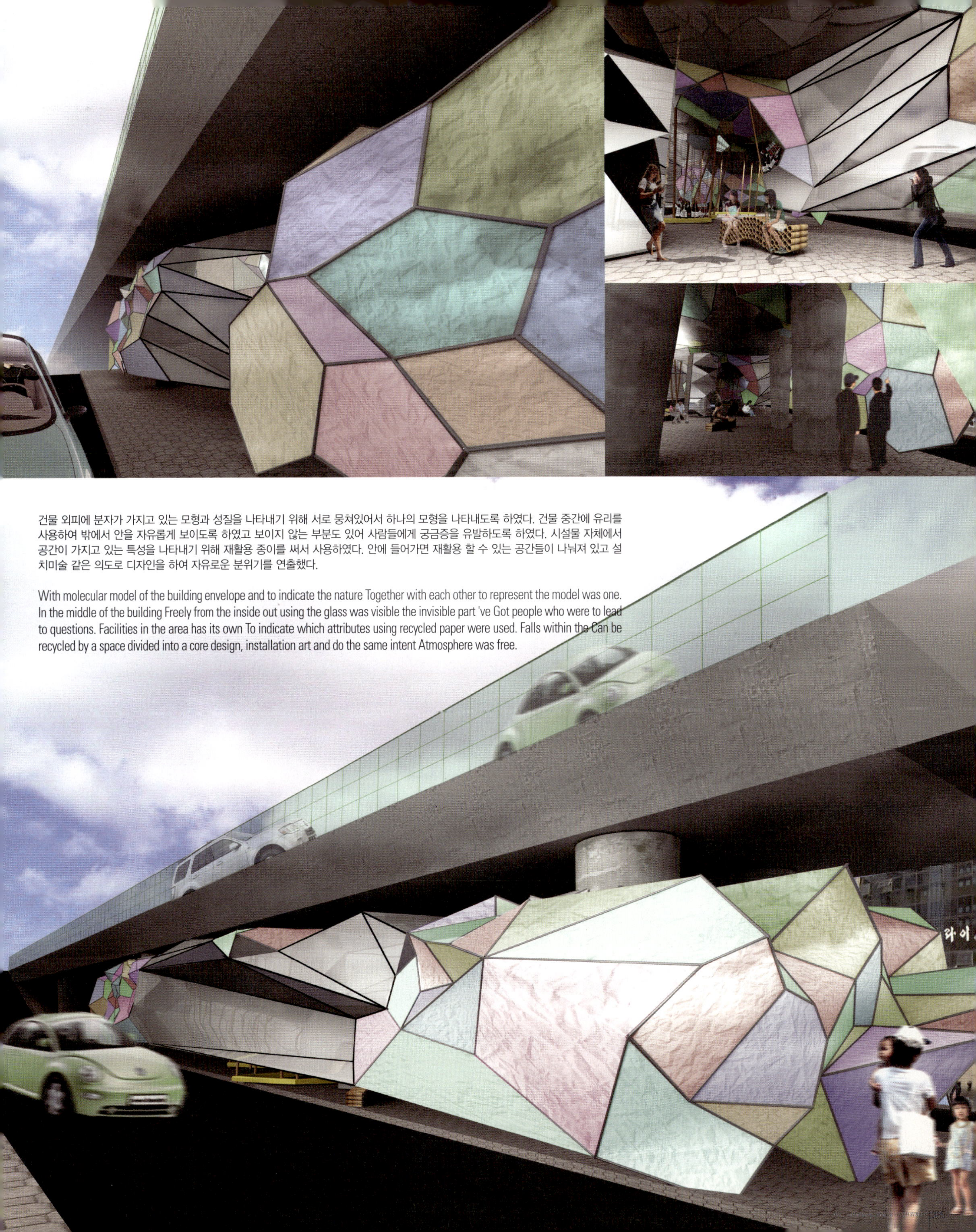

건물 외피에 분자가 가지고 있는 모형과 성질을 나타내기 위해 서로 뭉쳐있어서 하나의 모형을 나타내도록 하였다. 건물 중간에 유리를 사용하여 밖에서 안을 자유롭게 보이도록 하였고 보이지 않는 부분도 있어 사람들에게 궁금증을 유발하도록 하였다. 시설물 자체에서 공간이 가지고 있는 특성을 나타내기 위해 재활용 종이를 써서 사용하였다. 안에 들어가면 재활용 할 수 있는 공간들이 나눠져 있고 설치미술 같은 의도로 디자인을 하여 자유로운 분위기를 연출했다.

With molecular model of the building envelope and to indicate the nature Together with each other to represent the model was one. In the middle of the building Freely from the inside out using the glass was visible the invisible part 've Got people who were to lead to questions. Facilities in the area has its own To indicate which attributes using recycled paper were used. Falls within the Can be recycled by a space divided into a core design, installation art and do the same intent Atmosphere was free.

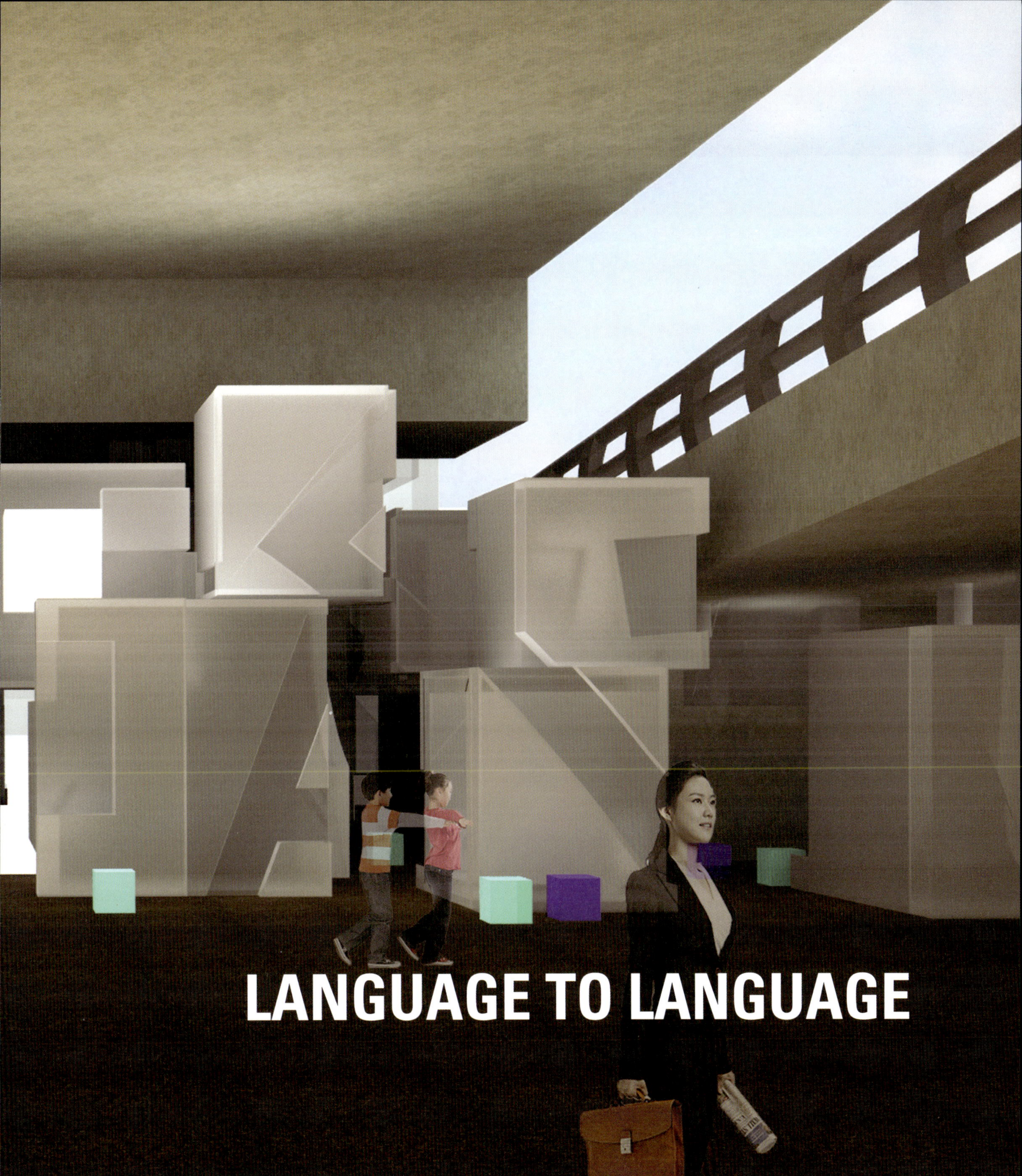
LANGUAGE TO LANGUAGE

BACKGROUND

글로벌 시대에 맞추어 세계적으로 한국의 위상이 높아지고 있는 만큼, 한국은 외국인 거주자 120만 명 시대를 앞두고 있다. 이 중 외국인 근로자의 수가 가장 많은 안산시는 3만 명에 달하는 외국인들이 거주하고 있다. 하지만 이러한 위상과는 맞지 않게 그들을 위한 복지 공간은 부족하다. 특히 집에 혼자 남게 되는 외국인 근로자 2세들을 위한 복지 공간은 찾아보기 힘들다.

Korea keeps having high prestige and foreign settlement's number is up to 1,200,000. Among them foreign workers is a lot of in An-san, (about 30,000). But, for them, welfare system is insufficient unlike our prestige. Specially it is difficult to find out welfare space for children of foreign workers.

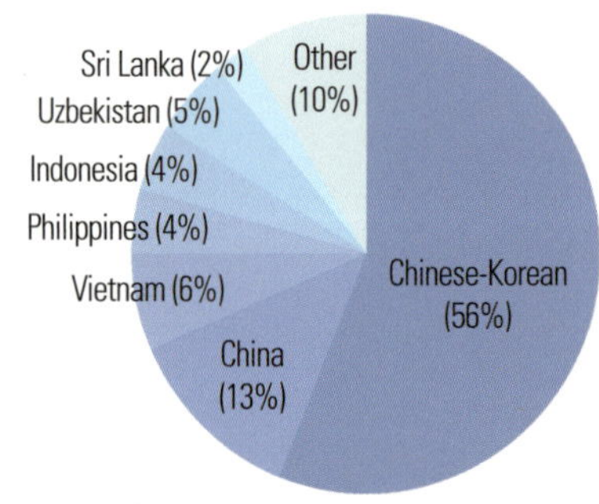

단원구 등록 외국인 국적별 현황
Danwon registered foreigners by nationality status

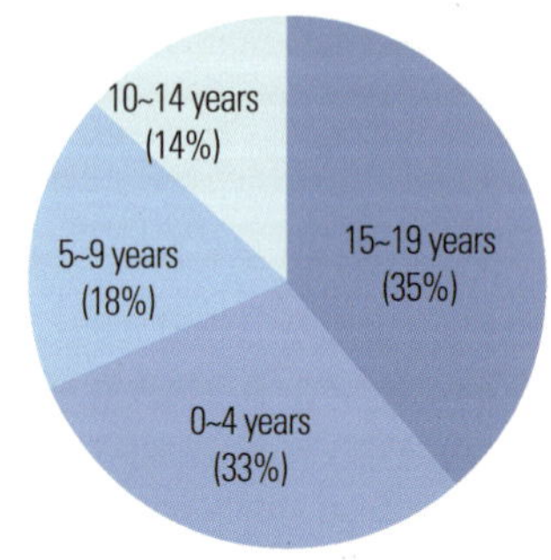

단원구 등록 외국인 근로자 2세 연령별 현황
Danwon registration status by age of the children of foreign workers

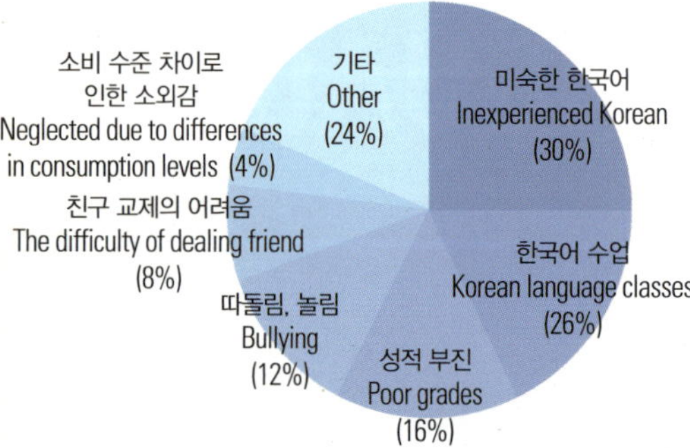

외국인 근로자 2세들이 겪는 고충
Grievances suffered by children of foreign workers

SITE ANALYSIS

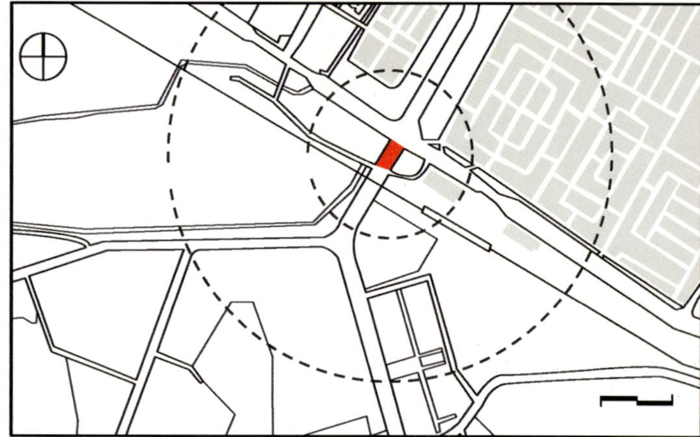

경기도 안산시 단원구 원곡동 원곡교
Wongokdong, Danwon-gu, Anshan, Gyeonggi-do

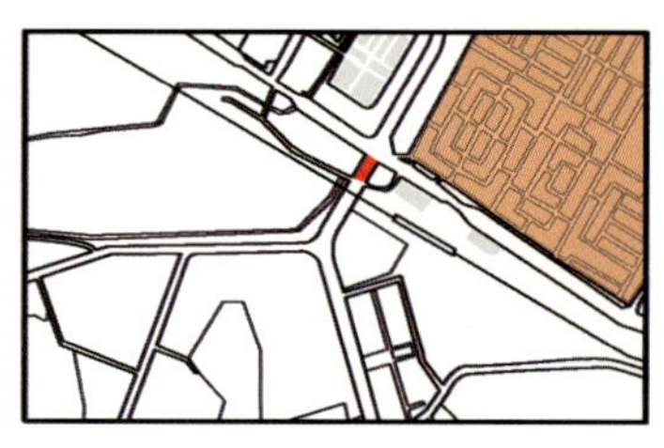

국경 없는 마을
Village without borders

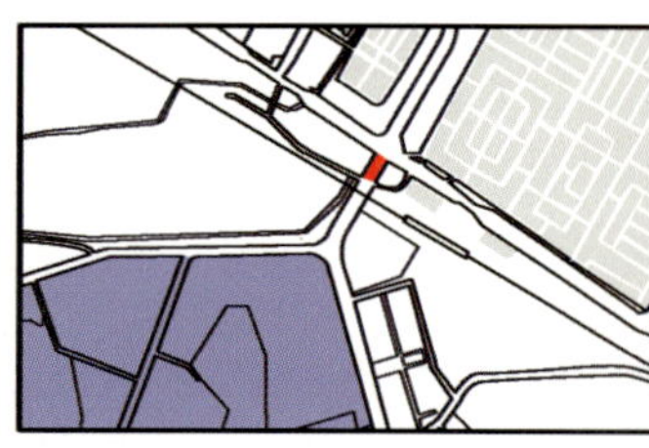

반월 공업 단지
Banwol Industrial Park

교통편
Traffic

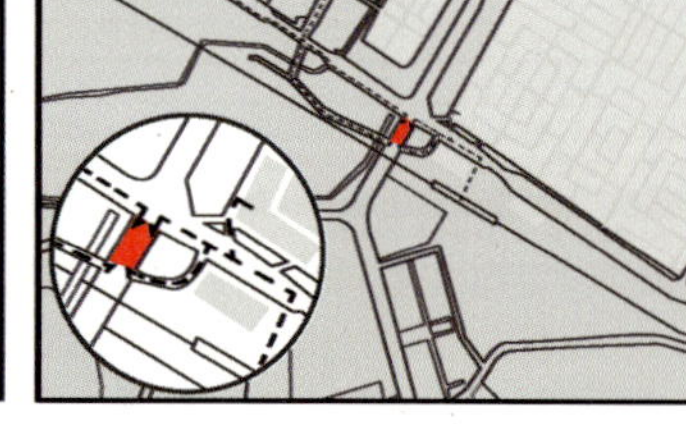

접근성
Access

TARGET

'다문화 가정의 아이들을 위한 놀이터' 가 아닌 '모든 아이들이 어울려 놀 수 있는 놀이터' 를 형성하여 외국인 근로자와 우리 사이의 보이지 않는 경계선을 완화시켜보고자 한다. 대체로 어른보다는 어린 아이들이 친목을 다지기 쉽고, 경계심이 덜하여 어울리는 것에 오랜 시간이 걸리지 않는다. 아이들의 친목을 통해 어른들의 친목 또한 자연스럽게 유도하여 그들과 우리의 사이를 나누지 않고, 함께 즐길 수 있는 공간을 형성한다.

I think making not playground for children of a multicultural family but for all children is good for broking border between foreign workers and us. Generally, it is easy to promote friendship for young children than adults because young children have soft sense of security. Through children's friendship, it is purpose to lead adult's friendship too. then it will be place to enjoy each other.

CHARACTER & ISSUE

다리 밑의 공간과 외국인 근로자는 사람들의 무의식 속에 부정적인 존재로 자리 잡고 있다. 다리 밑이라는 어두운 공간을, 외면당하고 소외당하는 외국인 근로자의 포용 공간으로 형성하여 긍정적인 사고를 조금이나마 부여하고자 한다.

Space under the bridge and foreign labors is becoming negative to people unconsciously. I want to change dark space that is under the bridge into a broad-minded space to convey people positive thought.

STORYBOARD

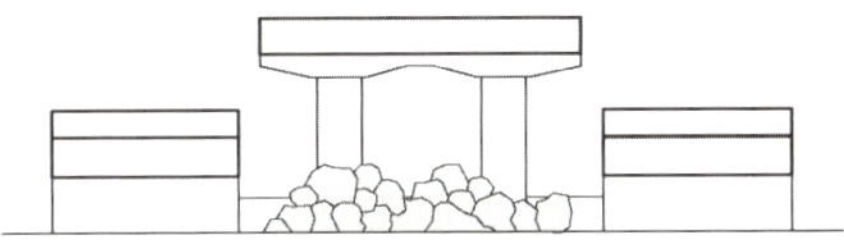

명목상의 자재창고로 쓰이던 다리 밑의 버려진 공간. 어둡고 음침한 분위기 때문에 주된 통로임에도 불구하고 사람들이 마음 놓고 다니지를 못한다.

Because of thrown away, dark, and gloomy space under the bridge, even though that is a core passage, people can't come and go without anxiety.

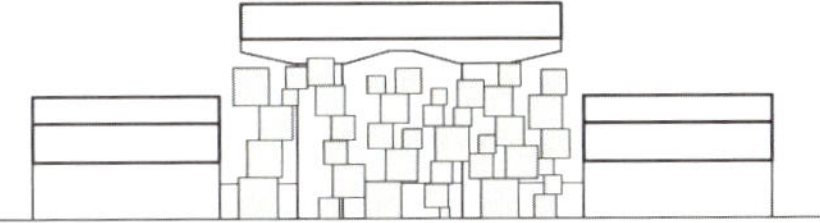

어두운 다리 밑의 공간이 다문화 가정 어린이들을 위한 공간으로 재탄생 했다.

Then, dark space under the bridge is made newly for children of multicultural family.

집에 홀로 남게 되는 다문화 가정 어린이들이 하나 둘 모여들기 시작하고, 일을 나가야 하는 부모도 마음 놓고 일을 할 수 있게 되었다.

After that, children who remain alone flock from various quarters and their parents can go to their workplace comfortably.

모여 든 아이들은 서로 어울려 재미있게 논다. 때때로 한국 유치원 친구들이 놀러와 함께 놀기도 한다. 저녁이 되면 보육 교사와 함께 공부를 한다.

In there, children hang out interestingly. sometimes children who is korean hang out together. At night, they study with nurturer teacher.

퇴근 시간이 되어 부모가 찾아오면 아이들은 집에 돌아간다.

On finishing work, their parents come back and go to home together.

EXPLODED VIEW

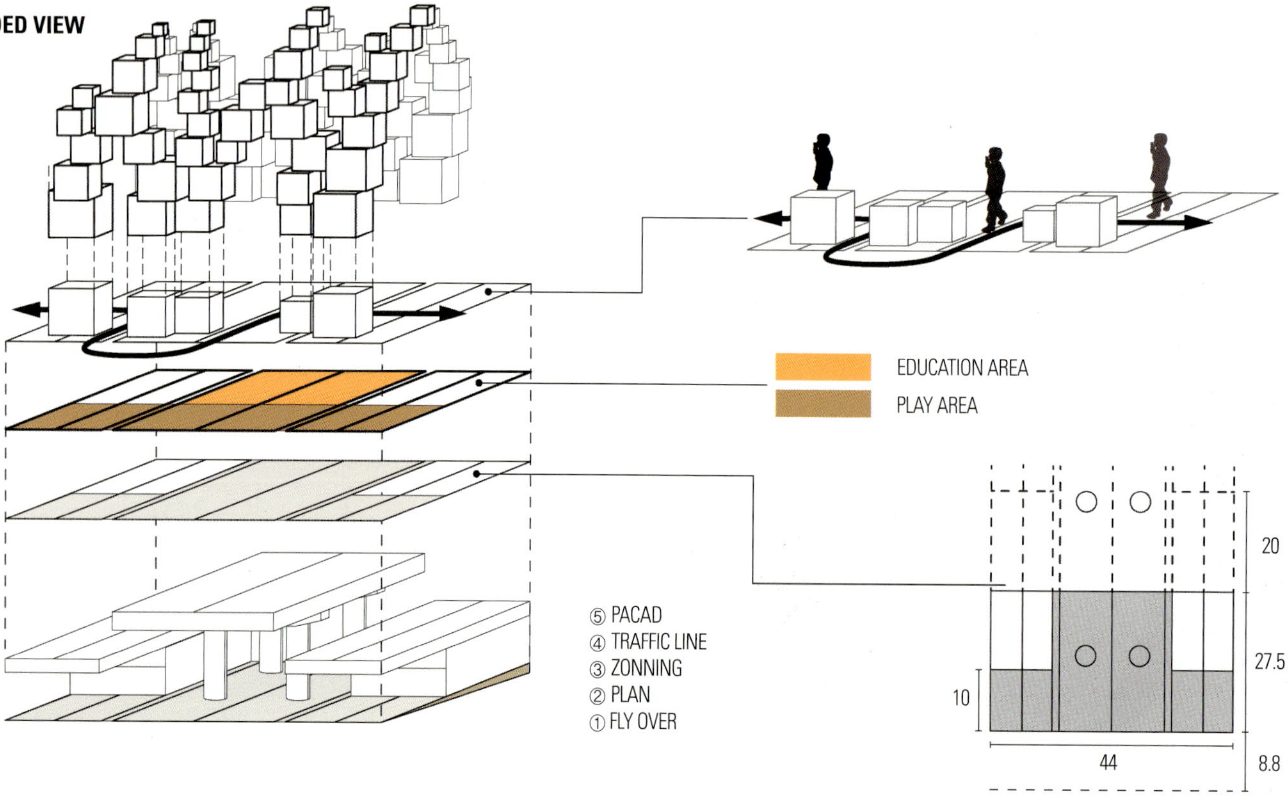

EDUCATION AREA

PLAY AREA

⑤ PACAD
④ TRAFFIC LINE
③ ZONNING
② PLAN
① FLY OVER

20

27.5

10

44

8.8

CONCEPT

언어

사람의 생각과 느낌을 표현하거나 전달하기 위한 기호. 음성 또는 문자.
언어는 커뮤니케이션의 가장 기초가 되는 수단인 반면 높은 장벽이 되기도 한다. 언어의 극복을 통해 '너와 나' 사이가 아닌 '우리' 사이를 만들고자 한다.

Language

sign, spoken language, and word to convey or express people's thought and feeling.
language is basic tool to communicate but it is high border. Throu gh overcoming difference of language, I want to make relation not 'you and I' but 'we'.

MOTIVE IMAGE

매스로 공간에 의미부여
Give meaning to the mass in space

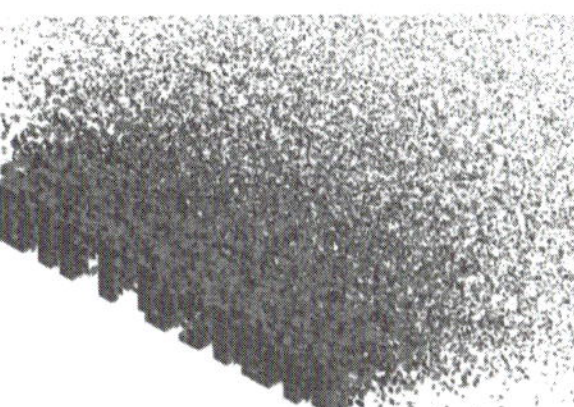

맞잡은 손
Holding hands

문자의 재조합
Character of the recombination

문자의 깨짐
Character of the broken

부분적인 색깔
Partial COLOR

PROCESS 1

각 나라의 언어를 모은다.
I collect individual language.

언어를 추상화 시킨다.
I make language abstract.

추상화 된 언어를 배치한다.
Became the language of abstraction should be placed.

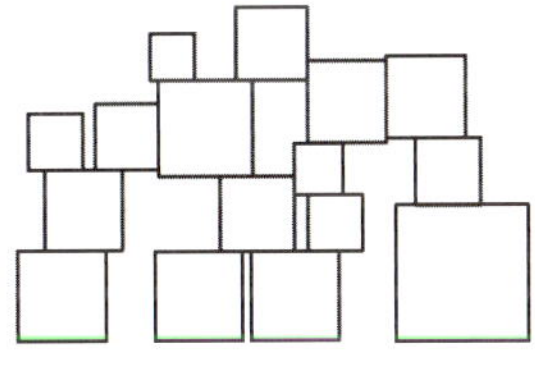

추상화 된 언어의 형태를 큐브에 적용시킨다.
Abstraction makes the language applicable to the form of a cube.

각 나라의 언어를 모은다.
I collect individual language.

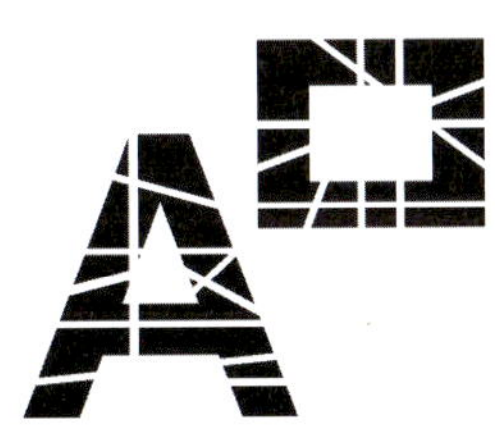

언어를 깨뜨린다.
I break language.

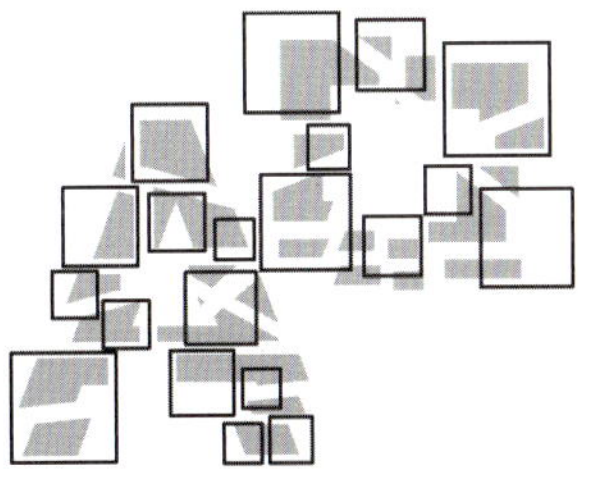

깨어진 언어 조각들에 의미를 부여한다.
I grant meanings to broken language.

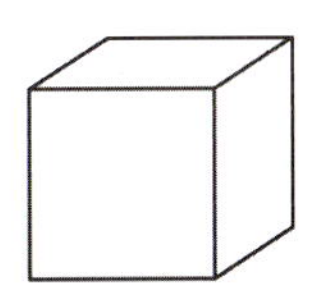

의미가 부여된 깨진 언어 조각들은 큐브가 되어 공간을 형성한다.
The broken language that is granted for meaning becomes cube and forms space.

PROCESS 2

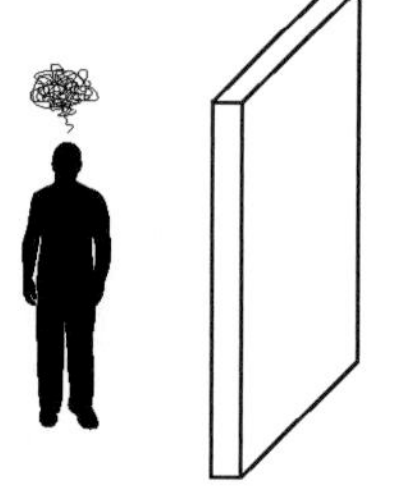

서로 다른 언어로 인해 장벽이 생긴다. 언어의 장벽으로 인해 그들과 우리 사이의 거리는 점점 멀어진다.

Because of different language, border is made and for this, relation between them and us gradually is estranged.

그들과 우리 사이에 장벽을 만드는 각각의 언어를 깨뜨린다. 깨어진 언어의 조각들을 다시 합쳐 새로운 의미, 또는 형태를 만든다.

I break each language which makes border between us. I collect and form broken language again.

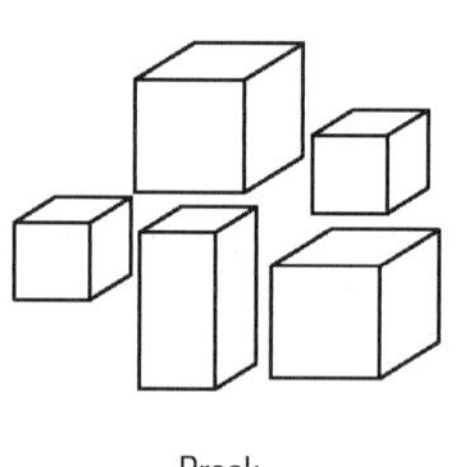

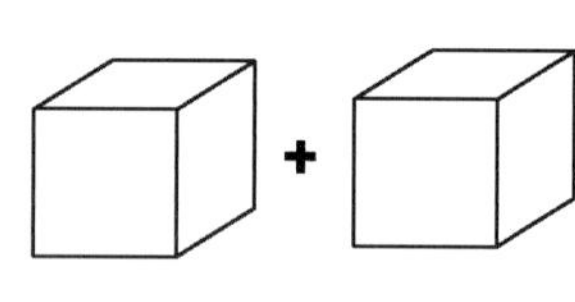

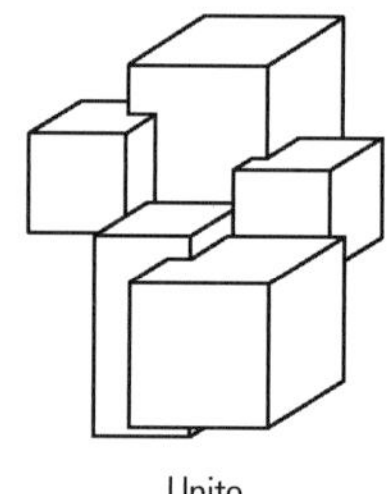

PROCESS 3

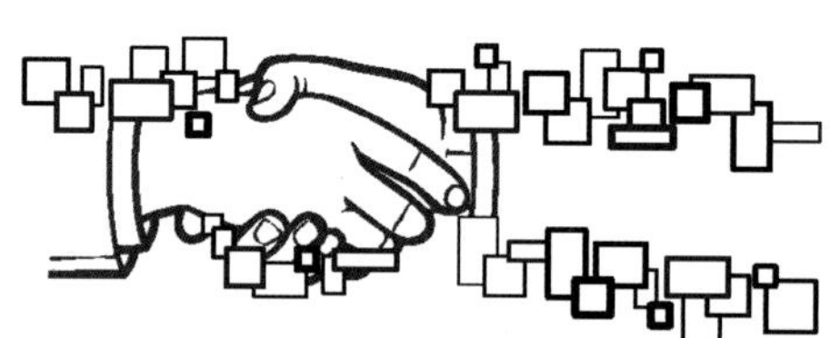

언어에는 문자와 음성뿐만 아니라 몸짓이나 행동도 포함된다. 아이들이 서로 어울려 노는 모습, 손을 잡는 모습, 포옹 하는 모습 등도 하나의 언어가 될 수 있다.

Because of different language, border is made and for this, relation between them and us gradually is estranged.

어린 아이들이 서로 어울려 놀며, 맞잡게 되는 손의 형태를 모티브로 공간을 디자인 한다. 큐빅 형태의 오브제로 공간을 형성하며, 오브제에는 각 나라의 언어라는 의미를 부여한다.

Therefore, form of cruising, holding hands becomes motive and design using that.

맞잡은 손의 형태를 따라 오브제를 쌓는다. 쌓아진 오브제는 하나의 공간을 형성한다.

I pile up objects following form of holding hands. Objects pilled up make space.

DETAIL FUNCTION

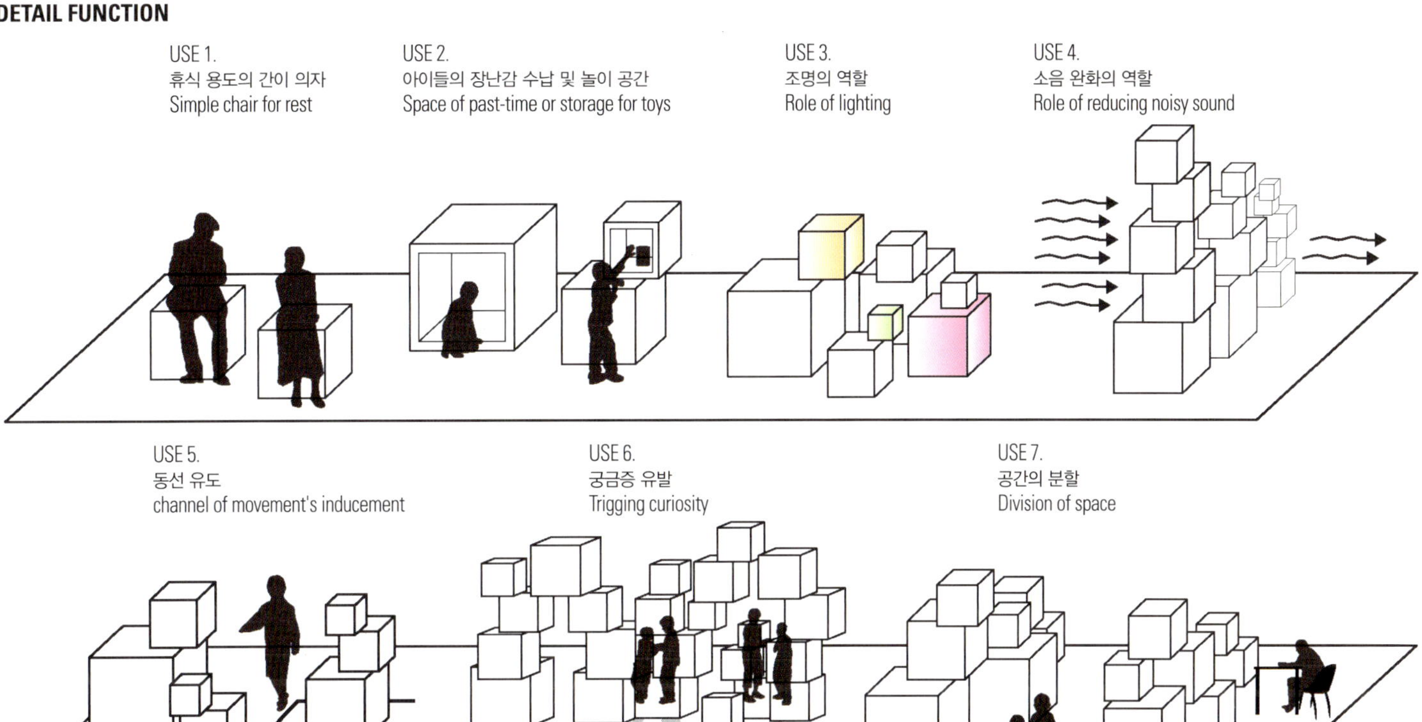

Flyover의 입.출구로 동선의 유도와 놀이장소등 여러 기능을 한다.

Flyover of the inlet and outlet to the induction of copper and several features are turned off nolyijang.

2012

INTERNATIONAL DESIGN EXCHANGE PROJECT

The 3rd Hidden Space Project

HIGH STREET

LONDON Metropolitan University
Sir John Cass Department of Art, Media and Design
Interior Design and Technology BA (Hons)
UK

GACHON University
Department of Interior Architecture
Rep. of KOREA

DONGYANG MIRAE University
Department of Interior Design
Rep. of KOREA

FHHS 2010 - *The 1st Hidden Space Project* **"LIVING BRIDGES"**

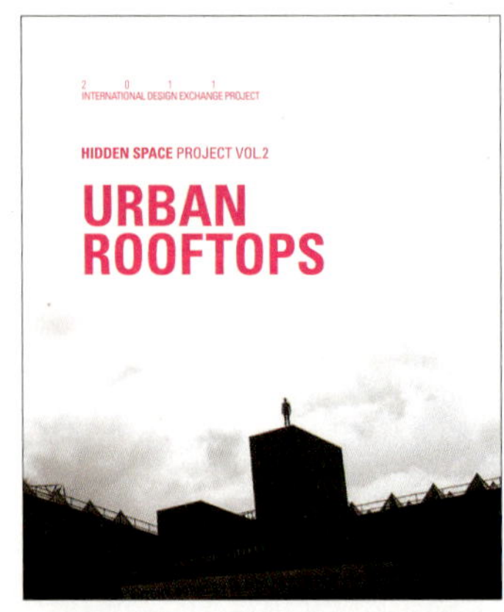

FHHS 2011 - *The 2nd Hidden Space Project* **"URBAN ROOFTOPS"**

FHHS 2012 - *The 3rd Hidden Space Project* **"HIGH STREET"**

* FHHS (Future Horizon for Hidden Space)

2010년 [LIVING BRIDGES]주제로 영국 런던 메트로폴리탄대학과 한국의 경원대학(현-가천대학), 한국의 동양미래대학이 공동으로 런던 아키텍 페스티벌 참가하면서 시작된 **FHHS**는 우리시대 숨겨진 공간들을 10년간 10개의 주제로 교류전을 기획, 3회까지 진행했으며, 이후 주제를 현재 기획중이다.

올해는 2011년에 이어서 [HIGH STREET]를 주제로 3번째 작품집이 발간되며, 또 2012년 6월25일부터 7월중 런던 아키텍트 페스티벌 프로그램의 일환으로 전시가 진행될 예정이다.

FHHS는 이후 계속해서 우리주변의 숨겨진 공간들을 찾아 열정과 지혜, 그리고 성찰의 노정을 계속해 나갈 예정이다.

2012

INTERNATIONAL DESIGN EXCHANGE PROJECT

The 3rd Hidden Space Project

HIGH STREET

2012년 7월 1일 1쇄 발행

Published By 2012 HIGH STREET Project
(Kaye Newman / Janette Harris / Park, Young Tae / Ahn, Sung Hee)

Total Design
Park, Young Tae | DONGYANG MIRAE University
만든이 | 박영태
편집,기획, 표지 디자인 | 박영태

Assistant
Park, Hye Jin / Lim, So Young / Kim, Hye Jin | DONGYANG MIRAE University
편집 보조 | 박혜진, 임소영, 김혜진 – 동양미래대학 실내디자인과

발행처 | 도서출판 조경
TEL. | 031-955-4966~8
인쇄 | 백산인쇄
출력 | 한결그래픽스

정가 | 30,000원